METALLURGICAL ENGINEERING

VOLUME I

Engineering Principles

METALLURGICAL ENGINEERING

VOLUME I
Engineering Principles

REINHARDT SCHUHMANN, JR.

Associate Professor of Metallurgy

Massachusetts Institute of Technology

ADDISON-WESLEY PUBLISHING COMPANY

READING, MASSACHUSETTS · MENLO PARK, CALIFORNIA
LONDON · AMSTERDAM · DON MILLS, ONTARIO · SYDNEY

The past ten years have brought general recognition of the fact that the traditional college training in the established arts of extracting the common metals from their ores no longer is the soundest basis for a successful career in the metallurgical profession. One major factor in this awakening has been the rapidly expanding contributions of chemical thermodynamics to the solutions of practical metallurgical problems. It is a little disturbing to find that many of these contributions have been made by chemists with little or no formal metallurgical education. Another significant trend has been the increasingly prominent role of chemical engineers in metallurgical activities. Although graduate chemical engineers have devoted little or no time in college to the study of specific metallurgical arts, their point of view and mastery of basic engineering principles, and their skill in applying these principles quantitatively to practical problems, have often enabled them to compete on a better than even footing with metallurgical graduates. The advantages of chemical engineering training have appeared most pronounced in the more rapidly growing and developing fields, for example, in the extractive metallurgies of titanium, magnesium, and uranium and in developments relating to by-product recovery from metallurgical operations.

The teaching of metallurgical engineering has been deeply influenced in various ways by the evolution and maturation of the science of physical metallurgy. This younger science has in some ways outgrown and overshadowed its sister science of chemical metallurgy, while both have remained under the same roof. In other words, the scope of the field of metallurgy as a whole has been changing. If undergraduate metal-

lurgical education is to resist the temptation to narrow and specialize in areas of short-range attractiveness, it is clear that both branches must be taught in terms of basic scientific and engineering principles. Consequently there is not time in the undergraduate curriculum for the student to become very well acquainted with current arts, equipment, and industrial practices in either physical or chemical metallurgy.

There has been general agreement as to the need for reorganization of instruction in extractive metallurgy to meet the changing outlook of the metallurgical profession. Moreover, there has been considerable further agreement as to the general directions and objectives of such reorganization. Three objectives in particular have seemed most compelling to me. In the first place, the subject matter of metallurgical engineering should be organized according to basic principles and unit processes and not according to metals treated. Second, the important underlying science of physical chemistry should be more intimately interwoven with metallurgical engineering instruction and not left too much apart in separate courses. Third, the primary emphasis should be placed on analysis and quantitative solution of engineering problems rather than on descriptive knowledge of current practices.

This book has been prepared from lecture notes worked up during several years of teaching courses in process metallurgy and metallurgical engineering to Juniors and Seniors in Metallurgy at the Massachusetts Institute of Technology. During this period several rearrangements of the subject matter have been made and a revised undergraduate curriculum in metallurgy has been placed in operation. In this curriculum, Metallurgical Engineering constitutes a sequence of three one-semester courses, starting in the second semester of the Junior year. The organization of these courses stems from the thesis that the most logical basis of classifying the activities of metal extraction, and of classifying knowledge pertaining thereto, is according to the unit processes, which are the individual chemical steps used in extracting metals from their ores. Accordingly, the unit processes are the focal points of the plan. However, the unit processes themselves are not the fundamental principles of metallurgical engineering; each unit process involves coordinated application of a variety of scientific and engineering principles. Some mastery of these principles is an essential prerequisite to the study of the unit processes, even though the ultimate objective is to apply the principles to specific unit processes to obtain results of engineering value. Hence, the three undergraduate courses in metallurgical engineering at M.I.T. cover, respectively, (1) engineering principles common to all the unit processes of extractive metallurgy, (2) the unit processes, and (3) complete flowsheets and integrated processes for various metals. This volume is the text for the first course and Volume II will be the text for the second. For the third course, C. R. Hayward's *Outline of Metallurgical Practice* and J. L. Bray's *Nonferrous Production Metallurgy* and *Ferrous Production Metallurgy*, supplemented by current periodicals, are satisfactory textbooks.

In my experience, this order of presentation of engineering principles and unit processes meshes well with the student's simultaneous studies of physical chemistry. It is clear that physical chemistry can mean more to the student if he is given opportunities to make use of it, while still fresh, in solving relatively concrete technical problems. In this volume are extensive applications of the

First Law of Thermodynamics, and these are presented to show as clearly as possible the bridge between rigorous scientific statements of the principles and the frequently less rigorous engineering applications. Similarly in Volume II, considerable emphasis will be placed on equilibrium data for the chemical reactions and changes in state involved in all the unit processes, and the effort will be made to show the practical importance of such data.

Problem solving is a most vital approach to the study of metallurgical engineering. Accordingly a great deal of care has been taken in devising problems which will stimulate and develop analytical thinking and engineering judgment on the part of the student. Many of the problems were first used in one-hour open-book examinations and thus should aid in demonstrating to the student that metallurgical education is not learning formulas and information, but is learning to think and to judge.

As can be seen by examining the references listed at the end of each chapter, I have leaned heavily on the literature of chemical engineering. In the chapter on steady heat flow in particular, I made free use of the concepts and methods of W. H. McAdams and H. C. Hottel, as presented in the well-known *Heat Transmission*. Another important reference book whose organization and development of the problem approach to metallurgical engineering have greatly influenced my point of view is *Metallurgical Problems* by Allison Butts. I hope that students and other readers will be stimulated to become familiar with these and other reference books cited at the end of each chapter, since in most instances the references give much fuller presentations of important topics.

So many of my metallurgical colleagues and friends have given me assistance and encouragement during the preparation of this manuscript that I cannot begin to make adequate individual acknowledgments. To John Chipman, Head of the Metallurgy Department, and to my fellow staff members and students at M.I.T., I am indebted not only for specific help and inspiration but also for an environment well suited for developing the ideas of metallurgical engineering education on which this book is based. The following metallurgists examined all or part of the first seven chapters in a preliminary multigraphed edition and made valuable suggestions and comments: J. B. Cunningham, P. L. deBruyn, H. A. Doerner, J. F. Elliott, K. L. Fetters, F. A. Forward, A. M. Gaudin, J. L. Gregg, E. T. Hayes, R. R. Jones, B. M. Larsen, J. S. Marsh, S. Mellgren, W. O. Philbrook, A. K. Schellinger, A. W. Schlechten, S. Skowronski, J. P. Spielman, D. H. Whitmore, and O. Zmeskal. I also wish to express my thanks to Miss Agnes R. Friend, to Miss Sibyl Warren, and to the staff of Addison-Wesley Press who have helped in many ways in the preparation of the manuscript. Finally, I am grateful to my wife, whose constant encouragement and cheerful toleration of odd working hours made possible the completion of the writing.

R. Schuhmann, Jr.

February, 1952

CONTENTS

CHAPTER 1. THE UNIT PROCESSES OF CHEMICAL METALLURGY 1

Engineering fundamentals of the unit processes. Classification of the unit processes. Metallurgical systems.

CHAPTER 2. STOICHIOMETRY 12

Stoichiometric principles. Modes of operation of unit processes. Weighing, sampling, and analysis. Stoichiometric calculations.

CHAPTER 3. THE HEAT BALANCE 43

Conservation of energy. The heat balance.

CHAPTER 4. METALLURGICAL FUELS 69

Classification of fuels. Pulverized coal. Coke. Gaseous fuels. Fuel oil.

CHAPTER 5. COMBUSTION OF FUELS AND HEAT UTILIZATION 109

A fuel-fired metallurgical furnace. Combustion stoichiometry. Heat utilization in furnaces. Heat recovery from flue gases; regenerators and recuperators. Fuel utilization in the open-hearth furnace.

CHAPTER 6. FLUID FLOW 143

Energy balances. Friction. Flow measurement. Draft. Compressed air and blast.

CHAPTER 7. STEADY HEAT FLOW 191

Conduction. Convection. Radiation. Steady heat flow in some metallurgical systems.

CHAPTER 8. UNSTEADY HEAT FLOW 259

CHAPTER 9. PHASES IN PYROMETALLURGICAL SYSTEMS 284

Typical multiphase systems. Metals and metal solutions. Slags. Mattes and other products.

CHAPTER 10. REFRACTORY MATERIALS 325

Refractory properties. Metallurgical refractories. Metallurgical furnace construction.

APPENDIX 368

INDEX 379

CHAPTER 1

THE UNIT PROCESSES OF CHEMICAL METALLURGY

Metallurgy is the science and art of extracting metals from their ores, refining them, and preparing them for use.

Metals are extracted from the earth's crust by sequences of operations and processes. First the metallic minerals must be found in the earth's crust in deposits of sufficient size and metal content, and with other characteristics which will make profitable the extraction of the metal. The metal is mined as an ore or mineral mixture, which usually contains sizable proportions of waste minerals mixed and intergrown with the valuable metallic minerals. For most ores the next step, carried out in mineral dressing plants, is to separate the valuable minerals. The rejection of the waste minerals is accomplished by first crushing and grinding the ore to sever the minerals from each other and then concentrating the ore by such processes as gravity concentration, froth flotation, and magnetic separation. These dressing operations generally do not involve changes in the physical or chemical identities of the minerals which are separated.

The further steps of extracting the metals from the mineral concentrates and refining them are necessarily chemical in nature, and require the carrying out of a wide variety of chemical reactions on a large scale. Many of these chemical processes are conducted in furnaces at high temperatures, and some in water solutions at ambient temperatures; still others utilize electricity to produce chemical changes. Usually a series of such chemical steps is involved in producing the final refined metal from the mineral raw material.

After the metal is extracted from the ore and refined, it undergoes further treatment to adapt it to its ultimate use. By alloying, working, and heat-treating, the physical metallurgist controls the properties of metals which determine their utility. The property-controlling steps usually are integrated closely with the final metal-fabricating steps that yield the metal end products and structures.

Among the varieties of engineering activity encompassed in following the metal from the earth's crust to the final metal product, we may delineate *chemical metallurgy* as that field of engineering which deals with the extraction of metals from their raw materials and with the refining of metals by chemical processes on an industrial scale.

Unit operations and unit processes. The individual sequential steps in producing metal from ore may be called *unit operations* and *unit processes.* These terms were first used by chemical engineers to designate the unit actions in the manufacturing processes of the chemical industry. In a broad sense, the extraction of metals may be considered as *chemical engineering,* since the unit operations and unit processes in the two fields are similar and in many cases identical.

The sequences of unit operations and unit processes are different for every metal; in fact, there is usually a choice of procedures for extracting a given metallic element. As a result, literally scores of procedures are used to extract metals. When these procedures are examined carefully, however, it is found that they all consist of different permutations and combinations of a relatively small number of unitary steps. Because the total number of these unit operations and unit processes is small and the basic principles of the unit are the same regardless of the particular materials or metals treated, the unitary concept is appropriate and logical as the basis for organizing the knowledge of the arts of extracting and refining metals.

Table 1–1 lists the principal unit actions for metal extraction. These are divided into two groups, corresponding approximately to the division of labor between mineral dressing and chemical metallurgy which has been customary in the past. Since this division is roughly equivalent to the chemical engineer's distinction between unit *physical* operations and unit *chemical* processes, the corresponding designations are used in Table 1–1. However, a feature of the classification which has greater scientific importance is that *the unit processes primarily involve chemical reactions and/or changes in state of aggregation,* while the *unit operations primarily do not involve either bulk reactions or changes in state.* Other bases of classifying might be used, or precise definitions might even be avoided, but the one just given will be followed throughout this book.

Integrated processes. An integrated process may be defined as a coordinated sequence of unit operations and unit processes which is in some respects self-complete. It may be self-complete in that it represents all the actions in a plant, or in a section of the plant in which a particular job is done. It is sequential and coordinated in that materials flow from one step to the next in an orderly fashion; the product of one step serves as the feed for the next. Thus integrated processes are described conveniently by diagrammatic flowsheets.

Typical flowsheets. Typical flowsheets for the extraction of copper, aluminum, and iron from their ores are shown in Figs. 1–1, 1–2,

TABLE 1-1

Principal Unit Operations and Unit Processes
for Extracting Metals

Unit Operations	Unit Processes
Comminution	Gas-solid processes (including
Screening	roasting, calcining, gaseous
Classification	reduction and drying)
Separation of solids from	Sintering and pyroagglomeration
fluids	Reduction of metal oxides
Heavy-fluid separation	Retorting
Jigging	Simple smelting
Flowing film concentra-	Blast furnace smelting
tion and tabling	Converting
Flotation and agglomera-	Refining of liquid metals
tion	Melting and liquation
Magnetic separation	Casting and solidification
Agitation and mixing	Distillation and sublimation
Materials handling	Hydrometallurgical processes
	Electrolytic processes

and 1–3 respectively. These flowsheets are much condensed to show only the essential and major steps and paths of material flow. Each step is indicated by a rectangle; the rectangles for the unit operations are dotted to distinguish them from the unit processes. Detailed flowsheets, listing all of the individual items of equipment and showing flows to and from each item, would fill several pages for any one of the three metals. Chemical analyses and relative tonnages of materials at various stages are given so that the main flow of materials through the process can be followed. These data show a common characteristic of processes of metal extraction, namely, the step-by-step *separation* of the metal from other chemical elements. The need for several kinds of unit operations and unit processes in sequence arises from the fact that no *single* unit operation or unit process is capable of separating the metal from all of the elements associated with it in the ore. Even if a unit process could accomplish such separation in one pass, its usefulness probably would be limited because it would cost more than a series of steps, each designed to operate efficiently on a given part of the separation job.

The flowsheets for copper and aluminum extraction include both steps involving chem-

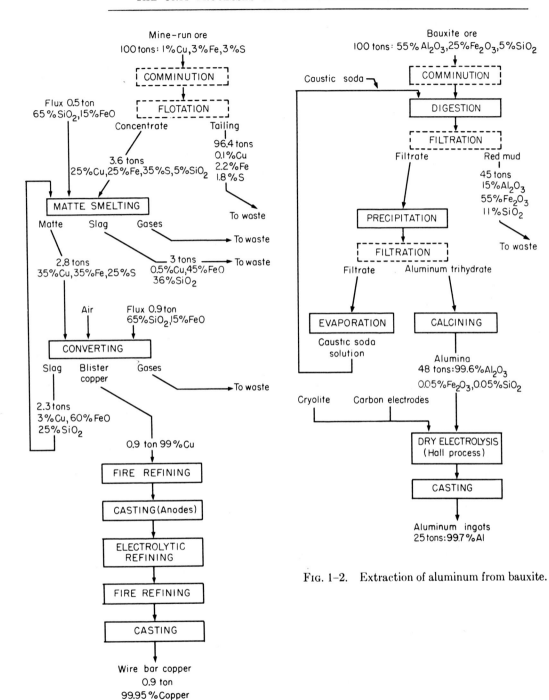

FIG. 1-1. Extraction of copper from a low-grade ore.

FIG. 1-2. Extraction of aluminum from bauxite.

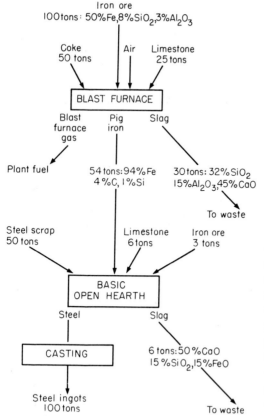

Iron ore
100 tons: 50%Fe,8%SiO₂,3%Al₂O₃

Coke   Air   Limestone
50 tons   25 tons

BLAST FURNACE

Blast   Pig   Slag
furnace   iron
gas

Plant fuel   54 tons:94%Fe   30 tons: 32%SiO₂
4%C, 1%Si   15%Al₂O₃,45%CaO

To waste

Steel scrap   Limestone   Iron ore
50 tons   6 tons   3 tons

BASIC
OPEN HEARTH

Steel   Slag

CASTING   6 tons:50%CaO
15%SiO₂,15%FeO

Steel ingots
100 tons   To waste

FIG. 1–3.   Extraction of iron from iron ore.

cal change (*mineral dressing*); (2) chemical processing of concentrates to yield metal (*chemical metallurgy, smelting, process metallurgy*). However, many flowsheets, like that for aluminum, depart to advantage from this order of treatment.

The work of separating a given chemical element from the metal may be divided among several unit operations and unit processes in a given flowsheet, and the division of labor may be varied over wide limits to meet special circumstances. For example, in the copper flowsheet, sulfur is separated from the copper in four steps: first, in the flotation tailing as pyrite (FeS_2); second, as sulfur dioxide in the gases from matte smelting; third, as sulfur dioxide in the gases from converting; and fourth, as sulfur dioxide in the gases evolved during fire refining. Complete separation of sulfur from copper is still technically feasible when either of the first two steps is omitted; in fact, under some conditions ores are fed directly to matte smelting without previous concentration, or flotation concentrates are fed directly to converting without passing through the matte-smelting step.

Thorough understanding of metallurgical flowsheets in the scientific sense starts with thorough understanding of the individual unit operations and unit processes. Thorough understanding in an engineering sense additionally requires translation of technical characteristics of the unit actions into terms of costs, raw materials, power, and labor. In the final analysis the choice among flowsheets and among different methods of operation must be based largely on over-all cost comparisons.

Chemical metallurgy — a branch of engineering. A chemist may demonstrate in the laboratory that a certain chemical reaction or set of reactions will extract metal from an ore or separate an impurity from a metal.

ical changes and steps in which minerals and other substances are separated without changes in their physical or chemical identities. The copper flowsheet is representative of that for many metals, in that unit physical operations predominate in the earlier steps while the later steps are mainly unit chemical processes. In the aluminum flowsheet, on the other hand, the unit operations and unit processes are well intermixed. To a considerable degree, the procedures long in use for the common metals show the sequence represented by the copper flowsheet: (1) dressing of raw materials to produce mineral concentrates, without bulk chemi-

Further, he may obtain complete scientific data as to the conditions of temperature, pressure, composition, equilibria, and rates for the reactions. Even with this complete technical and scientific data available, the work of the metallurgical engineer has only begun. In order that this knowledge may become useful to man, *the engineer must design, erect, and operate equipment and plants in which the reactions can be carried out to produce metal in tonnage quantities and at a profit.*

Successful operation of an industrial process requires intimate knowledge of the relations of the large-scale process to its surroundings and careful coordination of those relations with each other and with the internal characteristics of the process. Some of the more important items to be considered are:

Supplies of raw materials, fuels, water, reagents.

Supply and application of energy, whether heat or mechanical energy.

Manpower requirements and direction.

Transportation, storage, and handling of raw materials and products.

Waste disposal: tailings, slags, gases, etc.

Instrumentation and control.

These represent the costs of a process, and the engineer not only must understand how these process requirements are met but also must know their costs and keep the costs under control to achieve maximum profit.

The science of chemical metallurgy. Since chemical metallurgy deals with reactions and changes in state, the chief underlying science is chemistry. Of particular concern are energy and heat effects accompanying chemical changes; equilibrium relationships; kinetics, mechanisms, and rates of chemical changes; and surface chemistry and physics. A feature of metallurgical chemistry is the emphasis on high-temperature, nonaqueous reactions.

Twenty-five years ago the metallurgist's knowledge of the chemistry of a unit process was summed up commonly by a list of chemical equations thought to occur in the process. The over-all reactions usually were known fairly well, since they could be determined by analytical data on the compounds present in the feed to and the products from the process. Also, data on heats of formation were used with those over-all reactions to obtain heat evolutions or absorptions of the process as a whole. However, the individual chemical equations thought to represent reactions occurring within the process frequently rested on scanty evidence or on no evidence at all. Knowledge of the temperature and pressure conditions favoring specific metallurgical reactions was also limited, although metallurgists skilled in the art knew by experience the range of conditions under which the over-all process could be conducted.

Today the metallurgist is finding it to his advantage to become more and more of a practicing physical chemist. He is learning to apply thermodynamic data to predict whether or not a given reaction is possible, to predict the effect of changing temperature on the course of a reaction, and to achieve quantitative scientific understanding of many important phenomena in those metallurgical arts and processes which heretofore have evolved mainly by trial and error. One gauge of the value of this approach is the fact that many schools in recent years have introduced advanced courses in metallurgical thermodynamics, required of all metallurgy undergraduates, and metallurgists practicing in industry have approved this move.

Although the rapidly expanding application of chemical thermodynamics to metallurgical systems represents the major scientific work now in progress in chemical metallurgy, interest is developing in applying other branches of chemical science. The kinetics of chemical changes, dealing with the motions, collisions, and interactions of atoms and molecules, is of particular importance in relation to the rates at which metallurgical reactions occur. The chemistry and physics of surfaces also is recognized as playing an important role in some processes, although little work has yet been done in applying this science to metallurgical systems. Processes for extracting and refining metals by the use of electricity lead the metallurgist into the science of electrochemistry.

Since chemical metallurgy is at once a science, a branch of engineering, and an empirical art, a rational balance among these three should be achieved. The metallurgist must develop and operate processes whether or not he understands them completely in a scientific sense. Were he limited to using processes completely understood scientifically, little metal would be produced. On the other hand, the practicing metallurgist should exploit and expand science to the fullest, if for no other reason than that it pays financially to do so. He should strive to avoid the habit of classification into "practical" and "theoretical," with separate compartments for the two.

Engineering fundamentals of the unit processes

While the unit operations and unit processes comprise the units of organization of the engineering activities of extracting metals, they are not the fundamentals. Just as all flowsheets and integrated processes can be broken down into a few unit processes, so the unit processes can be analyzed in terms of a few engineering fundamentals which they all have in common. Table 1-2 lists these common features of the metallurgical unit processes. These fundamentals represent a logical starting point for the study of chemical metallurgy, and most of them are taken up in Volume I of this text.

TABLE 1-2

Engineering Bases of the Metallurgical Unit Processes

1. Stoichiometry
2. Thermochemistry and thermophysics; heat balances
3. Fuels and combustion
4. Pyrometry
5. Flow of heat
6. Flow of fluids
7. Phases and phase equilibria in metallurgical systems
8. Refractories and furnaces
9. Chemical equilibria
10. Rates of reaction
11. Instrumentation and control

A brief consideration of a typical unit process will show the need for mastering the separate engineering principles and features before attempting detailed study of the unit process itself. The unit process, *simple smelting*, is a key step in the copper-extraction flowsheet already considered (Fig. 1-1). In this step, the feed materials are cold granular ores and liquid slag containing copper, iron, sulfur, and oxygen in various chemical combinations, along with gangue oxides such as silica, alumina, magnesia, etc. This mixture is heated to temperatures in excess of 1200°C, so that everything is molten. Two liquid phases result: (a) the copper matte, consisting of copper, iron, and sulfur, and (b) the slag, a solution of iron oxide, silica, alumina, magnesia, and other oxides, but low in copper content. These two phases separate under the influence of gravity, the heavier matte accumulating in the bottom of the furnace and the slag layer on the top. These two liquid products are tapped out separately and are the two main products. In addition, some reactions yield sulfur dioxide and other gases which pass out with the flue gases. Figure 1-4 shows the essential features of a fuel-fired reverberatory furnace, typical apparatus for conducting this process.

Each of the fundamental subjects listed in Table 1-2 contributes in an important way to the engineer's understanding of the simple smelting process and hence to its operation, as is shown below:

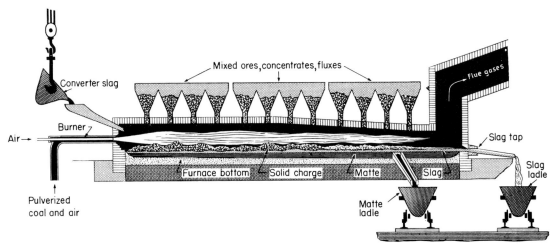

FIG. 1–4. Simple smelting in a reverberatory furnace.

1. *Stoichiometry.* The total amount of any chemical element, for example, sulfur, entering the furnace in fuel, air, and charge must equal the total leaving in the products plus accumulation in the furnace. When new compounds form or old ones decompose, the quantities of the respective elements involved are simply related in terms of the combining weights, which are simple multiples of the atomic weights. Thus the formation of 64 lb of SO_2 in the gas requires 32 lb of sulfur and 32 lb of oxygen. Weights of gaseous products are converted to volumes by means of the gas laws. These elementary principles relate the weights, volumes, and chemical analyses of the incoming and outgoing substances, and hence are the basis of many calculations the metallurgical engineer needs to make. These include calculations of proper amounts of various charge components; calculations of weights, volumes, or analyses of feed components or products not directly measured; and estimates of recoveries or distributions of the elements among the products.

2. *Thermochemistry and thermophysics.* Large quantities of heat are involved in high-temperature processes like the simple smelting process. The heat account sheet or the heat balance ranks close in importance to the materials balance sheet. Chemical reactions, changes in state, and simple temperature changes are accompanied by evolutions or absorptions of heat which can be evaluated easily from fundamental thermochemical and thermophysical data, such as heats of formation, latent heats of melting, specific heats, etc. The basis of all such calculations is the First Law of Thermodynamics, i.e., the Law of Conservation of Energy. The heat supply to the simple smelting process in Fig. 1–4 comes chiefly from combustion of the fuel and from the hot slag charge. This heat intake is accounted for as heat contained in the liquid and gaseous products, as heat loss to the surroundings, and as heat absorbed by certain reactions within the furnace. Since the cost of the fuel is a large item for this simple smelting process, quantitative heat accounting is an obvious part of intelligent operation.

3. *Fuels and their combustion.* The copper-smelting reverberatory furnace shown in Fig. 1–4 might burn about 150 tons of powdered

coal for the 1000 tons of charge smelted in 24 hours. The properties and behavior of the fuel directly affect the process in many ways. Both the quantity of heat supplied by the fuel and the temperature level at which it is supplied must be considered. Also the products of combustion, including the oxygen in the excess air admitted to obtain good combustion, may react with the charge. The metallurgical engineer's approach to fuel and combustion problems rests mainly on stoichiometric and thermochemical principles.

4. *Flow of heat.* Heat energy produced by burning the fuel is transferred from the flame to the furnace charge and furnace walls. The walls and charge exchange heat by radiation. Heat flows out of the furnace through the furnace walls and is lost in the surroundings. Heat is conducted from the surface to the interior of the solid charge, and through layers of liquid slag and matte. Some of these heat transfers are essential to the process, and the engineer seeks to provide conditions most favorable to them. Other transfers, such as the heat loss to the surroundings, represent waste and are reduced as far as possible, consistent with other requirements of the process.

5. *Pyrometry.* Many different temperatures are of interest in the reverberatory furnace in Fig. 1–4. Flame temperatures, flue gas temperature, wall temperatures (inside and out), and matte and slag temperatures are of obvious importance in the intelligent and efficient utilization of heat and in their effects on the chemical reactions and state changes carried out within the furnace. Pyrometry, the measurement of these temperatures, is a difficult art but one which the metallurgist must use to the fullest extent.

6. *Flow of fluids.* The combustion of 150 tons of coal per day requires about 40,000,000 cu ft of air and yields a much larger volume of hot flue gases which must be drawn through the furnace, carried in flues, and eventually disposed of. For smooth, successful operation of the furnace this large volume of flow must be under close control at all points. The designs of the furnace chamber, gas inlets and outlets, flues, chimneys, and draft controls are in large measure fluid-flow problems.

7. *Chemical equilibria.* The materials charged to any one matte-smelting furnace comprise a great variety of chemical compounds, especially metal oxides, sulfides, sulfates, and silicates, and thus many chemical reactions may take place as the charge is smelted. The following are a few of the many possible reactions:

$$Cu_2O + FeS \longrightarrow Cu_2S + FeO$$
$$3Fe_3O_4 + FeS \longrightarrow 10\ FeO + SO_2$$
$$2FeS + 3O_2 \longrightarrow 2FeO + 2SO_2$$

It is of obvious importance that the conditions of temperature, pressure, and phase composition under which a given reaction takes place and the conditions under which it does not take place be known quantitatively. In fact, the very basis of the process is the fact that the first of the three reactions written above can be made to proceed to the right so that substantially all the copper added to the system is changed into sulfide and therefore is concentrated in the matte or liquid sulfide phase. Much useful information of this kind can be derived from the study of physicochemical data on equilibrium, such as equilibrium constants.

8. *Phases and phase equilibria.* A large number of solid phases are included in the furnace charge, and the chief furnace products are two liquid phases, slag and matte. Both the physical and the chemical properties of these liquid phases depend on composition and temperature, and any given phase is stable only through certain ranges of composition and temperature. Information of this kind is available through the study of

"constitution diagrams" or "equilibrium diagrams" which depict quantitatively the conditions under which a given phase or combination of phases is stable. Often the physical and chemical behavior of a given phase can be correlated with the information furnished by the constitution diagram. For example, the principal components of the copper-smelting slag are Fe, O, and SiO_2, so that an understanding of the phases and phase equilibria in this ternary system should contribute directly to an understanding of slag properties. Similarly, the Cu-Fe-S system is a starting point for the study of mattes.

9. *Kinetics and reaction rates.* Whereas study of equilibrium constants and phase equilibrium data may tell what will occur in the reverberatory furnace if the materials all are held in the furnace long enough to reach equilibrium, in actuality equilibrium is not attained completely under operating conditions. Reactions occurring at the surface between slag and matte can occur only as fast as the reacting atoms are able to reach this surface by diffusion, convection, or other mechanisms. Particles of solid charge melt only as fast as the necessary heat can be transferred to them. When two kinds of atoms collide, only a certain proportion of the collisions may result in reaction. In the final analysis, the kinetics determine the rate of smelting and the capacity of the furnace.

10. *Refractory materials and furnaces.* Steel and other metals, glass, and rubber are satisfactory construction materials for most chemical laboratory apparatus and apparatus for low-temperature chemical processes. However, these and other common materials of construction fail at temperatures hundreds of degrees lower than the 1200° to 1300°C which prevail in the reverberatory furnace. The refractory bricks of which the furnace is constructed must withstand not only high temperatures but also chemical erosion, mechanical abrasion, mechanical stresses, and stresses arising from cooling and heating. They must have low enough heat conductivity so that heat is retained in the furnace. The service requirements are different for different parts of the furnace structure, and for each kind of service a choice must be made based on knowledge of the properties and limitations of the various available refractory materials. The refractories represent a major item both of first cost and of operating cost in the copper smelting process.

11. *Instrumentation and control.* A furnace such as the one shown in Fig. 1–4 can be operated by an experienced man without special instruments and control devices. However, in a modern plant, such a furnace probably would have automatic indication and recording of temperatures at various points, combustion control, draft control, and other instruments and devices to eliminate much of the guesswork from furnace operation.

Most of the eleven engineering fundamentals discussed above are involved in each unit process. Moreover, the specific practical problems which come up in the development and application of a unit process usually can be resolved into terms of one or more of these eleven engineering fundamentals, plus economic considerations. The effective solution of such problems requires the ability on the part of the engineer to recognize the roles of stoichiometry, thermochemistry, heat flow, fluid flow, and so on, and the ability to deal with each in a quantitative fashion.

Classification of the unit processes

A classification of the unit processes for extracting metals was given as part of

Table 1-1. The purpose of this, as of other classifications, is to group like with like. The principal differentiation among these unit processes is on the basis of the kinds of phases involved in the process. For example, melting is a solid-liquid transformation; simple smelting involves production of two or more liquid phases; refining of molten metals involves the addition of reagents (usually solid or gaseous) to impure liquid metal with the removal of the impurity to another phase (gas, liquid, or solid); hydrometallurgical processes involve reactions of aqueous solutions and solids. Reduction of metal oxides is carried out in systems of various combinations of phases and overlaps the other unit processes, but is listed separately because of the important group of principles concerned.

For any one of the unit processes, the principles are generally the same regardless of the chemical compositions, temperature, pressure, size of system, etc. Thus, melting copper and melting lead are the same in principle, although the temperature, heat requirement, atmospheric requirement, and other quantitative variables have quite different values for the two metals.

A further useful and significant result of classifying unit processes according to the important phases handled is that processes carried out in similar apparatus are grouped together. For example, any simple smelting operation can be conducted in a reverberatory furnace, an arc furnace, or a crucible. A gas-solid reaction, whether calcining, desulfurization roasting, or gaseous reduction, can be carried out in a multiple-hearth furnace, a rotary kiln, a muffle, a fluidized-solid system, a flash system, or a combustion boat in a tube furnace.

The unit processes are taken up in Volume II of this text.

Metallurgical systems

Whether he is dealing with a principle, a process, one aspect or part of a process, or a problem, an engineer either consciously or unconsciously focuses his attention on a limited group of relevant objects or materials and excludes irrelevant information. The portion of the universe under consideration in any given case is the *system*. The rest of the universe is the *surroundings*, but usually only immediate surroundings are of interest in dealing with a given system. These concepts of system and surroundings are themselves simple and self-evident. Their importance lies in the absolute necessity for careful delineation and specification of the system and its boundaries when dealing with scientific and engineering relationships. Except in cases where no doubt is possible, statements of such relationships should be prefaced by precise statements or descriptions of the system under consideration.

Plant metallurgical processes are designed to produce metal and not to illustrate scientific principles. Accordingly, they generally turn out to be very complex from a purely scientific point of view. A simple, direct, and quantitative application of elementary principles to yield practically useful information is rarely possible. One all too frequent outcome is that the metallurgist may be overcome by the apparent complexity to the extent of relying fully on previous operating experience and rule of thumb, while dismissing principles as impractical, too theoretical, or as not taking everything into account. To avoid this outcome, the first and frequently the most difficult hurdle to be surmounted is the analysis of the process into terms of a number of simple systems, to each of which elementary principles are readily applied. In other words, systems are the common denominators which must be found to relate principles to processes.

The important systems to be considered in analyzing a given process may include:

The contents of the apparatus, or of any zone or geometric part thereof.

The apparatus, or parts, exclusive of contents.

Apparatus and contents.

One phase (e.g., the gas phase) or various combinations of phases (e.g., slag and metal).

A single particle or a specified group of particles.

A limited group of the substances or chemical elements involved in the process.

A moving fluid with confining walls or included stationary objects.

The above is not intended as a complete listing of all the possible kinds of systems, but is illustrative of the kinds of analyses which are most frequently used.

Logical though this systematization may seem when expressed in general terms, the fact is that skill in the analysis of complex problems into terms of simple systems is not easily and quickly acquired. The student is urged to make a conscious effort to develop this skill while solving the problems in the subsequent chapters. To this end, the first step in the written solution of a problem should be a specification of the system or systems under consideration. Often this specification can take the form of a rough sketch or box diagram of the system with the contents labeled and with labeled arrows to show relations between the system and its surroundings. Such a sketch is desirable even if a written specification is included, because the sketch aids in mental visualization of the system. Solving problems without clearly visualizing the systems is "formula-punching," a procedure of little engineering value.

CHAPTER 2

STOICHIOMETRY

Stoichiometry deals with relations between the quantities and analyses of the materials entering, leaving, and participating in a process. These stoichiometric relations are themselves very simple and well within the scope of high school chemistry, but they are the bases of such a wide variety of engineering calculations that the metallurgical engineer cannot develop too much ingenuity and proficiency in their use. The chief fundamental principles are the Law of Conservation of Elements, the Law of Combining Weights, and the Gas Laws. Most of the applications of these principles to metallurgical systems are for the purpose of calculating weights, volumes, or analyses of materials handled in a process. For a process in operation, these calculations give data which are very difficult, sometimes impossible, to obtain by direct measurement in the plant. Also, such calculations frequently are necessary in planning how a process is to be carried out, for example, in determining the proper quantities of the various materials to be fed into the process to obtain the desired products.

The materials balance or metallurgical balance for an operating process is an essential and routine plant record, in much the same way as an income and outgo financial account is a part of operating a business. The metallurgical balance shows the weights and analyses of input and output materials, and the calculated inputs and outputs of each of the important metals, elements, and compounds. This accounting also serves as a check on plant data in that the various totals of input and output should be equal.

The metallurgical balance is the starting point for a great many other important calculations, including cost calculations and various technical calculations presented in subsequent chapters.

The wide practical usefulness of stoichiometric relationships for all unit processes is made possible by the fact that they require little or no detailed understanding of the internal workings of the process. They need only weight and assay data on the materials entering and leaving the process.

Stoichiometric principles

Conservation of elements. Matter is composed of atoms, which are units consisting of a compact, positively charged nucleus, accounting for most of the weight, and of negatively charged electrons which may be regarded as moving in definite orbits about the nucleus. The chemical behavior of an atom is determined by its *atomic number*, which is the number of positive charges on the nucleus and which in an electrically neutral atom must be equal to the total number of electrons in orbits around the nucleus. The nucleus and its atomic number are unchanged when the atom participates in all ordinary chemical and mechanical processes, for example, in all the unit operations and unit processes of extractive metallurgy. Thus, the number of atoms of a given atomic number, that is the number of atoms of a given chemical element, is neither decreased nor increased by any chemical process. Since in electrically neutral atoms and combinations of atoms each atomic nucleus is accompanied by electrons

12

equal in number to the atomic number, the Law of Conservation of Elements may be stated as follows: *The quantity of any element is neither decreased nor increased by chemical or physical processes taking place in systems that contain the element.**

Atomic weights. Although for a given element the charge on the nucleus (atomic number) is a fixed number, the constitution and weight of the nucleus may be different in different atoms of the same element. Atoms of the same atomic number but of different weights are isotopes. Isotopic atoms behave so identically in chemical processes that it is extremely difficult to separate them from each other. Accordingly, the proportions in which the various isotopes of a given element occur in nature are maintained throughout the chemical processes of nature and of man. The persistence of the natural isotopic proportions makes possible the determination of an average atomic weight for each element. In determining atomic weights, that for oxygen is set arbitrarily at 16.000 and the atomic weights for the other elements are determined relative to this value for oxygen. The quantity of an element represented by the atomic weight in grams (gram atomic weight) contains 6.02×10^{23} atoms.

Table A–1 in the Appendix gives atomic weights for the chemical elements found in nature. For many purposes, especially in engineering work, the accuracy of the measured data does not justify the extra labor of using precise values for the atomic weights. Therefore, it is often convenient to round off the atomic weights to the nearest unit or half unit. Rounded figures will be used

---

* This statement is rigorous for all systems to be considered in this book, but does not apply in systems where nuclear reactions take place.

almost exclusively for the subsequent calculations and problems in this book.

Combining weights. Atoms combine to form molecules. Such combinations are brought about by rather specific interactions involving principally the electrons in the outermost orbits of the combining atoms. These interactions are chemical bonds and include electron sharing, electron transfer, resonant bonds, etc. Owing to the specific character of these bonds and to geometrical limitations to the number of atoms which can fit around another atom, the molecular groupings generally contain a small number of each kind of atom. Thus, the relative weights in which elements combine or react with each other can be expressed accurately by the products of the atomic weights and small whole numbers.

The principle of combining weights is taken for granted in the everyday use of formulas and equations. Thus the formula Fe_3O_4 represents 1 molecule of a compound containing 3 atoms of iron to 4 atoms of oxygen or 166.52 weight units of iron to 64.000 weight units of oxygen. Similarly, a chemical equation represents quantities, as shown below:

	Fe_3O_4	$+ 4H_2 =$	$3Fe$	$+ 4H_2O$
MOLECULES OR MOLS:	1	4	3	4
ATOMS:	3 of iron 4 of oxygen	8 of hydrogen	3 of iron	8 of hydrogen 4 of oxygen
RELATIVE WEIGHTS:	230.52	8.06	166.52	72.06

The tabulation of numbers of atoms above shows for each given kind of atom the same total number (and therefore the same total weight) on both sides of the equation. The Law of Conservation of Elements requires that the equation be balanced in this way.

The relative weight corresponding to a formula obviously can be expressed in any desired system of units, provided consistency is maintained. Thus, it can be decided that "Fe₃O₄" represents 230.52 g of iron oxide. This quantity is one "gram formula weight." Or it may be more convenient to define one formula weight as 230.52 kg, or 230.52 lb, or even 230.52 tons. These units are also called the gram mol, kilogram mol, pound mol, and ton mol, respectively.

Gas laws. Boyle's Law states that at constant temperature the volume of a given mass of gas is inversely proportional to the pressure. Gay-Lussac's Law states that at constant pressure the volume is proportional to the absolute temperature. These statements are strictly true only for an ideal gas, and lead to the ideal-gas equation:

$$pV = NRT, \qquad (2\text{--}1)$$

in which p = pressure, V = volume, N = number of mols of gas, R = gas constant, and T = absolute temperature. With p in atmospheres, V in liters, N in g mols, and T in °K (°C + 273), the gas constant R has the value 0.082 liter-atm per degree. Real gases closely approach the ideal behavior represented by the above equation at low pressures but may deviate considerably at high pressures. The errors introduced by assuming ideal behavior for several real gases at different pressures are shown by the data in Table 2–1. Since very few metallurgical systems involve gases at high pressures, it can be seen that no significant errors are introduced by assuming ideal behavior for real gases in nearly all metallurgical engineering calculations.

Although calculations of gas volumes are easily made using the ideal-gas equation,

TABLE 2-1

Corrections to Ideal-Gas Behavior*

0°C

Gas	Pressure, atmospheres			
	1	10	50	100
Air	-0.06%	-0.57%	-2.27%	-3.11%
N_2	-0.05	-0.43	-1.57	-1.57
O_2	-0.09	-0.96	-4.40	-7.75
H_2	+0.06	+0.63	+3.15	+6.45
CO	-0.05	-0.45	-2.25	-2.80
CO_2	-0.33		-89.57	-79.93

*To obtain actual volume, correct the volume calculated from ideal-gas equation by percentages tabulated above.

engineers find a different method of calculation advantageous. The engineering type of calculation is based upon use of constants representing the volume of 1 mol of ideal gas at standard conditions, that is, at 0°C (32°F) and 1 atm. The figures most commonly used are:

1 g mol = 22.4 liters at standard conditions,
1 kg mol = 22.4 m³ at standard conditions,
1 lb mol = 359 cu ft at standard conditions.

The abbreviation (STP) is used to indicate standard conditions; for example, "the quantity of gas liberated was 12 cu ft (STP)."

Often it is convenient to carry through calculations and present stoichiometric data with all gas quantities expressed in terms of volume at standard conditions, even though the gases involved may be handled in the process under conditions far from standard. Such a procedure simplifies calculations and is particularly suitable when the gas temperatures are not accurately known or when several different gas temperatures are involved.

When the volume under conditions other than standard is desired, the calculation follows the relation

$$V = NV_o \times \frac{p_o}{p} \times \frac{T}{T_o} \qquad (2\text{--}2)$$

in which V, p, T, and N represent actual volume, absolute pressure, absolute temperature, and number of mols ($=$ mass/molecular weight); $V_o = 22.4$ if V is in liters and N in g mols or if V is in m³ and N in kg mols, or $V_o = 359$ if V is in cu ft and N in lb mols; and T_o and p_o represent standard temperature and pressure. An advantage of this type of volume calculation is that p_o and p can be expressed in atmospheres, millimeters of mercury, or pounds per square inch, and T and T_o in either °K (°C $+$ 273) or °R (°F $+$ 460) without introducing more new constants to remember (of course, both p and p_o must be in the same units in any one calculation and must be absolute, not gauge pressures; likewise T and T_o must be in the same units in any one calculation). On the other hand, the many values of the gas constant R for the various combinations of these units defy remembrance and may be difficult to find in handbooks.

The *law of combining volumes* follows from the ideal gas laws and affords a further simplification of engineering calculations. According to this principle, the relative volumes of the gaseous reactants involved in a definite chemical reaction and measured at the same temperature and pressure are small whole numbers. If the formulas used for the gaseous reactants represent the actual molecular formulas, these relative combining volumes are the same as the numbers of mols reacting. For example, if 100 cu ft (STP) of methane reacts completely with oxygen to form carbon dioxide and steam, the reaction is

$$CH_4(gas) + 2O_2(gas) \longrightarrow$$
$$CO_2(gas) + 2H_2O(gas).$$

By inspection, 200 cu ft (STP) of O_2 will be required, and 100 cu ft (STP) of CO_2 and 200 cu ft (STP) of H_2O gas will be formed.

The ideal-gas equation ($pV = NRT$) applies to the component gases of ideal gaseous mixtures, when p is the partial pressure of the component, N is the number of mols of the component, and V is the volume occupied by the gaseous mixture. This is the basis of the following often used relation between different expressions of gas analyses:

$$\text{volume } \% = \text{mol } \% = \frac{\text{partial pressure}}{\text{total pressure}}.$$
$$(2\text{--}3)$$

Modes of operation of unit processes

In the laboratory and in a number of full scale metallurgical processes, *batch* operation is used. A definite batch of material is placed in the apparatus and processed through a definite sequence of conditions and steps to obtain the products. A characteristic of batch operation is that the conditions in the system vary with time.

For large scale processes, *continuous* operation possesses many advantages. Continuous streams of feed materials enter the apparatus and continuous streams of products leave the apparatus. Two varieties of continuous operation can be distinguished; in one the system is said to operate continuously in a *steady state*, in the other the operation is in an *unsteady state*. For a system operating continuously in the steady state, the variables expressing conditions in the system are constant and do not vary with time, and the flows of materials and energy in and out of the system are at uniform rates. The system in the steady state is in many respects the simplest system to analyze quantitatively, as well as the easiest to operate practically. A simple example of a steady state system is a combination of fuel burner and brick combustion chamber, operated with constant fuel rate and constant air supply in surroundings that are at a fixed temperature and pressure. During the starting-up period, as the walls of the combustion chamber are being heated, the system is in an unsteady state. However, a steady state is reached after a period of heating up, the time required to reach this state depending mainly on the heat capacity of the system and the rate of generating heat in the flame. When the steady state prevails, such variables as temperature, temperature gradient, pressure, pressure gradient, gas velocity, gas composition, specific reaction rate, and many others which measure the conditions at any fixed point in the system are constant. The relations between these variables derived from chemical and physical principles can be applied quantitatively almost as if the system were static.

The absence of the necessity for dealing with time as a primary variable in steady systems greatly facilitates the quantitative engineering calculations which can be made in applying stoichiometric, thermochemical, heat-flow, fluid-flow, and other principles to the system. Accordingly, many of the illustrative calculations and problems given subsequently in this book refer to systems in the steady state. All actual operating systems are unsteady to some degree because of accidental or uncontrolled variations in operating conditions. However, even when wide variations occur, the purposes of most engineering calculations are best served by assuming a steady state and then using averaged data. Care and judgment must of course be exercised to obtain average data which fairly represent an average state of operation and not a temporary or unusual set of conditions. The additional accuracy which might theoretically be obtained by a rigorous analysis of unsteady operation rarely justifies the additional complications brought into the calculations.

Cyclic operation is a special case of unsteady state operation in which the system continually and repeatedly goes through a certain sequence of conditions. For example, in a heat regenerator, hot flue gas is passed through a chamber filled with a checkerwork of bricks until the temperature of the bricks approaches that of the incoming flue gas. Then the hot flue gas is diverted and cold air is passed through the bricks for heating. After the bricks are cooled by the air, the air flow is diverted and hot flue gas again brought in to start the next cycle. In this mode of cyclic operation, the regenerator system operates in the unsteady state but passes through the same sequence of conditions in each cycle. The temperature at a given point in the regenerator will move up and down cyclically, perhaps covering a very wide range. Also, the quantity of heat energy stored in the regenerator will vary cyclically over a correspondingly wide range. Engineering calculations for this kind of unsteady state can become extremely complicated and difficult because time must be considered an important primary variable.

From the above discussion, it will be evident that in applying stoichiometric and other principles to metallurgical systems, not only must the system under consideration be defined but also the mode of operation of the system must be taken into account. Thus, for a batch process the data usually cover a period of observation of one batch or an integral number of batches; for steady state operation, the data may be based arbitrarily on a definite time interval, as one hour or one day, or on a given quantity of feed or product, such as 1 ton or 100 tons; and for cyclic operation the data may cover one cycle or an integral number of cycles. If steady state operation is to be assumed, the validity of the assumption should be checked.

The *materials balance* of an unsteady system may be stated *for a given chemical element* as follows:

$$input = output + accumulation. \quad (2-4)$$

A steady system, by definition, will contain a fixed and unvarying total quantity of any given element so that the element balance simplifies to:

$$input = output. \quad (2-5)$$

This simpler relation also applies to batch, cyclic, and other unsteady state systems if input and output are measured for a period with the system in the same condition at the beginning and end of the period.

Rate of treatment, capacity, and retention time. One of the items of stoichiometric data used most often on a unit process is naturally the rate of treatment or tonnage. The specification of rate of treatment may be in terms of quantity of feed material, quantity of metal put through, or quantity of the important product, and for continuous processes usually is expressed in tons per 24-hour day. For example, a 1000-ton blast furnace is one which produces 1000 short tons (2000 lb) of pig iron in 24 hours; on the other hand a copper reverberatory furnace is rated in terms of the tons of solid charge smelted per 24 hours. Equipment operated batchwise usually is rated in terms of quantity fed or quantity produced per batch; the rate of treatment then is this figure multiplied by the number of batches treated per day. Thus, a 100-ton open-hearth furnace might tap 100 tons of steel per heat, averaging between two and three heats per day to give a rate of treatment of over 200 tons per day.

Capacity ratings, though indicative of the maximum rate of treatment, are in most cases nominal figures based on standard conditions, and may be unattainable for some conditions and easily exceeded in others. A common practice is to express rate of treatment as a percentage of nominal rated capacity.

The retention time, also called time of treatment or time of residence, is related closely to the rate of treatment. Thus improvements in a process which reduce the necessary time of treatment result in increased rate of treatment. In batch processes the retention time of materials in the system can be measured directly. In systems operating in the steady state with continuous flows of materials in and out, the method of estimating retention time depends very much on the internal workings of the process. The *mean retention time* (θ) defined below may be taken as a suitable gross measure of retention for any chemical element.

$$\theta = \frac{quantity\ in\ system}{rate\ of\ treatment}. \quad (2-6)$$

Any consistent set of units may be used; for example, if the quantity in the system is in pounds, the rate of treatment might be in pounds per hour and θ in hours. However, the retention calculated in this way is an average figure with different meanings for different kinds of material flows through the system. Thus, the passage of materials through the apparatus may involve one or a combination of the following three mechanisms: *displacement, mixing,* and *short-circuiting*.

An example of perfect *displacement* is a single file of ball bearings rolling in a *V*-trough through a furnace, Fig. 2–1. For such a system, the mean retention time (θ) calculated by the above relation is correct for each and every ball; that is, if the furnace chute holds 100 balls and the rate of passage is 10 per minute, the retention time for each ball is 10 minutes. Displacement is the predominant mechanism in a number of con-

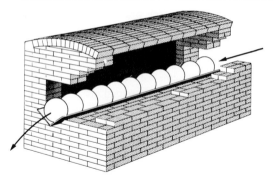

FIG. 2-1. Perfect displacement: ball bearings flowing through V-trough.

tinuous metallurgical processes, for example, in gas-solid reactions carried out as the solids move through the length of a rotary kiln.

Figure 2-2 is a simple system illustrating the effect of *mixing* on retention time. A copper sulfate solution and water are fed to an agitating tank, operated under ideal conditions of perfect mixing. Under these conditions, each drop of feed solution is uniformly distributed throughout the tank the instant after it enters; the solution throughout the tank and the effluent solution have exactly the same composition.

Since the concentration of copper sulfate is constant through the system, the mean retention time calculated as indicated above is V/v in which V is the volume of solution in the system and v is the volume of solution discharged from the agitator per unit time. To measure the retention time of Cu atoms in the tank, let us assume that at time $t = 0$ an injection of copper sulfate with n_o radioactively tagged copper atoms (Cu*) is added with the feed. The quantity of radioactive copper should be small enough not to have an appreciable effect on the total copper concentration, but large enough so that the radioactive atoms can be counted as they discharge from the tank. Now let n equal the number of Cu* atoms in the tank at time t ($n = n_o$ when $t = 0$). Then the

concentration in the tank and in the effluent at time t is n/V. The number leaving the tank in a time increment dt then is $-dn$, which must equal the product of the effluent concentration n/V and effluent volume $v\,dt$,

$$-dn = \frac{n}{V} v\,dt. \qquad (2\text{-}7)$$

Integrating, and remembering that $n = n_o$ when $t = 0$,

$$\frac{n_t}{n_o} = e^{-\frac{v}{V}t}. \qquad (2\text{-}8)$$

Since V/v is the mean retention time θ defined previously and n_t/n_o may be defined as r, the fraction still remaining in the system at time t,

$$r = e^{-\frac{t}{\theta}}. \qquad (2\text{-}9)$$

The fraction of the Cu discharging in the time interval from t to $t + dt$ is $-dr$; in other words $-dr$ is the fraction of the Cu atoms with retention times between t and $t + dt$. Differentiating the previous equation then,

$$-\frac{dr}{dt} = \frac{e^{-\frac{t}{\theta}}}{\theta}. \qquad (2\text{-}10)$$

The plot of $-dr/dt$ against t in Fig. 2-3 based on this equation shows graphically

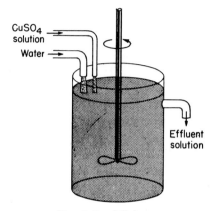

FIG. 2-2. Mixing.

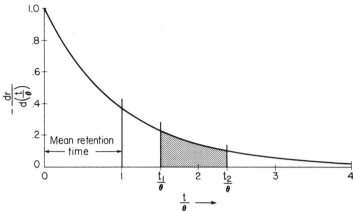

FIG. 2-3. Distribution of retention times for perfect mixing in a continuous agitator.

the distribution of retention times. The total area under this curve from $t = 0$ to $t = \infty$ is unity. The area under the curve between any two times t_1 and t_2 represents the fraction of the atoms with retention times between t_1 and t_2. Using Eq. (2-9), it can be shown that the fraction of the copper retained in the agitator for less than the mean retention time θ is $1 - 1/e$ or 0.63. Another characteristic of mixing shown by the curve in Fig. 2-3 is that the most probable retention time is zero; that is, a larger fraction of the Cu* leaves the agitator in the first second after mixing than in any subsequent one-second interval.

When mixing predominates over displacement as the characteristic of flow through an apparatus, the above analysis shows that a substantial part of the feed material is retained for a very short time while another substantial part is retained for times much larger than the mean retention time. Accordingly, if a definite time of retention is required, for example, to complete a chemical reaction, some substances may pass through unreacted while others are retained longer than necessary. One way to meet this problem is to divide the apparatus into cells or compartments with a series flow of materials from cell to cell, but with good mixing only within each cell.

Retention time not only measures duration of treatment, but for continuous processes also gives the engineer a measure of the time the system requires to respond to changes. The time of response is commonly referred to as the lag. If, for example, a material flows through a furnace by perfect displacement, a change in feed composition does not show up in the product until a time θ after the change in feed, where θ is the retention time. On the other hand, if the system operates with perfect mixing of the contents, the products start to respond at once to a change in feed, but after a time θ equal to the mean retention time, the response is only 63% complete. After time 2θ the change is 86.5% complete, and after time 3θ it is 95% complete. This general relationship is readily illustrated in terms of the mixing system shown in Fig. 2-2, as follows: Assume first that $CuSO_4$ and water are fed steadily at rates to produce a volume v per unit time of effluent solution containing c_1 grams $CuSO_4$ per liter. Then, at $t = 0$, let the feed rates be changed to be equivalent to a solution at the same volumetric rate v but at a new concentration c_2. The actual concentration c of $CuSO_4$ in the effluent solution will change from c_1 toward c_2 in accordance with the following relations, which are based on a simple materials balance for the time interval dt, during which the concentration in the mixing tank and in its effluent changes by dc. V is the volume of the tank, so that the mean retention time θ is V/v, as in the previous derivations.

$$c_2 v\, dt \;=\; cv\, dt \;+\; V\, dc \qquad (2\text{--}11)$$
$$\text{(input)} \quad \text{(output)} \quad \text{(accumulation)}$$

Integrating, from $t = 0$ when $c = c_1$, and substituting $\theta = V/v$,

$$\frac{c_2 - c}{c_2 - c_1} = e^{-\frac{t}{\theta}}. \qquad (2\text{--}12)$$

The fraction on the left of Eq. (2–12) is a relative measure of the uncompleted change at time t compared to the total change to be accomplished, so that we can say, for the *perfect mixing* system,

$$\% \text{ completion of change} = 100(1 - e^{-\frac{t}{\theta}})$$
$$= 100(1 - 10^{-\frac{t}{2.30\theta}}). \qquad (2\text{--}13)$$

To take a numerical example, assume that a 1% alloy is being melted continuously in a furnace with good mixing and a retention time θ of $1\frac{1}{2}$ hr, and that the feed is then changed to a 1.3% alloy. The total change to be accomplished is 0.3% in content of alloying element. After 3 hr ($t = 2\theta$), Eq. (2–13) indicates that the change will be 86.5% complete (since $e^{-2} = 0.135$) so that the alloy leaving the furnace will contain 1.26% of alloying element.

An appreciation of the nature and magnitude of the time lag of a system in responding to changes is important to the engineer who is attempting to control or adjust the process. The human impulse when adjustments are needed is to twirl the valves until things are right, but conservative adjustments, each followed by a period of patient waiting proportional to the retention time, are less likely to lead to disaster.

Short circuiting may occur from a number of causes, including location of feed entrances too close to product exits. The short-circuited material has a short retention time, but the reduction in retention time is likely to be more serious than that obtained as a result of mixing because it usually involves poorer exposure to conditions in the system.

Another factor affecting retention time in some practical metallurgical systems is that part of the contents of the system may be static and not in reality a part of the flow through the system. In a hearth-type furnace which has mechanical rakes to move granular solids over the hearth from feed to discharge, the material under the rakes may remain indefinitely as a dead load. Similarly, deposits built up on furnace walls or on furnace bottoms may constitute another kind of dead load. When retention times are calculated for such circumstances by Eq. (2–6), usually the quantities of materials tied up in the dead load should be omitted from the quantity in the system.

Since the metallurgical unit processes generally involve multiphase systems, with heterogeneous reactions proceeding at various rates, retention times for different chemical elements may differ widely in the same process. Moreover, the relative roles of displacement, mixing, and short-circuiting in the flow of materials through the system may be very difficult to assess. For these reasons, a quantitative application of the relations presented in the preceding paragraphs may prove difficult or impossible.

Weighing, sampling, and analysis

The most important data on a practical

metallurgical system are the stoichiometric data which tell the kinds, quantities, and compositions of materials entering and leaving the system. Since the objective in a unit process or sequence of unit processes is to extract metal and to separate chemical elements from each other, it is obvious that the stoichiometric data must furnish the primary yardsticks of metallurgical performance.

Weighing. Batch weighing involves individual weighing of loaded railroad cars, skips, trucks, hoppers, etc., and taking the net weight of material as the gross less the tare weight. This is a common practice in metallurgical plants, and requires the installation of suitable track scales or platform scales. When proper care is taken by the operator in obtaining both gross and tare weights individually and the scales are standardized, the weights should be correct to within 0.5%.

Solid materials moved by belt conveyors are weighed continuously by apparatus such as the Merrick weightometer, which is a weighing and integrating device supporting a short section of the conveyor belt.

Where no provisions are made for weighing large tonnages, various methods of weight estimation may be used for approximate results. Many of these methods are essentially volumetric, and involve counting the number of carloads, shiploads, or ladles and multiplying by a specially determined average weight per load. Or, the number of cubic feet in a stockpile or storage bin, for example, may be estimated by geometric methods and multiplied by the weight per cubic foot determined from actual measurement or from experience. Table 2-2 gives some typical figures used for this purpose. Still another source of approximate weight data involves calibration of feeding devices, such as belt feeders, pan feeders, in terms

TABLE 2-2

Bulk Densities of Various Materials

Material	Condition	Bulk density lb/ft^3
Coal	Broken, loose	45 to 60
Coke	Loose	23 to 32
Gravel	Dry	100 to 120
Iron ore (hematite)	Loose	150
Limestone	Broken, loose	100
Sand	Dry, loose, unsized	117
Slag	Granulated	53 to 60
Slag	Bank, crushed	80
Stone, average	Crushed	95 to 100
Stone, heavy (trap, greenstone)	Crushed	107

For fine-crushed ores and similar materials, the bulk density of the crushed but unsized material is about 57% of the density of the original solid mass.

of weight delivered per revolution and then counting revolutions with an automatic revolution counter.

As will be seen later, direct weighing of all feed and product materials for each processing step is not necessary and usually is not feasible. Many of the quantities can be calculated as accurately as they can be measured directly. However, direct weighing may be necessary for important plant intake materials, particularly if these materials are expensive or are purchased from other companies, and for the final salable products of the plant. Also, in making up furnace charges from ores and other solid materials, weighing is commonly used.

For ores and granular solids in general, the determination of moisture content at the time of weighing is as important a step as the weighing itself. Chemical analyses and many stoichiometric calculations are based on "dry weights," and an error in percent of moisture is just as serious as a direct weighing error.

Sampling. Poor sampling practices are a prime cause of inaccuracies in stoichiometric data. Unfortunately, accurate sampling of large quantities of ores, metals, and metallurgical products is often a difficult task, even a costly one. The cost of sampling facilities and manpower is direct and tangible, but the benefits from accurate sampling, though indirect, less tangible, and long-range, still can represent many dollars per month. Since numerical analyses can be obtained just as easily on poor samples as on good samples, a feeling of false security is easily engendered in the mind of the metallurgist in the office when he sees only the numerical results and does not observe how the samples were taken in the plant.

Composition of material moving in a process is characteristically nonuniform, except in certain special cases. Ideally, of course, a continuous process operating in a steady state should give uniform flow of materials, but such a state cannot be assumed; sampling and analysis are necessary to establish the existence of a steady state. The variations in composition may be variations with time or variations with position in the stream or in the transporting container; usually both time and position variations must be taken into account. For example, ore loaded into a railroad car is likely to be segregated in the car; conse-quently, the sample should be taken as a top to bottom section or increment, not just off the top. Then the frequency of sampling (number of sections per car and proportion of cars sampled) depends on the time uniformity of the material. Similarly, in sampling ingots, the pattern of taking drill samples from any one ingot must take into account segregation during solidification, while the number of ingots sampled will depend on the uniformity of metal composition as a function of time of pouring the ingots.

Deviations of the sample composition from the true material composition can be classified into two kinds: systematic and random. Systematic errors can be eliminated or allowed for only by study of the specific system being sampled and by appropriate design of sampling technique. Thus, segregation of alloy components during solidification may leave a lower-melting composition in the part of the ingot to solidify last, and a study of the mechanism and geometry of segregation will lead to a pattern of sampling giving a representative sample. Granular solids segregate systematically according to particle size and density. This leads to the practice of sampling "all of the stream part of the time" rather than "part of the stream all of the time." Again, in sampling suspensions such as solids in water, dust carried in a gas stream, metal containing inclusions, etc., careful study is necessary to eliminate systematic sampling errors.

The normal probability relation applying to random errors is useful in deciding on the size and frequency of samples, if some statistical data can be collected. For example, consider a continuous stream of material which is being sampled by taking all of the stream part of the time. Further, assume that variations in composition along the stream are random and nonsystematic and assume that the chemical analysis of each sample is accurate. Under these conditions take a series of n equal-sized samples according to a definite schedule and assay this set of samples accurately. From this data, the probable error of any single sample cut can be calculated:

probable error of analysis of one sample

$$= 0.6745 \sqrt{\frac{\sum \left(\begin{array}{c}\text{deviations of individual}\\ \text{analyses from mean}\end{array}\right)^2}{\text{number of samples}}}.$$

$$(2\text{--}14)$$

The probable error is defined so that statistically half the measurements will have greater errors than the probable error and half will have smaller errors.

The probable error of the mean analysis for n cuts is the same as the probable error of the assay of a composite sample made up by mixing n cuts, and is equal to the probable error for a single cut divided by $\sqrt{n}$. Thus, the number of individual sample cuts which must be mixed to obtain a composite sample with a desired probable error is given by:

number of cuts composited (n)

$$= \left(\frac{\text{probable error of single cut}}{\text{probable error of } n\text{-cut composite}}\right)^2. \tag{2-15}$$

If no systematic errors are introduced, the increase in effective number of cuts can be obtained by increasing the size of the cuts as well as by increasing the number of cuts. That is, the probable error is inversely proportional to the square root of the quantity of sample taken.

The probable-error calculations described above are based on first obtaining a set of statistical data by an experimental sampling of the system. During the period of experimental sampling, the true assay for the system sampled should not vary systematically with time. For example, if experimental sampling data are taken on a stream of crushed ore over a 24-hr period while the ore assay is increasing, the calculations will not be valid since they apply only when the variations in assay are random.

In sampling granular and lump solids, the size of sample required to obtain a given precision of sampling will depend on the particle size of the material and on the variability in composition from particle to particle. To take an extreme case, if the material being sampled is a mixture of two kinds of particles A and B, all the same size, one kind pure A and the other kind pure B, the sample required for given precision would involve a given total number of particles; therefore the sample weight would be proportional to the cube of the particle size. However, in crushed ores, slags, and other granular materials the compositions of the particles do not show such extreme variability, especially in the coarser pieces. Also, the variability of particle composition depends a great deal on the nature of the material, whether it is "spotty," uniform, finely disseminated, coarsely disseminated, etc. Accordingly, a purely theoretical prediction of proper sample size is hardly feasible, and the choice of sample size must be based in some way on experience or experiments with the type of material being sampled. Various authorities have published tables giving sample weights for ores, fuels, and other materials as a function of particle size. In many of these tables, the recommended sample weights follow the relation

$$W = kD^a, \tag{2-16}$$

in which W is the sample weight, D is the maximum size of particle, k is a constant depending upon the nature of the material and the precision desired, and a is a constant with a value usually between 2 and 3. For most purposes, it appears to have been satisfactory to use $a = 2$; that is, for a given precision the sample weight is made proportional to the square of the maximum particle size. Thus, if a 50-lb sample is considered necessary on $\frac{1}{4}$-inch material, an 800-lb sample should be taken on 1-inch material.

From the original stream of material to the gram or so used for chemical analysis, not one but usually a sequence of sampling steps are taken. Ordinarily the first step is the most difficult and the source of the largest errors, so that the subsequent steps of working up the sample and finally weighing out a small portion for analysis can be designed to have errors negligible by comparison.

Sampling practices. Many pitfalls await the unsuspecting sampler, and it is not always possible to foresee them in any given system. For example, in the drill sampling of solid metals, different analyses may result from drilling up and from drilling down, because in drilling down more of the fine powder produced at the start may be retained in the interstices of the metal. Then, too, in drill sampling there is always the question of how much surface dirt and surface oxidation should be included in the sample. Wet and dry drilling can give different results under some circumstances. Proper location of holes on the drilling template to avoid segregation errors is of obvious importance, but at times can be a treacherous problem.

Each system to be sampled presents its own problems, so that no positive rules can be laid down for avoiding systematic errors of various kinds. Table 2–3 summarizes information on typical sampling practices. The reader is advised to consult the references listed for details.

Analyses. The practicing metallurgical engineer is plagued occasionally by difficulties in obtaining prompt and accurate analyses, and even may find himself giving advice to analytical chemists or attempting analyses himself. Close liaison between metallurgist and analytical chemist and good understanding of each other's problems are essential to efficient plant operation. Nevertheless, a discussion of analytical chemistry and analytical practices is outside the scope of this book.

The last decade or so has been characterized by far-reaching developments in automatic and instrumental analyses and their application to industrial processes.

These methods are proving of great dollar-and-cents value in metallurgical plants, not only through manpower savings but also through the improvements in operating efficiency which result from better control. Thus, in Chapter 1, Instrumentation and Control was included in the list of the chief engineering aspects of metallurgical processes.

Most analytical results on solid samples are reported as weight percent on a dry basis. The drying is commonly an hour or more at 105–120°C. On the other hand, weight data in the plant are most likely obtained on an undried, "as is" basis. Thus, the percentage of moisture in the undried material is an important item of stoichiometric data, one in which error or entire neglect may have serious consequences. Assuming that the moisture sample is truly representative of the condition of the material at the time of weighing, obviously the drying procedure used in determining the percentage of moisture should be the same as, or equivalent to, that used in drying the main sample before chemical analysis.

Metallurgical calculations involve several different kinds of analyses, such as ultimate or elemental analyses, rational analyses, mineralogical analyses, proximate analyses, etc. Ordinary chemical analyses consist of determinations of the percentages of individual chemical elements, and are known as *ultimate* or *elemental* analyses. However, for many purposes analyses in terms of the various compounds present in the material are of interest. Such analyses are known as *rational* analyses. Where information of this kind is essential, chemical methods are sometimes applicable for the determination of one compound or group of compounds in the presence of other compounds containing the same elements. Considerable effort has been expended, for example, in attempting

TABLE 2-3

Typical Sampling Practices

Materials	Name of sampling method	Description
Granular solids: ores, concentrates, fluxes, etc., in heaps, railroad cars, boats, wheelbarrows	Grab sampling	Samples taken by hand with scoop or shovel, usually according to a fixed geometric pattern over the surface of the material.
Fine granular solids in cars, boats, bins, stockpiles	Pipe sampling	Pipe with longitudinal slot is forced into material, twisted until filled, and then withdrawn with sample. Pipe samples are taken at fixed intervals over the lot of ore being sampled.
Granular solids	Shovel sampling	During manual unloading or transferring by shovel, every alternate (or every fifth or tenth) shovelful is taken for the sample.
Granular solids in lots of less than 50 tons	Coning and quartering	Material is carefully piled in form of cone, the cone is flattened out and divided into four quarters. Two opposite quarters are taken as sample. Procedure repeated to obtain smaller cut.
Granular solids in continuous streams, slurries, ore pulps, etc.	Automatic mechanical sampling	Ore stream is caused to fall short distance under gravity, and sample is automatically taken at regular time intervals by a mechanical cutter designed to take "all of the stream part of the time."
Fine granular solids, small lots	Riffle sampling	Material is poured over equally spaced parallel riffles arranged so that half the stream is diverted into the sample and half into the reject. Riffling repeated as necessary to obtain desired cut.
Ore shipments, especially to custom metallurgical plants	Sampling mills	Ore is passed through a continuous plant in which it is first crushed, then sampled mechanically. The first sample cut is recrushed and resampled, and these operations are repeated to obtain an accurately representative sample generally weighing a fraction of a percent of the original ore.
Water solutions, fine slurries, pulps, nonvolatile liquids	Dip sampling	Mix thoroughly in agitator and dip out sample.
Petroleum products, other liquids, in tanks, tank cars		Lower a stoppered bottle to desired level, release stopper. An all-level sample may be taken by lowering bottle to bottom and then raising it to top at such a rate that it is just filled when it reaches the surface.
Liquid slags, mattes, metals, etc.	Chilled-rod sampling	A cold metal rod or bar is passed quickly through a stream of the molten material or is quickly inserted and then withdrawn from a batch held in the furnace, ladle, or crucible. The frozen shell sticking to the rod is taken as the sample.
Liquid slags, mattes, metals, etc.	Spoon or ladle sampling	A sample is taken with a metal spoon or small sampling ladle and quickly poured into a specially designed sample mold or onto a cold steel plate. Final sample for chemical analysis may be obtained by drilling hole through sample ingot. With some materials, sample may be granulated by pouring from ladle into water.
Cast metal, ingots, anodes, etc.	Drill and template sampling	Holes are drilled through pieces at specified points. A template may be used to systematize location of drill holes. The drill cuttings are collected as the sample.
Large metal sections: blooms, billets, slabs, squares, etc.	Drill sampling	Drill parallel to long axis of piece at a point halfway between center and outside on a diagonal.
Small metal sections: pipe, bar, rod, etc.	Machining	Mill off a complete cross section at right angles to the long axis.
Dusty air, other dust-laden gases	Dust sampling	Draw sample of known volume of gas through device which collects dust. Dust may be separated from gas by various means, including filtration, settling, electrostatic separation, impingement, etc.

REFERENCES

Liddell, Donald M., Handbook of Nonferrous Metallurgy. Vol. I, Principles and Processes. New York: McGraw-Hill, 1945.

Chemists, U.S. Steel Corporation, Sampling and Analysis of Carbon and Alloy Steels. New York: Reinhold, 1938.

ASTM Standards. Philadelphia: American Society for Testing Materials, 1946 (published triennially).

Taggart, Arthur F., Handbook of Mineral Dressing. New York: John Wiley, 1945.

by rational analyses to find the compounds present in coal. More often, however, for ores and metallurgical products, rational analyses are calculated from ultimate analyses, through the use of collateral knowledge or reasonable assumptions about the compounds present. A *mineralogical* analysis is a kind of rational analysis, in that the composition is given in percentages of the various minerals present. A *proximate* analysis gives the results of a standardized analytical or testing procedure such as might be used in commercial specifications, and does not necessarily give the true chemical composition. For example, a proximate analysis of coal gives percentages of "fixed carbon," "volatile matter," "ash," and "moisture," which cannot be defined in terms of the elements or compounds present in the coal.

One common but sometimes misleading practice is the expression of ultimate or elemental analyses in terms of the calculated percentages of compounds rather than of elements. Thus, the results of Al, Fe, Si, P, and other element determinations may be reported as % Al_2O_3, % Fe_2O_3 or % FeO, % SiO_2, % P_2O_5, etc., respectively, even for materials where the oxides are not present as such or even when the original material does not contain the necessary amount of oxygen. Stoichiometric calculations frequently require conversion of such data back to element percentages. In any event, analyses reported in terms of compounds should not be taken as evidence for the presence of the compounds, unless other reasons or evidence are available.

Gas analyses. Gas analyses are generally given on a dry, dust- and fume-free basis, as percent by volume, and are given in percentage of the molecular species actually comprising the gas. The volume percent is equal to mol percent, well within the accuracy of the usual analytical data. Many industrial gas analyses, especially those of products of combustion, are made with the well-known Orsat apparatus (Fig. 2–4) or some modification thereof. The common Orsat gives volume percentages directly;

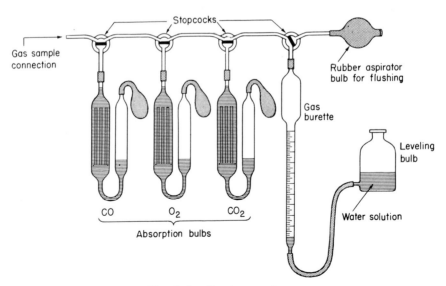

FIG. 2–4. Orsat apparatus.

and because the gases are handled over water and in the apparatus are always saturated with water regardless of original water content, the Orsat percentages give actual analysis on a dry basis. The common Orsat gives % N_2, % O_2, % CO_2, and % CO, but additions can be made to determine other gases.

Determination of moisture is usually a separate determination. In many cases, however, it is possible to omit a direct moisture determination and to obtain the moisture content by calculation from other data. The water content of a gas can be expressed as volume percent, but other units are also common in metallurgical calculations, including:

percent by volume;

percent by weight;

humidity, or pounds H_2O per pound dry gas (= grams H_2O per gram dry gas, etc.);

dew point, in °C or °F;

relative humidity, percent of saturation;

grains moisture per actual cubic foot;

grams moisture per actual cubic meter;

partial pressure, millimeters.

Table A–2 in the Appendix gives vapor pressures of water at temperatures from 0 to 100°C, which are needed for stoichiometric calculations when moisture content is expressed in terms of dew point or relative humidity.

Dry air at sea level analyzes by volume, 20.99% O_2, 78.03% N_2, 0.94% A, and 0.04% other gases, chiefly CO_2. Since both N_2 and A are inert in most processes, and in fact are determined together in gas analyses, the analysis of air is usually taken to be 79.0% N_2, 21.0% O_2. Another useful figure in stoichiometric calculations is the average molecular weight of dry air, 29.0. These slightly rounded-off figures will be used

hundreds of times in the problems in this book.

Constantly increasing use is being made of various methods of automatically indicating and recording gas analysis. Instruments involving physical measurements of thermal conductivity and measurements of infrared absorption are finding many applications around metallurgical operations.

Stoichiometric calculations

Metallurgical balances. When complete and accurate stoichiometric data are available for a process, a table is easily drawn up to show the balance of materials and the balance for each important element. A typical metallurgical balance is given in Table 2–4. The primary data are the weights and analyses of the materials entering and leaving the process — the other columns are calculated. Obviously, the metallurgical balance contains the most essential items of information for judging and controlling metallurgical performance, just as the balance sheet of a business is essential in determining profit or loss and in controlling the business.

In making over-all comparisons and appraisals of metallurgical performances, the metallurgist commonly abstracts one or two figures from the balance sheet. For example, the percentage recovery of the valuable metal in the primary product is a widely used performance yardstick. Frequently, too, recovery and grade figures will be abstracted jointly, for example, "97% of the copper was recovered in a 40% Cu matte." In still other circumstances, attention may center on minimizing metal losses, with percentage loss or metal content of a waste product serving as the measure of performance.

Details of accounting vary from plant to plant and from company to company, but

TABLE 2-4

Metallurgical Balance for an Iron Blast Furnace*

Materials	Weight lb	Iron %	Iron lb	Silica %	Silica lb	Alumina %	Alumina lb	Lime %	Lime lb	Magnesia %	Magnesia lb	Phosphorus %	Phosphorus lb	Manganese %	Manganese lb	Sulfur %	Sulfur lb
Feed Materials																	
Ore No. 1	15,978	49.80	7,957	5.8	927	3.0	479	0.2	32	0.2	32	0.068	10.87	0.77	123	0.010	2
Ore No. 2	1,653	46.80	773	8.2	135	2.3	38	0.2	3	0.2	3	0.064	1.06	1.12	19	0.062	1
Ore No. 3	7,893	47.00	3,710	10.6	837	2.3	182	0.2	16	0.2	16	0.078	6.16	0.65	51	0.014	1
Ore No. 4	1,026	53.30	547	3.2	33	2.3	24	0.1	1	0.2	2	0.072	0.74	0.43	4	0.016	
Cinder, scale, scrap	3,450	61.25	2,113	7.9	272	1.3	45	1.5	52	0.4	14	0.155	5.35	0.59	20	0.112	4
Coke	15,505			5.3	822	3.1	480	0.3	47	0.2	31	0.021	3.26			0.921	143
Limestone	7,733			2.7	209	1.5	116	51.5	3982	1.3	100	0.031	2.40			0.070	5
TOTALS			15,100		3235		1364		4133		198		29.84		217		156
Products																	
Pig iron: calculated	16,064	94.00	15,100	2.14**	344							0.185	29.84	0.903	145	0.03	5
actual				2.68**								0.158		1.03		0.03	
Slag: calculated	8,846			32.7	2891	15.4	1364	46.7	4133	2.2	198			0.82†	72	1.71	151
actual				33.8		15.7		47.2		2.0				0.78			
TOTALS (calculated from feed)			15,100		3235		1364		4133		198		29.84		217		156

*After Camp, J. M., and Francis, C. B., The Making, Shaping, and Treating of Steel. Pittsburgh: Carnegie-Illinois Steel Corp., 1940.

**Present in pig iron as Si.

†Present in slag as MnO.

complete balances are always kept on input and output of the plant as a whole, of important divisions of the plant, and of some of the individual unit processes. These may be on a daily basis, but arranged so that weekly, monthly, and yearly balances are readily composited. Figure 2–5 gives an example of a daily balance.

Several factors operate to prevent exact balances between input and output. Accumulation or depletion of material in process may account for large differences between input and output over a short term. Periodic inventories and cleanups, however, furnish a direct check on accumulation and depletion. Spillages and unaccountable losses also show up when the materials balance is calculated. Often, however, consistent differences between input and output are the result of errors in the primary stoichiometric data, and it can become a serious and embarrassing problem to determine whether failure to balance is due to unex-

plained losses, accumulations in equipment, weighing errors, or analytical errors.

Calculation of missing stoichiometric data. Direct and accurate measurement of the quantities of all the products of a given unit process is difficult, sometimes practically impossible. Particular difficulties arise in attempting to measure the large quantities of hot gases which are major products of many processes — one such measurement in a 10-ft diameter flue may constitute a long research problem in itself. Accordingly, calculation of weights and volumes of some of the products is common practice and in many cases gives more accurate information than can be hoped for from any direct measurement. As will be seen, the readily measurable data may make possible two or more independent calculations of the same quantity and give an opportunity for checking.

Since sampling and analysis are often feasible where quantity measurements are

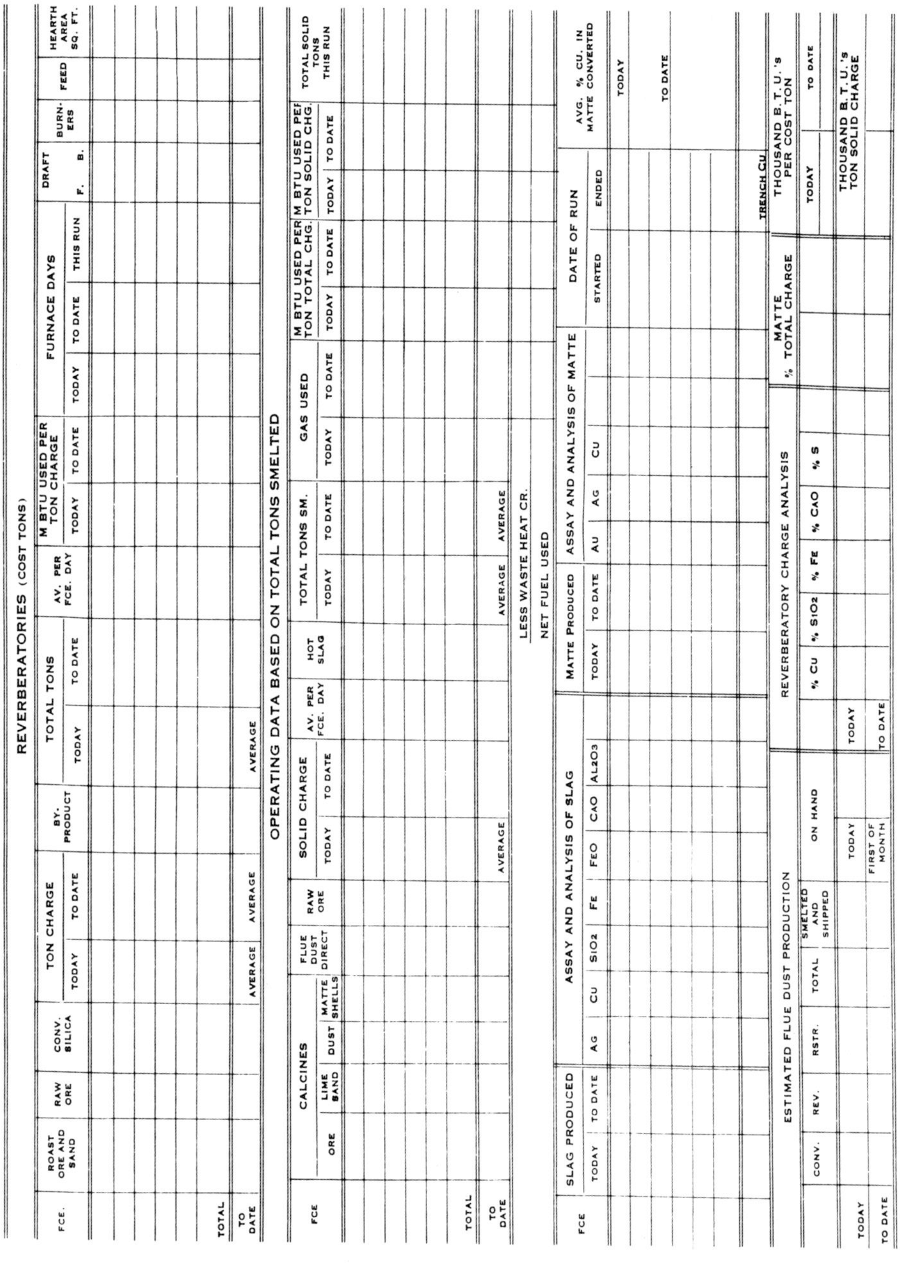

Fig. 2–5. Typical form for daily balance record.

not, less dependence is placed on calculated analyses except for checking. However, when the materials fed to a process are very erratic and spotty in metal content or when their metal content is so low that analytical accuracy is poor, the calculated feed analysis may be more reliable than the chemical analysis.

In principle, all the quantities entering and leaving a continuous process in the steady state, or a batch process with no accumulation or loss, can be calculated if complete analyses on all materials and the

FIG. 2–6. Steady state process.

quantity of one of the materials are known. Thus, for the process indicated in Fig. 2–6, let A, B, and C equal weights of feed materials and X, Y, and Z equal weights of products. Feed and products are all analyzed for elements 1, 2, 3, 4, etc. Let a_1, a_2, a_3, etc., equal respective percentages of each element in product A; b_1, b_2, b_3, etc., equal respective percentages in product B; etc. Then the principle of conservation of elements gives:

$$a_1A + b_1B + c_1C - x_1X - y_1Y - z_1Z = 0;$$
$$a_2A + b_2B + c_2C - x_2X - y_2Y - z_2Z = 0;$$
$$a_3A + b_3B + c_3C - x_3X - y_3Y - z_3Z = 0;$$
$$a_4A + b_4B + c_4C - x_4X - y_4Y - z_4Z = 0.$$

As many of these equations can be written as there are elements analyzed for in all the materials. Also,

$$A + B + C - X - Y - Z = 0.$$

These simultaneous equations can be solved for the unknown quantities; if n independent element balances can be written, n unknowns can be solved for. Though these equations are written in a form to suggest that the unknowns are weights (A, B, C, X, Y, Z, etc.), one or more of the individual analyses may be among the n unknowns to be solved for. If there are more independent equations than unknowns, two sets of answers can be calculated and checked against each other.

Solutions of simultaneous equations set up from practical data sometimes are grossly in error because the solution can greatly magnify small errors in the original data. Even impossible negative answers may be obtained. Such results are due to the fact that solutions for unknowns in simultaneous equations are necessarily obtained by a method of difference, and the large errors arise when at some stage in solving the equations a small quantity is determined as a difference between two large quantities. Care must be taken to recognize and avoid troubles of this kind.

In practice it is rarely necessary to set up formally and to solve a list of simultaneous equations such as that indicated above. Brief study of the problem usually discloses simpler procedures for a given type of process. These are based on the fact that all or substantially all of a given element enters or leaves the process in one stream. For example, N_2 may enter entirely with air and leave entirely in the gaseous product; if the percentage of N_2 in the flue gas is available and the volume of either air or flue gas is available, the other volume is readily calculated without regard for other stoichiometric data on the process. Similarly, if all C enters in the fuel and leaves in the flue gases, data on fuel rate plus analyses of the fuel and of the flue gases for total C permit direct calculation of the quantity of flue gases. In smelting processes producing liquid slag as a major product, all SiO_2, CaO, Al_2O_3, MgO, etc., may leave in the slag product, or near enough to all so that no large error is introduced by ignoring other products in balances for these oxides.

It should be emphasized that stoichiometric calculations for a given system do not depend on a detailed knowledge and understanding of the process within the system (except for information on accumulation in batch and unsteady state systems). What is required is only data on what enters and what leaves the system. Thus the principles discussed in this chapter suffice for the solution of a wide variety of problems on all the unit processes which the student can solve at this stage without concerning himself with internal process details. However, it is easier to work on a problem with some conception of its physical counterpart in mind, so many of the stoichiometric problems on specific unit processes will be given in later chapters.

Charge calculations. Charge calculations are necessary in advance of process operation to determine what quantities of materials must be fed to the process to obtain the desired products. For example, a smelting furnace may be fed a mixture of three or four different ores, scrap metal, and various materials returned from other steps in the plant. The furnace products may be liquid metal and liquid slag. In smelting operations of this kind, smooth operation requires that the slag composition be held within rather narrow limits. Accordingly, the proportions of each of the available feed materials to be mixed in the charge are calculated carefully in advance.

Stoichiometric calculations made before the process is carried out obviously require some knowledge or assumptions as to the nature of the products and as to the way in which the various elements will be distributed among the products. For a given unit process the behavior of some of the elements becomes self-evident when the process

is examined or described, while for other elements empirical stoichiometric rules based on previous operating experience may guide the calculations. For example, in making a charge calculation, it might be assumed from previous experience that 96% of a given metal in the charge would report in the metallic product of the furnace, 3% in the slag, and 1% in the flue gas. In the same smelting process, all the SiO_2, Al_2O_3, and CaO in the charged materials could be expected to enter the slag. Often the charge calculations are much simplified by intelligent use of approximations without affecting the value of the results. Referring to the example above, the charge calculations might be greatly simplified by assuming all the metal reported in the liquid-metal product instead of just 96%; this assumption probably would not have any effect of practical significance on the estimation of quantities of materials to be fed to the process.

A general procedure for solving stoichiometric problems. Stoichiometric problems are rarely difficult, but can be long and tedious and usually offer many opportunities for arithmetic and other errors which are not easy to find and correct. Accordingly, from the very start the student should deliberately follow a logical procedure of problem solving, clearly writing out all steps so that the solution can be quickly retraced and checked by another man after the problem is cold. *The point of view that arithmetic, decimal point, and dimensional errors are not serious if the principle of the calculation is correct cannot be tolerated in engineering calculations, for obvious reasons.*

The procedure outlined below will be found suitable for solving most stoichiometric problems, subject of course to modification in individual cases.

(1) Make a simple line sketch of the system under consideration. Add labeled

arrows for each feed and product material. A simple square or rectangle to represent the system, with feed arrows on the left and product arrows on the right, may be sufficient if the system is easily visualized.

(2) Add the known stoichiometric data (quantities and analyses) to the sketch if space permits or tabulate it separately if necessary.

(3) Note the stoichiometric assumptions and relations characteristic of the unit process which can serve as a basis for the solution.

(4) Choose a convenient basis for the calculations. This may be a unit of time (e.g., 1 hr or 24 hr), a single batch in a batch process, or a fixed quantity (e.g., 1 lb, 100 lb, 1 ton) of one important feed or product material. At the same time select a consistent set of units to be followed throughout the calculations which will require a minimum of conversion factors.

(5) Write out the balanced chemical equations which will account for quantitative relations between specific compounds in the feed and product materials. Do not attempt to write all the equations to represent the details of reactions and reaction mechanisms within the system.

Refer to the given data and the selected calculation basis, and insofar as possible determine the quantities of reactants involved in the chemical equations. Write these quantities under the equations.

In solving many problems, a complete set of equations to represent the stoichiometry of the process is not necessary, since the equations representing element balances [see (6), below] give most of the needed information.

(6) Write the equations representing the element balances which are independent of balances already set up by the chemical

equations under (5). Indicate the unknowns by appropriate symbols or words. In the same way write out equations to represent the stoichiometric assumptions listed under (3).

(7) Inspect the relations in (5) and (6) and decide on a procedure for solving for the unknowns. Frequently a procedure can be found which involves solving one equation at a time for one unknown, then using that answer to make possible the solution of the next equation, etc. Such a procedure is preferable to setting up a group of simultaneous equations and attempting to solve them as a group.

Carry out the solution as planned.

(8) Make up a metallurgical balance table for the system from the given and calculated weights and analyses. The preparation of this table affords a good way of checking the answers, since the input and output of individual elements can be calculated independently of the previous calculations.

(9) Check the solution to see that all given conditions are fulfilled, using computations independent of those originally made in solving the problem.

Examples of stoichiometric calculations. As already mentioned, stoichiometric calculations involve data on quantities and compositions of materials entering and materials leaving a metallurgical system and generally do not involve information as to what is actually occurring within the system. Accordingly, examples of calculations for several unit processes can now be considered, even though the unit processes themselves will not be described in detail until later chapters. It might be noted also that the over-all stoichiometric relations for a unit process, such as those described in the following examples, generally represent the best starting point for studying a unit process.

Example 1

Preliminary design calculations are being made for a calcining operation to be carried out continuously in a rotary kiln. The kiln will receive a damp filter cake of aluminum hydroxide, analyzing 55% Al_2O_3 and 45% total H_2O (free and combined), and will deliver pure Al_2O_3 as the solid product. The fuel consumption is estimated to be 0.2 lb of fuel oil per lb of Al_2O_3 produced, the fuel oil analyzing 85% C and 15% H. The air for combustion will be controlled so that 20% excess air is supplied over the theoretical quantity for perfect combustion of the fuel. Find the volume of gases leaving the kiln per ton of Al_2O_3 produced and find also the wet and dry analyses of the flue gases.

Solution

Figure 2–7 shows the system schematically, with arrows indicating the flow of feed and products.

ASSUMPTIONS:

Complete decomposition of $Al(OH)_3$ into Al_2O_3 and H_2O.
Complete combustion of fuel to CO_2 and H_2O.
20% excess air for combustion.

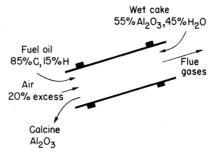

BASIS OF CALCULATION:

1 short ton (2000 lb) Al_2O_3; English units.

CHEMICAL REACTIONS:

(1) $2Al(OH)_3 \longrightarrow Al_2O_3 + 3H_2O$.
(2) $C \text{ (fuel)} + O_2 \longrightarrow CO_2$.
(3) $H \text{ (fuel)} + \frac{1}{4}O_2 \longrightarrow \frac{1}{2}H_2O$.

ELEMENT BALANCES:

N_2 in air = N_2 in flue gases.
Al in feed = Al in calcine.

FIG. 2–7. Calcining aluminum hydrate in a rotary kiln.

CALCULATIONS:

The H_2O in the flue gases comes from the free and combined water in the feed material and from the combustion of the fuel.

$$\text{Weight } H_2O \text{ from feed} = \frac{45}{55} \times 2000 \text{ lb} = 1640 \text{ lb}.$$

$$\text{Volume } H_2O \text{ from feed} = \frac{1640}{18} \times 359 \text{ cu ft} = 32{,}600 \text{ cu ft (STP)}.$$

$$H_2O \text{ from fuel} = \frac{0.15 \times 0.2 \times 2000}{2} \text{ lb mols} = 30 \text{ lb mols}$$

$$= 30 \times 359 \text{ cu ft} = 10{,}800 \text{ cu ft (STP)}.$$

$$\text{Total } H_2O \text{ in flue gases} = 32{,}600 + 10{,}800 = 43{,}400 \text{ cu ft (STP)}.$$

$$CO_2 \text{ in flue gases} = \frac{0.85 \times 0.2 \times 2000}{12} \text{ lb mols} = 28.3 \text{ lb mols}$$

$$= 28.3 \times 359 \text{ cu ft} = 10,160 \text{ cu ft (STP)}.$$

To find the O_2 and N_2 in the flue gases, it is necessary first to calculate the air consumption. From equation (2) above,

cu ft O_2 to burn C in fuel = cu ft CO_2 formed = 10,160 cu ft (STP).

From equation (3),

cu ft O_2 to burn H in fuel = $\frac{1}{2}$ cu ft H_2O formed = 5400 cu ft (STP).

Total O_2 theoretically required for combustion = 10,160 + 5400 cu ft

$$= 15,560 \text{ cu ft (STP)}.$$

20% excess oxygen = 0.2 × 15,560 = 3110 cu ft (STP) to flue gas.

$$N_2 = \frac{79}{21} \times \text{total } O_2 = \frac{79}{21} \times (15,560 + 3110) = 70,230 \text{ cu ft (STP)}.$$

Now quantities of all the flue gas constituents can be tabulated and totaled, and the wet and dry analyses calculated:

	Cu ft (STP)	Volume %	
		Wet basis	Dry basis
CO_2	10,160	8.0	12.2
N_2	70,230	55.3	84.1
O_2	3,110	2.5	3.7
H_2O	43,400	34.2	—
Totals	126,900	100.0	100.0

EXAMPLE 2

A melting furnace burns 10 tons per day of pulverized coal having the following analysis:

C	H	O	N	H_2O	Ash	Total
65%	5%	12%	1%	5%	12%	100%

The Orsat analysis of the flue gases is:

N_2	CO_2	CO	O_2	Total
81%	15%	0%	4%	100%

Calculate the following: (a) the volume of dry air used for combustion and the percent excess over theoretical air, (b) the wet and dry volumes of flue gases leaving the furnace, and (c) the percentage of H_2O in the flue gases.

Solution

The combustion system is indicated schematically in Fig. 2–8.

ASSUMPTIONS:

None of the coal is left unburned.

BASIS OF CALCULATIONS:

24 hr of operation, 20,000 lb of coal; English units.

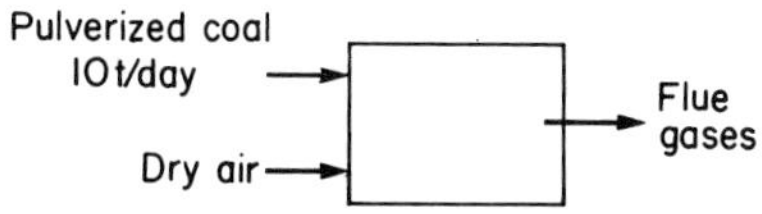

CHEMICAL REACTIONS:

(1) $C \text{ (coal)} + O_2 \longrightarrow CO_2$.

(2) $H \text{ (coal)} + \frac{1}{4}O_2 \longrightarrow \frac{1}{2}H_2O$.

Pulverized coal
10 t/day

Dry air

Flue
gases

Fig. 2–8.   Combustion system for melting furnace.

ELEMENT BALANCES:

N_2 in coal (as combined N) + N_2 in air = N_2 in flue gases.
C in coal = C in flue gases.
Total H in coal (as combined H and H_2O) = H in flue gases (as H_2O).
O_2 in coal (as combined O) + O_2 in air = total O_2 in flue gases (as O_2, H_2O, and CO_2).

Inspection shows that the stoichiometric relations represented by chemical equation (1) and the carbon balance can be solved directly to find the dry volume of flue gases, after which the other relations can be solved.

CALCULATIONS:

From equation (1) and the carbon balance,

$$\text{cu ft of } CO_2 \text{ in flue gases} = \frac{20{,}000 \times 0.65}{12} \times 359 = 389{,}000 \text{ cu ft (STP)}.$$

The CO_2 in the flue gases is 15% by volume on the dry basis, so:

$$\text{dry volume flue gases} = \frac{389{,}000}{0.15} = \underline{2{,}590{,}000 \text{ cu ft (STP)}}.$$

$$H_2O \text{ from combustion of H in coal} = \frac{0.05 \times 20{,}000}{2} \times 359 = 180{,}000 \text{ cu ft (STP)}.$$

$$H_2O \text{ from moisture in coal} = \frac{0.05 \times 20{,}000}{18} \times 359 = 20{,}000 \text{ cu ft (STP)}.$$

Therefore,

$$\text{wet volume flue gases} = 2{,}590{,}000 + 180{,}000 + 20{,}000$$
$$= \underline{2{,}790{,}000 \text{ cu ft (STP)}},$$

and

$$\% \, H_2O \text{ in flue gases} = \frac{200{,}000}{2{,}790{,}000} \times 100 = 7.2\%.$$

Now the volume of air supplied can be calculated from the nitrogen balance:

$$N_2 \text{ in flue gases} = 0.81 \times 2{,}590{,}000 = 2{,}100{,}000 \text{ cu ft (STP)}.$$

$$N_2 \text{ from fuel} = \frac{0.01 \times 20,000}{28} \times 359 = 2560 \text{ cu ft (STP), which can be neglected.}$$

$$\text{Volume of dry air supplied} = \frac{2,100,000}{0.79} = \underline{2,660,000 \text{ cu ft (STP)}}.$$

The oxygen balance affords a check on the calculations thus far:

$$O_2 \text{ from air} \quad = 2,660,000 \times 0.21 \qquad = 559,000 \text{ cu ft}$$

$$O_2 \text{ from dry fuel} = \frac{0.12 \times 20,000}{32} \times 359 \qquad = 27,000 \text{ cu ft}$$

$$O_2 \text{ from fuel moisture} = \frac{0.05 \times 20,000}{18 \times 2} \times 359 = \underline{10,000 \text{ cu ft}}$$

$$\text{Total } O_2 \text{ input} = 596,000 \text{ cu ft (STP)}$$

$$O_2 \text{ in flue gases as } CO_2 \qquad = 389,000 \text{ cu ft}$$

$$O_2 \text{ in flue gases as } H_2O = \frac{200,000}{2} \qquad = 100,000 \text{ cu ft}$$

$$O_2 \text{ in flue gases as } O_2 \quad = 0.04 \times 2,590,000 \quad = \underline{104,000 \text{ cu ft}}$$

$$\text{Total equivalent } O_2 \text{ output} = 593,000 \text{ cu ft (STP)}.$$

This can be considered a satisfactory check on the oxygen balance.

$$\begin{array}{c} \text{Theoretical } O_2 \text{ required for} \\ \text{combustion of fuel} \end{array} = \frac{(O_2 \text{ for burning C}) + (O_2 \text{ for burning H})}{-(O_2 \text{ from fuel})}$$

$$= 389,000 + \frac{180,000}{2} - 27,000$$

$$= 452,000 \text{ cu ft (STP)}.$$

$$\% \text{ excess } O_2 \text{ supplied in air} = \frac{559,000 - 452,000}{452,000} \times 100 = \underline{23.5\%}.$$

EXAMPLE 3

Iron ore, limestone, and coke of the analyses given below are to be charged to an iron blast furnace:

Analysis, %

	Fe	SiO_2	CaO	MgO	Al_2O_3	H_2O	Mn	P	C	S
Ore	50.0	8.0			3.0	10	2	0.05		
Limestone		4	50	2	1	1				
Coke	1	5			3	2			86	1

The furnace is operated to produce a pig iron analyzing 94% Fe, 4% C, and 1% Si, and a slag in which % CaO + % MgO = % SiO$_2$ + % Al$_2$O$_3$. Previous experience indicates that the coke consumption will be 1800 lb per 2000 lb of pig iron produced and that the quantity of air in the blast will be 80% of that theoretically required to burn the C in the coke to CO. Figure 2–9 shows the system schematically.

Calculate the following: (a) the weights of ore and limestone to be charged per 2000 lb of coke, (b) the volume of dry air to be supplied in ft^3/min (STP) for a furnace producing 1000 short tons of pig iron per 24 hours, (c) the composition of the blast furnace gas and the volume produced per day in a 1000-ton furnace.

SOLUTION

ASSUMPTIONS:

Assume all iron from the ore reports in pig iron and neglect iron from coke. Neglect Mn in calculations. Assume air is dry. Assume all H$_2$O in ore, stone, and coke is evaporated to enter blast furnace gas without decomposition. Coke consumption = 1800 lb per 2000 lb pig iron. In slag, % CaO + % MgO = % SiO$_2$ + % Al$_2$O$_3$. Air 80% of equivalence to burn C to CO. Neglect dust loss.

CHEMICAL EQUATIONS:

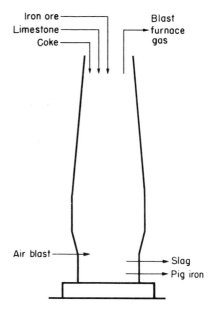

Iron ore ——
Limestone ——
Coke ——

Blast furnace gas

Air blast ——

Slag
Pig iron

FIG. 2–9. Iron blast furnace.

(1) $C + \frac{1}{2}O_2 \longrightarrow CO$.

(2) $3CO + Fe_2O_3 \longrightarrow 2Fe + 3CO_2$.

(3) $2C + SiO_2 \longrightarrow 2CO + Si$ (in pig iron).

(4) $CO_2 + C \longrightarrow 2CO$.

(5) $CaCO_3 \longrightarrow CaO + CO_2$.

(6) $MgCO_3 \longrightarrow MgO + CO_2$.

ELEMENT BALANCES:

C in coke + C in CaCO$_3$
    = C in pig iron + C in gases.

N$_2$ in air = N$_2$ in blast furnace gases.

H$_2$O in ore, stone, and coke
    = H$_2$O in blast furnace gases.

Fe in ore = Fe in pig iron (see assumptions).

CaO in limestone = CaO in slag.

MgO in limestone = MgO in slag.

Si in ore, stone, and coke
    = Si in slag + Si in pig iron.

Al$_2$O$_3$ in ore, stone, and coke = Al$_2$O$_3$ in slag.

(For a process as complex as that in the iron blast furnace, a variety of combinations of stoichiometric equations and element balances are possible, providing each combination sums up to the same over-all results insofar as materials leaving and materials entering are concerned.)

(a) *Charge calculation.* Basis, 2000 lb of coke; English units. Inspection of the available stoichiometric relations shows that the quantity of ore can be calculated directly from the iron balance and the given coke consumption, after which the quantity of limestone can be calculated by setting up the equation given to represent the desired slag composition.

$$\text{Lb of pig iron} = \frac{2000}{1800} \times 2000 = 2220 \text{ lb.}$$

$$\text{Lb of Fe} = 0.94 \times 2220 = 2090 \text{ lb.}$$

$$\text{Lb of ore} = \frac{2090}{0.50} = \underline{4180 \text{ lb.}}$$

Let x = weight of limestone to be charged per 2000 lb of coke.
Slag constituents:

SiO_2 in slag = SiO_2 in ore + SiO_2 in stone + SiO_2 in coke − equiv. SiO_2 (as Si) in pig iron

$$= 0.08 \times 4180 + 0.04x + 0.05 \times 2000 - 0.01 \times 2220 \times \frac{60}{28}$$

$$= 334 + 0.04x + 100 - 48 = 386 + 0.04x \text{ (lb)}.$$

Al_2O_3 in slag = Al_2O_3 in ore + Al_2O_3 in stone + Al_2O_3 in coke

$$= 0.03 \times 4180 + 0.01x + 0.03 \times 2000$$

$$= 125 + 0.01x + 60 = 185 + 0.01x.$$

$$\text{CaO in slag} = \text{CaO in stone} = 0.50x.$$

$$\text{MgO in slag} = \text{MgO in stone} = 0.02x.$$

Now the given relation, $\text{CaO} + \text{MgO} = \text{SiO}_2 + \text{Al}_2\text{O}_3$, can be set up in terms of x:

$$0.50x + 0.02x = 386 + 0.04x + 185 + 0.01x.$$

Solving this relation,

$$x = \underline{1215 \text{ lb of limestone.}}$$

(b) *Blast requirement per minute.* The volume of air can be calculated using chemical equation (1) and the assumption that 80% of the C in the coke reacts.

$$\text{1000 tons of pig iron per day} = \frac{1000 \times 2000}{24 \times 60} \text{ lb/min}$$

$$= 1390 \text{ lb of pig iron per min.}$$

$$\text{Lb of coke per min} = 1390 \times \frac{1800}{2000} = 1250 \text{ lb of coke per min.}$$

$$\text{Cu ft of O}_2 \text{ per min} = \frac{0.8 \times 0.86 \times 1250}{12} \times \frac{1}{2} \times 359$$

$$= 12,870 \text{ ft}^3/\text{min (STP)}.$$

$$\text{Cu ft of air per min} = \frac{12,870}{0.21} = \underline{61,200 \text{ ft}^3/\text{min (STP)}.}$$

(c) *Blast furnace gas.*   Basis, 2000 lb of coke.   Inspection of the available stoichiometric relations shows that the quantities of N_2 and H_2O in the blast furnace gas can be calculated straightforwardly, whereas calculation of the CO and CO_2 quantities will require working out the stoichiometry represented by chemical equations (1) to (6).

$$N_2: \text{cu ft of } N_2 = \frac{79}{21} \text{ cu ft } O_2 \text{ per 2000 lb coke.}$$

The oxygen requirement is 80% of theoretical to convert C in coke to CO.

$$\text{Cu ft of } N_2 = \frac{79}{21} \times 0.8 \times \frac{0.86 \times 2000}{12} \times \frac{359}{2} = 77{,}500 \text{ cu ft (STP).}$$

$$H_2O: \text{lb of } H_2O \text{ in blast furnace gas} = \text{lb in ore} + \text{lb in stone} + \text{lb in coke}$$
$$= 0.1 \times 4180 + 0.01 \times 1215 + 0.02 \times 2000$$
$$= 470 \text{ lb.}$$

$$\text{Cu ft of } H_2O \text{ in gases} = \frac{470}{18} \times 359 = 9370 \text{ cu ft (STP).}$$

CO and CO_2: the number of lb mols for each of the reactions (1) to (6) can be calculated first to determine the stoichiometric behavior of carbon.

Reaction (1): $\text{mols C} = \text{mols CO} = 0.8 \times 0.86 \times \frac{2000}{12} = 114.5 \text{ lb mols.}$

Reaction (2): $\text{mols CO} = \text{mols } CO_2 = \frac{3}{2} \times \text{mols Fe} = \frac{3}{2} \times \frac{0.5 \times 4180}{56} = 56.0 \text{ lb mols.}$

Reaction (3): $\text{mols C} = \text{mols CO} = 2 \times \text{mols Si} = 2 \times \frac{0.01 \times 2220}{28} = 1.6 \text{ lb mols.}$

Reaction (4): C for reaction (4) = total C in coke − C consumed in reactions (1) and (3) − C in pig iron.

$$\text{Mols C for reaction (4)} = \frac{0.86 \times 2000}{12} - 114.5 - 1.6 - \frac{0.04 \times 2220}{12} = 19.7.$$

Mols CO_2 consumed = 19.7.

Mols CO produced = 2 × mols C = 39.4.

Reaction (5): $\text{mols } CO_2 = \text{mols CaO} = \frac{0.5 \times 1215}{56} = 10.8 \text{ lb mols.}$

Reaction (6): $\text{mols } CO_2 = \text{mols MgO} = \frac{0.02 \times 1215}{40} = 0.6 \text{ lb mol.}$

Now the net lb mols of CO and CO_2 produced in the process can be found by algebraic addition of the quantities involved in reactions (1) to (6):

$$\text{Lb mols of CO} = 114.5 - 56.0 + 1.6 + 39.4 = 99.5 \text{ lb mols.}$$
$$\text{Lb mols of } CO_2 = 56 - 19.7 + 10.8 + 0.6 = 47.7 \text{ lb mols.}$$

Cu ft of CO = 99.5 × 359 = 35,700 cu ft (STP).

Cu ft of CO_2 = 47.7 × 359 = 17,100 cu ft (STP).

Tabulating the calculated quantities of the constituents of the blast furnace gas:

	Cu ft (STP)	Volume %	
		Wet	Dry
N_2	77,500	55.5	59.5
H_2O	9,370	6.7	—
CO	35,700	25.6	27.4
CO_2	17,100	12.2	13.1
Total volume per 2000 lb coke	139,670	100.0	100.0

Since 1800 lb of coke are used per 2000 lb of pig iron and the furnace produces 1000 tons of pig iron per day, the coke consumption in 24 hr is 900 tons. Accordingly, the quantity of blast furnace gas produced per day is 900 × 139,670 or 126,000,000 cu ft (STP), on a wet basis.

SUPPLEMENTARY REFERENCES

BUTTS, ALLISON, *Metallurgical Problems*. New York: McGraw-Hill, 1943.

HOUGEN, O. A., and WATSON, K. M., *Chemical Process Principles. Part I, Material and Energy Balances*. New York: John Wiley, 1943.

PERRY, JOHN H., *Chemical Engineers' Handbook*. New York: McGraw-Hill, 1949.

PROBLEMS

2–1 Metallic iron is made by heating a mixture of magnetite and carbon at 1100°C in a crucible. The gases evolved during the process average 80% CO, 20% CO_2. Calculate (a) carbon consumption, in lb C reacting per lb of Fe produced, (b) volume of gases produced, in cu ft (STP) per ton of Fe produced, (c) volume of gases produced, in ft³/ton of Fe, measured at 1100°C and 740 mm total pressure.

2–2 A copper matte analyzes 30% Cu. Calculate the % Fe and the % S for each of the following stoichiometric assumptions. (a) The matte is a mixture of Cu_2S and FeS. (b) The matte is 95% Cu_2S and FeS, balance silica and other compounds not containing Cu, Fe, or S. (c) The matte is 10% Fe_3O_4, 85% Cu_2S and FeS, balance compounds not containing Cu, Fe, or S.

(d) The matte is a mixture of Cu, Fe, and S, but with only 90% of the S theoretically required to form Cu_2S and FeS.

2–3 The complete charge to a matte-smelting furnace contains Cu, Fe, and S in the proportions Cu:Fe:S = 1:4:1.2. In the smelting process 98% of the Cu and 75% of the S are recovered in the matte product. Assuming that the matte is a stoichiometric mixture of Cu_2S and FeS, calculate (a) the matte grade in % Cu and (b) the percentage of the Fe recovered in the matte.

2–4 A mixture of 50% CO and 50% CO_2 is passed into a laboratory furnace at the rate of 100 ml/min, and deposits C by the reaction

$$2CO \longrightarrow CO_2 + C.$$

The gas leaving the furnace is 51% CO_2, 49% CO.

Calculate the rate of C deposition in the furnace, in mg/hr.

2–5 100 tons of hard lead (98% Pb, 2% Sb) are melted in a steel kettle and then treated with 2 tons of PbO. The products are (1) a slag consisting of PbO and Sb_2O_3, analyzing 20% Sb, (2) a Pb–Sb alloy of lowered Sb content and negligible oxygen content. Calculate the % Sb in the final alloy.

2–6 A slag analyzing 55% FeO, 10% Fe_2O_3, and 35% SiO_2 is melted in a platinum crucible under an inert atmosphere, and it is found that 100 g of the slag loses 1 g of Fe which dissolves as Fe in the solid Pt. There is no loss of Si or O. Calculate the slag analysis after melting, as % FeO, % Fe_2O_3, and % SiO_2.

2–7 A zinc flotation concentrate, substantially pure ZnS, is roasted to ZnO without fuel. The roaster gases contain 6% SO_2 and 1% SO_3, balance N_2 and O_2. Calculate the quantity of dry air fed to the roasting apparatus, in standard ft^3/short ton of ZnS.

2–8 50 short tons/24 hr of a flotation concentrate is roasted continuously with dry air. The concentrate contains 80% ZnS, 5% PbS, 5% FeS_2, and the balance inert gangue minerals. Assume that roasting converts the ZnS to ZnO, the PbS to PbO, and the FeS_2 to Fe_3O_4, all the S from these compounds entering the gaseous product as SO_2 and SO_3. The gases leaving the system analyze 5% SO_2 and 1.5% SO_3. Calculate (a) the quantity of air entering the roaster, in ft^3/min (STP), (b) the excess air, as percent of the quantity theoretically required for the reactions, (c) the complete analysis of the gases leaving the roaster, (d) the quantity of H_2SO_4 which might be made from the gases, assuming conversion of all the S, in short tons/24 hr.

2–9 Pyrite (FeS_2) is burned in a furnace with an excess of air to produce Fe_2O_3 and SO_2. The flue gas from the furnace contains 6.3% SO_2, balance N_2 and O_2. Calculate (a) theoretical air consumption, m^3 (STP) per metric ton of pyrite, (b) actual air consumption, m^3 (STP) per metric ton of pyrite, (c) % excess air, (d) volume of flue gas (STP) per metric ton of pyrite.

2–10 A mixture of chalcopyrite ($CuFeS_2$), pyrite (FeS_2), and chalcocite (Cu_2S) analyzes 30% Cu, 30% Fe, and 40% S. Calculate the mineralogical analysis.

2–11 A continuous agitator of 50-gal capacity is fed with solid copper sulfate and water, and overflows continuously 10 gal/min of a solution containing 0.5 lb $CuSO_4$ per gal. One day the copper sulfate feeder is accidentally shut off for 20 min. Prepare a graph showing the variation of $CuSO_4$ concentration with time in the effluent solution, for the period from the shutoff until the concentration returns to substantially its normal value.

2–12 The Orsat analysis of a gas gives the following percentages by volume: N_2, 75%; CO_2, 15%; O_2, 10%. The dew point of the gas is 30°C, measured when the total pressure of the gas is 745 mm. Calculate (a) complete gas analysis, volume percentages on a wet basis, (b) complete gas analysis, weight percentages on a wet basis, (c) gas density at STP, in lb/ft^3, (d) gas density at 500°F, 745 mm pressure.

2–13 Moisture in the air is determined by measuring the dew point and the barometric pressure, which are found to be 16.5°C and 752 mm, respectively. The air temperature is 240°C. Calculate (a) % H_2O by volume in the air, (b) humidity, lb H_2O/lb dry air, (c) relative humidity, %, (d) grains moisture per actual cu ft.

2–14 A lead sinter to be smelted in a blast furnace analyzes 8% CaO, 25% SiO_2, and 15% Fe. In smelting this sinter, a slag is desired with CaO, FeO, and SiO_2 in the proportions

$$CaO:FeO:SiO_2 = 4:6:7.$$

To obtain this slag, iron ore and limestone of the following analyses are added to the charge:

	% Fe	% CaO	% SiO₂
Iron ore	45.0	0.5	15.0
Limestone	2.0	46.0	8.0

In addition, coke is to be added to the charge as fuel in an amount equal to 12% of the weight of sinter + iron ore + limestone. The coke contains 2.5% Fe_2O_3 and 12.0% SiO_2. Assume that all the lime, iron, and silica enter the slag as CaO, FeO, and SiO_2. Calculate the weights of iron ore, limestone, and coke to be used per 2000 lb of lead sinter.

2–15 Ten 1-lb samples of a copper ore taken from a large lot of minus 3-mesh ($\frac{1}{4}$-inch) ore have the following analyses in % Cu:

0.85, 1.02, 1.18, 0.94, 1.03, 1.09, 0.82, 1.07, 0.90, 1.14.

(a) Estimate the weight of sample which should be taken of this 3-mesh lot of ore in order to obtain an analysis with a probable error of 0.01% Cu. (b) Estimate the weight of sample which should be taken of this same ore, but of a different lot crushed only to $1\frac{1}{2}$ inches, to obtain an analysis with a probable error of 0.02% Cu. Note your assumptions.

2–16 A shaft furnace consisting of a vertical cylinder 15 ft in diameter and 50 ft high is fed continuously at the top at the rate of 25 tons/hr of lump solids. The solids move down the shaft countercurrent to an upward gas stream which flows at the rate of 12,000 ft³/min (STP). At the bottom of the shaft the gas is at 2500°F and 5 psi gauge pressure, and at the top it is at 500°F and atmospheric pressure. The bulk density of the solids in the shaft is about 60 lb/ft³, and it is estimated that the void space available to gas flow is 45% of the total volume.

Calculate (a) retention time of solids in the shaft, hr; (b) rate of downward movement of charge, inches/min; (c) retention time of gases in the shaft, sec (assume average temperature and pressure); (d) gas velocities in void space at top and bottom, ft/sec; (e) tons of gases per hr (assume average molecular weight = 30).

2–17 A charge of 200 g of roasted ores and fluxes mixed is heated to 1300°C in a fireclay crucible, and after an hour it is found that the crucible contains two homogeneous liquid phases, a slag and a matte. The crucible and contents are then cooled and broken up, and samples of the two phases are analyzed for Cu, Fe, and S. Analyses of the charge and products are as follows:

	% Cu	% Fe	% S
Original charge	9.17	32.6	10.5
Products:			
matte	32.5	36.1	25.3
slag	1.21	28.5	1.3

Assume the matte and slag contain all the Cu and Fe from the charge, and that no Cu or Fe is picked up from the crucible. The charge dissolves some of the crucible, but also loses some weight by volatilization.

Calculate (a) weights of matte and slag, in grams, and (b) sulfur loss by volatilization, in grams.

Draw up a table giving the complete metallurgical balance for this experiment, including for each of the three elements the distributions in grams and percentages among the products.

CHAPTER 3

THE HEAT BALANCE

All chemical processes involve various kinds of energy interchanges with the surroundings. In most of the unit processes for extracting and refining metals, heat energy is the principal form of energy exchanged between the metallurgical system and its surroundings. The supply and utilization of heat rank in importance with the supply and utilization of raw materials in determining costs and in determining the success or failure of a process. For example, the availability of fuel or low-cost electric power has in numerous cases been the primary factor in the choice of plant location or the choice of process. Accordingly, a system of energy accounting or an energy balance showing input and output of heat and other forms of energy, like the materials balance and the profit-and-loss statement, represents one of the indispensable tools the metallurgical engineer uses constantly and even routinely.

Just as the principle of conservation of matter affords a simple, straightforward approach in setting up a materials balance, so the principle of Conservation of Energy or the First Law of Thermodynamics affords a sound basis for setting up an energy balance. In both cases, a relatively complete accounting can be made from a knowledge of what goes into the system and what comes out, with little or no need to consider the complexities and mechanisms of the process within the system.

Conservation of energy

The law of conservation of energy can be stated in various ways, but the following statement will serve for the present: "Whenever a quantity of one kind of energy is produced, an exactly equivalent amount of another kind (or other kinds) must be used up." Some of the *kinds of energy* produced and consumed in metallurgical operations are described briefly below.*

Heat is the kind of energy that passes from one body to another solely as a result of a difference in temperature.

Mechanical work is done when a force acts on a body and a displacement of the body occurs in the direction of application of the force. The quantity of work is measured by the product of force times displacement.

Electrical work is done by the passage of an electric current; in a simple direct-current circuit it is measured by the product of voltage and current.

Heat and work thus involve energy in transit and might be called transient kinds of energy. Of equal importance are the various nontransient kinds of energy which are stored within a body or within a system. Some of these are the following:

Internal energy is stored within a system by virtue of the relative motions, forces, and arrangements of the atoms and molecules in the system. The temperature is one evidence of internal energy; that is, some of the internal energy can be withdrawn as heat when the temperature falls. The pressure and volume also represent internal energy which can be drawn upon by allowing

---

* Texts on physics and thermodynamics should be referred to for a rigorous and more complete discussion of the various kinds of energy.

43

the system to expand. Also, part of the internal energy of a system may be released as heat or work by allowing a chemical reaction to occur in the system.

Kinetic energy is possessed by a body by virtue of its relative motion, and is measured by $\frac{1}{2} mu^2$, where m is mass and u is velocity. This kind of energy is particularly important in the flow of gases and liquids.

Potential energy is possessed by a body by virtue of its position and the force of gravity, and is converted to other kinds of energy when the body falls.

The descriptions of various kinds of energy given above are not entirely definitive, but as these energy concepts are applied later to specific systems and specific quantitative relationships their meanings will become more exact. Thus, the mathematical statement of the First Law given below can be considered as a definition of internal energy.

The Law of Conservation of Energy is commonly stated in the form of an energy balance or energy equation for a process occurring within a system. Before the process the system is said to be in *state 1* or the *initial state* and after the process the system is said to be in *state 2* or the *final state*. The process may thus be referred to as a *thermodynamic change in state*, or simply a *change in state*, not to be confused with the special case of change in state of aggregation which refers to melting, vaporization, etc. A system is in a definite state when all of its properties are defined, for example, by specifying the temperature, pressure, kinds and quantities of matter, and states of aggregation for the system.

The First Law equation for a process occurring in a system is

$$E_2 - E_1 = Q - W, \qquad (3\text{--}1)$$

where E_1 and E_2 are the *energy contents* (or internal energies) of the system in state 1 and state 2, respectively, Q is the *heat absorbed by the system* from the surroundings, and W is the *work done by the system* on the surroundings.

The energy content E of a system is a thermodynamic property of the system; that is, when the state of the system is fixed, the value of E is also fixed. If this were not so, it would be possible to devise a cyclic process which could be carried out in the system, whereby the system would supply energy to the surroundings and yet be returned to its original state. Such a process is impossible because it would violate the Law of Conservation of Energy. Since E_2 and E_1 depend only on the initial and final states of the system, it follows that $E_2 - E_1$ in Eq. (3–1) is independent of the path taken by the system between state 1 and state 2. This fact accounts for the great engineering usefulness of Eq. (3–1), since it makes possible the setting up of energy balances for very complex processes without the necessity for knowing all about the path or for having information about the mechanism of the process.

Although $E_2 - E_1$ depends only on initial and final states, and therefore the difference $Q - W$ also depends only on initial and final states, Q and W individually depend on the path of the change in state. That is, Q and W represent energy exchanges with the surroundings and are not thermodynamic properties of the system. For a given change in state, then, a path by which the heat, Q, absorbed from the surroundings is greater will also be a path by which more work, W, is being done on the surroundings.

Fortunately Q and W for a given change in state are fixed in their values by relatively simple specifications of the path; complete knowledge of the path and the process mechanism is not at all necessary. Some of the paths treated in thermodynamics are the following: constant pressure, constant volume, adiabatic $(Q = 0)$, and reversible. Almost all the processes of extracting metals follow constant-pressure paths, in fact they are carried out at substantially 1 atmosphere. For constant-pressure processes, a more useful form of the First Law can be worked out, as discussed below.

The heat content or enthalpy of a system, H, is a thermodynamic property (its value is fixed when the state of the system is fixed), defined as follows:

$$H = E + pV, \qquad (3\text{--}2)$$

where E is the energy content as already defined, p is the pressure, and V the volume of the system. Combining with Eq. (3–1), the First Law can be written for *any* thermodynamic change in state:

$$H_2 - H_1 = Q - W + p_2V_2 - p_1V_1. \quad (3\text{--}3)$$

For a constant-pressure process, in which all work done by the system on the surroundings is the work of expansion, that is, $\int_1^2 p\,dV$, $p_2 = p_1 = p$ and $W = p(V_2 - V_1)$. (3–4)

Combining Eqs. (3–3) and (3–4),

$$H_2 - H_1 = Q_p \quad \text{or} \quad \Delta H = Q_p, \quad (3\text{--}5)$$

where Q_p is the heat absorbed by the system in changing from state 1 to state 2 by a *path of constant pressure in which only expansion work is done on the surroundings.*

If the change in state is at constant pressure, but involves work other than expansion work on the surroundings,

$$\Delta H = Q_p - W', \qquad (3\text{--}6)$$

where W' is the work done by the system on the surroundings in the constant-pressure process, but does not include expansion work. For example, in an electric furnace process at constant pressure, $-W'$ equals the input of electrical work to the system.

Since most metallurgical processes follow paths of substantially constant pressure, the statements of the First Law in Eqs. (3–5) and (3–6) are very useful. Their importance and usefulness lie in the fact that they relate in simple fashion a quantity of great practical significance, the heat effect for a process (Q_p), to a thermodynamic property of the system, the heat content (H). The change in heat content (ΔH) depends only on the initial and final states and not on the path of the process; hence the heat absorptions or evolutions of practical processes can be evaluated from data on the properties of the practical system before the process and after the process, data usually not too difficult to obtain.

Heat-content data. Heat content is an important thermodynamic property, not only in energy balances such as those discussed in this chapter, but also in many other useful chemical calculations. Accordingly, physical chemists have made relatively comprehensive experimental measurements, directly and indirectly, of the heat contents of pure substances in various states. Moreover, further effort has gone into the compiling, simplifying, and systematizing of the great accumulation of this kind of information, to make it easy to find and easy to use. For the heat-content data used in this book, the primary sources are (1) K. K. Kelley, *Contributions to the Data on Theoretical Metallurgy, X. High Temperature Heat-Content, Heat-Capacity, and Entropy Data for Inorganic Compounds.* Bulletin 476, U. S. Bureau of Mines (1949); (2) *Selected Values of Chemical Thermodynamic Properties.* U. S. Bureau of Standards (1949).

Difficulties arise in attempting to assign absolute values to the heat content H because the quantities generally determined experimentally are changes in heat content, ΔH, for various thermodynamic state changes. However, the lack of absolute values of H does not affect the practical usefulness of the data, since, as will be seen, relative values of H and values of ΔH for specific changes in state suffice for all ordinary purposes.

Systematic enthalpy data are readily available for the following simple thermodynamic changes in state, all at constant pressure:

(1) Temperature changes in pure substances.

(2) Changes in state of aggregation in pure substances (including allotropic changes in solids).

(3) Formation of compounds from the elements at 25°C.

These three kinds of primary enthalpy data, combined in various ways, make possible substantially complete heat accounting for most metallurgical processes. However, another kind of thermodynamic state change in many metallurgical processes is:

(4) Formation of solutions, dilution of solutions.

Enthalpy changes have been measured for dissolution of common substances in water. On the other hand, very few measurements have been made of enthalpy changes associated with formation of various nonaqueous solutions at high temperatures, such as slags which are solutions of various oxides in each other. Fortunately, these heats of solution at high temperatures are probably small in most cases, compared with other enthalpy changes in the process, so that they can be neglected or merely estimated without introducing serious errors into the over-all heat accounting.

As has been indicated repeatedly above, the relations under discussion apply only to constant-pressure processes, and moreover much of the available thermodynamic information is for processes at 1 atmosphere pressure. No significant errors are introduced into energy calculations for pyrometallurgical processes if pressure changes even of a substantial fraction of an atmosphere are ignored, because W (expansion work) is generally small in comparison with Q. On the other hand, in dealing with high-pressure processes, behavior of steam, compressed gases, fluid flow, and related systems, the relations derived for constant-pressure processes cannot be applied. Doubtful cases should be resolved by referring back to the First Law for any process, Eq. (3–3).

Effect of temperature. Changes in heat content with changes in temperature for pure substances may be presented systematically in graphs, tables, or empirical equations, and may be presented directly in terms of heat content or indirectly in terms of heat capacity. Anyone making frequent use of thermodynamic data will of necessity learn to use all forms of data. However, in this book the calculations and calculation methods will be based primarily on the empirical equations for heat contents and heat capacities, and on the tabulated values of heat contents given by K. K. Kelley in U. S. Bureau of Mines Bulletin 476.

The variation of heat content with temperature is expressed adequately for most substances by the empirical relation

$$H_T - H_{298} = aT + bT^2 + cT^{-1} + d. \quad (3\text{–}7)$$

In this equation, $H_T - H_{298}$ is the increase

in heat content, in calories per mol,* as the substance is heated from 298°K (25°C) to $T°K$. The temperature 298°K or 25°C (77°F) is known as the *base temperature*, and $(H_T - H_{298})$ is then the *heat content above the base temperature*. Temperature T must be expressed in °K (= °C + 273), so that temperature conversion usually is necessary in engineering calculations. However, this temperature conversion is a small price to pay for the privilege of utilizing the most recent and accurate basic information in the form in which physical chemists are accustomed to present it.

Using Eq. (3–7), the heat content above the base temperature of 298°K is readily calculated from given values of a, b, c, and d as long as no change in state of aggregation occurs between 298°K and T. However, if the heat content is to be calculated for the same substance in a different state of aggregation, another equation with different values of a, b, c, and d must be used. In all cases, the empirical equation gives $H_T - H_{298}$ or the heat content above a base temperature of 298°K. Also, in all cases H_{298} is the heat content of the substance in its most stable state at 298°K, so that $H_T - H_{298}$ represents the increase in heat content from the base temperature and reference state of aggregation to the actual temperature and actual state of aggregation. The quantity $H_T - H_{298}$ is also called the *sensible heat*, especially in engineering calculations and in process heat balances.

The molal heat capacity at constant pressure, usually just called the *heat capacity* C_p, is defined by

---

* Gram calories per gram mol, kilogram calories per kilogram mol, or pound calories (Centigrade heat units) per pound mol. To obtain British thermal units per pound mol, multiply the number of calories per mol by 1.8.

$$C_p = \left(\frac{\partial H}{\partial T}\right)_p. \qquad (3–8)$$

The empirical relation between C_p and T is found by differentiating Eq. (3–7):

$$C_p = (a) + (2b)T - (c)T^{-2}. \qquad (3–9)$$

Thus, if values of the empirical constants a, b, c, and d are available for a substance, either the heat-content or the heat-capacity equations can be written without calculation except for multiplying b by 2 and c by -1 in the heat-capacity equation. Also, it should be noted that the heat capacity has the same numerical value expressed in several different systems of units:

$$\frac{(\text{gram calories})}{(\text{gram mol})(°C \text{ or } °K)} = \frac{(\text{kilogram calories})}{(\text{kilogram mol})(°C \text{ or } °K)}$$

$$= \frac{(\text{pound calories})}{(\text{pound mol})(°C \text{ or } °K)}$$

$$= \frac{(\text{British thermal units})}{(\text{pound mol})(°F \text{ or } °R)}.$$

It is clear that the pound calorie is a convenient unit if quantities are expressed in the English system of units and temperatures in °C; the British thermal unit is convenient if quantities are expressed in the English system and temperatures in °F. On the other hand, the pound calorie is not an appropriate unit if temperatures are in °F.

Integration of Eq. (3–8) gives

$$H_2 - H_1 = \int_{T_1}^{T_2} C_p \, dT. \qquad (3–10)$$

The change in heat content with change in temperature can also be expressed in terms of the mean heat capacity, C_m:

$$H_2 - H_1 = C_m(T_2 - T_1). \qquad (3–11)$$

The value of C_m depends on both T_2 and T_1, as can be seen by comparing Eqs. (3–10) and (3–11):

$$C_m = \frac{\int_{T_1}^{T_2} C_p \, dT}{T_2 - T_1}. \qquad (3\text{–}12)$$

A common practice, especially in approximate engineering calculations, is to use heat capacities per unit weight, usually called *specific heats,* instead of molal heat capacities. Such calculations follow the relation

$$H'_2 - H'_1 = c_m(t_2 - t_1), \qquad (3\text{–}13)$$

where $H'_2 - H'_1$ is the change in heat content per unit weight, c_m is the *mean specific heat* (at constant pressure) over the temperature range t_1 to t_2, and $t_2 - t_1$ is the temperature change. Strictly speaking, Eq. (3–13) is just as exact as Eq. (3–10), and some authors have compiled tables and empirical equations of mean specific heats in relation to temperature. However, since the latest and most accurate data are now usually in the forms discussed earlier, it is recommended that Eq. (3–13) be used primarily for estimations and rapid calculations. Also, this relation is convenient for water (specific heat = 1), for various mixtures and solutions (for example, "stone," slag, brick, cork, etc.), and for use of handbook data. It should be noted that specific heat has the same numerical value for the several different combinations of heat units, weight units, and temperature units already indicated for heat capacity.

The engineer's most frequent uses of heat-content data are for determining sensible heats and the heat quantities evolved or absorbed when the temperature of a substance is changed between two known levels. For such calculations it is convenient to use heat-content data in tabulated form, so that all that is necessary to find the sensible heat of a unit quantity of material at a given temperature is to make a simple arithmetic interpolation between heat-content values tabulated at even temperatures. The heat evolved or absorbed in a specified temperature change at constant pressure is found simply as the difference between the heat contents at the two temperatures. Accordingly, Table A–3 in the Appendix gives $H_t - H_{77}$ in Btu/lb mol at even hundreds of °F for most of the pure substances commonly appearing in metallurgical calculations. For the calculations in which empirical heat-content or heat-capacity equations are necessary, Table A–3 also includes values of the constants a, b, c, and d for Eqs. (3–7) and (3–9). Table A–4 gives approximate values of mean specific heat (c_m) for various common materials which are also useful in engineering calculations.

Changes in state of aggregation. After a solid is heated to its melting point, additional heat must be supplied to melt it. The heat required for melting at constant pressure is equal to the increase in heat content from the solid to the liquid and is known as the *heat of fusion* $(\Delta H_{\text{fusion}})$. An equal quantity of heat is liberated during solidification, so that $\Delta H_{\text{solidification}} = - \Delta H_{\text{fusion}}$. Similarly, heat effects accompanying vaporization and allotropic changes in solids are measured by *heats of vaporization* and *heats of transformation.*

Values of ΔH for changes in state of aggregation vary with the temperature and pressure under which the change is carried out. Accordingly, the values usually reported are those for the normal melting temperature, boiling temperature, or transformation temperature at 1 atm pressure. The small variation of ΔH with temperature is usually of no consequence in heat-balance calculations, except for the use of different values for the heat of vaporization of water. Thus, the heat of vaporization of water at its boiling point, 100°C, is 542 cal/g (975 Btu/

lb), whereas at 25°C it is 583 cal/g (1050 Btu/lb).

As already noted in the previous section, the empirical equations for heat content above the reference state at 298°K take into account the heat-content changes associated with changes in state of aggregation. Accordingly, additional data on heats of fusion, vaporization, transformation, etc., are not necessary for ordinary energy balance calculations. However, the values of ΔH for changes in state of aggregation can be calculated if desired from the heat-content data in the Appendix as the difference between $(H_T - H_{298})$ for the two states of aggregation at the temperature T at which the change in state occurs.

EXAMPLE

The melting point of aluminum is 659°C or 932°K.

$$\text{For solid Al: } H_T - H_{298} = 4.94T + 1.48 \times 10^{-3}T^2 - 1605$$

$$= 4.94 \times 932 + 1.48 \times 10^{-3}(932)^2 - 1605$$

$$= 4285 \text{ cal/mol, at the melting point.}$$

$$\text{For liquid Al: } H_T - H_{298} = 7.00T + 330$$

$$= 7.00 \times 932 + 330$$

$$= 6854 \text{ cal/mol at the melting point.}$$

$$\text{Therefore, } \Delta H_{\text{fusion}} = 6854 - 4285 = 2569 \text{ cal/mol}$$

$$= 1.8 \times 2569$$

$$= \underline{4624 \text{ Btu/lb mol.}}$$

(Using the numerically tabulated heat contents for solid and liquid Al at the melting point (1217°F), a rounded figure of $12,330 - 7700 = 4630$ Btu/lb mol is obtained.) Dividing by the atomic weight of Al, the heat of fusion per unit weight is found:

$$\Delta H'_{\text{fusion}} = \frac{4624}{26.97} = 171 \text{ Btu/lb.}$$

Heats of formation. When a chemical compound is formed from its elements, heat is either liberated or absorbed. If the reaction of formation is carried out at constant pressure and with only expansion work on the surroundings, by the First Law the quantity Q_p of heat absorbed must equal the increase in heat content ΔH of the system resulting from the reaction. This quantity ΔH is fixed when the quantities and thermodynamic properties of the reacting elements (the initial state) and of the compound produced (final state) are fixed. It may be convenient to include the necessary thermodynamic data on initial and final states in the chemical equation. Thus the formation of CO_2 from the elements at 298°K (25°C) can be represented by:

$$\text{C (graphite)} + O_2 \text{ (gas, 1 atm)}$$
$$= CO_2 \text{ (gas, 1 atm),}$$

$$\Delta H_{298} = -94,050 \text{ cal/mol.}$$

The information in parentheses specifies the states of the reactants and reaction products, and the equation itself represents quantities of reactants.

Ordinarily it is not necessary to include detailed information on thermodynamic states of all the reactants in the equation, because even without specification the states are clearly understood. Thus, in the above equation for formation of CO_2 at 25°C, it is not necessary to say that O_2 and CO_2 are gases at 25°C. Also the pressure specification is not necessary, first, because 1 atm is generally assumed in the absence of specification, and second, because moderate pressure variations do not affect ΔH appreciably insofar as most metallurgical calculations are concerned. On the other hand, the state of aggregation of the C should be indicated in the equation, since C can exist as diamond, as graphite, or as amorphous carbon, and each of these three gives different heats of formation for CO_2. Also, the temperature should always be specified, because ΔH varies with temperature.

For systematic thermochemical calculations, such as metallurgical heat balances, tables of heats of formation give ΔH (or $-\Delta H$) per mol for the reactions of forming compounds at the base or reference temperature of 25°C, when the reacting elements and the resulting compound are each in their standard or most stable states at this temperature and at 1 atm pressure. Where there is the possibility of two reference states, as there is for H_2O (liq) and H_2O (gas), the appropriate designation is usually given.

Heats of formation at 25°C for most of the compounds involved in metallurgical calculations are given in Table A–5 in the Appendix.

Heats of reaction. As demonstrated previously, ΔH for any process depends only on the initial and final states and not on the path. Accordingly, if a process is divided into several steps and ΔH is determined for each step, the algebraic sum of the ΔH values for all the steps must equal ΔH for the original direct process. Therefore, when a chemical reaction is equal to the algebraic sum of two or more individual reactions, the ΔH for that reaction is equal to the algebraic sum of the ΔH values for the individual reactions. Naturally, the ΔH values per mol for the individual reactions must be multiplied by the appropriate number of mols before addition, corresponding to the multiplication of the chemical equations themselves before addition. Also, it should be kept in mind that a reaction in reverse will have a ΔH of the same numerical value but of opposite sign.

Using the principle outlined in the previous paragraph, the heat of reaction at a given temperature for any reaction can be found by algebraic addition of the heats of formation (ΔH_f) at that temperature of the reaction products and subtraction of the heats of formation of the reactants. That is,

$$\Delta H = \Sigma(\Delta H_f \text{ of reaction products})$$
$$- \Sigma(\Delta H_f \text{ of reactants}). \quad (3\text{--}14)$$

EXAMPLE

The values of ΔH_f per mol for the formation of the reactants and products from the elements, multiplied by the number of mols of each, can be written below the chemical equation, as follows:

$$Fe_3O_4 \ + \ 4CO \ \longrightarrow \ 3Fe \ + \ 4CO_2 \quad \text{at 25°C.}$$
$$-267{,}000 \quad 4(-26{,}420) \qquad 0 \qquad 4(-94{,}050)$$

Applying Eq. (3–14),

$$\Delta H_{25°C} = 4(-94{,}050) - (-267{,}000) - 4(-26{,}420)$$
$$= -3{,}520 \text{ cal/mol Fe}_3\text{O}_4.$$

Complex processes. Methods have been presented in the previous sections for evaluating ΔH for the simple constant-pressure processes of heating and cooling, changes in state of aggregation, formation of compounds at 25°C, and chemical reactions at 25°C. Practical processes are, of course, far more complex than any of these. However, it is usually feasible to secure reasonably complete data on the initial and final states even for the most complex processes. These data consist primarily of the materials balance, giving quantities and kinds of substances entering and leaving the process, plus information on temperatures and states of aggregation of these input and output substances. Data on process temperatures, reaction details, phases, etc., within the process itself are not necessary in describing initial and final states. When complete data on initial and final states are available, an ideal process can be set up on paper to accomplish the same over-all change in state and therefore to have the same value of ΔH. This ideal process generally has little resemblance to the actual process, beyond the required identity of initial and final states with the actual process. However, the ideal process is set up as the sum of a number of individual simple processes, for each of which ΔH is easily calculated by methods already discussed.

In calculating ΔH for a complex process, a schematic diagram facilitates analysis of the problem and is helpful in avoiding errors and omissions. Figure 3–1, for example, represents the following batch process: Liquid Cu_2S at 1200°C is charged into a converter and then blown with cold air (25°C) to produce blister copper. All the oxygen from

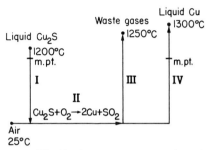

Fig. 3–1. Idealized process, converting Cu_2S to Cu.

the air reacts with the Cu_2S, and at the end of the blow the converter contains liquid Cu at 1300°C. The average temperature of the gases leaving the converter during the blow (N_2 and SO_2) is 1250°C. In the diagram, initial states for all entering substances are shown at the left and final states for the products are shown at the right. Temperature is the ordinate. Each simple step in the ideal process is represented separately by an arrow, heating and cooling by vertical arrows and reactions by horizontal arrows. These steps obviously do not correspond at all to the way the process is carried out in practice, but the ideal process accomplishes the same thermodynamic change in state as the actual process, so that $Q_{\text{process}} = \Delta H_{\text{process}}$ $= \Delta H_{\text{I}} + \Delta H_{\text{II}} + \Delta H_{\text{III}} + \Delta H_{\text{IV}}$.

Ideal processes for calculating ΔH, such as the one represented in Fig. 3–1, are generally set up so that the chemical reactions occur at the base temperature (25°C) for which tabulated heats of formation and reaction are available. Also the reactions are between substances in their most stable or reference states at the base temperature. Accordingly, the calculation of ΔH for a complex process involves algebraic addition of

ΔH values for the following three kinds of steps.

(1) Change all input substances from actual entering temperatures and states to base temperature and reference states at base temperature.

(2) Carry out reactions at base temperature.

(3) Change all reaction products and output materials from base temperature and reference states to actual final temperatures and states.

In using the various First Law equations and heat-content data, a common difficulty is keeping the positive and negative signs straight for the thermodynamic variables. It is, of course, obvious that errors in sign are often much more consequential than ordinary arithmetic errors. The answers to any uncertainties about sign are to be found in a clear understanding of the definitions of the thermodynamic quantities. However, the tabulation below may be useful as an amplification and review of the definitions which have been given previously.

Signs of ΔH and Q_p

(Constant-pressure processes, expansion only form of external work)

POSITIVE (+)

Heat absorption from surroundings
Heating, temperature increase
Melting
Vaporization
Endothermic, heat-absorbing chemical reactions
Most dissociation reactions

NEGATIVE (−)

Heat evolved to surroundings, heat loss
Cooling, temperature decrease
Freezing

Condensation
Exothermic, heat-evolving chemical reactions
Most reactions of formation from the elements

Signs of W

POSITIVE (+)

System does work on surroundings
Expansion, volume increase
System delivers work to shaft
System generates electricity

NEGATIVE (−)

Surroundings do work on system
Compression, contraction, volume decrease
System receives work through shaft
System uses electricity, electrolysis, electric heating

Heats of reaction at high temperature. Figure 3–2 indicates the steps involved in calculating heat of reaction at a temperature above the base temperature. That is,

$$\Delta H_T = \Delta H_{298} + \Sigma \Delta H \text{ for cooling reactants,}$$
$$T \text{ to } 298°K + \Sigma \Delta H \text{ for heating reaction products, } 298°K \text{ to } T. \quad (3\text{–}15)$$

If there are no changes in state of aggregation of reactants or reaction products between $298°K$ and $T°K$, the last two terms often are approximately equal numerically and of course have opposite signs, so that $\Delta H_T = \Delta H_{298}$ serves as a useful rough approximation.

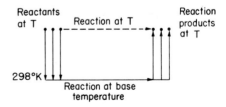

FIG. 3–2. Ideal process for calculating heat of reaction at other than base temperature.

The heat balance

The heat balance for a process is a statement of the energy balance in a form which is very convenient practically, and has, for engineering purposes, the virtue of being less abstract than the formal thermodynamic treatment of the same facts. The heat balance, like the materials balance, accounts for heat quantities in two categories, input and output, whose totals must be identical. All items in the heat balance are positive, so if the simple rules are followed, the confusion of plus and minus signs often experienced with thermodynamic calculations is minimized.

No new principles are introduced in the heat balance; it is simply another way of stating the First Law of Thermodynamics for a constant-pressure process. However, a few conventions and definitions are involved to obtain positive heat terms only and to insure their proper classification into input and output.

In the first place, in setting up a heat balance, a single *base* or *reference temperature* must be chosen and specified. This base temperature is preferably 25°C and is a temperature for which heats of formation are easily available; however, in special cases a temperature characteristic of the process may be used. As will be seen, the individual terms are quite different in heat balances set up at different base temperatures for the same process. Hence, heat balances set up for different base temperatures are not to be compared directly, term by term.

After selecting the reference temperature, reference states at this temperature must be understood for each substance in the input and output materials. For most substances there is only one stable state at the reference temperature, so that specification is not needed. Water appears in one form or another in many processes and specification is required, since the reference state can be either H_2O (gas) or H_2O (liq).

Obviously the heat balance must refer to initial and final states of a definite system, which may be a single batch of material, a unit weight of throughput, the throughput for a fixed time interval as a day or an hour, etc. As in the case of the materials balances, the heat balance for a continuous process in the unsteady state would take the form

heat input = heat output
+ heat accumulation.

To avoid difficulties in evaluating the last term, therefore, considerable effort is worthwhile to obtain data for substantially steady state operation.

The *sensible heat* in an input or output substance is the positive quantity of heat required to change the substance from reference temperature and reference state to the actual temperature and state in which it is present in the input or output of the process. Thus, if the reference temperature is 25°C, and a substance is fed to the process at 25°C and in its reference state, the substance brings no sensible heat into the process; on the other hand if the substance is still hot from previous treatment it brings sensible heat into the process. The calculation of sensible heat is a calculation of ΔH for the heating from reference state to actual state.

Heats of reaction must have the values corresponding to the reference temperature, and not to the process or to other temperatures, and must be for the reactions between substances in their reference states at the base temperature.

Using the conventions and definitions discussed above, the items appearing as *heat input* will include:

(1) Sensible heats in input materials.

(2) Heats evolved in exothermic reactions.

(3) Heat supplied from outside system (e.g., electrically).

Items appearing in the *heat output* tabulation will include:

(4) Sensible heats in output materials.

(5) Heats absorbed in endothermic reactions.

(6) Heats absorbed in bringing input materials to reference temperature and state.*

(7) Heat loss to surroundings.

The sum of the heat input items should equal the sum of the heat output items. As already mentioned, all items will be positive. Also, each item is readily visualized as a definite quantity of either heat input or output without necessity for keeping track of plus and minus signs and their meanings.

Figure 3–3 shows the close relationship between the thermodynamic statement of the First Law for a constant-pressure process and the conventional arrangement of a heat balance. Essentially the heat balance is a rearrangement of the First Law equation to give an equation in which all the numbers have positive signs. The input versus output statement has a number of advantages for engineering purposes, because the engineer's operating problems relate usually to supplying heat and utilizing heat, not to the measurement of changes in an abstract thermodynamic property like the heat content. Also, common sense tells whether a given item belongs in heat input or output, so that the possibility of confusion of signs is smaller

than with the formal thermodynamic statement.

Typical heat balances. The heat balance, like the materials balance, furnishes for a unit process an accounting which is essential to economical and efficient operation. This does not mean that a continuous accounting, day by day and week by week, is necessary in the case of a heat balance, as it is for a materials balance. A reasonably accurate determination of the heat balance for a plant-scale operation may constitute in itself a major research problem, to be attempted only at infrequent intervals. Accordingly, the heat accounting for a process is often rather qualitative and incomplete. Even then it is an important basis of process control and development.

Typical heat balances for three different metallurgical processes are given in Tables 3–1, 3–2, and 3–3.

The first obvious fact about any heat balance is that it shows simply what the important sources of heat energy are and what their relative contributions to the total are. In *autogenous* processes, the heat evolved in the metallurgical reactions themselves, plus the sensible heats in the input materials, account for the heat input. In *nonautogenous* processes, the heat is supplied wholly or in part by heat evolved from combustion of fuel, heat supplied electrically, or heat otherwise supplied from outside the system. Thus, the heat balance for Zn-ore roasting in Table 3–2 is representative of an autogenous process, whereas that for reverberatory smelting in Table 3–1 represents a nonautogenous process. Generally some or all of the sources of heat are amenable to control, and the heat balance serves as one guide in appraising the effects on the process of changes in these sources of heat. For example, Table 3–1 shows that the charging of roasted ore while still hot from

---

* For example, if H_2O (gas) is the reference state and H_2O (liq) is present in feed materials (a very common case), this term will include the heat of vaporization for bringing H_2O (liq) to the reference state H_2O (gas).

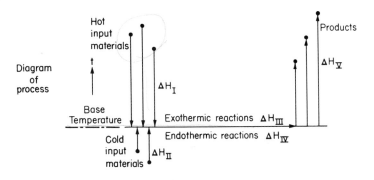

Diagram of process

$$\text{Thermodynamic } \Delta H_{total} = Q_p - W' = \Delta H_I + \Delta H_{II} + \Delta H_{III} + \Delta H_{IV} + \Delta H_V$$

Thermodynamic statement of First Law

Heat balance:

Input	Output
1. Sensible heats, input materials, $(-\Delta H_I)$	4. Sensible heats, output materials (ΔH_V)
2. Heats evolved in exothermic reactions, $(-\Delta H_{III})$	5. Heats absorbed in endothermic reactions, (ΔH_{IV})
3. Heat supplied electrically, $(-W')$	6. Heats absorbed bringing input materials to ref. states (ΔH_{II})
	7. Heat to surroundings $(-Q_p)$

$$\text{Totals} \quad -\Delta H_I - \Delta H_{III} - W' \quad = \quad \Delta H_V + \Delta H_{IV} + \Delta H_{II} - Q_p$$

Fig. 3-3. Heat accounting for constant-pressure process.

TABLE 3-1

Heat Balance of Coal-Fired Reverberatory Furnace Smelting Roasted Copper Ore

Heat input	Btu per day	%
Heat from combustion of coal	1,827,676,000	94.6
Sensible heat in roasted ore	103,680,000	5.4
TOTAL	1,931,356,000	100.0
Heat output		
Sensible heat in slag	335,820,000	17.4
Sensible heat in matte	96,000,000	5.0
Heat absorbed in decomposing limestone ($CaCO_3 \rightarrow CaO + CO_2$)	96,567,800	5.0
Sensible heat in flue gases	927,050,900	47.9
Heat losses to surroundings	475,917,300	24.7
TOTAL	1,931,356,000	100.0

TABLE 3-2

Heat Balance of Roasting Furnace Treating Zinc Concentrates

Heat input	lb-cal/hr	%
Heat evolved in oxidizing sulfides (e.g., $ZnS + 3/2\ O_2 \rightarrow ZnO + SO_2$)	13,371,000	100.0
Heat output		
Sensible heat in calcined product	1,403,000	10.5
Sensible heat in roaster gases	8,980,000	67.1
Heat loss to surroundings	2,988,000	22.4
TOTALS	13,371,000	100.0

TABLE 3-3

Heat Balance for an Iron Blast Furnace

Heat input	Btu/ton of pig iron	%
1. Heat evolved, combustion of C to CO	5,189,000	68.9
2. Sensible heat in blast	2,145,000	28.5
3. Heat evolved, gaseous reduction ($Fe_2O_3 + 3CO \rightarrow 2Fe + 3CO_2$)	192,000	2.6
TOTALS	7,526,000	100.0
Heat output		
4. Sensible heat in pig iron	1,020,000	13.7
5. Sensible heat in slag	874,000	11.6
6. Sensible heat in blast furnace gas	568,000	7.5
7. Heat absorbed, evaporation of water in charge	613,000	8.1
8. Heat absorbed, decomposition of $CaCO_3$	627,000	8.3
9. Heat absorbed, decomposition of H_2O in blast	213,000	2.8
10. Heat absorbed, reduction of SiO_2 to Si	211,000	2.8
11. Heat absorbed, reduction of MnO to Mn	79,000	1.0
12. Heat absorbed, reduction of P_2O_5 to P	23,000	0.3
13. Heat absorbed, "solution loss" reaction ($CO_2 + C \rightarrow 2CO$)	280,000	3.7
14. Sensible heat removed in cooling water	672,000	8.9
15. Heat losses to surroundings (by difference)	2,346,000	31.3
TOTALS	7,526,000	100.0

a previous operation contributes 5.4% of the heat input. If the roasted ore were cooled before charging to the reverberatory furnace, this portion of the heat input would have to be supplied by burning additional fuel. Table 3-3 shows that the preheat in the air blast furnishes a very substantial portion of the total heat input (28.5%) to the blast furnace, and shows in part why no modern iron blast furnaces are operated without preheating the air.

Similarly, a heat balance shows in simple and concise fashion how the heat is used. Some items of heat output are inherent in the metallurgical process itself and thus must be considered as essential and useful applications of energy. Both the copper smelting furnace (Table 3-1) and the iron blast furnace (Table 3-3) must deliver liquid products (matte and slag, and pig iron and slag, respectively) hot enough to flow freely out of the furnace, so the sensible heats in these products are essential. When the process includes endothermic chemical reactions, some usually represent reactions essen-

tial to the process (for example, the decomposition of limestone in Table 3-3) while others are side reactions to be minimized as much as possible (for example, the "solution loss" reaction in Table 3-3). Some sensible heat items of heat output cannot be classified as strictly essential to the metallurgical process itself, but represent various kinds of inefficiencies which, theoretically at least, might be reduced or eliminated by alterations in furnace design or in the method of utilizing heat. For example, in Table 3-1 the large sensible heat in the flue gases is characteristic of the use of fuels to supply heat for high temperature processes, but could be nearly eliminated if the heat were supplied electrically. In Table 3-3, the sensible heat in the blast furnace gases is contained in a very large quantity of gas and at such a low temperature (about 300°F) that a substantial reduction in this item of heat output is not practical. Heat losses to the surroundings are an inevitable item in all heat balances for high temperature processes, since heat always flows from regions of high temperature to those of lower temperature. However, if the heat flow to the surroundings appears excessive in the heat balance, furnace insulation or other measures to reduce heat loss may be found worthwhile.

Careful study of the heat balance often discloses possible lines of improvement in the process, especially improvements leading to savings in fuel. If modifications are to be considered for a going process, for example, modification in compositions of input materials, in fuel, in rate of treatment, or in process temperatures, a preliminary appraisal of the probable effects of such modifications on the heat balance may help in preparing for operating difficulties resulting from the changes.

Broadly speaking, the heat balance indicates the thermal efficiency of a chemical process. Various single percentage figures, such as the ratio

$$\frac{\text{useful heat output}}{\text{total heat input}} \times 100,$$

may be calculated and actually called thermal efficiencies, but these single efficiency figures are likely to oversimplify to the point of being misleading.

Procedure in calculating a heat balance. Even after the necessary data are collected, the calculation of a heat balance may prove long and arduous, and beset by possibilities for mistakes. Accordingly, a logical and standard procedure of calculation is desirable, even if it occasionally requires repetition of the obvious. The following procedure is suggested.

(1) Work out the complete stoichiometry and materials balance, as outlined in steps (1) to (9), pages 31–32, Chapter 2.

(2) Add to the stoichiometric data the temperatures at which all materials enter and leave the system.

(3) Fix and specify the basis of the heat balance, including quantity throughout and and period of time covered, reference temperature (usually 25°C), and reference state for water (and for other substances when not clearly understood).

(4) Calculate sensible heat for each input and output material.

(5) Calculate heats of reaction for the quantities of all the chemical reactions which are necessary to account for the over-all stoichiometry of the process.

(6) Calculate heats required, if any, to bring input materials up to the reference states. Also, calculate heat supplied electrically or otherwise from surroundings, if any.

(7) List and total input and output items, finding heat loss by difference, so that input and output totals are equal. The heat loss may be checked independently by estimation based on the principles of heat flow (Chapter 7).

(8) Check the calculations.

Choice of reactions. There often arises the question of which set of chemical equations should be used to account for the over-all chemical change. Thus reduction of Fe_2O_3 to Fe with C, the C going to CO, might be described by any one of the following three methods:

I. $Fe_2O_3 + 3C \longrightarrow 2Fe + 3CO$

$\Delta H_{25°} = 117{,}240 \text{ cal/mol } Fe_2O_3$

II. $Fe_2O_3 \longrightarrow 2Fe + \frac{3}{2}O_2$

$\Delta H_{25°C} = 196{,}500 \text{ cal/mol } Fe_2O_3$

$\frac{3}{2}O_2 + 3C \longrightarrow 3CO$

$\Delta H_{25°C} = -79{,}260 \text{ cal/mol } Fe_2O_3$

III. $Fe_2O_3 + 3CO \longrightarrow 2Fe + 3CO_2$

$\Delta H_{25°C} = -6390 \text{ cal/mol } Fe_2O_3$

$$3CO_2 + 3C \longrightarrow 6CO$$

$$\Delta H_{25°C} = 123,630 \text{ cal/mol Fe}_2O_3$$

Since all three methods represent the same change in state, the total ΔH for the chemical change described is found to be 117,240 cal. Accordingly, the *net* contribution to a heat balance would be 117,240 cal/mol Fe_2O_3 as heat output or heat consumption. Representing the chemical change by I, this would be accounted for as a single item of heat output. By method II, 196,500 cal would appear in the heat output column and 79,260 cal in the heat input column, and the total heat input and output for the process as a whole would each be 79,260 cal larger than by method I. Also, by method III, 123,630 cal would appear as heat output and 6390 cal as heat input, making the total input 6390 cal greater than by method I. By all three methods, however, the heat loss to the surroundings calculated by difference will still come out the same, because in all three the net contribution to the heat balance is the same.

Thermodynamically, all three of the above are equally correct, since reaction mechanisms in carrying out a given change in state are not considered. Clearly, however, the choice affects the outlook of the heat balance insofar as heats of reaction are concerned. Also using the combination of reactions which gives the largest input and output totals places greater relative emphasis on reaction heats compared with the other items. For example, the combination giving largest totals makes the heat loss appear the smallest proportion of the total heat input, and apparently (but not in actuality) gives a higher thermal efficiency for the process.

Some have followed the custom in the past of writing reactions for decomposing input compounds into elements, and reactions for combining elements into the out-

put compounds (corresponding to method II, above). Since most heats of formation (ΔH) are negative, this procedure is equivalent to including in the heat input the items corresponding to heats of formation of all compounds formed in the process and in the heat output the items corresponding to heats of formation of compounds consumed in the process. Though not objectionable on thermodynamic grounds, this procedure should be discouraged because it can give a very unrealistic conception of the actual heat effects within the process. In particular, it has the effect of "padding" the heat balance, making the reaction heats appear more important than they are and making such items as the heat loss appear small.

Although the choice may be arbitrary in some cases, it is recommended that a set of equations be used and heat effects calculated which represent insofar as possible real heat evolutions or absorptions in the process. This practice was followed in setting up the blast furnace heat balance in Table 3–3. Thus the heat evolved from combustion of coke is computed for the burning of C to CO, as it is known to occur above the tuyères, and the reduction of Fe_2O_3 to Fe is primarily an exothermic gaseous-reduction reaction occurring in the stack of the furnace. Older blast furnace heat balances given in the literature on the subject do not follow these practices, some showing, for example, the major part of the C burning to CO_2 or showing the direct reduction of Fe_2O_3 by solid C as a strongly endothermic reaction.

When heat balances for the same or for similar processes are to be compared, the basis of writing the chemical equations should correspond.

Reference temperatures and states. In addition to a choice of equations, choices of reference temperature and reference states are also arbitrary and affect the numerical makeup of the heat balance. The temperature 25°C (77°F) is the most convenient reference temperature for calculations because good heat-content data and heats of formation are directly available for this reference temperature. Also 25°C is particularly appropriate as a reference level because it is almost always a good approximation of ambient temperature. In the past, other reference temperatures near ambient temperature, such as 0°C (32°F), 18°C (64.4°F), etc., have been used, but these are so close to each other and to 25°C that for most purposes direct comparisons of heat balances can be made without significant errors. In particular, heats of reaction generally vary so slowly with temperature that the change over 25°C is almost always less than the probable error of the thermodynamic data on heats of formation.

For some processes it is of interest to calculate the heat balance by using a characteristic process temperature as the reference temperature. Thus steelmaking might be referred to 1600°C, or copper converting to 1200°C or 1300°C. This results in a description of the process from an entirely different point of view. To illustrate this difference, a simple process of blowing white metal (Cu₂S) to copper may be considered. Liquid Cu_2S at 1300°C is charged into a converter and air is blown through the liquid to carry out the following over-all reaction:

$$Cu_2S \text{ (liq)} + O_2 \longrightarrow 2Cu \text{ (liq)} + SO_2.$$

Assume that the air consumption is the theoretical quantity required for the reaction and that the air enters at 25°C. Also, assume that the copper and the effluent gases are all at 1300°C. Table 3-4 gives heat balances for this process, calculated for two different reference temperatures, 25°C and 1300°C. Whereas the heat balance based on 25°C shows the heat input to consist of a large sensible heat contribution from the hot charge of Cu_2S plus the heat of reaction, the heat balance based on 1300°C shows the heat of reaction to be the only item of heat input, with no input of sensible heat. Also, whereas the 25°C tabulation of heat output shows the heat loss and large

TABLE 3-4

Effect of Choice of Reference Temperature on Heat Balance

Process: Cu_2S (liq, 1300°C) + O_2 (25°C) + $\frac{79}{21} N_2$ (25°C) $\longrightarrow$

$2Cu$ (liq, 1300°C) + SO_2 (1300°C) + $\frac{79}{21} N_2$ (1300°C)

Basis of calculations: 1 mol Cu_2S

Heat Balance A — Reference Temperature 25°C			
Heat input	Cal	Heat output	Cal
Heat of reaction at 25°C	51,960	Sensible heat in Cu	23,550
Sensible heat in Cu_2S	33,190*	Sensible heat in gases	52,390
		Heat to surroundings (diff.)	9,210
TOTAL	85,150		
		TOTAL	85,150

Heat Balance B — Reference Temperature 1300°C			
Heat input	Cal	Heat output	Cal
Heat of reaction at 1300°C	55,940	Heat to bring air to 1300°C	46,730
		Heat to surroundings (diff.)	9,210
TOTAL	55,940		
		TOTAL	55,940

*Found by extrapolation: For solid Cu_2S, $H_T - H_{298} = 20.32T - 4275$ (Table A-3). Heat of fusion has been estimated to be 5500 cal/mol. Hence, for Cu_2S (liq), $H_T - H_{298} = 20.32T + 1225$ (estimate).

sensible heats in gaseous and liquid products, the 1300°C tabulation shows the same heat loss and a large heat requirement for heating cold air to the reference temperature. In addition, it should be noted that the totals of input and output are quite different for the two reference temperatures. If the one tabulation represents the point of view of a man outside the converter, the other equally valid point of view is that of a genie inside the apparatus.

Practical difficulties. Textbook problems are likely to give a misleading conception of the ease of setting up and solving a heat balance for a furnace operation in the plant. In practice, the difficulties involved in obtaining data for a satisfactory heat balance are such as to discourage all but the most stouthearted metallurgist.

One of the most serious difficulties with a continuous process is finding a period of steady operation during which the flow of all materials through the process is uniform both in rate and composition and the other variables such as the temperatures are constant. Accumulation and storage of heat in the system are more intangible and more difficult to assess than accumulation and storage of materials. The same kind of uncertainty arises even more with batch and cyclic processes.

Temperature measurements on furnace products can account for large uncertainties. When liquids are tapped from a furnace (e.g., liquid metal, slag, etc.), the temperature is likely to vary during the period of tapping, this in addition to the difficulty of measuring the temperature at any one time. The flue gas, whose sensible heat is a large item in most heat balances, may flow through irregularly-shaped ducts, several feet in width, with large variations in temperature across the flow. These variations, together with unsteadiness of flow at a given position, can make a good temperature average almost impossible of attainment.

Some uncertainties in the thermodynamic data, such as heat capacities, heats of reaction, heats of solution, etc., are troublesome at times. However, this source of error is not often serious, and most of the thermochemical and thermophysical information available is more accurate than the other information used in the heat balance calculations.

Heat loss to the surroundings is usually found by difference, and like other determinations by difference may be grossly in error, or may even come out negative. A rough check can be made by direct calculation of heat flow to the surroundings, using methods discussed in Chapter 7.

The greatest practical difficulties encountered in making a balance for a going furnace operation cannot justify failure on the part of the metallurgist to think and analyze in terms of the heat balance, using the best data at hand. At least a qualitative or semiquantitative picture of the distributions of energy input and output is indispensable to intelligent operation. Perhaps no heat balance measurement will afford greater difficulties than that for the open-hearth furnace used in steelmaking. This is a batch operation, characteristically an unsteady state operation, but the attempt to set up quantitative heat balances has proved valuable.

EXAMPLE

Calculation of heat balance for an iron blast furnace. In this example, the calculations will be given for the heat balance already referred to in Table 3–3. The operating data are taken from a paper by T. L. Joseph and K. Neustaetter (*Blast Furnace and Steel Plant,*

Vol. 35, 944–948, August, 1947), but the calculations given by these authors have been modified slightly to bring them into accord with the thermodynamic data and methods of calculation used in this book.

Figure 3–4 shows the system schematically with arrows indicating the flow of materials in and out.

The following are average data for a 20-day period of operation of this blast furnace during which the average production rate was 507 short tons of pig iron per day.

Coke consumption:	1527 lb per ton of pig iron.
Coke analysis:	95.2% C; 1.0% N on dry basis.
	Actual moisture, 5.28%.
Limestone:	820 lb per ton of pig iron; 43.75% CO_2.
Moisture in charge:	583 lb H_2O per ton of pig iron, added in ore burden and during charging.
Moisture in air:	67% relative humidity at 72°F.
Slag produced:	971 lb per ton of pig iron.
Cooling water:	8036 gal per ton of pig iron, raised 10°F in temperature.
Analysis of pig iron:	3.9% C, 1.1% Si, 1.8% Mn, 0.21% P.
Analysis of top gas:	16.6% CO_2, 21.3% CO, 2.2% H_2, 59.9% N_2.
Temperatures:	Blast, 1300°F.
	Top gases, 300°F.

Assume all entering materials except blast are at the reference temperature 77°F (25°C).

STOICHIOMETRY

A convenient basis for the calculations is 1 ton of pig iron. Inspection of the data given shows that the computation of the heat balance will require stoichiometric calculations of the quantities of blast and of top gases and also calculations of the quantities involved in each of the important chemical reactions listed below.

(1) $C + \frac{1}{2}O_2 \longrightarrow CO$.

(2) $Fe_2O_3 + 3CO \longrightarrow 2Fe + 3CO_2$.

(3) $MnO + C \longrightarrow Mn + CO$.

(4) $SiO_2 + 2C \longrightarrow Si + 2CO$.

(5) $P_2O_5 + 5C \longrightarrow 2P + 5CO$.

(6) $CaCO_3 \longrightarrow CaO + CO_2$.

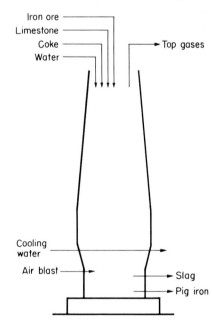

Fig. 3–4. Flow of materials, iron blast furnace.

(7) $C + H_2O$ (blast) $\longrightarrow CO + H_2$.

(8) $CO_2 + C \longrightarrow 2CO$.

MATERIALS BALANCES:

C in coke + C in limestone = C in pig iron + C in top gas.

N in air + N in coke = N in top gases.

H_2O in charge = H_2O in top gases.

The carbon balance can be solved to find the dry quantity of top gas:

Dry weight of coke = 1527 × 0.9472 = 1446 lb.

C in coke = 1446 × 0.952 = 1377 lb.

C in limestone = $820 \times 0.4375 \times \dfrac{12}{44}$ = 98 lb.

Total C input = 1377 + 98 = 1475 lb.

C in pig iron = 2000 × 0.039 = 78 lb.

C in top gas = 1475 − 78 = 1397 lb

= 116.4 lb mols.

Since the top gas analyzes 16.6% CO_2 and 21.3% CO on a dry basis,

$$\text{lb mols dry top gas} = \frac{\text{lb mols C in top gas}}{\text{mol } \% \ CO_2 + \text{mol } \% \ CO} = \frac{116.4}{0.166 + 0.213} = 307.1 \text{ lb mols.}$$

Accordingly, on a dry basis, the top gas contains:

CO_2; 16.6% × 307.1 = 51.0 lb mols.

CO; 21.3% × 307.1 = 65.4 lb mols.

H_2; 2.2% × 307.1 = 6.7 lb mols.

N_2; 59.9% × 307.1 = 184.0 lb mols.

The top gas actually contains 583 lb of H_2O from the charge or,

$$\frac{583}{18} = 32.4 \text{ lb mols.}$$

Now the N_2 balance can be solved to find the quantity of dry air in the blast:

lb mols N_2 in air = lb mols N_2 in top gas − lb mols N_2 in coke

$$= 184.0 - \frac{0.01 \times 1446}{28}$$

$$= 183.5 \text{ lb mols.}$$

$$\text{lb mols } O_2 \text{ in air} = \frac{21}{79} \times 183.5 = 48.7 \text{ lb mols.}$$

Total lb mols air = 232.2.

The quantity of H_2O in the blast can be calculated from the humidity and temperature of the outside air. The vapor pressure of water at 72°F is 20 mm Hg, so that at a relative humidity of 67%,

$$\text{lb mols H}_2\text{O per mol of hot air} = \frac{20}{760} \times 0.67 = 0.0177,$$

$$\text{lb mols H}_2\text{O in blast} = \frac{0.0177}{0.9823} \times 232.2 = 4.2 \text{ lb mols H}_2\text{O}.$$

The above calculations complete the necessary accounting for quantities of input and output materials. Now it is necessary to determine the quantities involved in each of the chemical reactions.

Reaction (1): lb mols C = lb mols CO = 2 × lb mols O_2 in blast = 97.4 lb mols.

Reaction (2): lb mols CO = lb mols CO_2 = $\frac{3}{2}$ × lb mols Fe in pig iron = $\frac{3}{2} \times \dfrac{2000 \times 0.93}{55.8}$ = 50.0 lb mols.

Reaction (3): lb mols C = lb mols CO = lb mols Mn in pig iron = $\dfrac{2000 \times 0.018}{55}$ = 0.7 lb mol.

Reaction (4): lb mols C = lb mols CO = 2 × lb mols Si in pig iron = $\dfrac{2 \times 2000 \times 0.011}{28}$ = 1.6 lb mols.

Reaction (5): lb mols C = lb mols CO = $\frac{5}{2}$ × lb mols P in pig iron = $\frac{5}{2} \times \dfrac{2000 \times 0.0021}{31}$ = 0.3 lb mol.

Reaction (6): lb mols CO_2 = $\dfrac{820 \times 0.4375}{44}$ = 8.2 lb mols.

Reaction (7): lb mols C = lb mols CO = lb mols H_2O in blast = 4.2 lb mols.

Reaction (8): The quantity of carbon in this reaction is found by difference: lb mols C in reaction (8) = lb mols in coke − lb mols in pig iron − lb mols consumed in reactions (1) to (7)

$$= \frac{1377}{12} - \frac{2000 \times 0.039}{12} - (97.4 + 0.7 + 1.6 + 0.3 + 4.2)$$

$$= 4.1 \text{ lb mols C}.$$

Thus the "solution loss" reaction (8) consumes 4.1 lb mols CO_2 and produces 8.2 lb mols CO.

As a check on the above calculations, the quantities of CO and CO_2 in the top gases can be calculated and compared with the quantities calculated previously from other data.

$$\text{mols CO} = 97.4 - 50 + 0.7 + 1.6 + 0.3 + 4.2 + 8.2 = 62.4,$$
$$\text{mols CO}_2 = 50 + 8.2 - 4.1 = 54.1.$$

Compared with the quantities previously calculated from the top gas analysis, the CO is low and the CO_2 is high, each by about 3 mols. This discrepancy may be accounted for by the assumption that part of the iron in the charge entered as Fe_3O_4 rather than Fe_2O_3 [see equation (2), page 61].

THERMAL CALCULATIONS

Based on the above stoichiometric data and on available thermochemical data, the heat balance can now be calculated item by item. The results are tabulated in Table 3–3 (page 56), and the calculations are given below.

Item 1. Heat evolved, combustion of C to CO.

$\Delta H = -29,600$ cal/mol (amorphous C)
$\quad\ = -53,280$ Btu/lb mol.

Total heat evolved $= 53,280 \times 97.4 = \underline{5,189,000\text{ Btu.}}$

Item 2. Sensible heat in blast.
Blast temperature is 1300°F.

Gas	$H_{1300} - H_{77}$ Btu/lb mol (Table A–3)	Lb mols	Btu
O_2	9,455	48.7	460,000
N_2	8,935	183.5	1,640,000
H_2O	10,810 *	4.2	45,000
	Total sensible heat in blast $= \underline{2,145,000\text{ Btu}}$		

Item 3. Heat evolved, gaseous reduction [Reaction (2)].

$\Delta H = -6390$ cal/mol Fe_2O_3
$\quad\ = -11,500$ Btu/lb mol Fe_2O_3.

Total heat evolved $= \frac{1}{3} \times 50 \times 11,500 = \underline{192,000\text{ Btu.}}$

Item 4. Sensible heat in pig iron.

Mathesius has estimated that 1 lb of pig iron carries about 510 Btu of sensible heat from the furnace.
Sensible heat in pig iron $= 2000 \times 510 = \underline{1,020,000\text{ Btu.}}$

Item 5. Sensible heat in slag.

Mathesius has estimated that 1 lb of slag carries about 900 Btu from the furnace.
Sensible heat in slag $= 971 \times 900 = \underline{874,000\text{ Btu.}}$

Item 6. Sensible heat in top gas.

Top gas temperature is 300°F.

---

* Reference state, H_2O (gas) at 77°F.

| Gas | $H_{300} - H_{77}$ | | |
	Btu/lb mol	Lb mols	Btu
CO_2	2160	51.0	110,000
CO	1565	65.4	102,000
H_2	1550	6.7	10,000
N_2	1560	184.0	287,000
H_2O	1820	32.4	59,000

Total sensible heat in top gas = 568,000 Btu

Item 7.   Heat absorbed, evaporation of water in charge.

The heat of vaporization of water at 298°K (77°F) is 10,520 lb cal/lb mol.

Heat absorbed, evaporation of water = $32.4 \times 10,520 \times 1.8$

$$= 613,000 \text{ Btu.}$$

Item 8.   Heat absorbed, decomposition of limestone [Reaction (6)].

$\Delta H = 42,500$ lb cal/lb mol.

Total heat absorbed, decomposition of $CaCO_3 = 8.2 \times 42,500 \times 1.8$

$$= 627,000 \text{ Btu.}$$

Item 9.   Heat absorbed, decomposition of water in blast [Reaction (7)].

$\Delta H = 28,200$ lb cal/lb mol.

Total heat absorbed, decomposition of H_2O in blast = $4.2 \times 28,200 \times 1.8$

$$= 213,000 \text{ Btu.}$$

Item 10.   Heat absorbed, reduction of SiO_2 [Reaction (4)].

$\Delta H = 146,200$ lb cal/lb mol.

Total heat absorbed, reduction of $SiO_2 = \dfrac{1.6}{2} \times 146,200 \times 1.8$

$$= 211,000 \text{ Btu.}$$

Item 11.   Heat absorbed, reduction of MnO [Reaction (3)].

$\Delta H = 62,400$ lb cal/lb mol.

Heat absorbed, reduction of $MnO = 0.7 \times 62,400 \times 1.8$

$$= 79,000 \text{ Btu.}$$

Item 12.   Heat absorbed, reduction of P_2O_5 [Reaction (5)].

ΔH = 212,000 lb cal/lb mol.

Heat absorbed, reduction of $P_2O_5 = \dfrac{0.3}{5} \times 212,000 \times 1.8$

$$= \underline{23,000 \text{ Btu.}}$$

Item 13.   Heat absorbed, solution-loss reaction [Reaction (8)].

ΔH = 38,000 lb cal/lb mol.

Heat absorbed, solution loss = 4.1 $\times$ 38,000 $\times$ 1.8

$$= \underline{280,000 \text{ Btu.}}$$

Item 14.   Sensible heat removed in cooling water.

Cooling water = 8036 gal
= 67,200 lb.
Heat removed = 67,200 $\times$ 1 $\times$ 10 = $\underline{672,000 \text{ Btu.}}$

Item 15.   Heat losses and unaccounted for.

Total heat input, items 1 to 3	7,526,000 Btu
Total heat accounted for, items 4 to 14	5,180,000 Btu
Heat losses and unaccounted for =	2,346,000 Btu.

Supplementary References

Butts, Allison, *Metallurgical Problems.* New York: McGraw-Hill, 1943.

Glasstone, S., *Thermodynamics for Chemists.* New York: Van Nostrand, 1947.

Hougen, O. A., and Watson, K. M., *Chemical Process Principles. Part I, Material and Energy Balances.* New York: John Wiley, 1943.

Kelley, K. K., *Contributions to the Data on Theoretical-Metallurgy. X. High Temperature Heat-Content, Heat-Capacity, and Entropy Data for Inorganic Compounds.* U. S. Bureau of Mines, Bull. 476, 1949.

Kubaschewski, O., and Evans, E. Ll., *Metallurgical Thermochemistry.* New York: Academic Press, 1951.

Lewis, G. N., and Randall, M., *Thermodynamics and the Free Energy of Chemical Substances.* New York: McGraw-Hill, 1923.

National Bureau of Standards, *Selected Values of Chemical Thermodynamic Properties*, 1949.

Noyes, A. A., and Sherrill, M. S., *A Course of Study in Chemical Principles.* New York: Macmillan, 1938.

Perry, John H., *Chemical Engineers' Handbook.* New York: McGraw-Hill, 1949.

Steiner, L. E., *Introduction to Chemical Thermodynamics.* New York: McGraw-Hill, 1948.

Wenner, Ralph R., *Thermochemical Calculations.* New York: McGraw-Hill, 1941.

Problems

3–1 In carrying out a given process, 500 Btu of heat are supplied to a system and the system increases in volume by 4 cu ft. The pressure on the system is constant at 1 atm. Calculate (a) W in ft-lbs, (b) ΔH in Btu, (c) ΔE in Btu.

3–2 A dc motor operates steadily at 115 v and

25 amp and delivers 3.2 hp through its shaft. How much heat is dissipated by the motor to its surroundings, in Btu/hr?

3-3 15,000 ft³/min (STP) of a flue gas analyzing 74% N_2, 14% CO_2, 10% H_2O, and 2% O_2 passes through a waste-heat boiler which cools it from 2050°F to 560°F. Calculate the horsepower equivalent of the heat given up by the flue gas in the boiler.

3-4 A batch of 20 tons of liquid copper at 2350°F is cooled to 2200°F by adding solid copper at 77°F. Assume there is no heat loss. (a) What quantity of solid copper should be used? (b) What temperature would be reached if 2 tons of solid copper were added instead?

3-5 Calculate the mean specific heat of dry air for the following temperature ranges: (a) 77°F to 2000°F, (b) 77°F to 1000°F, (c) 1000°F to 2000°F.

3-6 Calculate ΔH for the following reaction at 25°C, as cal/mol of FeS:

$$3FeS + 5O_2 \longrightarrow Fe_3O_4 + 3SO_2.$$

Is this reaction endothermic or exothermic?

3-7 How much heat is absorbed by the following reaction at 25°C and 1 atm, in Btu/100 lb of Zn produced?

$$ZnO + C \longrightarrow Zn + CO.$$

3-8 Hematite (Fe_2O_3) is reduced to metallic iron by carbon, and the resulting gaseous reaction product is a mixture of CO and CO_2, with two parts CO to one part CO_2 by volume. Calculate ΔH for this process at 25°C in the following units: (a) Btu per short ton of Fe produced, (b) kg cal per metric ton of Fe_2O_3 reduced.

3-9 A sample of tungsten carbide (WC) is burned to WO_3 and CO_2 (gas) using O_2 in a closed bomb, so that the combustion occurs at constant volume. The bomb and contents are at 25°C before combustion and are cooled back to the same temperature afterwards. The total heat evolved from the bomb during the combustion and subsequent cooling is 6080 joules/g WC. Assume that both CO_2 and O_2 behave as ideal

gases, for which (1) the ideal-gas equation of state is valid and (2) H and E are independent of pressure. Calculate (a) ΔH for the combustion reaction at 25°C and 1 atm, in cal/mol WC, (b) the heat of formation of WC from tungsten metal and graphite at 25°C (ΔH_f), in cal/mol.

3-10 Calculate ΔH for the following reaction at 1000°C, in cal/mol:

$$CO_2 + H_2 \longrightarrow H_2O \text{ (gas)} + CO.$$

3-11 Derive an equation giving ΔH in cal/mol as an algebraic function of T in °K for the following reactions:

(a) $CO + \frac{1}{2}O_2 \longrightarrow CO_2$
(b) ZnO (s) + C (graphite) $\longrightarrow$
$$Zn \text{ (gas)} + CO \text{ (gas)}.$$

[Note: Zn (gas) has the same molal heat capacity as argon. Also, for Zn (liq) $\longrightarrow$ Zn (gas), $\Delta H = 27{,}430$ cal/mol at the boiling point, 1180°K.]

3-12 Air at 300°F and carbon at 2000°F react to yield a mixture of N_2, CO, and excess C, the mixture having a temperature of 2400°F. A 100% excess of C is present. The process is conducted at a constant pressure of 1 atm. Calculate Q for this process in Btu/lb of C initially present in the system. Is this quantity of heat absorbed or evolved?

3-13 Calculate and tabulate the heat balance for the operation of calcining aluminum hydroxide described in Example 1, page 33. As a basis for the heat balance, take one short ton (2000 lb) of Al_2O_3. Express the heat balance items in Btu. Use 77°F as the reference temperature and H_2O (gas) as a reference state at 77°F. Assume that all input materials enter at 77°F. The calcine leaves the kiln at 1000°F and the flue gases leave the kiln at 400°F. [Note: In the absence of further information on the fuel oil, it may be assumed that the heat of combustion of the fuel oil is the same as that of a mixture of elemental C (amorphous) and H_2 of the same analysis (85% C; 15% H_2 by weight).]

3-14 2000 kg/hr of a wet zinc concentrate (90% ZnS, 10% H_2O) is roasted with dry air to produce solid ZnO and a gaseous product containing (dry basis) 5% SO_2, balance N_2 and O_2. All the sulfur from the entering ZnS is converted to SO_2. The input materials are at 25°C and the products (ZnO and gases) are at 427°C. What is the rate of heat loss from the furnace to the surroundings, in kg cal/hr?

3–15 Calcium carbonate is decomposed to make CaO and CO_2 by heating in a rotary kiln, using as fuel a natural gas consisting of pure methane (CH_4). The CO_2 formed by decomposition of the calcium carbonate mixes with the products of combustion to form the flue gas product.

Fuel consumption is 8000 ft³/ton (STP) of CaO produced.

The average flue gas analysis *on a dry basis* is 75% N_2, 4% O_2, and 21% CO_2.

Temperatures are as follows: entering calcium carbonate, fuel, and air all substantially 77°F, calcium oxide product leaves at 2200°F, flue gases leave furnace at 1100°F.

Calculate the heat balance.

3–16 A fuel gas (CO + H_2) is manufactured continuously by passing oxygen and steam through a reactor containing a fluidized bed of coke (assume pure carbon). The coke and oxygen are fed at 77°F; the steam is low-pressure steam entering at about 212°F. The reactor is held at 1850°F by controlling the steam/oxygen ratio, and the fuel gas leaves the reactor at 1850°F. Assume no heat loss from the reactor to the surroundings. Assume reactions go to completion, so that the gas consists entirely of H_2 and CO. (a) Calculate the weight ratio of steam to oxygen for operation under the conditions described. (b) What will be the analysis of the fuel gas?

3–17 5000 ft³/min (STP) of a gas analyzing 80% N_2, 15% CO_2, and 5% H_2O is to be cooled from 1400°F to 600°F by a water spray installed within the hot-gas flue. The water is brought in at 60°F and is all evaporated in the main gas stream. Calculate (a) the water requirement, in gal/min, (b) the % H_2O in the cooled gas.

3–18 ZnO and C in stoichiometrically equivalent amounts are charged cold (77°F) into a retort and heated to 2400°F. The following reaction occurs, Zn vapor and CO leaving the retort at 2400°F:

$$ZnO + C \longrightarrow Zn \text{ (gas)} + CO.$$

Calculate: (a) the quantity of heat required to bring the reactants to 2400°F, in Btu/lb of Zn; (b) the quantity of heat to be supplied to the retort at 2400°F, to carry out the reaction; (c) a heat balance for the process, reference temperature 77°F; (d) a heat balance for the process, reference temperature 2400°F; (e) a heat balance for the process, reference temperature 77°F and based on the following reactions:

$$ZnO + CO \longrightarrow Zn \text{ (gas)} + CO_2$$
$$CO_2 + C \longrightarrow 2CO.$$

The Zn-CO mixture from the retort passes into a condenser where it is cooled to 1000°F, the Zn being condensed as a liquid and the CO passing on out of the condenser at this temperature.

Calculate: (f) the quantity of heat to be removed from the condenser, in Btu/lb of Zn; (g) the sensible heat in the CO leaving the condenser plus the heat of combustion of the CO to CO_2, in Btu/lb of Zn. Compare this figure with the result of part (b) above.

CHAPTER 4

METALLURGICAL FUELS

Metallurgical fuels are the substances which are burned to supply heat at usable temperature levels for metallurgical operations on a commercial scale. Few of the unit processes of chemical metallurgy are autogenous; most of them consume large quantities of heat which must be supplied either by burning fuels or electrically. In most parts of the United States, fuels represent a cheaper source of heat than electricity; in other parts of the world the situation is sometimes reversed, because of lack of fuel, availability of cheap hydroelectric power, or other factors. On the whole, however, the metallurgical industry is strongly dependent on fuels and consumes quantities of fuels which are almost commensurate with the quantities of metals produced. For example, the coke consumption in the iron blast furnace approaches a ton per ton of pig iron, and additional fuel is consumed in making steel from pig iron.

From the point of view of the fuel industry, the metallurgical industry is one of the largest consumers. This is especially true of coal, with metallurgical uses accounting for about one-fifth of the total tonnage of coal mined in the United States. As a result the two industries are closely integrated in many areas, and the metallurgical engineer may find himself concerned with supply as well as with utilization of fuels.

As will be seen in subsequent chapters, fuels are much more than just a means of supplying the heat quantity shown in the heat balance. Metallurgical reactions require high temperatures, and the attainment of these temperatures often is a more difficult fuel problem than the supply of the right number of Btu. In some processes, for example blast furnace processes, the fuel serves as a reducing agent and as an important process reactant, as well as a means of supplying heat. In these and other processes, the products of combustion of the fuel come into direct contact with furnace contents and are mixed with the furnace products, so that the fuel participates actively in the process, beyond its primary function of supplying heat.

Classification of fuels

Fuels may be classified as indicated in Table 4-1. At least one in each of the classes of fuels indicated in the tabulation is important metallurgically.

TABLE 4-1

Classification of Fuels*

Type	Natural or primary fuels	Manufactured or secondary fuels
Solid	Anthracite coal Bituminous coal Lignite Peat Wood	Coke Charcoal Briquets
Liquid	Petroleum	Tar Petroleum distillates Petroleum residua Alcohols Colloidal fuel
Gaseous	Natural gas	Illuminating gas, city gas Coal gas, coke-oven gas Producer gas Water gas Oil gas Blast furnace gas Acetylene

*After Haslam, R. T., and Russell, R. P., Fuels and Their Combustion. New York: McGraw-Hill, 1926.

Of the solid primary fuels, bituminous coal represents the largest single source of energy for metallurgical purposes. However, the largest portion of the coal is not burned directly, but serves as raw material for manufacturing the secondary fuels: coke, tar, and artificial gases. The coal burned directly in metallurgical furnaces is now all finely pulverized before burning, although a few decades ago metallurgical furnaces were heated almost entirely by grate firing of coarse coal, to the exclusion of other methods. The other solid primary fuels, such as anthracite, lignite, peat, and wood, are used very little as metallurgical fuels and only in unusual circumstances.

Over 100 million tons of the bituminous coal mined annually in the United States is used to make metallurgical coke. Thus coke might well be regarded as a special-property fuel made for metallurgical purposes. The unique combination of physical and chemical properties possessed by coke is essential to a number of processes which either will not operate at all or will not operate as economically with any of the other kinds of fuel shown in Table 4-1. Blast furnaces in particular require coke. Charcoal is similar to coke in some ways, but is more expensive and has rather limited use.

Gaseous fuels, both natural and artificial, are much used and possess some advantages for heating metallurgical furnaces. Natural gas is an economical fuel of high quality which is important for plants near the gas fields or near the network of gas distribution lines recently constructed in some parts of the country. The artificial gases consumed metallurgically are made largely from coal and coke, some as by-products from other uses of coal.

Liquid fuels, or fuel oils, compete with gas and pulverized coal for heating furnaces. The kind of fuel oil which can be used for a given installation is primarily a matter of costs and economics; in some places even primary crude petroleum may be used, but usually it is advantageous to use by-product residual tars and distillates which are not valuable for other purposes.

Summarizing, the principal fuel products used in metallurgical processes can be classified as follows:

(1) Pulverized coal.
(2) Coke.
(3) Gaseous fuels.
(4) Fuel oils.

Of these, pulverized coal, gases, and fuel oils compete for firing furnaces, while coke is a special-purpose fuel and metallurgical raw material.

Coal

Origin and rank. Coal is formed in the earth by prolonged action of geological forces and conditions on accumulations of plant and vegetal matter. Nature's process of making coal can be traced easily in qualitative outline, since various stages of the coalification process can be observed in the earth's crust at the present time. The recognized sequence is as follows:

wood $\longrightarrow$ peat $\longrightarrow$ lignite
$\longrightarrow$ bituminous coal
$\longrightarrow$ anthracite coal $\longrightarrow$ graphite

One important evidence of the correctness of this sequence is that traces of the original wood structure are readily observed in many coals along with other more or less identifiable plant remains. These structures become less and less distinct as the coalification proceeds.

The formation of peat beds involves a combination of woody growth in wet, swampy places with favorable biochemical conditions. The dead vegetal matter accumulates and slowly decays to form peat. Observations of present-day peat bogs indicate a rate of accumulation of compact peat averaging about one foot per century. Since the peat is further compressed during subsequent coal formation, a 10-foot peat accumulation might correspond eventually to 2 or 3 feet of bituminous coal. The transformation of peat into coal begins after geological changes which result in the burial of the peat under thick beds of sedimentary material. Great pressures from the weight of the overlying sedimentary material then play an important role. Also high temperatures may be important in later stages of coalification, especially in forming graphite.

As coalification proceeds in the sequence discussed above, major changes occur in both physical and chemical character. Noticeable physical changes are in color, strength, hardness, density, and structure. The chemical changes are even more important to the fuel user. The oxygen content (on an ash- and moisture-free basis) decreases in round numbers from 40% or more for wood to 30% for peat, 20% for lignite, 5% for bituminous coal, and 2% for anthracite coal. Likewise the percentage of volatile matter (driven off by heating at 950° C in the absence of air) shows a progressive decrease from about 70% for wood down to 5% or less for anthracite (ash- and moisture-free basis). As the contents of oxygen and volatile matter decrease, the percentage of fixed or nonvolatile carbon increases from about 30% for wood and peat to almost 100% for anthracite coal.

The differences in physical and chemical properties corresponding to various stages of the coalification sequence determine the behavior of the material as a fuel. Accordingly, it is useful to rank the fuel in terms of the extent of coalification that has taken place. Thus lignite is of higher rank than peat, anthracite coal is of higher rank than bituminous coal, etc. The major ranks are further subdivided. For example, sub-bituminous, bituminous, and semibituminous specify ranks of bituminous coal. The commercial ranking of coals is based on a few specific chemical and physical tests which suffice to place the coal properly in the coalification sequence.

Composition and constitution. Coals are extremely complex substances, undoubtedly mixtures of many different chemical compounds. This complexity is not surprising in view of the vegetal origin of coal. The chemistry of wood itself is very complex to start with. Cellulose, $(C_6H_{10}O_5)_n$, has been established as a major constituent. Among the other important compounds in wood are lignin, various complex carbohydrates and sugars, gums, resins, oils, etc. The fermentation and geological metamorphosis of coalification modify these compounds, destroying some rapidly and some very slowly, and causing new chemical combinations (e.g., ulmins, humus acids, hydrocarbons). Owing to the loss of mass, minor constituents in the vegetal matter may be concentrated during the process. Accordingly, about all that can be said about the constitution of coal without getting onto controversial ground is that coals are amorphous mixtures containing carbon and carbon compounds, especially hydrocarbons, carbon-hydrogen-oxygen compounds, and some organic nitrogen and sulfur compounds. Furthermore, since coals are noncrystalline and are of infinitely variable compositions, they should be classed as rocks, not as minerals.

The study of the individual constituents and compounds in coal is called *rational analysis*. Rational analyses are based on interpretation of data from a variety of tests, such as leaching the coal with organic solvents, destructive distillation at different temperatures, chemical attack, and microscopic study. Most of these investigations show what compounds can be derived from the coal but do not necessarily prove the presence of these compounds in the original coal. Although in several instances correlations have been made or attempted between data on kinds and quantities of various substances in the coal and the use properties of the coal, rational analyses lie in the field of basic research and play no role in the day-to-day use of fuel.

Petrographic analysis, along with other types of microscopic study, has been widely used in research on coal constitution. Under the microscope, coal is found to be made up of several constituents of different optical properties, which have been given names such as fusain, clarain, vitrain, anthraxylon, and others. Although these constituents appear to be the same from one coal to another, they are not considered to be pure substances or single chemical compounds.

The foregoing discussion of the constitution of coal has dealt with what might be called the coal substance itself. In addition, commercial coals contain inorganic matter of various kinds, varying in intimacy of mixing from large pieces of shale and rock and pyrite crystals to submicroscopic matter deposited in the coal structure itself. The inorganic mineral matter accounts for the ash obtained when the coal is burned.

Proximate analysis. The practical classification of coals and their appraisal for a given application are based to a large extent on the proximate analysis. This is an empirical analysis, in that the results depend on the analytical procedure used, and therefore a closely specified and standardized procedure must be followed. However, the data obtained, percentages of "fixed carbon," "volatile matter," "ash," and "moisture,"

have considerable significance in relation to the behavior and use characteristics of the coal.

The accepted standard procedures for the proximate analysis and other common tests on coal and coke have been published by the American Society for Testing Materials. Close adherence to these standards is essential, so that tests by different laboratories, by buyer and seller, etc., will be comparable. Hereafter, brief outlines of these tests will be given, together with the number of the ASTM standard in which the further details necessary for proper execution of the tests will be found.

None of the tests to be discussed gives significant data unless carried out on *truly representative* samples. Obvious though this is, poor sampling practice still is probably the most common and most overlooked source of error. One of the secondary troubles in sampling coal is allowing for changes in moisture content during handling, so that all analyses can be referred to a uniform basis or sample condition, which might be "as received," "air dried," "dry," etc. Thus the understanding and practice of standard sampling methods (see ASTM D-492) is fully as important as following the proper analytical procedure.

The proximate analysis (ASTM D-271) consists of the following determinations:

Moisture (M): obtained by drying 1-g sample for 1 hr at 105°C.

Volatile matter (VM): loss in weight of 1-g sample heated for 7 min at 950°C in absence of air, less the moisture.

Ash (A): residue after complete combustion in muffle at 700–750°C.

Fixed carbon (FC): percentage fixed carbon is 100 less sum of percentage moisture, volatile matter, and ash.

Since moisture content will be subject to some more or less accidental variations, comparisons may be facilitated by recalculating the proximate analysis to a *moisture-free* basis, making volatile matter + ash + fixed carbon = 100%. Similarly, it may be of interest to compare coals on the basis of the character of the organic coal substance itself, apart from the inorganic and mineral matter which may be mixed in varying proportions. For such purposes, the proximate analysis is recalculated to an *ash- and moisture-free* basis, so that volatile matter + fixed carbon = 100%.

Ultimate analysis. The chief chemical elements in coal (apart from the associated inorganic mineral matter) are C, H, O, N, and S. The direct chemical determinations of these elements require more work than the proximate analysis, but yield data of considerable value to the user of coal and furnish part of the basis for appraising the suitability of the coal for a given application. Also, as will be seen, the ultimate analysis in terms of the elements is necessary for accurate calculation of materials balances and serves for calculation of calorific power in the absence of laboratory measurement.

For the ultimate analysis (ASTM D–271) C, H, N, and S are determined by chemical analysis and expressed on a moisture-free basis. The ash is determined in the same way as for the proximate analysis and also is calculated to the moisture-free basis. Then % O is found empirically as 100 − (% C + % H + % N + % S + % ash). The oxygen content found by difference in this way obviously will reflect all the errors in the other determinations, as well as further errors in the assumptions that the ash contains no C, H, N, or S and that the ash-forming portion of the coal does not change weight during the ash determination. Nevertheless, as a measure of O in the coal, not including O in the

ash, the figure obtained by difference is adequate for most purposes.

When the fuel ash enters a process and mixes with other process materials, as occurs in many metallurgical furnaces, chemical analysis of the ash is necessary as a basis for charge calculations and materials balances. Ash analysis may include SiO_2, Fe_2O_3, CaO, Al_2O_3, and other metal oxides, and sometimes S and P as well.

Calorific power. The calorific power or heating value of a fuel is the quantity of heat evolved by complete combustion of a unit quantity of the fuel under specified combustion conditions. Unfortunately, calorific values of different fuels are reported on slightly different bases; that is, they are referred to several different combustion conditions. Ideally the basis of reporting calorific values should be consistent with the most common basis of reporting heats of reaction, because calorific values are in reality specific cases of heats of reaction. If this practice were followed, the calorific value of a fuel would be equal to $-\Delta H$ for the process of complete combustion of a unit of fuel carried out with fuel and oxygen (or air) initially at 25°C and 1 atm and with the products of combustion brought to the same temperature and pressure. To complete this definition, the reference state for water in the products of combustion must be chosen, as discussed in the next paragraph.

Most fuels contain hydrogen and/or water and therefore the choice of a reference state for water in the products of combustion is necessary in giving calorific power. If H_2O (liq) is the reference state, the calorific value includes heat evolved (latent heat of vaporization) in condensing water in products of combustion at 25°C and is known as the *gross* or *high* calorific power. If H_2O (gas) is the reference state for water in the products of combustion, latent heat of vaporization is not included and the calorific value is known as the *net* or *low* calorific power. The difference between gross and net calorific powers, the heat of vaporization of water at 25°C, is 584 kg cal/kg or 1050 Btu/lb of water in the products of combustion. The net calorific value might be considered more appropriate for comparative purposes, since none of the heat derived from condensation of water is available for practical purposes. On the other hand, the laboratory determination of calorific power involves condensation of water in combustion products and gives the gross calorific power. Care must be exercised in using reported figures, since the specification as to which figure is being cited is often omitted. However, when no specification is made, the gross heating value, including heat of condensing water in combustion products, is usually understood.

Gross heating values of solid and liquid fuels are determined experimentally in an oxygen-bomb calorimeter, so that reported values are of heat evolved in a constant-volume combustion at 20°C (68°F) and several atmospheres pressure and with H_2O condensed to liquid in the combustion products (ASTM D–271, D–407). By an analysis based on the First Law it can be shown that the largest correction to be made to convert the oxygen-bomb values to heats of combustion at 25°C and 1 atmosphere is the correction from a constant-volume process to a constant-pressure process. This correction amounts only to about $+ 30$ Btu/lb of water formed by burning hydrogen,* so that the oxygen-bomb figure for a coal will be

---

* In ASTM D–407, the heat of vaporization of water is taken as 1055 Btu/lb and it is specified that the *net calorific power at constant pressure* should be calculated from *gross calorific value at constant volume* (the bomb heating value) by subtracting 1030 Btu/lb of water.

within 0.1% of the gross calorific value on a constant-pressure basis. Accordingly, in the interests of simplification, henceforth in this book it will be assumed that reported gross calorific values represent gross heats of combustion at constant pressure and at 25°C.

For fuels which are pure chemical compounds (or elements) or known mixtures of compounds, the calorific power can be calculated accurately from the known heats of formation of the compounds in the fuel and of the oxides formed in combustion. For fuels like coal and fuel oil, which are of unknown or indefinite chemical constitutions, calorimetric determinations are essential to obtain fully reliable and accurate values; however, by making certain assumptions calorific values accurate to within 2 or 3% can be calculated from the ultimate analyses.

Dulong's formula is widely used in calculating calorific power of coal from the analysis:

low calorific power (kg cal/kg or lb cal/lb) =

$$81C + 340 \left(H - \frac{O}{8} \right) + 22S$$
$$- 5.84 (9H + M), \qquad (4-1)$$

or in English units:

low calorific power (Btu/lb) =

$$146C + 610 \left(H - \frac{O}{8} \right) + 40S$$
$$- 10.5 (9H + M). \qquad (4-2)$$

In these equations, C, H, O, S, and M represent the percentages of the various elements and of moisture as determined by the standard procedures already discussed. The analyses should, of course, be expressed on the same basis as that desired for the calorific power, that is, for the coal "as received," "dry," or "air dry," as the case may be. These equations can be derived from the following empirical assumptions:

Carbon is present as "amorphous" carbon with a heat of oxidation to CO_2 of 8100 kg cal/kg (compared with 7831 kg cal/kg for carbon in the form of graphite).

Hydrogen is present as H_2 and as H_2O, the quantity of H_2 being found by subtracting the amount equivalent to the oxygen percentage from the total hydrogen found in the ultimate analysis. The heat of oxidation of H_2 to H_2O (liq) is rounded off to 34,000 kg cal/kg H_2 (compared with 33,890 kg cal/kg from heat of formation of water).

Sulfur is assumed to be present as elemental S, and the heat of oxidation to SO_2 is rounded off to 2200 kg cal/kg.

The last term in the equation, which is omitted in calculating gross calorific power, represents the latent heat of vaporization for water in the fuel plus that formed by burning hydrogen.

Calorific powers may be reported on an "as received" basis or on a "dry" basis. The gross calorific power is reduced by the presence of moisture in the proportion of the wet weight to the dry weight, but the net calorific power is reduced additionally by moisture because of the subtraction of latent heat of vaporization for the moisture. In comparing and classifying fuels, the calorific power is often recalculated to an "ash- and moisture-free" basis.

Classification of coals. Table 4-2 gives typical proximate analyses, ultimate analyses, and gross calorific values for various ranks of coal, along with similar data on wood and peat for comparison.. The solid fuels are arranged in order of increasing rank, from top to bottom, so that the progressive changes in properties characteristic of the coalification sequence can readily be seen. In particular, the trends in volatile matter, fixed carbon, moisture, oxygen content, and calorific power should be noted.

The wide ranges in properties of commercial coals together with the wide differences in properties desired in different applications make necessary a definite system of classification for coals. That is, once a satisfactory practice is established using a given class of coal, it should be possible to recognize, specify, and purchase this class of coal from any source without being forced to depend on one coal bed and without excessively detailed purchase specifications. Since the properties vary systematically with the ranking in the coalification sequence, commercial classification generally is based on rank. However, since rank is a geological criterion not measurable by a simple yardstick, the problem of classification becomes one of using the results of standard analyses and

TABLE 4-2
Typical Analyses and Calorific Values of Wood, Peat, and Coals*

Kind of fuel	Proximate analysis, %				Air-drying weight loss, %	Ultimate analysis,%					Gross calorific value, Btu/lb
	Volatile matter	Fixed carbon	Ash	Moisture		S	H	C	N	O	
Wood							6.3	49.5	1.1	43.1	5,800
Peat	26.1	11.2	6.0	56.7	53.4	0.6	8.3	21.0	1.1	63.0	3,586
Lignite	35.3	22.9	7.2	34.6	15.5	1.1	6.6	42.4	0.6	42.1	7,090
Sub-bituminous coal	27.6	44.8	3.3	24.3	16.2	0.4	6.1	55.3	1.1	33.8	9,376
Bituminous coal	27.1	62.6	7.1	3.2	1.8	1.0	5.2	78.0	1.2	7.5	13,919
Semibituminous coal	14.5	75.3	8.2	2.0	1.4	2.3	4.1	80.0	1.2	4.2	14,081
Semianthracite coal	8.5	76.6	11.5	3.4	2.6	0.6	3.6	78.4	1.0	4.9	13,156
Anthracite coal	1.2	88.2	7.8	2.8	1.5	0.9	1.9	84.4	0.6	4.4	13,298

*Data are on the "as received" basis. After E. S. Moore, Coal. New York: John Wiley, 1940.

TABLE 4-3

Classification of Coal by Rank*
ASTM D-388

Legend: FC = fixed carbon, VM = volatile matter, Btu = British thermal units.

Class	Group	Limits of fixed carbon or Btu mineral-matter-free basis	
I. Anthracitic	1. Meta-anthracite	Dry FC, 98% or more (dry VM, 2% or less)	Nonagglomerating†
	2. Anthracite	Dry FC, 92-98% (dry VM, 2-8%)	
	3. Semianthracite	Dry FC, 86-92% (dry VM, 8-14%)	
II. Bituminous	1. Low-volatile bituminous coal	Dry FC, 78-86% (dry VM, 14-22%)	
	2. Medium-volatile bituminous coal	Dry FC, 69-78% (dry VM, 22-31%)	
	3. High-volatile A bituminous coal	Dry FC, less than 69% (dry VM, more than 31%); moist‡ Btu, 14,000§ or more	
	4. High-volatile B bituminous coal	Moist‡ Btu, 13,000-14,000§	
	5. High-volatile C bituminous coal	Moist Btu, 11,000-13,000§	Either agglomerating or nonweathering¶
III. Sub-bituminous	1. Sub-bituminous A coal	Moist Btu, 11,000-13,000§	Both weathering and nonagglomerating
	2. Sub-bituminous B coal	Moist Btu, 9,500-11,000§	
	3. Sub-bituminous C coal	Moist Btu, 8,300-9,500§	
IV. Lignitic	1. Lignite	Moist Btu, less than 8,300	Consolidated
	2. Brown coal	Moist Btu, less than 8,300	Unconsolidated

*This classification does not include a few coals which have unusual physical and chemical properties and which come within the limits of fixed carbon or Btu of the high-volatile bituminous and sub-bituminous ranks. All these coals either contain less than 48% dry, mineral-matter-free fixed carbon or have more than 15,500 moist, mineral-matter-free Btu.

†If agglomerating, classify in low-volatile group of the bituminous class.

‡Moist Btu refers to coal containing its natural bed moisture but not including visible water on the surface of the coal.

§Coals having 69% or more fixed carbon on the dry, mineral-matter-free basis shall be classified according to fixed carbon, regardless of Btu.

¶There are three varieties of coal in the high-volatile C bituminous coal group, namely, variety 1, agglomerating and nonweathering; variety 2, agglomerating and weathering; variety 3, nonagglomerating and nonweathering.

It is recognized that there may be noncoking varieties in each group of the bituminous class.

"Moist" basis, Parr formulas: (1) Dry, Mm-free FC $= \dfrac{FC - 0.15\ S}{100 - (M + 1.08\ A + 0.55\ S)} \times 100$

(2) Dry, Mm-free VM $= 100 -$ Dry, Mm-free FC

(3) Moist, Mm-free Btu $= \dfrac{Btu - 50\ S}{100 - (1.08\ A + 0.55\ S)} \times 100$

tests (such as the proximate analysis and the calorific value) as the basis of ranking. Some differences of opinion exist as to the proper relationship between specifications and rank, and a few coals have unusual combinations of properties. Hence, a number of different classifications have been used in the past. The classification given in Table 4-3 is the result of a great deal of study by various interested groups and committees, and is generally known as the ASTM classification (ASTM D-388). The special basis of calculation should be noted. In particular, the proximate analyses and calorific values are recalculated to a "mineral-matter-free" basis. This kind of specification was originally developed and tried out by Parr, as a better comparative measure of the properties of the combustible coal substance ("unit coal") than the simpler "ash-free" specification. Parr's formulas, given in Table 4-3, are expressed in terms of the results obtained in a standard proximate analysis (M, A, FC, VM, and % S) and in a determination of the gross calorific value on coal containing its natural bed moisture but not including visible water on the surface of the coal. The following considerations are the basis of the formulas:

Assuming that all the S in the coal is present as FeS_2* and that FeS_2 is converted to Fe_2O_3 in the ash, a correction of $\frac{5}{8}$ of the weight of the S is needed to compensate for loss in weight from FeS_2 to Fe_2O_3. Also, on the same stoichiometric assumptions, since $\frac{10}{8}$S represents the weight of Fe_2O_3 obtained in the ash from FeS_2, the weight of ash exclusive of Fe_2O_3 from FeS_2 is $(A - \frac{10}{8}S)$. It is assumed that the earthy, FeS_2-free mineral matter (e.g., shale) carries a constant moisture content ("water of hydration") of 8% (not evolved in drying but evolved in the ash determination at higher temperature). Accordingly, the original mineral matter in the coal is considered to be the weight of the ash corrected both for the loss of weight in oxidizing FeS_2 and for the loss of water of hydration, or

total mineral matter
$$= A + \tfrac{5}{8}S + 0.08 (A - \tfrac{10}{8}S),$$

which can be simplified and rounded off to

total mineral matter = 1.08A + 0.55S.

The combustible coal substance, or "unit coal," is the total weight less the moisture and mineral matter, or

combustible coal substance
$$= 100 - (M + 1.08A + 0.55S).$$

In the usual determination of volatile matter, FeS_2 probably decomposes to FeS, so that $\frac{1}{2}$ of the S is included in the volatile

---

* This assumption is not fulfilled in actuality, since in some coals the major part of the S may be in organic compounds.

matter. On the other hand, in the ash determination the loss of weight from FeS_2 to Fe_2O_3, as already discussed, is $\frac{5}{8}$ of the S. Hence, the fixed carbon found by difference will be in error by $\frac{1}{8}$ of the weight of the S. Therefore, Parr subtracts 0.15S from the measured fixed carbon to obtain a more accurate measure of the fixed carbon in the coal substance itself.

The subtraction of 50S from the calorific value is a correction for the heat evolved in burning FeS_2 to Fe_2O_3, since this part of the calorific power has no bearing on the nature of the "unit coal."

The classification depending on mineral-matter-free specifications should not be influenced by accidental inclusion of shale and other extraneous matter and likewise should not be influenced by use or nonuse of coal preparation and washing. Obviously, these factors have no bearing on the rank of the coal.

An important criterion in distinguishing bituminous and higher ranking coal from sub-bituminous and lower ranking coal is resistance to weathering. The *weathering* coals tend to disintegrate on exposure, whereas the *nonweathering* coals are resistant to exposure. A quantitative, accelerated weathering test may be used to determine the *weathering index*, which is the percentage fines produced by a standard exposure.

Another criterion in the ASTM classification is the *agglomerating* or *nonagglomerating* character of the coal. Agglomerating coals are those coals which in the standard determination of volatile matter make an agglomerate button that will support a 500-g weight or a button that shows swelling or cell structure. The agglomeration or nonagglomeration of a coal is not to be confused with *coking* and *noncoking*, although the two are related. The classification of coals as between coking and noncoking is based on whether or not the coals are suitable for making coke in commercial equipment, and no sure-fire laboratory tests are available for making this distinction.

A more qualitative and more descriptive summary, given by Fieldner and Davis, of the distinguishing physical and chemical characteristics of various ranks of coals is given in Table 4–4. The decrease in moisture content with increase in rank applies only to the normal moisture content as mined, and should not be considered as an accurate criterion. The ratio of fixed carbon to volatile matter was one of the first criteria proposed for classification of coals (W. R. Johnson, 1844). This ratio is sometimes called the "fuel ratio" and is still used as a simple means of comparing coals. Thus, the fuel ratio gives a rough indication of the type of flame obtained in burning the coal on grates, a high ratio of fixed carbon to volatile matter corresponding to a short intense flame and a low ratio to a long, luminous flame.

Coal quality. Grading of coals. It should be emphasized that the ranking and classi-fication of coals do not measure coal quality. Specifically, higher rank should not be considered as higher quality. Although the layman sometimes thinks of anthracite as of higher quality than bituminous, this feeling results from its greater suitability for certain domestic purposes, and there are many applications for which bituminous coals are more suitable.

In different applications, different criteria of quality may be emphasized. Thus, in making coke for the iron blast furnace, the sulfur content of the coal is one of the important factors, but in other metallurgical processes the sulfur does little harm and percent sulfur becomes only a secondary criterion of coal quality. For some purposes, the particle size distribution and, in particular, the percentage of fines are significant. Thus the term "degradation" may be used to refer to the size reduction and disintegration of the coal into fines which occur as the coal is handled. The calorific value (as received) and the percent ash almost always are important criteria of quality, as might be expected. Also, the composition and properties of the

TABLE 4-4

Physical and Chemical Characteristics of Coals*

| Coal | Chemical characteristics | | Physical characteristics |
	Approximate % normal moisture	Ratio of fixed carbon to volatile matter	
Lignite	30–45		Distinctly brown; either markedly claylike or woody in appearance; falls into pieces on exposure to weather.
Sub-bituminous	18–30		Black; no distinct woody texture; disintegrates and loses moisture on exposure to weather, but less rapidly than lignite.
Bituminous	3–15	3	Slightly affected by exposure to weather.
Semibituminous	3–6	3–7	Slightly affected by exposure to weather.
Semianthracite	3–6	6–10	Slightly affected by exposure to weather.
Anthracite	2–3	10–60	Slightly affected by exposure to weather.

*Fieldner, A. C., and Davis, J. D., "Modern Views of the Chemistry of Coals of Different Ranks as Conglomerates." *Trans. AIME* 71, 229, 1925.

ash frequently must be taken into account in metallurgical processes. Obviously fixing the rank of a coal fixes none of these quality factors.

The ASTM has set up standard designations for grading coal (ASTM, D–389) which are useful for making specifications in buying and selling.

Although coal rank is not changed by crushing, washing, sizing, concentration, and other operations of mechanical preparation, coal quality is improved by mechanical preparation. Thus, by preparation, sulfur, ash, and extraneous mineral matter are rejected and the calorific value is increased. Moreover, variations in the mined coal are smoothed out so that quality becomes more uniform. In fact, a uniform product can be made and at the same time low-cost, non-selective mining methods (e.g., strip mining) can be adopted. Accordingly, with depletion of high-grade deposits, increased quality requirements, and competition with other fuels, mechanical preparation of coal is already widely practiced and undoubtedly will be used in a larger and larger proportion of the total coal production as time goes on.

Coal deposits. Although coal is a metallurgical raw material of relatively low unit cost, large quantities are used. As a result, cost of transportation of coal from mine to metallurgical plant constitutes a sizeable fraction of gross fuel cost and frequently represents a major cost item of metal extraction. Therefore countries without their own coal deposits are likely to turn to other fuels, if available, or to electric power in preference to importing coal. The United States fortunately is blessed with ample deposits and large reserves of coal, in fact the largest of any country in the world. However, even within the borders of the United States, transportation costs of fuel from source to point of use have much influence on plant

location and metallurgical economics in general. The well-known and outstanding example is the location of the United States iron and steel industry in relation to coal deposits.

Figure 4–1 shows the distribution of known coal deposits in the United States, based on data compiled by the U. S. Geological Survey. The coal areas are divided into six provinces, as follows:

Eastern province
Interior province
Gulf province
Northern Great Plains province
Rocky Mountain province
Pacific Coast province

Each province in turn is divided into *regions.* For example, the Eastern province includes the Appalachian, Anthracite, and Atlantic Coast regions. Further geographical subdivisions are made as a matter of convenience in referring to coal occurrence, the regions being divided into coal *fields,* and coal fields being divided into *districts.* The district is the smallest division, and generally refers to areas in which the coal is mined from a single continuous bed or a group of closely related beds. The buyer of coal may desire the coal from a given district because its properties and behavior are particularly suited to his needs in ways not gauged entirely by rank, grade, and other commercial specifications.

The Eastern province is the most important coal-producing area and accounts for a large part of United States resources of high-rank coal. Nearly all the anthracite mined in this country has come from the Pennsylvania Anthracite region. The Appalachian region has accounted for a major fraction of the high-rank bituminous coal mined and in large measure has been the basis of the great industrial development of the Eastern United

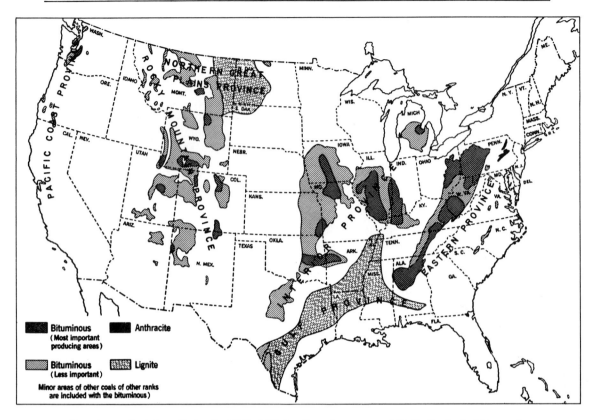

Fig. 4–1.   Coal deposits in the United States (based on maps of U. S. Geological Survey).

States and of the great iron and steel industry of that area.   Some of the best coking coals have come from western Pennsylvania, near Pittsburgh (e.g., Connellsville coke).

The bulk of the coal in the other five provinces is of lower rank than the coal in the Eastern province, although high-rank bituminous and anthracite coals are concentrated in a few districts of industrial importance. Coals suitable for making metallurgical coke are mined in sufficient quantity in the West to take care of the needs of the western smelters and metallurgical plants.

The lignites and low-ranking bituminous coals, of which there are extensive deposits, particularly in the Gulf and Great Plains provinces, are not mined on a large scale because these materials cannot compete with the higher ranking coals and other desirable fuels which are still available in ample supply for years to come.

Although the United States coal reserves of all ranks appear almost inexhaustible in quantity at present and anticipated rates of consumption, the long-range trend is toward gradual depletion of the best grades and highest quality coals.   Thus, the gradual decrease in quality of metallurgical coke has become of serious concern to the iron and steel industry because of the direct relation between coke quality and the efficiency and economy of blast furnace operation.   To counterbalance this trend, removal of impurities and closer control of quality by widespread adoption of modern coal preparation methods is to be expected. To date, progress in coal preparation has not been as rapid as might be desired because of the small but appreciable cost and because of the long-ingrained habit in the coal industry of looking for better quality in the ground.

Coal storage. Thermodynamically *any* mixture of fuel with air or oxygen is unstable at ambient temperatures. However, in most cases the rate of reaction of fuel with oxygen is low or negligible at ambient temperatures and some kind of deliberate ignition is necessary to start active burning or explosion. If the rate of oxidation is low, the heat of reaction is readily dissipated to the surroundings and no temperature rise occurs. On the other hand, if the rate of oxidation is high enough to start with, heat accumulates, the fuel mixture increases in temperature, and the rate of reaction increases further in an accelerating and self-propagating fashion. Thus spontaneous combustion occurs at a threshold of conditions for which the reaction rate is just sufficient to generate heat at a rate faster than it is dissipated to the surroundings. Solid coals and air react rapidly enough at low temperatures so that this threshold is easily crossed, especially with coals containing fine particles or fine pyrite, and coals stored in large piles or bins with poor dissipation of heat. *Finely pulverized coal, containing large percentages of minus 200-mesh material, cannot be stored safely in contact with air.*

Various rules and practices have been followed in coal storage to avoid spontaneous combustion, but coals vary so much in their properties that each situation should be considered on its own merits. Storage under water is sometimes used, but requires some special facilities and leaves the coal very wet.

Capping the pile with a more or less airtight cover of asphalt prevents entrance of air and is useful for long-term storage. When coal is stored in piles or in bins, it is important to prevent segregation of sizes in different parts of the mass since segregation produces spots where entrance of air, reactive fines, and poor heat dissipation are particularly favorable to spontaneous combustion.

A common procedure is to maintain a regular check on the temperature of coal in storage. Temperature measurements may be made in iron pipes inserted near the bottom of the pile and left in place. A temperature of 120°F is considered a warning that a close watch is needed, but a rise to 160°F or so is sometimes permitted. Usually the coal is moved when it becomes too hot.

Pulverized coal

Pulverized coal is widely used for heating melting and smelting furnaces of all kinds in metallurgical plants. Some decades ago coarse coal was burned on grates as the chief source of heat for pyrometallurgical processes. Grate firing with various kinds of mechanization (for example, stokers) is still important for domestic heating and for many industrial purposes. However, for metallurgical heating, coal is burned in quantity only in pulverized form. Among the reasons for widespread adoption of pulverized coal firing for metallurgical furnaces are the flexibility and easy adjustment to process requirements, high combustion efficiency, low costs, and adaptability to wide variations in coal rank and quality.

Whereas the technique of coal firing on grates has little in common with gas firing and fuel-oil firing, pulverized coal firing is similar to them in many respects; consequently, the three are to a considerable extent interchangeable without major changes in furnace design. Finely ground coal settles slowly in air and can be handled as a thick suspension in air which flows much like a liquid. At the furnace, the mixture of coal dust and air is blown into the combustion chamber through a simple and relatively compact burner and forms a flame in the same way as does a mixture of air and atomized oil, or a mixture of air and fuel gas. Of course pulverized coal is not as clean to use as gas or oil; part of the finely divided ash settles in the furnace chamber and joins the materials in process and another part is carried out in suspension in the flue gases.

Mechanism and rate of burning coal particles. A 200-mesh particle of coal burns in substantially the same way as a large lump of coal, but the process is more easily observed with a lump. Three consecutive stages are recognized. First, the particle is dried and heated until ignition occurs. In the second stage volatile matter distills out and burns. Finally the residual coke and fixed carbon burn by a gas-solid reaction. The technique of heating with pulverized coal is based on the fact that the rates of all three steps of burning increase markedly with decrease in particle size, so that for particles 200-mesh and finer burning is completed in a fraction of a second as the particle is carried through the combustion chamber in a high velocity gas stream.

The times required for ignition and for distillation and burning of the volatile matter of fine coal particles are very small, of the order of a few thousandths of a second. The major part of the time of combustion is the time required for burning the residual solid coke, the third stage of combustion. The rate of this gas-solid reaction will be proportional to the solid-gas surface area and will depend also on the partial pressure of oxygen in the gas, on the diffusion rates of oxygen and gaseous reaction products to and from the particle surface, and on the chem-

ical reactivity of the solid itself. Theoretical analysis leads to relations showing the time of combustion to be proportional to a power n of the particle size, the value of n varying from about 0.5 to 2 depending on the assumptions made in the analysis. A few experimental data indicate very roughly a value of n in the vicinity of unity. The combustion times for particles in the 100- to 200-mesh size range fall in general between 0.1 and 1 second, varying with the coal and the conditions of burning.

The powdered coal flame consists of a moving cloud of burning particles. Hence, by combining the information on rates of reaction of individual particles discussed in the preceding paragraph with data on the size distribution of the coal, an understanding of the flame itself and an estimate of its volumetric space requirements can be obtained. Hottel and Stewart have made a detailed analysis of this problem in "Space Requirements for the Combustion of Pulverized Coal," *Ind. Eng. Chem.* **32,** 719–730 (1940), which should be consulted for details.

As in other processes where gas-solid reactions determine rates, the specific rate of reaction at a given instant, expressed, for example, as the fraction of the solid carbon consumed per second, should be proportional to the specific surface of the solid in cm^2/g of solid. Thus, in expressing the fineness of powdered coal it is appropriate to use the specific surface as a single-figure criterion of fineness. On this basis, the pulverizing operation can be considered as a surface-making operation.

Equipment. A major consideration in the arrangement of equipment for powdered coal firing is that the powdered coal cannot be stored safely and conveniently for any length of time. Also precautions must be taken to avoid accidental accumulations of explosive mixtures of air and coal dust in and around equipment and buildings auxiliary to the burner installation. Other troublesome problems have been encountered and overcome in developing the art of pulverized coal firing. Coals as received at the plant commonly carry too much moisture for smooth handling in pulverizers and burners so that drying facilities must be incorporated. The pulverizers must be capable of delivering coal of 80 to 90% minus 200-mesh over long periods without mechanical breakdown. Close control of coal feed rate and of the fuel-air ratio must be provided so that efficient combustion is obtained over the whole range of operating conditions.

Many different arrangements are used for preparing and burning pulverized coal, since for each of the unit operations of preparing the coal (that is, feeding, drying, grinding, sizing, conveying) a variety of equipment is available. For example, the grinding may be conducted in a ball mill, a roller mill, a ball-race mill, or an attrition mill. A separate dryer may be provided, or drying may be accomplished by sweeping the pulverizer and auxiliary equipment with air preheated in a separate air heater. However, regardless of the choice of equipment, the common arrangements fall into two categories which may be called the "unit system" and the "storage system," respectively.

The *unit system* is illustrated in Fig. 4–2. With this system, each burner installation involves a self-complete unit for preparation and pulverizing of the coal, with a continuous flow of coal from coarse coal bin to burner and no storage of fine coal at any point. The entire unit is located at the furnace.

The *storage system* is illustrated in Fig. 4–3. With this arrangement, a closed storage bin of several hours' capacity is provided for the pulverized coal so that the steps of preparing the pulverized coal are not so rigidly coupled to the operation of the combustion system

and each can be controlled independently. The storage system is suited for large plants with pulverized coal firing of a number of furnaces, since the coarse-coal unloading and storage facilities and the pulverizing plant can be located centrally and conveniently. The finely pulverized coal is readily moved around the plant in pipes to the various burners. This arrangement is also known as the "central system."

Pulverized coal burners for metallurgical furnaces are usually relatively simple, consisting mainly of a jet or pipe which spouts the coal-air mixture into the furnace chamber at a suitable velocity. In some cases all the combustion air is mixed with the coal dust ahead of the burner; in other cases the burner jets a suspension of the coal in part of the air (primary air) while the rest of the air (secondary air) is fed to the combustion chamber through openings around the burner. Separate feeding of primary and secondary air is necessary if the secondary air is highly preheated, because coal becomes sticky on heating.

The pulverizer and burner with necessary auxiliary equipment constitute a relatively elaborate installation compared to the facilities required for firing fuel oil or gas. Hence, although coal is generally cheaper, against the saving in fuel cost must be balanced the larger capital costs and operating costs (mainly power for pulverizers) and added maintenance for the pulverized coal installation. Accordingly, pulverized coal is most attractive for large furnaces and power plants, where the coal consumption is a matter of several tons per hour.

Coke

Coke is the coherent cellular residue from the destructive distillation of coal in the absence of air. The destructive distillation process by which coke is produced is called

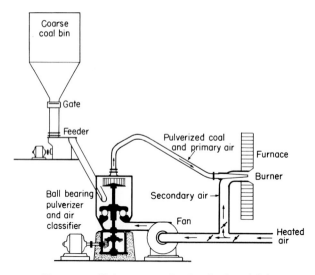

FIG. 4–2.   Unit system of pulverized coal firing.

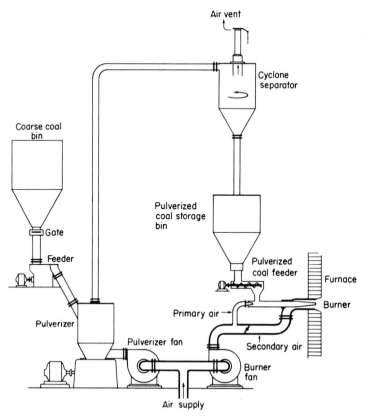

FIG. 4–3.   Storage system of pulverized coal firing.

84

coking or *carbonization*. During coking the volatile matter of the coal is driven off, leaving a product high in fixed carbon. Chemically, then, coke might be considered an impure commercial form of solid carbon to be classed with charcoals, sugar carbon, petroleum coke, graphite, and other commercial carbons. However, the wide metallurgical use of coke as a fuel and reducing agent is the result not only of its cheapness as a source of carbon but also of its special combination of physical and chemical properties, possessed by no other form of carbon. In fact, a number of important processes are virtually founded on the special properties of coke.

Mechanism of coking. As indicated above, coke is made by heating coal in the absence of air until substantially all the volatile matter is removed. The complex organic compounds which make up the coal substance are decomposed in steps, and many varieties of compounds are volatilized, different gases being evolved at different stages of the pyrolysis. The volatilized compounds range from simple gases such as CO, CO_2, H_2O, H_2, N_2, CH_4, H_2S, S_2, and NH_3 to various complex hydrocarbons and other organic compounds, some of them containing nitrogen and sulfur. As will be discussed, the substances volatilized from coal as by-products of coking are important and valuable raw materials for the chemical industry.

During the decomposition the mass of coal fuses and becomes plastic. At the same time it swells and expands. As the coking proceeds to completion, the mass slowly solidifies. The fusion, swelling, and solidification account for the characteristic structure of coke. Moreover, differences in the behavior of different coals during these successive stages account for important structural and property differences in the resulting coke. To take an extreme example, non-coking coals may decompose without becoming

plastic at any stage. Considerable control over coke properties can thus be exercised by mixing different coals and controlling the proportions in the mixture.

Coke ovens. Metallurgical coke is made in two types of ovens, *beehive* ovens and *by-product* ovens. Good quality coke can be made in either type, but the beehive oven wastes the valuable by-products of coking, whereas they are recovered profitably in the by-product process. Another definite advantage of the by-product oven is its greater flexibility in handling poorer coals and in controlling coke properties. Accordingly, the beehive process now accounts for a relatively small percentage of metallurgical coke in the United States, and beehive ovens are used mainly as a facility of low capital cost for emergency use or standby capacity. During World War II some new beehive ovens were constructed and put into operation as the quickest and cheapest method of getting the necessary increase in coke production.

The beehive oven (Fig. 4–4) is about 12 ft in diameter, 6 ft high, and takes a charge of 6 to 8 tons of coal as a bed about 2 ft deep. The walls are of brick, usually banked on the outside with dirt or clay. A hole at the top serves for charging and for exit of gases. The door opening at one side gives access to the oven for leveling the charge, introducing water for quenching, and removal of coke. During the coking period, the door opening is bricked nearly to the top, leaving only a small opening for entrance of air. One operating cycle is as follows. The coal is charged into the oven and leveled. The door is bricked to about $1\frac{1}{2}$ inches from the top. With the oven hot from the preceding charge, volatile matter from the coal starts to come off and burns with the air entering the slot in the door. Throughout the coking period (2 or 3 days) heat is supplied by the combustion of gases distilled from the coal, and the coking proceeds gradually from the top of the charge to the bottom of the oven. At the end of the coking period, the door is broken down and the coke is quenched by spraying with water in the oven. Finally, the coke is drawn by hand and the oven is ready for the next charge. Obviously, making coke in beehive ovens is an art in which the human element and the coke burner's skill have a major bearing on the results.

In the by-product process, the coal is charged batchwise into silica-brick or fire-brick retorts and these retorts are heated externally, using a gaseous fuel (Fig. 4–5). The matter volatilized from the coal is piped from

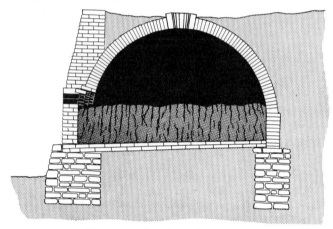

Fig. 4–4. Beehive coke oven.

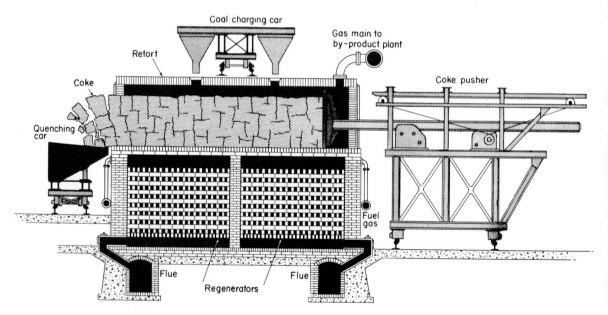

Fig. 4–5. By-product coke oven.

the retorts to the by-product recovery plant, where the by-products are separated from each other. The retorts are generally about 40 ft long, 13 ft high, and $1\frac{1}{2}$ ft wide, the narrow width making possible the uniform penetration of heat from the outside of the retort to the center of the charge. Coal is charged through holes along the top of the retort (about 15 tons or more per retort). At the end of the coking period, about 18 hours after charging, coke is discharged from the end of the retort by a mechanical pusher which moves through the whole length of the retort. The retort is made with a slight longitudinal taper to facilitate this kind of discharge. The hot coke falls from the retort into a car, where it is quenched with water.

By-product ovens are constructed in batteries of from 30 to 90 retorts, parallel and closely spaced so that they can be enclosed and heated in one furnace unit and so that a single traveling pusher serves all the retorts in a battery. Although each retort individually operates batchwise on a cycle of less than 24 hr, the charging and discharging of the retorts in a battery is carried out around the clock, so that the battery as a whole operates substantially in a continuous fashion. The fuel may be part of the coke-oven gas obtained as one of the by-products. Since coke-oven gas is a relatively high-grade fuel gas, in some districts it is profitable to heat the coke ovens with a lower grade fuel, such as producer gas or blast furnace gas, and to sell or use the coke-oven gas elsewhere. Whatever fuel is used, the coke-oven batteries are built with regenerators to recover heat from the flue gases after passage over the retorts and to return this heat to the system.

Temperature is an important variable in the carbonization process, affecting not only the rate of carbonization but also the yields of coke and various by-products and the properties of the coke. In *low-temperature coking*, the maximum temperature is in the range 450–700°C. In *high-temperature coking* the temperature is carried to 800–1000°C or higher. Only high-temperature coke has the properties desired for metallurgical purposes, and within the high-temperature coking range considerable control over coke properties can be exercised by regulating the heating rate and temperature.

Typical yields from one ton of coal by the by-product process are as follows:

Coke	1400 lb
Coke-oven gas, calorific value of 550 Btu/ft³	11,000 cu ft
Ammonium sulfate	25 lb
Tar	10 gal
Light oil	3 gal

The by-products have many uses and are the starting points for several industries. Gas and tar are frequently used as fuels in the coke plant and in the steelworks with which the coke plant is affiliated. By-product ammonia represents one of the major raw materials for industries using chemical nitrogen. On refining by distillation and other processes, the tar and oil yield a great variety of solvents, oils, and industrial chemicals, including such products as benzols, toluol, naphthalene, anthracene, dyes, cresol, phenol, road tar, creosote, etc.

Coking coal. Coking coal has been defined for practical purposes as a coal which "will yield a merchantable coke when carbonized by a commercial method in an existing type of coke oven." This definition involves several related questions: Will the coal yield coherent coke when heated? Is the behavior of the coal during carbonization suitable for existing apparatus and coking procedures? Assuming coke can be made with existing facilities, will it be merchantable?

So far no way has been found for predicting with any degree of certainty the behavior
of coals during pyrolysis from the rank, analyses, or other data readily obtained in the
laboratory. Destructive distillation of some
coals leaves a powdery residue with no evidence of plasticity at any stage, while other
coals of similar specifications might yield
hard coke. Low-volatile and high-volatile
bituminous coals behave differently during
coking. For example, the high-volatile coals
generally contract during coking while the
low-volatile coals expand, although this relation cannot be taken as a fixed rule since the
contraction or expansion depends on factors
other than the amount of volatile matter.
The contraction or expansion during coking
has considerable direct importance in the by-
product oven, since expansion may lead to
sticking in the retort and damage to the oven
structure. A common practice, therefore, is
to blend low- and high-volatile coals to obtain
a more favorable combination of expansion
behavior and coke quality than is obtainable
by one coal alone.

Even if a coal or coal mixture can yield a
coke of satisfactory structure and mechanical
properties and can be carbonized without
difficulty in standard equipment, it may still
be unusable because of high contents of impurities deleterious in metallurgical processes. In particular, limits are placed on
percent ash, sulfur, and phosphorus.

Although United States reserves of bituminous coal are very large, the portions of
these reserves which can be considered as good
coking coal are being depleted rapidly. During and since World War II, the metallurgist
has had to use poorer quality coke, often with
serious drops in efficiency and production
rate.

Properties of coke. As already indicated,
the physical and mechanical properties of
coke are just as critical metallurgically as the
coke analysis and chemical properties. The
major tonnage use of metallurgical coke is in
the blast furnace process, and a full discussion of the properties required in coke should
be based on a detailed study of this process.
However, certain obvious features may be
pointed out. The solid coke, mixed with
lump ores and fluxes, is subjected to heavy
loads and abrasion in moving down the shaft
of the furnace. Since the process depends on
the passage of huge gas volumes up through
the shaft, the coke should not contain or
break down into fines which increase the resistance to gas flow and also increase the dust
loss. Chemically, the coke is utilized to a
large extent in reactions with gases (e.g., O_2,
CO_2) so that it must have the right degree of
reactivity. Coke porosity favors rapid reaction. The ash portion of the coke must be
removed in the furnace slag, and additions to
the furnace charge are generally required to
flux the constituents of the ash. In the iron
blast furnace, the phosphorus and part of the
sulfur from the coke are picked up by the iron
and eventually must be removed before the
iron is made into steel.

Chemical analyses of coke are made by
the same procedures as for coal. The usual
coke specifications include the proximate
analysis, %P, %S, and the analysis of the ash.
The ranges of variation in metallurgical cokes
are about as follows:

Fixed carbon   83–90%
Volatile matter 0.5–4% (preferably   2%)
Ash   4–15% (preferably <10%)
Moisture   <5% (preferably <3%)
S   0.5–3% (preferably <1%)
P   <0.04%

Several kinds of tests are used to appraise the physical properties of coke. The
porosity, or "volume of cell space," is determined by measuring the apparent specific
gravity of the lump coke and the true specific

gravity of the coke exclusive of voids (ASTM D–167). These data fall generally in the following ranges:

Apparent specific gravity	0.8–1.1
True specific gravity	1.8–2.0
Porosity	41–55%

Resistance of coke to breakage is measured in a standard "shatter test" in which a 50-lb sample of +2-inch coke is dropped 4 times onto a steel plate. The percentage of the original coke still retained on a 2-inch screen after the drops measures the resistance to breakage (ASTM D–141). The range for blast furnace coke is from 70 to 87% retained on a 2-inch screen. A "tumbler test" (ASTM D–294) is sometimes used as a measure of resistance to abrasion. Also, measurements of the compressive strength are occasionally reported.

Closely sized coke is preferred for blast furnaces and at least the fines must be screened out. The minus $\frac{3}{4}$-inch or minus 1-inch fines separated from the coarse coke are the "coke breeze," which has a variety of uses in metallurgical plants.

The reactivity of coke is a property difficult to define and to measure, but nevertheless a property which is important in all uses of coke. No standard laboratory test has been developed on which the coke user can rely to appraise coke reactivity or combustibility. Instead the appraisal has necessarily been based on how well the metallurgical process works when the given coke is used. When the various uses of coke are examined, it is evident that the chief reactions of coke are reactions with gases, particularly O_2, CO_2, and CO. Accordingly, the experimental studies of coke reactivities generally have involved measurements on coke-gas reactions under controlled conditions. Two kinds of measurements have been used: (1) reaction rates of CO_2 with coke ($CO_2 + C \longrightarrow 2CO$)

and (2) ignition temperatures of coke in O_2 or air. Undoubtedly more use will be made of tests of this kind in the future.

Gaseous fuels

Gaseous fuels possess a number of advantages over other fuels: cleanliness and absence of ash, ease of handling, ease of control, flexibility, and good combustion characteristics. These advantages lead to the widespread use of gaseous fuels wherever availability and costs permit. It is even desirable in some operations to manufacture gas in the plant from available solid and liquid fuels of less desirable characteristics. Gases made from coal in particular are widely used. In some cases, the fuel gases are recovered as by-products of other plant operations (blast furnace, coke oven, zinc retort, etc.).

Constituents of gaseous fuels. Most fuel gases are mixtures, the different fuels consisting for the most part of different proportions of a small number of common constituents. These common constituents may be classed in two groups as follows:

Combustible	*Diluents*
Hydrogen (H_2)	Nitrogen (N_2)
Carbon monoxide (CO)	Carbon
Methane (CH_4)	dioxide (CO_2)
Ethane (C_2H_6)	Water (H_2O)
Ethylene (C_2H_4)	
Other hydrocarbons	

In addition to these principal constituents, small percentages of other volatile organic compounds, sulfur compounds, etc., may be present. Thus the properties of commercial fuel gases depend primarily on the relative proportions of combustible and diluent gases and on the nature of the combustible compounds.

Table 4–5 gives a comparison of the pure combustible gases in terms of volume of oxygen required for combustion, volume of combustion products, and low calorific power. On these bases the combustible gases can be roughly classified into two categories: first, the two *base* gases, H_2 and CO, of calorific powers around 300 Btu/ft³ and with low volumes of oxygen consumption and combustion products; and second, the *enriching* or hydrocarbon gases with calorific values of about 1000 Btu/ft³ for methane, above 1500 Btu/ft³ for higher hydrocarbons, and with much higher volumes of oxygen consumption and combustion products.

Commercial fuel gases. The fuel gases used metallurgically cover a wide compositional range which is best examined in terms of the relative proportions of diluent, base, and enriching gases. Table 4–6 gives typical analyses and Btu values for the most common fuel gases. It will be noted that calorific values in Table 4–6 are reported as Btu/ft³ of gas measured dry at 60°F and 30 inches Hg pressure, whereas the data in Table 4–5 involved measurement of gas volumes at 32°F and 29.92 inches Hg (STP). Another common basis is Btu/ft³ of gas at 60°F and 30 in-

TABLE 4-5

Properties of Pure Combustible Gases

Gas	Volumes O₂ required per volume of fuel*	Volumes flue gas per volume of fuel,* using theoretical air	Net calorific power	
			Btu/ft³ (STP)	kg cal/m³ (STP)
Hydrogen (H_2)	0.5	2.88	290	2,582
Carbon monoxide (CO)	0.5	2.88	341	3,034
Methane (CH_4)	2.0	10.52	962	8,560
Ethane (C_2H_6)	3.5	18.16	1,698	15,110
Propane (C_3H_8)	5.0	25.80	2,433	21,650
Ethylene (C_2H_4)	3.0	15.28	1,627	14,480
Acetylene (C_2H_2)	2.5	12.40	1,510	13,440

*All volumes STP.

TABLE 4-6

Analyses and Calorific Values of Typical Fuel Gases*

	% by volume								Calorific values, Btu/dry ft³ (60°F, 30 inches Hg)	
	H_2	CO	CH_4	C_2H_6	C_2H_4 and illuminants	N_2	CO_2	O_2	Gross	Net
Blast furnace gas	3.7	26.3	0.0	0.0	0.0	57.1	12.9	0.0	96.7	94.8
Producer gases:										
Pittsburgh coal, best	12.7	27.5	3.4	0.0	0.8	50.8	4.6	0.2	176.5	167.1
Pittsburgh coal, average	15.0	24.7	2.3	0.0	0.8	52.2	4.8	0.2	154	142
Illinois coal	11.0	25.4	2.5	0.0	0.6	54.0	5.2	0.1	155	146
Coal gas	48.5	7.4	32.2	1.0	5.6	2.3	2.2	0.8	620	564
Coke-oven gas	57.0	5.9	29.7	1.1	3.4	0.7	1.5	0.0	592	523
Blue water gases:										
from coal	52.2	32.7	3.1	0.2	0.2	6.8	4.5	0.3	311	284
from coke	49.7	39.8	1.3	0.1	0.0	5.5	3.4	0.2	306	278
Carburetted water gas	34.4	31.3	12.0	0.9	5.6	8.2	2.9	0.8	554	507
Oil gas	50.8	10.2	27.6	-	3.5	5.1	2.6	0.2	548	481
Natural gas	0.0	0.0	84.5	11.5	0.0	3.8	0.2	0.0	1040	940
Natural gas	0.0	0.0	94.5	0.5	0.5	4.0	0.2	0.3	963	868

*After Camp, J. M., and Francis, C. B., *The Making, Shaping, and Treating of Steel*. Pittsburgh: Carnegie-Illinois Steel Corp., 1940.

ches Hg, but saturated with water (p_{H_2O} = 0.52 inch Hg at 60°F). The lowest grade combustible gases are high in the diluent constituents and contain only CO and H_2 as combustible matter. Reduction in diluent content leads to the successively higher calorific values of producer gas and then water gas. To obtain calorific values much above 300 or so Btu/ft³, the gas must be enriched with hydrocarbons. Values near 1000 or more Btu/ft³ are obtained only with gases composed mainly of hydrocarbons.

With fuel gases of low calorific power, obviously large volumes of gas must be supplied and burned to meet a given Btu requirement, and thus the cost of handling and distribution per Btu goes up rapidly with decrease in calorific power. Accordingly, low grade gases such as blast furnace gas and producer gas are burned very near the point of manufacture, as a rule within the same plant. Coke-oven gas and enriched gases of 500 Btu/ft³ and higher can be distributed economically as city gas. The natural gases of high calorific value are piped long distances cross-country from the gas and oil fields to industrial centers.

Brief descriptions of the sources of the common fuel gases are given below.

Blast furnace gas is produced in large quantities as a by-product of the iron blast furnace, and is an important fuel only around iron blast furnace plants. Part of it is burned for heating the air blast to the blast furnace and the rest is used for powering gas engines, heating coke ovens, and other uses where high-grade fuels or very high temperatures are not required.

Producer gas is made by treating coal or coke with a mixture of air and steam, usually in an apparatus called a gas producer located in the plant near the furnace where the gas is burned. Use of producer gas as fuel can be regarded as a cheap method of obtaining the advantages of gaseous fuel when coal is the only fuel readily available at the plant. The calorific value of the producer gas represents a large fraction of the calorific value from the original coal, generally 70% or over.

Coal gas and *coke-oven gas* are the noncondensable gases obtained from the destructive distillation of coal. Coke-oven gas, as the name indicates, is the by-product of coke plants, of which the primary product is blast furnace coke. To a limited extent coals are carbonized in gas retorts to give coal gas, the gas being the primary product and the coke constituting a by-product usually too friable for metallurgical uses.

Water gas (*blue gas*) is made by the reaction of steam with a hot bed of coal or coke. Since the reaction is endothermic, a cyclic process must be used, with alternate production of water gas and heating by passing air through the bed. *Carburetted water gas* is water gas enriched with hydrocarbons obtained by thermal decomposition of petroleum oils. Carburetion increases the calorific value of the water gas, and therefore city gas often is of this type.

Oil gas is used as city gas in some localities, especially where oil is cheap and plentiful and coal or coke relatively expensive. Since the thermal decomposition of oil into gaseous hydrocarbons is endothermic, a cyclic process is used in which some of the oil is burned to supply heat. Lampblack is a by-product of oil-gas manufacture.

Natural gas is found in the earth, trapped at great depths and under high pressure. Frequently, but not always, the gas occurs with petroleum and thus is released as a by-product. In the past, in fact, a great deal of natural gas has been burned at the oil wells as a waste product. However, increasing costs and decreasing quality of other fuels have led in recent years to the construction of long pipelines from the fields to urban and industrial areas. The largest production of natural gas is in Texas, Louisiana, California, and Oklahoma, which together account for about three-quarters of the United States production, but substantial quantities are also produced in many other states, including West Virginia, Kansas, Pennsylvania, Kentucky, Ohio, New Mexico, and Montana.

Gasification of coal and coke. Several of the fuel gases just described are manufactured from coal or coke. When coal is the raw material for gasification, the first stage in gasification is thermal decomposition of the coal compounds with evolution of the volatile matter and formation of coke. These thermal decomposition reactions are very complex, occur over a wide temperature range, and yield a complex gaseous mixture containing tars, oils, and ammonia as well as the combustible gases separated in coal gas or coke-oven gas. The calorific value of the gas obtained from coal by destructive distillation, however, corresponds to only a fraction of the calorific value of the original coal. The further recovery of the heating value of coal in the form of a gaseous fuel involves reacting the coke with gases such as O_2, H_2O, and CO_2. These reactions are very important to metallurgists, not only because of their utilization in coal gasification but also because they are vital reactions within the many metallurgical processes in which carbon is an important reactant.

A common system in gasification is to feed the coal or coke continuously into the top of a vertical column, arranged so that the solids flow down through the column countercurrent to the flow of gases. Air, steam, oxygen, carbon dioxide, or mixtures of these are passed in at the bottom, and the fuel gas leaves the top of the column. When the column is operated properly, the solids leaving the bottom consist of the ash portion of the solid fuel. If the column is fed uni-

formly, it should reach a steady state such that the conditions at any point in the column (temperature, gas analysis, etc.) will not vary with time. As will be seen, such a countercurrent, steady state system possesses several characteristics which make it efficient for the gasification reactions. In particular, the action of the column as a heat exchanger conserves sensible heat in the apparatus, the cold raw fuel acting to cool the product gases before they leave the column and the cold gas acting to cool the ashes, and both mechanisms carrying heat into the high-temperature zones of the column where it is utilized most effectively.

The simplest gasification process is the system shown in Fig. 4–6, consisting of passing air up through a bed of coke. Starting at the bottom of the column, the air is preheated in passing through the hot ash. Leaving the ash zone, the air comes in contact with hot carbon and the oxygen reacts rapidly to form CO_2:

$$C + O_2 \longrightarrow CO_2. \qquad (1)$$

This combustion reaction predominates over all other reactions in the lower section of the coke.* The oxygen content of the gas is substantially all consumed in this section, as the equilibrium is far to the right and the reaction rate is high at elevated temperatures. As the CO_2 content of the gas builds up in passing through the combustion zone, the

----

* The detailed reaction mechanism has not been definitely established. Considerable evidence exists that CO is formed first (e.g., $C + \frac{1}{2}O_2 \longrightarrow CO$). As long as an excess of O_2 is present in the gas, the CO reacts immediately to form CO_2 (e.g., $CO + \frac{1}{2}O_2 \longrightarrow CO_2$) under most conditions, and the net result is the same as if reaction (1) occurred directly. However, with extremely high gas velocities the direct formation of CO can be observed without the intermediate formation of CO_2.

CO_2 starts to react with hot carbon to form CO:

$$CO_2 + C \longrightarrow 2CO. \qquad (2)$$

Reaction (2) is slower than reaction (1), and at moderate temperatures (e.g., less than 900°C) equilibrium is reached with only incomplete reaction of CO_2 to CO. Accordingly, reaction (2) starts in the active combustion zone, continues as the gas moves up, and is the principal reaction in the zone above the combustion zone. In passing through the zone where reaction (2) predominates, the gas cools, owing to heat absorption by the reaction, and reaches a temperature at which the reaction practically ceases (or even reverses to some extent). Above this point, the gas is cooled further in preheating the downward flow of raw coke. This description of the process of gasifying coke with air is based on experimental observations, particularly on analyses of gas samples from various levels in the coke beds. Figure 4–7 gives curves which summarize the results of an extensive investigation of this kind. In any given case, of course, the shapes and relative positions of these curves will vary with conditions in the bed, especially with temperature, flow rate, and coke reactivity. It is possible to operate a coke column with upward flow of air to yield a gas very low in CO_2, and consisting therefore mainly of N_2 and CO. Such a gas might be termed "ideal air-blown producer gas" and is approached in coke beds operated at high temperatures favorable to reaction (2).

Figure 4–8, based on data by Clement, Adams, and Haskins, shows the extent of reduction of CO_2 to CO by reaction (2) as a function of the temperature of the coke bed and the time of contact of the coke with the gas. The gas fed to the coke is the mixture of N_2 and CO_2 (21% CO_2) obtained by converting all the oxygen in air to CO_2. The upper asymptote at 34.6% CO corresponds to complete conversion of the O_2 in the air into CO. Starting at the left of the graph, the curve for infinite time of contact gives the %CO calculated from the equilibrium constant for reaction (2). From this curve it is clear that temperatures over 900°C are required for high yields of CO even at the lowest gas velocities. As shorter and shorter reaction times are considered in the other

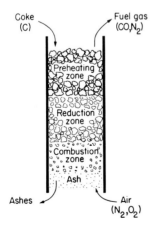

Ashes

FIG. 4–6. Gasification of coke with air.

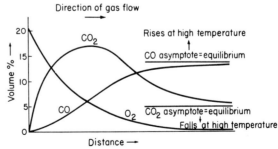

FIG. 4–7. Gas compositions in air-blown coke bed (after Walker, Lewis, McAdams, and Gilliland).

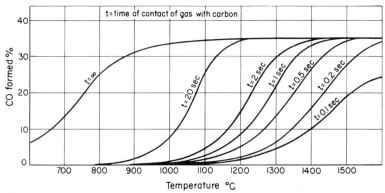

FIG. 4–8. Effect of contact time and temperature on extent of reaction:
$CO_2 + C \longrightarrow 2CO$ (after Clement, Adams, and Haskins).

curves, the CO yield at any given temperature becomes smaller and smaller. With contact times on the order of one second in commercial gas producers, temperatures of at least 1300°C are required to get good conversion of CO_2 to CO. With the exception of the equilibrium curve ($t = \infty$), the positions of the curves for given contact times will depend on the kind of carbon used. The curves for a more reactive coke or for charcoal will be to the left of those shown, while for a less reactive coke they will be to the right.

The combustion reaction (1) in the air-blown coke column is strongly exothermic, so that the combustion zone is characterized by a high temperature and large evolution of heat. On the other hand, reaction (2) is endothermic, so that part of the heat generated in the combustion zone and carried up the column (by conduction and gas flow) is consumed in the reduction zone. This effect, plus the cooling effect of new coke, leads to the gradual drop in temperature from the top of the combustion zone to the top of the coke bed. The over-all result of the two reactions in the bed is a large heat evolution. Thus, when such a column is operated continuously, as in an air-blown gas

producer, either the products and the coke bed will reach high temperatures or some auxiliary cooling must be provided.

The process represented in Fig. 4–6 occurs in a bed of coal being burned on a grate, as well as in the column of coke in a blast furnace. In both cases the net heat evolution in the process is necessary and desirable. On the other hand, when the process is conducted to manufacture a fuel gas, as in an air-blown gas producer, some or all of the heat evolved in the gas producer may be lost and troubles may be experienced from excessively high temperatures developed in the gas producer.

When steam is blown through a bed of hot coke, it reacts as follows:

$$C + H_2O \longrightarrow CO + H_2, \qquad (3)$$

and

$$C + 2H_2O \longrightarrow CO_2 + 2H_2. \qquad (4)$$

Both these reactions proceed simultaneously, and the CO_2 formed by (4) reacts subsequently with more C:

$$CO_2 + C \longrightarrow 2CO. \qquad (2)$$

Figure 4–9 summarizes experimental data on passing steam through carbon beds at 900 to 1200°C, showing the course of these reactions

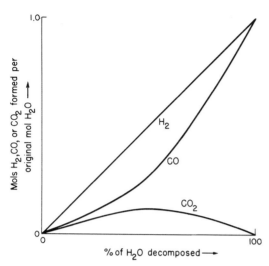

FIG. 4–9. Decomposition of steam in coke beds (after Haslam, Hitchcock, and Rudow).

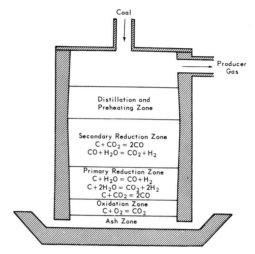

FIG. 4–10. Reaction zones in a gas producer (after Haslam and Russell).

as the gases pass through the bed. If the bed conditions are such that the reactions can go substantially to completion, reaction (3) represents the over-all result of the process and the final product in an ideal system is water gas, 50% CO and 50% H_2.

The over-all water-gas reaction is endothermic, so that heat must be supplied from the outside or from other reactions in the bed if the process is to be carried out continuously. The common procedure in water-gas manufacture is a cyclic one, in which the bed of coke is alternately blown with air to heat it and then run with steam to make water gas. The gases from the air blow are discharged to waste. In making producer gas, the air and steam are blown through the bed simultaneously and continuously, the proportions of the two being adjusted so that the heat evolution in the reaction of O_2 with coke balances the heat absorption of the endothermic reactions and the heat loss from the apparatus.

Figure 4–10 shows the principal reactions and reaction zones when a mixture of air

and steam is passed up through a bed fed with raw coal at the top. At the very bottom in the ash zone, the air and steam are preheated to some extent by contact with the hot ash leaving the active zone of the producer. In the oxidation zone, the O_2 of the gas is largely consumed by reaction (1), which generates the heat required for the process. Heat flows up from this zone and is carried by the gas into the reduction zone, where the endothermic reactions (2), (3), and (4) of reducing the CO_2 and H_2O predominate. Above and overlapping the reduction zone is the fuel preheating zone, in which the fuel absorbs heat from the hot gases. As the coal passes down through this zone and is heated, the volatile matter distills out and joins the main gas stream, and the coal becomes coke. Hence, the upper part of the bed is also called the distillation zone.

Other reactions may play some role in the gas-producing process of Fig. 4–10. The homogeneous gaseous reaction,

$$CO + H_2O \longrightarrow CO_2 + H_2, \qquad (5)$$

probably occurs as part of the process in the reduction zone, and tends to go to the right with decreasing temperature. As the gas cools in the upper part of the bed, and in the gas space above the bed, carbon deposition may occur by reversal of reaction (2):

$$2CO \longrightarrow CO_2 + C \qquad (6)$$

Reactions (5) and (6) tend to lower the calorific power of the gas.

The general arrangement of a typical gas producer is shown in Fig. 4–11. The body is a cylindrical steel shell lined with firebrick. A steel ashpan supports the bed of fuel and is kept full of water which acts as a gas seal to the body of the producer. Ash is quenched in the water and discharged over the edge of the ashpan. Air and steam are introduced through a pipe passing up through the center of the ashpan. The top of the producer carries the fuel feeder, the gas offtake, and a device for leveling and stirring the bed. Generally there is another water seal between top and body of the producer. In one com-

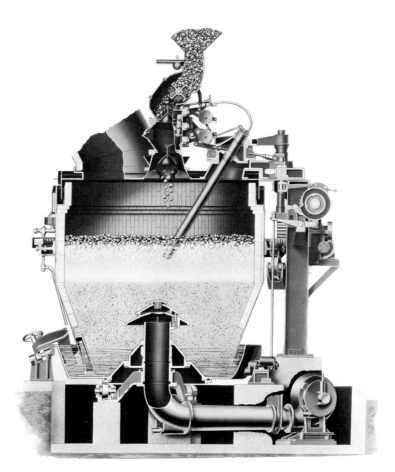

FIG. 4–11. Wellman mechanical gas producer (courtesy of The Wellman Engineering Co.).

mon arrangement, the top is stationary while the body and ashpan rotate slowly (about 1 revolution in 10 minutes), an adjustable plow serving to discharge ash over the edge of the ashpan. Approximate ranges of sizes and operating characteristics of commercial producers are as follows:

Depth of fuel bed	4–10 ft
Diameter	6–12 ft
Capacities	0.5–2 tons fuel/hr
Steam rate	0.2–1 lb/lb of fuel
Gas yield	55–80 ft³/lb of fuel

Either coke or coal, or mixtures of the two, may be used in gas producers, coke giving the easiest operation. Coke breeze and screenings separated from blast furnace coke are often used. When coal is used, the volatile matter evolved in the producer contains a considerable amount of tar, which may represent as much as 10% of the heating value of the raw gas. If the gas is not stored or distributed, but is led while still hot into a furnace adjacent to the gas producer, cleaning and tar removal are not necessary. Another problem which may arise in coal-fed producers is the fusion and clinkering of the ash, which can seriously limit the temperatures and firing rate of the producer.

Other techniques of coal gasification have been under development in recent years and may prove eventually to be of commercial importance. In one mine in Alabama the coal is being converted into a fuel gas underground, without actually mining the coal. Another promising technique is gasification in a fluidized bed of fine coal, the bed being kept in an agitated, fluid condition by the controlled upward flow of gas through the bed. With this system, reaction rates are very high and close control over temperature and other reaction conditions can be maintained. Also, this system may permit the use of O_2 instead of air, with the attendant improvement in gas quality.

Heat balances of gasification processes; recovery of calorific value. In a conventional heat balance for a chemical process at constant pressure, the input includes $-\Delta H$ for the exothermic reactions and the output includes ΔH for the endothermic reactions. The two sets of reactions should correspond to the complete stoichiometry of the process. When solid or liquid fuels are burned to supply heat to a process, the chemical reactions are not known because the chemical compounds in the fuel are not known, so the contribution of all the fuel combustion reactions to the heat balance is lumped into one heat-input item readily calculated from the calorific value of the fuel. Gross heating value is used for heat balances with H_2O (liq) as the reference state, net heating value if H_2O (gas) is the reference state. Now, when gasification processes are considered, a different point of view must be taken. The ΔH values for the gasification reactions themselves cannot be evaluated satisfactorily because the fuel compounds are unknown. Also, the calorific value of the coal is not a realistic item of heat input because the coal is not burned. However, by a simple analysis in terms of the First Law, it can be shown that the net contribution of gasification reactions to the heat balance is correctly represented if the heat input includes calorific value of the coal fed to gasification and the heat output includes calorific values of the gas and other combustible products. Thus, strongly endothermic gasification reactions give products whose combined calorific values exceed the calorific value of the fuel gasified, but this increase is obtained at the expense of completely burning additional fuel to supply the heat required by the gasification reactions. This is well illustrated in water-gas manufacture. Conversely, an exothermic gasification reaction, such as gasification with air, requires no external heating and yields fuel products with less calorific power than the starting material. In a standard gas producer, the gasification reactions are balanced so that most of the calorific power of the solid fuel is recovered in the calorific power of the producer gas (65–80%), the calorific power not recovered being just enough to balance the heat loss of the apparatus (5–10%) and the sensible heats in the products (10–25%). If the gas producer is so situated that the producer gas can pass directly to the furnace, with little cooling, its sensible heat will represent an additional recovery of the calorific power of the coal being gasified.

Mechanism and characteristics of gaseous combustion. The simple oxidation reactions commonly written to represent gaseous combustion express the over-all results of combustion and afford a sound basis for stoichiometric and thermochemical calculations, but they do not represent actual reaction mechanisms. Thus, the reaction of hydrogen with oxygen probably involves at least the following principal reaction mechanisms:

$$OH + H_2 \longrightarrow H_2O + H$$
$$H + O_2 \longrightarrow OH + O$$
$$O + H_2 \longrightarrow OH + H$$

These are "chain reactions" whose propagation depends on the chain carriers, OH, H, and O, which are consumed in some reactions and produced in others. These chain carriers are very reactive, so that reactions at explosive rates are readily accounted for, even though the actual concentrations of the carrier radicals and free atoms at a given instant are so minute as to be negligible from a stoichiometric point of view. The concentration of these radicals can, however, be measured by spectroscopic methods.

Although the practical characteristics of gaseous combustion in flames and explosions are intimately related to the mechanisms of the reactions and specifically to the properties of chain reactions, the study and interpretation of reaction mechanisms is too complex and too highly specialized a field to concern the engineer who is interested primarily in supplying heat to a metallurgical process. However, in choosing, operating, and adjusting burners, some appreciation, at least from an empirical point of view, of the variables describing flame behavior is helpful.

The *ignition temperature* of a gas-air mixture is the lowest temperature at which rapid self-sustaining combustion occurs. Below this temperature slow combustion may occur, but in general the rate of heat loss to the surroundings exceeds the rate of heat generation by combustion, so that the combustion tends to die out. Since the factor of heat loss to surroundings is involved, obviously the ignition temperature is not a property of the fuel gas alone but depends on the arrangement of the combustion system. In a steady flame, ignition of incoming gas is accomplished by heat transferred back from the rapid combustion zone. Some typical ignition temperatures measured by Dixon and Coward for mixtures of air and various pure gases in laboratory apparatus are as follows: H_2, 580–590°C; CO, 644–658°C; CH_4, 650–750°C; C_2H_6, 520–630°C; C_2H_4, 542–547°C; and C_2H_2, 406–440°C.

The *limits of inflammability* are the limiting compositions of gas-air mixtures in which a flame is self-propagating. These limits serve not only as a guide to combustible mixtures, but also measure roughly the compositions which are explosive and must therefore be avoided or handled carefully around the plant. The *lower inflammability limit* is the minimum percent gas and the *upper inflam-*

mability limit is the maximum percent fuel gas in a gas-air mixture that will support combustion. Data for various gases are given in Table 4–7.

The *rate of flame propagation* or the *flame velocity* is the linear speed at which the flame will travel through a gas-air mixture. This speed is an important variable in burner design, in relation to the gas velocities within and adjacent to the burner. If the velocity of the gas-air mixture in the burner is less than the flame velocity, the flame will propagate back into the burner, that is, the burner will "backfire." On the other hand, if the gas issuing from the burner does so at a velocity greatly above the flame velocity, the flame may blow off. The rate of flame propagation varies over a wide range with the composition of the fuel gas, the properties of gas and air, and the size and shape of the container in which the flow is observed. The data in Fig. 4–12 for combustion of methane in air show maximum flame velocity for mixtures containing 10% methane, and a large increase in flame velocity with increase in

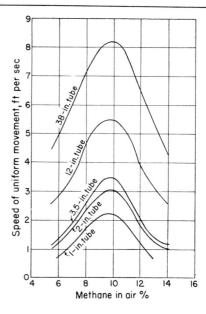

FIG. 4–12. Flame velocities in methane-air mixtures (from Haslam and Russell, *Fuels and Their Combustion*, McGraw-Hill).

diameter of the pipe through which the flame is propagated. The data for other gases, Fig. 4–13, show further the wide variations in behavior which must be allowed for in burner design.

Gas burners. Gas burners may be classified into three types, depending on the method of mixing fuel gas and air.

The simplest arrangement consists of admitting the gas and air into the combustion chamber through separate ports, usually adjacent to each other, and allowing the two to mix in the furnace. The rate of mixing and the size and shape of the flame are determined by port design, number and arrangement of ports, relative velocities of gas and air streams, angle of convergence, etc. This method of burning gas gives large, relatively slow-moving flames and has been widely used in firing open hearth steel furnaces with producer gas. Some of the conditions which favor the port type of burner are

TABLE 4-7

Limits of Inflammability of Gases in Air*

Gas	Limits as % by volume in air	
	% lower-limit mixture	% upper-limit mixture
Hydrogen (H_2)	6.2	71.4
Carbon monoxide (CO)	16.3	71.2
Coal gas	7	21
Coke-oven gas	7	21
Blue water gas	12	67
Blast furnace gas	36	65
Methane (CH_4)	5.8	13.3
Ethane (C_2H_6)	3.3	10.6
Ethylene (C_2H_4)	3.4	14.1
Ethyl alcohol (C_2H_5OH)	3.7	13.7
Benzene (C_6H_6)	1.4	5.5
Pentane (C_5H_{12})	1.3	4.9
Ether [$(C_2H_5)_2O$]	1.6	7.7

*After Haslam, R. T., and Russell, R. P., *Fuels and Their Combustion*. New York: McGraw-Hill, 1926.

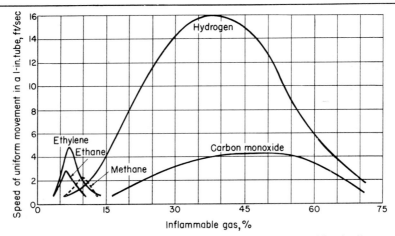

FIG. 4–13. Flame velocities in mixtures of fuel gases with air (from Haslam and Russell, *Fuels and Their Combustion*, McGraw-Hill).

the use of preheated fuel gas and air which ignite when mixed, dirty or tarry fuel gas, and the need for long slow flames.

The familiar Bunsen burner and also its domestic and industrial counterparts are of the *inspirator* type. In this type of burner, either the fuel gas or the air is delivered to the burner under pressure and is discharged from a nozzle or jet in such a way that its momentum is used in mixing gas and air and in delivering the mixture at a suitable velocity. Figure 4–14 is a cross section of a laboratory Bunsen burner utilizing the inspirator principle, with the gas under moderate pressure drawing in air at one atmosphere. A simple shutter controls the quantity of inspirated air. Not all the air for combustion is inspirated in the burner tube; part of it enters the flame from the atmosphere surrounding the flame above the burner. The former is called *primary* air, the latter *secondary* air; this division of the air supply is common to many kinds of burners. Figure 4–15 gives dimensions and shows other details of a large inspirator-type gas burner used for firing copper-smelting furnaces with natural gas at Anaconda. Five of these

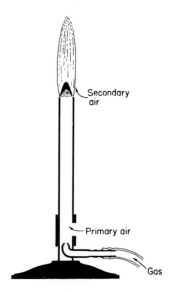

FIG. 4–14. Bunsen burner.

burners on one furnace burn a total of as much as 100,000 cu ft of natural gas per hour. A well-designed gas burner will operate efficiently over a considerable range of gas flow, about a 5 to 1 range for many industrial burners. Typical performance curves for an inspirator-type burner are shown in Fig. 4–

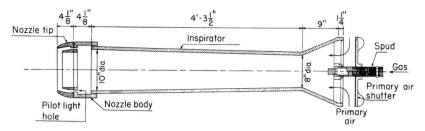

FIG. 4–15. Large inspirator-type gas burner used at Anaconda (from *Trans. AIME*).

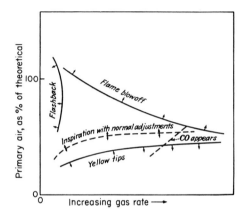

FIG. 4–16. Performance curves for inspirator-type gas burner.

16. The effects noted in the figure will be recognized by anyone who has adjusted a Bunsen burner.

A third common type of burner involves *proportioning* or *premixing* all or part of the air with the gas ahead of the burner. With this arrangement the burner itself may be a relatively simple nozzle designed to deliver the combustible mixture without backfire or without flame blowoff. A variety of arrangements are used for premixing. If both air and gas are under pressure the quantities may be controlled manually and the two brought together in a mixing tee. Mixing and proportioning may be accomplished in an inspirator fitting ahead of the burner, designed so that regulation of one flow automatically regulates the other in the proper

proportion. Another method of supplying a gas-air mixture involves a centrifugal fan with a shutter to regulate the air flow into the fan and the fuel pipe also feeding into the fan. Good mixing occurs in the fan, and the combustible mixture is delivered under pressure to the furnace. When any of the premixing methods are used, precautions may be necessary because the combustible mixture is an explosive mixture.

The rate of gaseous combustion is markedly increased at hot surfaces. This effect is utilized in tile-port burners, in which the burner port is made from a highly refractory tile and so arranged that the tile port is heated to incandescence. In other furnaces, the surface combustion phenomenon is applied by directing the flame against a furnace wall or roof or into a pile of crushed refractory. A large part of the combustion may occur at the solid surface so that the surface becomes a radiator of heat. Allowing any flame to brush a surface may aid in stabilizing the flame and in keeping it lit. Under some circumstances, however, surface combustion is disadvantageous, as it produces local overheating and rapid erosion of refractories or furnace parts projecting into the flame.

Fuel oil

Liquid fuels have proven advantageous in many metallurgical applications. Though fuel oils are usually somewhat higher in purchase cost than coal per Btu of calorific value, other characteristics may lead to cheaper over-all heating costs with fuel oils. In particular, fuel oils are easily handled, stored, and transported about the plant (in pipes). In connection with handling and storage, it is significant that fuel oils possess higher calorific values than any other commercial fuels, on either a weight or a volume basis. Also, like fuel gases, they are ashless and are very flexible in adjustment to widely varying operating requirements.

Petroleum and petroleum refining. Most of the liquid fuels used industrially are derived from crude petroleum. Petroleum is formed in the earth's crust from accumulated vegetal and animal matter by metamorphic processes probably similar in many ways to those of coalification. Also, petroleum contains the same predominant chemical elements as coal: carbon, hydrogen, oxygen, and sulfur, and sometimes nitrogen as well. The characteristic chemical compounds are hydrocarbons, with the organic sulfur, oxygen, and other compounds usually accounting for small percentages of the total.

Crude petroleums contain a great variety of hydrocarbons, of which only a relatively small number have been separated and positively identified. Both chain and cyclic or closed-ring hydrocarbon structures are abundantly represented. In each group are compounds with various carbon-to-hydrogen ratios, that is, of varying degrees of saturation of carbon with hydrogen. For example, the open-chain type hydrocarbons include the paraffin or saturated hydrocarbons starting with methane (CH_4) and having the general formula C_nH_{2n+2}, the olefins (C_nH_{2n}), and the acetylene series (C_nH_{2n-2}). Any crude petroleum contains hydrocarbons of all types, and each type may be represented by many individual compounds of widely varying molecular weights and numbers of carbon atoms per molecule. However, different types of hydrocarbons predominate in crudes from different sources, so that the crudes are classified as paraffin base (high in methane hydrocarbons), naphthene base (high in cyclic hydrocarbons), and mixed base.

The individual compounds in petroleum, considered separately, include gases, liquids, and solids. In general, the hydrocarbons of lowest molecular weights are gases (for example, methane, CH_4; ethane C_2H_6; ethylene, C_2H_4; acetylene C_2H_2), those of intermediate molecular weights are liquids (for example, octane, C_8H_{18}; octylene, C_8H_{16}; benzene, C_6H_6), while those of high molecular weights are solids. The gaseous hydrocarbons tend to separate from the petroleum when the pressure is released, hence natural gas is readily recovered as a separate product in many oil fields. The wide variations in volatility of the hydrocarbons in petroleum make fractional distillation the logical first step in separating the hydrocarbons from each other.

Crude petroleum is the raw material from which many important products are made, including gasolines, lubricating oils, fuel oils, solvents, and industrial chemicals. Gasoline and lubricating oils represent the major valuable products, whereas from the viewpoint of the oil refiner the fuel oils are less valuable by-products. Accordingly, in refining petroleum the objective is to obtain from the crude maximum yields of the most valuable products. Hence, industrial fuel oils in general represent a residuum from which it is not economical to manufacture more desirable products.

Petroleum refining has become a very complicated industry, so that no single procedure can be given as typical. At one time the principal step was fractional distillation of the crude to give the successively less volatile fractions: casinghead gasoline, raw gasoline, raw kerosene, gas oil, lube distillate, and residue. Each of the primary distillates was refined further for the market and the residue, constituting perhaps one-third of the original crude, was sold as fuel oil. Now, however, refining has become much more than separation of the compounds present in crude petroleum. Various of the primary distillation products are subjected to thermal decomposition ("cracking") and other chemical processes to produce greater quantities of hydrocarbons with the most desirable properties. Here again industrial fuel oils are obtained largely as residues or bottoms after distilling off the more volatile hydrocarbons.

Coal tar. Especially around steel plants, tar obtained as a by-product of coking may serve as fuel. The techniques of using coal tar as fuel are similar to those used with regular fuel oils obtained from petroleum, but tar is more viscous and more difficult to handle in burners and auxiliary equipment.

Chemical analyses of fuel oils. In spite of the wide variations in kinds and quantities of hydrocarbons in fuel oils, the compositions in terms of the chemical elements do not vary much from one fuel oil to another. The ultimate analyses give carbon contents usually between 83 and 88% and hydrogen from 10 to 13%, the heavier oils being highest in carbon and lowest in hydrogen. Sulfur is only a few tenths of a percent in many fuel oils, but may exceed 1% in the heavier residual fuel oils. When fuel oils are used for metallurgical processes, where sulfur pickup by the charge may be significant, special specifications may be made as to sulfur content. Oxygen may be on the order of 1%, and nitrogen usually is a fraction of 1%.

Water and sediment are generally determined as part of the purchase specifications and together may run as high as 2% in heavy fuel oils.

Properties of fuel oils. In buying and selling fuel oils, determinations are made of a number of properties which relate primarily to the measurement of fuel quantity and to the behavior of the fuel in burners and auxiliary handling equipment. As with coal specifications, standard methods are prescribed for making tests on fuel oils and are published by the ASTM.

Fuel oil is bought and sold on a volume basis, the standard unit being one U. S. gallon measured at 60°F. One barrel is 42 gallons. Since the temperature coefficient of expansion of oil is appreciable, the actual measurement of volume is accompanied by measurement of average temperature, and the volume is corrected to 60°F, using standard volume-correction tables. For the volume correction and for other purposes the specific gravity is also measured at the same time. The specific gravity is determined in degrees API with a specially marked hydrometer. The unit is defined as follows:

$$°API = \frac{141.5}{sp.\ gr.,\ 60°/60°F} - 131.5.$$

Readings at other than 60°F are corrected to this temperature in reporting properties. It will be noted that water has a specific gravity of 10.0° API, and fuel oils, being lighter than water, will give higher figures, the range being about from 10° to 50° API.

Other properties of fuel oils which affect their use and therefore are subject to specifications and standard tests include viscosity (ASTM D–88), pour point (ASTM D–97), flash point (ASTM D–93), distillation temperature (ASTM D–158), and carbon residue (ASTM D–524).

Grades of fuel oils. To facilitate buying, selling, and use of fuel oils, the various grades have been standardized on the basis given in Table 4–8. In general, from No. 1 fuel oil to to No. 6 fuel oil the volatility decreases, the

TABLE 4-8

Detailed Requirements for Fuel Oils*

ASTM D-396

Grade†	Flash point °F Min	Flash point °F Max	Pour point °F Max	Water and sediment percentage Max	Carbon residue percentage Max	Ash, percentage Max	Distillation temperatures, °F 10 percent point Max	90 percent point Max	90 percent point Min	End point Max	Viscosity, seconds Saybolt universal at 100°F Max	Saybolt universal at 100°F Min	Saybolt furol at 122°F Max	Saybolt furol at 122°F Min
1 Fuel oil, a distillate oil for use in burners requiring a volatile fuel	100 or legal	165	15‡	trace	0.05 on 10% residuum§		410			560‖				
2 Fuel oil, a distillate oil for use in burners requiring a moderately volatile fuel	110 or legal	190	15‡	0.05	0.25 on 10% residuum¶		440	600						
3 Fuel oil, a distillate oil for use in burners requiring a low-viscosity fuel	110 or legal	230	20‡	0.10	0.15 straight			675	600**		55			
5 Fuel oil, an oil for use in burners requiring a medium viscosity fuel	130 or legal			1.00		0.10						60	40	
6 Fuel oil, an oil for use in burners equipped with preheaters permitting use of a high-viscosity fuel	150			2.00††									300	45

*Recognizing the necessity for low-sulfur fuel oils used in connection with heat treatment, nonferrous metal, glass and ceramic furnaces, and other special uses, a sulfur requirement may be specified in accordance with the following table:

Grade of fuel oil	1	2	3	5	6
Sulfur, maximum %	0.5	0.5	0.75	No limit	No limit

Other sulfur limits may be specified only by mutual agreement between the buyer and seller.

†It is the intent of these classifications that failure to meet any requirement of a given grade does not automatically place an oil in the next lower grade unless, in fact, it meets all requirements of the lower grade.

‡Lower or higher pour points may be specified whenever required by conditions of storage or use. However, these specifications shall not require a pour point lower than 0°F under any conditions.

§For use in other than sleeve-type blue-flame burners, carbon residue on 10 percent residuum may be increased to a maximum of 0.12 percent. This limit may be specified by mutual agreement between the buyer and seller.

¶To meet certain burner requirements the carbon-residue limit may be reduced to 0.15 percent on 10 percent residuum.

**The minimum distillation temperature of 600°F for 90 percent may be waived if API gravity is 26 or lower.

††Water by distillation, plus sediment by extraction. Sum, maximum 2.0 percent. The maximum sediment by extraction shall not exceed 0.50 percent. A deduction in quantity shall be made for all water and sediment in excess of 1.0 percent.

‖The maximum end point may be increased to 590°F when used in burners other than sleeve-type, blue-flame burners.

viscosity increases, and the specific gravity increases (°API decreases). Also, the higher-numbered grades are cheaper. As might be expected, the more volatile, less viscous oils are more easily burned and therefore might be used with smaller furnaces (for example, domestic heating) and relatively specialized installations, whereas only the heavier grades would be used for large smelting furnaces.

Calorific value. Fuel oils of a given grade all have substantially the same calorific value, regardless of origin. In fact the variation from one grade to another is also relatively small. Gross calorific values are given either as Btu per pound or as Btu per gallon (at 60°F). Lighter grades of fuel oil have the highest heating values on a weight basis, but on a volume basis the heavier grades have the highest heating values. Thus the heavier fuel oils both are cheaper in cents per gallon and give the most Btu per gallon on combustion.

As with other fuels, precise determinations of calorific power are made in a calorimeter. However, for fuel oils it has been found that the gross heating value and the specific gravity are simply related, so that the heating value can be calculated from the specific gravity to a degree of accuracy sufficient for most practical purposes. For example, the following formula may be used:

Gross heating value, Btu/lb

$$= 18,650 + 40 \, (°API - 10).$$

The gross heating values of fuel oils fall generally in the range from 18,000 to 19,500 Btu/lb.

Oil burners. In burning liquid fuels, the actual combustion reaction generally is between the fuel vapor and the air. Thus, a prime requirement of an oil burner is that it should provide in some way for rapid and efficient vaporization of the oil. With some liquid fuels the vaporization can be accomplished satisfactorily within the burner, but the common arrangement in industrial oil burners involves *atomization* of the oil in the burner so that vaporization occurs quickly as the fine oil droplets are sprayed into the combustion chamber. At the same time the burner provides for thorough mixing of the atomized oil with air so that vaporization and combustion occur in quick succession.

Atomization of oil is a comminution process which consumes mechanical energy. Pressure atomization, common in domestic oil burners, involves forcing the oil under high pressure through a small orifice, the nozzle being designed so that a conical spray is produced. Steam or air atomization is generally used in the oil burners for metallurgical furnaces. Air is preferable to steam when thermal efficiencies alone are considered, but high-pressure steam is usually available around the plant and is cheaper than compressed air. Burners with steam or air atomization are made to handle the heaviest grades of fuel oils and tars.

Literally hundreds of burners of all degrees of complexity using steam or oil atomization have been designed. The atomization is accomplished in some burners by the rapid expansion of the steam or air when a mixture of oil and the atomizing agent is released from high pressure. Figure 4–17 illustrates a simple burner of this type. In other burners, the atomizing gas strikes at high velocity a stream or ooze of oil in such a way as to shear the oil into very fine droplets. Figure 4–18 shows the cross section of a burner based on this mechanism. The consumption of steam or air for atomization varies with the pressure of the atomizing agent and with the design of the burner. Typical figures for oil burners on open hearth

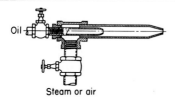

FIG. 4–17. Simple oil burner with steam or air atomization (reproduced by permission from W. Trinks, *Industrial Furnaces*, Vol. II, 2nd Ed., John Wiley and Sons, Inc., 1942).

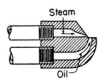

FIG. 4–18. Oil burner with steam atomization (reproduced by permission from W. Trinks, *Industrial Furnaces*, Vol. II, 2nd Ed., John Wiley and Sons, Inc., 1942).

furnaces are 4 lb of steam or 2.5 lb of air per gal of oil.

Coal tar and the heaviest grades of commercial fuel oil (for example, Nos. 5 and 6

fuel oils) are preheated before burning, to reduce the viscosity to a point at which the burner will perform satisfactorily.

SUPPLEMENTARY REFERENCES

ASTM Standards, Part III–A, Nonmetallic Materials. Philadelphia: ASTM, 1946 (published triennially).

BUTTS, ALLISON, *Metallurgical Problems.* New York: McGraw-Hill, 1943.

CAMP, J. M., and FRANCIS, C. B., *The Making, Shaping, and Treating of Steel.* Pittsburgh: Carnegie-Illinois Steel Corp., 1940.

CAMPBELL, M. R., *The Coal Fields of the United States.* U. S. Geol. Survey, Professional Paper 100, 1929.

CLEMENT, J. K., ADAMS, L. H., and HASKINS, C. H., *Essential Factors in the Formation of Producer Gas.* U. S. Bureau of Mines, Bull. 7, 1911.

GRISWOLD, JOHN, *Fuels, Combustion, and Furnaces.* New York: McGraw-Hill, 1946.

GUMZ, W., *Gas Producers and Blast Furnaces.* New York: John Wiley, 1950.

HASLAM, ROBERT T., and RUSSELL, ROBERT P., *Fuels and Their Combustion.* New York: McGraw-Hill, 1926.

JOHNSON, A. J., and AUTH, G. H., *Fuels and Combustion Handbook.* New York: McGraw-Hill, 1951.

LIDDELL, DONALD M., *Handbook of Nonferrous Metallurgy.* Vol. I, *Principles and Processes.* New York: McGraw-Hill, 1945.

LOWRY, H. H., COMMITTEE ON CHEMICAL UTILIZATION OF COAL, NATIONAL RESEARCH COUNCIL, *Chemistry of Coal Utilization*, two vols. New York: John Wiley, 1945.

MOORE, ELWOOD S., *Coal.* New York: John Wiley, 1940.

PERRY, JOHN H., *Chemical Engineers' Handbook.* New York: McGraw-Hill, 1949.

TRINKS, W., *Industrial Furnaces*, Vol. II. New York: John Wiley, 1947.

WALKER, W. H., LEWIS, W. K., McADAMS, W. H., and GILLILAND, E. R., *Principles of Chemical Engineering.* New York: McGraw-Hill, 1937.

PROBLEMS

4–1 A sample of mine-run coal gives the following data:

Proximate analysis, %

Moisture:	2.3
Volatile matter:	29.8
Fixed carbon:	58.7
Ash:	9.2
S:	0.5

Gross calorific power, Btu/lb: 13,780

(a) Calculate the proximate analysis and calorific power on a moisture-free basis. (b) Calculate the proximate analysis and calorific power on an ash- and moisture-free basis. (c) Find the classification of this coal by rank, according to ASTM D–388 (see Table 4–3).

4–2 The following data were obtained on a bituminous coal:

Ultimate analysis, dry basis, %

C:	77.6
H:	5.3
N:	1.5
O:	4.5
S:	2.8
Ash:	8.3

Moisture, 4%

Gross calorific power, dry basis, Btu/lb: 13,800

Calculate: (a) gross calorific value, moist basis, (b) net calorific value, dry basis, (c) net calorific value, moist basis, (d) gross calorific value, dry basis, using Dulong formula (compare with measured value).

4–3 A producer gas analyzes 50% N_2, 25% CO, 18% H_2, 6% CO_2, and 1% O_2.

Calculate: (a) net calorific power, Btu/ft³ (dry, STP), (b) net calorific power, kg cal/m³ (dry, STP), and (c) gross calorific power, Btu/ft³ (60°F, 30 inches Hg, saturated).

4–4 One short ton of coal of calorific power 13,500 Btu/lb yields as the main products on coking 0.75 ton of coke, calorific power 12,000 Btu/lb; 10,000 cu ft of coal gas, calorific power 550 Btu/ft³; and 75 lb of tar, calorific value 16,000 Btu/lb. The heating values of the other products may be ignored. 40% of the gas is burned to heat the oven.

Set up a heat balance for the coking process as a whole and calculate the total quantity of heat consumed as heat loss from the oven and sensible heats in the products (coke, gas, combustion products, etc.) in Btu/ton of coal. Express this heat consumption as a percentage of the original calorific value of the coal. What percentages of the original calorific power of the

coal are finally recovered in the coke, tar, and 6000 cu ft of coal gas?

4–5 (a) What is the composition of ideal air gas produced by gasifying pure C with dry air to obtain a mixture containing only N_2 and CO? (b) What percentage of the calorific power of the original carbon is represented in the calorific power of the air gas? Account for the remainder of the calorific power of the carbon.

4–6 What percentage of the calorific power of the carbon is represented in ideal water gas (50% CO, 50% H_2) produced by gasifying pure C with H_2O? Account for the fact that the total heating value of the gas is greater than that of the original carbon.

4–7 An ideal gas producer fed with pure C, air, and H_2O might be characterized as follows: (a) gasification reactions go to completion, so that the producer gas consists solely of N_2, H_2, and CO, (b) input materials and output materials contain no sensible heat (that is, are all at 77°F), and (c) the process is adiabatic.

For this ideal process, calculate (a) steam rate, lb/lb of C, (b) gas yield, ft³ (STP)/lb of C, (c) gas analysis, (d) calorific power of gas, Btu/ft³. Compare these results with data for commercial producers and producer gases.

4–8 Repeat the calculations of Problem 4–7 for an ideal gas producer fed with pure C, O_2, and H_2O.

4–9 The following data were obtained on the operation of a gas producer.

	Ultimate analyses, dry basis weight, %			
	Coal	Tar	Soot	Ashes
C	83.9	62.5	92.0	4.9
H	5.0	4.0		
S	0.7			0.4
N	1.4	} 21.0		
O	0.7			
Ash	8.3	12.5	8.0	94.7
	100.0	100.0	100.0	100.0

Producer gas analysis, dry basis
volume, %

CO	26.7
CH₄	3.5
H₂	13.9
CO₂	4.3
N₂	51.6
	100.0

Coal was fed to the producer at the rate of 4000 lb/hr (dry basis), and as fed contained 2% moisture. Estimated quantities of tar and soot were 120 lb and 80 lb per dry ton of coal fed, respectively. Steam was fed at the rate of 0.35 lb/lb of dry coal. Cooling water was used at the rate of 7 gal/min, and increased 60°F in temperature in passing through the producer. The following temperatures were measured or estimated:

Air:	77°F
Coal:	77°F
Steam:	281°F (50 psi)

Producer gas, tar, soot:	1400°F
Ashes:	150°F

Estimated specific heats are as follows: tar, 0.5, soot, 0.4, and ashes, 0.2.

Calculate:

1. (a) Net calorific power of coal, Btu/lb, dry (Dulong formula).
   (b) Net calorific powers of tar and soot, using Dulong formula.
   (c) Net calorific power of producer gas, Btu/ft³ (STP, dry).
   (d) Gross calorific power of producer gas, Btu/ft³ (60°F, 30 inches Hg, dry).
2. (a) Cu ft (STP) of dry air per lb of dry coal.
   (b) Cu ft (STP) of producer gas per lb of dry coal.
   (c) % H₂O by volume in producer gas.
   (d) % of steam decomposed to H₂ in producer.
3. Set up the heat balance for the gas producer, per dry lb of coal.

CHAPTER 5

COMBUSTION OF FUELS AND HEAT UTILIZATION

Engineering control of combustion and appraisal of efficiency in heat utilization follow the same principles whether the fuel is being burned in a house-heating furnace, an industrial steam boiler, or a metallurgical furnace. Combustion engineering in all cases is based primarily on simple stoichiometry and on the Law of Conservation of Energy, and this chapter deals chiefly with specific methods of applying stoichiometric principles and the First Law to combustion processes. Detailed illustrative calculations are given for a few representative systems, but the primary purpose of this chapter is to develop a point of view and a broad method of approach which will be useful in analyzing and solving the endless variety of practical problems which arise in the use of fuels to supply heat to metallurgical processes.

A fuel-fired metallurgical furnace

Pyrometallurgical processes are conducted in a great variety of furnaces, differing in size, shape, method of heating, method of charging and discharging, materials of construction, and other features. In general, each furnace is designed and constructed to meet the special requirements of a particular step or process. In spite of this diversity, however, fuel-fired furnaces all have much in common simply because they burn fuels. Once a furnace is designed and built, the furnace characteristics and the operating principles which guide furnace control and operation are those common to all fuel-fired furnaces. Accordingly, in presenting the common principles of combustion control

and fuel utilization in this chapter, it is sufficient to present them as applied to one particular kind of furnace, with the understanding that minor modifications and elaborations will be necessary in considering a different kind of furnace.

If any one kind of furnace could be considered a typical metallurgical furnace, this kind probably would be a *reverberatory* furnace. Reverberatory furnaces are used in one way or another in smelting, refining, heating, or melting virtually all the common metals. There is some justification for the statement that the reverberatory furnace is to the metallurgical engineer what the beaker is to the laboratory chemist. The essential features of a reverberatory furnace are that heat is supplied by combustion of a fuel and that the flame and combustion products pass directly over and are in direct contact with the materials being heated. That is, one chamber serves both as combustion chamber and as working chamber.

Figure 5–1 shows the main parts of a simple reverberatory furance. The furnace chamber is approximately rectangular in plan and is massively constructed of refractory brick and other forms of refractory materials on a solid foundation. The furnace roof is commonly a brick arch. The arch is kept under compression and the furnace walls are reinforced by a steel structure on the outside, consisting mainly of closely spaced vertical buckstays set into the foundation and horizontal, spring-loaded tie rods across the top between pairs of buckstays. The bottom of the furnace, or the

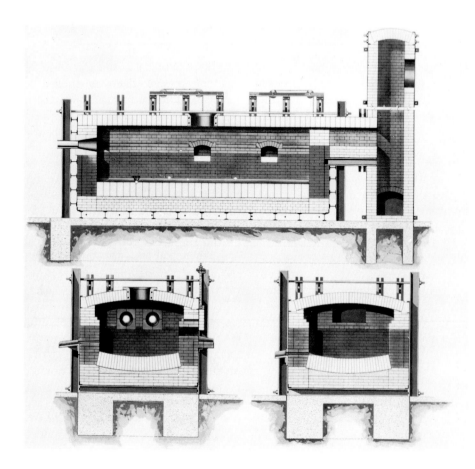

Fig. 5–1. A simple reverberatory furnace (courtesy of Harbison-Walker Refractories Company).

hearth, holds the material being heated, and may be designed to hold either solid materials or molten materials, depending on the particular application of the furnace. A variety of arrangements are used for charging, including doors in the sidewalls, charge holes in the roof, spouts in one end for liquids, and special "guns" for pulverized solids. If the furnace products are liquid, for example, liquid metals and slags, they are generally discharged through one or more tap holes. Reverberatory furnaces are used for both batch and continuous processes, but in the subsequent discussion of combustion principles a continuous, steady state operation will be assumed for the sake of simplicity.

Fuel and combustion air are introduced through openings at the firing end, using burner arrangements and air openings or shutters appropriate to the particular fuel being used. The flame generally occupies a substantial fraction of the volume of the furnace chamber over the charge. The gaseous products of combustion move from the firing end to the flue, the flow being maintained by a small suction pressure (a fraction of an inch of water) at the flue. As a result of the brick construction and the conditions to which it is subject, and as a result of the various openings into the furnace, the chamber is far from being a tight enclosure, so that a certain amount of air leakage into the furnace (or outward leakage of furnace gases) is inevitable and must be taken into account by the operator.

Reverberatory furnaces range in size from pilot-plant models handling a few hundred pounds of charge up to furnaces such as those used for copper smelting, which may treat more than 1000 tons of charge per day and are up to 30 ft wide and over 100 ft long. Some idea of the magnitude and scale of operation of reverberatory furnaces used in metallurgical processes is given by the data in Table 5–1.

Combustion stoichiometry

The starting point in studying combustion problems is the setting up of the materials balance, relating quantities and analyses of fuel, air, and flue gases. As indicated schematically in Fig. 5–2, the combustion system is essentially a steady state system with two entering streams, fuel and air, and one product stream, flue gas. For many metallurgical processes, the interchange of elements between the combustion products and the materials in process is small enough to be neglected in combustion calculations. For processes with appreciable interchange of materials between combustion products and charge, the combustion calculations must, of course, be modified, but it is still advantageous to deal with the combustion as a separate system rather than to calculate a single materials balance for combustion and metallurgical process combined.

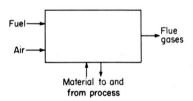

FIG. 5–2. Materials flow in steady combustion.

TABLE 5-1

Typical Reverberatory Furnaces

Process	Batch or continuous	Tonnage	Approximate process temperature	Size of furnace length × width, feet	Fuel		Gross Btu per furnace day
					Kind	Consumption	
Matte smelting, wet copper concentrates	Continuous	600 dry tons chg per day		102 × 25	Natural gas	3.4×10^6 ft^3 per day	3.7×10^9
Matte smelting, wet copper concentrates	Continuous	1000 dry tons chg per day		127 × 29	Pulverized coal	223 tons per day	5.7×10^9
Matte smelting, roasted copper concentrates	Continuous	1000 tons chg per day	2300°F	107 × 26	Natural gas	2.9×10^6 ft^3 per day	3×10^9
Copper refining	Batch	365 tons per day		46 × 17	Fuel oil	150 bbl per day	10^9

Some of the data for the combustion materials balance should be available by direct measurement, for example, the rate of fuel consumption, the analysis of the fuel, and the Orsat analysis of the flue gases. Other materials-balance data may be very difficult or impossible to measure directly and are usually calculated, for example, air consumption, air leakage, and quantity of flue gases. The missing data for the materials balance are found by solving a series of equations for the C balance, N balance, O balance, etc., for the combustion process. Usually the calculations are simpler if carried out in the order discussed below.

Carbon balance. The Orsat analysis gives the volume % CO and % CO_2 in the flue gases on a dry basis. The analysis of the fuel gives directly or indirectly the % C. Then, since the combustion air ordinarily contains a negligible amount of carbon, the carbon balance,

$$C \text{ in fuel} = C \text{ in flue gases,}$$

can be set up and solved to find the quantity of dry flue gas per unit of fuel. Combining this figure with the rate of fuel consumption gives the rate of production of dry flue gas.

The carbon balance, of course, becomes more complicated in some metallurgical furnaces when the charge itself contributes carbon, for example, when $CaCO_3$ is part of the charge and dissociates to contribute large quantities of CO_2 to the flue gases.

Nitrogen balance. In burning low-nitrogen fuels the nitrogen all enters with the air and leaves in the flue gases, so that the nitrogen balance, like the carbon balance, is usually a simple calculation:

$$N_2 \text{ in air} = N_2 \text{ in flue gases.}$$

If the fuel contains appreciable amounts of nitrogen, N_2 in the fuel must be included in the N_2 balance. After the flue gas quantity is found by the calculation of the carbon balance, the air consumption can be found by the nitrogen balance.

Air leakage into the furnace and flue system through charge doors, cracks in the brickwork, and other openings accounts for an appreciable fraction of the total air inflow in some cases. The air leakage between two points in the system (say between the furnace and the stack) can be calculated if representative flue gas samples can be obtained at the two points and analyzed by the Orsat method. With flue gas analyses available for both points, the total air consumption to each point may be calculated by a carbon balance followed by a nitrogen balance, as already discussed. The air leakage is then found as the difference between the total air consumptions at the two points.

For comparative purposes, air consumption is expressed as *percent of theoretical air*. The theoretical air requirement is based on the assumption of complete combustion of the fuel to CO_2, H_2O, SO_2, etc. To obtain efficient combustion, the air is usually adjusted to be a little greater than theoretical and the *percent excess air* is the excess over theoretical air, expressed as a percentage of the theoretical requirement. Thus, % excess air = (% of theoretical air − 100%).

Once a furnace practice is established, some simplifications may be desirable in the routine checking and control of combustion air. For example, it may be convenient to control combustion to obtain a predetermined % CO_2 in the flue gases, this percentage being recorded or indicated automatically and corresponding to a desired % excess air. Figure 5–3 illustrates such a relationship for oil-fired furnaces in which the furnace charge does not contribute any gases to the flue gas, so that a simple determination of % CO_2 is sufficient to determine excess air.

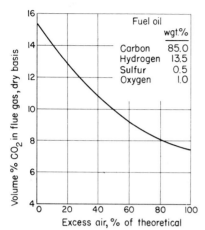

FIG. 5-3. CO_2 content of flue gases vs. % excess air, assuming complete combustion.

The percentage excess air can be checked by a calculation based only on the Orsat analysis, if the flue gases are derived entirely from the fuel and the air and not at all from the furnace charge, and if N_2 from the fuel is zero or negligible compared to that from the air. Under these conditions, the total quantity of oxygen supplied in the air per unit volume of flue gas can be calculated from the % N_2 in the flue gas, since each volume of N_2 originally accompanied a volume of 21/79 of O_2 in the air. The excess oxygen supplied per unit volume of flue gas is given by the O_2 percentage less the quantity of O_2 corresponding to unburned CO. That is,

$$\% \text{ excess air} = 100 \frac{\left(\% \, O_2 - \dfrac{\% \, CO}{2}\right)}{\left(\dfrac{21}{79}\right)(\% \, N_2) - \left(\% \, O_2 - \dfrac{\% \, CO}{2}\right)}.$$

Oxygen balance. The Orsat analysis gives the gas composition on a dry basis, so that an independent measurement or calculation is necessary to obtain the quantity of water in the flue gas and to account for all the H

and O. After the flue gas and air quantities are found by the carbon and nitrogen balances, the oxygen balance may be used to calculate the % water in the flue gas and then the volume of wet flue gas. For the oxygen balance, obviously, any oxygen introduced as moisture in the air or as CO_2, H_2O, or other compounds coming from the furnace charge must be taken into account. If the moisture entering the system with the fuel or furnace charge is unknown or uncertain, a direct measurement of the moisture content of the flue gas may be necessary.

Hydrogen balance. The element-balance calculations outlined above ordinarily suffice to complete the materials balance when the fuel analysis and the Orsat analysis of the flue gas are available and when data are not complicated by flue gas constituents coming from the charge. Then the hydrogen balance can be calculated as a check. Also, the hydrogen balance may be used to find % H_2O in the flue gases. In some circumstances, however, the hydrogen balance is useful in other ways. For example, the hydrogen content of the fuel may be calculated, or insufficient data on the quantities of gases contributed to the flue gases by the charge may introduce an unknown into the carbon or oxygen balance (or both) which makes the added hydrogen-balance equation essential to a solution of the complete materials balance.

Other stoichiometric calculations. When the materials in process in the furnace do not contribute elements to the flue gases or react with constituents of the gases, the C, N, O, and H balances, calculated in order, give the stoichiometry of combustion simply and without need of solving simultaneous equations. However, it is characteristic of some metallurgical processes that they contribute to the flue gas analysis in one or more ways. If adequate materials-balance data are available for the materials in process, it is a simple matter to add terms based on the known data to the element balances used for the combustion calculations. If such data are not available or cannot be estimated satisfactorily, full use of the C, N, O, and H balances still might give a solution, as illustrated below.

Oxygen consumption by the charge is common but does not affect the C, N, or H balances. These three balances applied to the fuel rate and Orsat analysis suffice to give flue gas quantity, total air consumption, and flue gas moisture. A direct moisture determination on the flue gas provides a check and then the O balance can serve to calculate oxygen consumption by the charge.

An unknown amount of CO_2 entering the furnace gases, for example from decomposition of carbonates, eliminates the direct calculation of dry flue gas quantity by the C balance. However, the C, N, O, and H balances can be set up as four simultaneous equations. Given the fuel rate and analysis, moisture content of combustion air, and Orsat analysis of flue gas, the equations can be solved to find quantity of dry flue gas, flue gas moisture, quantity of air, and the unknown quantity of CO_2 coming from the furnace charge.

Care must be exercised in using the results of complex stoichiometric calculations, since hidden in the solution of a set of simultaneous equations may be a calculation "by difference" not justified by the accuracy of the analytical data, especially the Orsat analysis. Wherever possible, enough stoichiometric data should be measured directly so that one element balance at least can be used to check the results of the others.

Heat utilization in furnaces

The complete heat balance of a furnace process shows in detail the sources of heat, including that obtained from combustion of fuel, and shows in detail the ways in which the heat is utilized or lost. However, in controlling combustion or in studying means for improving efficiency or economy of fuel utilization, it is helpful to consider separately the heat quantities directly related to the combustion of fuel and the heat quantities related to the metallurgical process proper. Also, in analyzing utilization of heat from combustion of fuels, the temperature levels at which the heat is made available to the process must be taken into account quantitatively. Particularly in high-temperature processes the operating variables relating to the attainment of a satisfactory temperature level are more critical than those relating to the supply of a certain total quantity of heat.

Adiabatic flame temperature (AFT). The maximum temperature theoretically attainable in burning a fuel-air mixture under given conditions is the temperature of the products of combustion when the combustion is carried out adiabatically. An adiabatic process is one in which the system neither loses heat to nor gains heat from the surroundings. In terms of the quantities used in the First Law equation, $Q = 0$, and if the process is both adiabatic and at constant pressure, $\Delta H = Q = 0$. In terms of the heat balance at constant pressure, the heat loss to the surroundings is nil for an adiabatic process, so that

sensible heat in combustion products = sensible heat in fuel and air + heat of combustion.

That is, for adiabatic combustion,

$\Sigma (H_T - H_{298})$ for combustion products

$= \Sigma (H_T - H_{298})$ for fuel and air

$- \Sigma \Delta H_{298}$ for all combustion reactions.

If the materials balance for the combustion process is known, as well as the temperatures of fuel and air, heat-content data and heat-of-reaction data can be used to set up an equation in which the temperature of the combustion products is the only unknown. Since this equation has terms in different powers of T, a successive-approximation type of solution is in order, as shown in the example below.

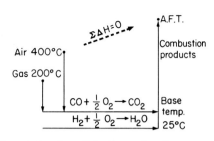

FIG. 5–4.   An adiabatic combustion process.

EXAMPLE

Calculate the adiabatic flame temperature for complete combustion of a producer gas with double the theoretical quantity of dry air. The producer gas analyzes 25% CO, 5% CO_2, 15% H_2, and 55% N_2 and is at 200°C. The combustion air is dry and is preheated to 400°C.

Figure 5–4 shows schematically the process for which the calculations are to be made.

STOICHIOMETRIC CALCULATIONS

Taking one mol of producer gas as the basis, the materials balance in terms of individual gas constituents is found to be as follows:

		mols
Fuel (200°C):	CO	0.25
	CO_2	0.05
	H_2	0.15
	N_2	0.55
	Total	1.00
Air (400°C):	O_2	0.40
	N_2	1.505
	Total	1.905
Combustion products	CO_2	0.30
	H_2O	0.15
	O_2	0.20
	N_2	2.055
	Total	2.705

Reactions: (1) $CO + \frac{1}{2}O_2 = CO_2$, 0.25 mol.

(2) $H_2 + \frac{1}{2}O_2 = H_2O$ (gas), 0.15 mol.

CALCULATION OF HEAT INPUT

Using the heat-content data in Table A–3 in the Appendix, the sensible heats in producer gas and air are found as follows:

$$(H_{473} - H_{298})_{\text{gas}} = 1260 \text{ cal,}$$
$$(H_{673} - H_{298})_{\text{air}} = 5150 \text{ cal.}$$

Finding the heats evolved in the two combustion reactions and adding,

$$- \Sigma \Delta H = 25,580 \text{ cal.}$$

Adding the three items of heat input,

$$\text{total heat input} = 31,990 \text{ cal/mol of producer gas.}$$

HEAT CONTENTS OF COMBUSTION PRODUCTS

The heat contents in the various constituents of the products of combustion can be expressed in terms of T as follows:

CO_2: $H_T - H_{298} = 0.30(10.55T + 1.08 \times 10^{-3}T^2 + 2.04 \times 10^5 T^{-1} - 3926)$,

H_2O: $H_T - H_{298} = 0.15(7.17T + 1.28 \times 10^{-3}T^2 - 0.08 \times 10^5 T^{-1} - 2225)$,

O_2: $H_T - H_{298} = 0.20(7.16T + 0.50 \times 10^{-3}T^2 + 0.40 \times 10^5 T^{-1} - 2313)$,

N_2: $H_T - H_{298} = 2.055(6.66T + 0.51 \times 10^{-3}T^2 - 2031)$.

Adding these and equating the sum to the heat input,

$$(H_T - H_{298})_{\text{Comb. prod.}} = 19.36T + 1.66 \times 10^{-3}T^2 + 0.68 \times 10^5 T^{-1} - 6149 = 31,990.$$

A successive-approximation solution of this equation for T can be based on the fact that the terms in T^2 and T^{-1} are small compared with the term in T.

For a first approximation, the terms in T^2 and T^{-1} can be neglected, to find $T = 1970°K$.

For the second approximation, substituting 1970 for T in the T^2 and T^{-1} terms and solving, $T = 1635°K$.

For the third approximation, substituting 1635 for T in the T^2 and T^{-1} terms and solving, $T = 1738°K$.

The successive approximations can be continued to obtain a solution of any desired precision as the answers continue to converge. Further approximations in this case are 1709, 1718, 1715, and 1716°K, so that $T = 1715°K$ to the nearest degree, or the adiabatic flame temperature = 1442°C.

Although the method of successive approximations may not appeal to a pure mathematician, it is of great engineering value and can be carried out quickly to obtain an answer of any desired precision if the calculator or slide rule is set up properly to start with.

The adiabatic flame temperature as defined and calculated above is a theoretical maximum which is not attained in actual flames. According to Trinks, actual flame temperatures are generally in the range from 48 to 75% of the calculated adiabatic flame temperatures. In the first place, complete combustion of CO to CO_2 and of H_2 to H_2O was assumed in the above example. At ordinary metallurgical furnace temperatures (say up to 1600°C or 3000°F), combustion reactions such as

$$CO + \tfrac{1}{2}O_2 \longrightarrow CO_2$$

and

$$H_2 + \tfrac{1}{2}O_2 \longrightarrow H_2O$$

go substantially to completion when the reactants are well mixed in appropriate quantities. At very high temperatures, however, chemical equilibrium is reached with appreciable quantities of CO and O_2 or H_2 and O_2 remaining unreacted. In other words, if the adiabatic flame temperature is first calculated on the basis of stoichiometrically complete combustion, a correction should be made for the dissociation of the products of combustion. The dissociation reactions are the reverse of the combustion reactions, absorbing rather than evolving heat, and therefore lowering the adiabatic flame temperature. The corrections for dissociation become appreciable at about 1600°C (3000°F) and large at temperatures above 2200°C (4000°F), so that published data on adiabatic flame temperatures sometimes are corrected for dissociation.

A second reason for the fact that actual flame temperatures are below calculated adiabatic temperatures is that actual flames are not adiabatic. Rapid though the combustion may be under favorable conditions, still it is not instantaneous. During the combustion, then, and before combustion is complete, the flame gives up heat to the surrounding furnace and furnace contents, and at positions in the flame where combustion is substantially complete, part of the heat generated in the flame has been transferred to the furnace and not retained in the products of combustion. For a given fuel-air combination, the highest actual flame temperature and the closest approach to the theoretical maximum are obtained under conditions of most rapid combustion. With gaseous fuels a perfect premixing of gas and air and provision of a catalytic solid surface favor rapid reaction, making possible temperatures within 100°C or so of the theoretical maximum under certain controlled laboratory conditions.

As the method of calculation in the above example showed, the adiabatic flame temperature is not a specific characteristic of the fuel, but depends on the conditions of combustion. In particular, the temperatures of fuel and air fed to combustion and the fuel-air ratio affect the calculation. Accordingly, a significant flame-temperature or calorific-intensity criterion for comparing fuels with each other is obtained when the AFT's are calculated for complete combustion with just the theoretical quantity of dry air and with air and fuel at a standard reference temperature. Table 5–2 gives such data for a number of fuels and includes calorific values for comparison. It can be seen that the theoretical adiabatic flame temperatures for a number of the common fuels are near or a little under 4000°F (2200°C). It should be noted that there is no over-all correlation between calorific power and flame temperature. For example, water gas gives the highest theoretical flame temperature of any of the fuels listed but has a lower calorific value than natural gas and coke-oven gas.

The actual flame temperature has considerable direct and indirect importance as a criterion of fuel utilization. An obvious requirement is that the flame temperature be above the temperature which must be reached in the furnace charge. The rate of transfer of heat from the flame to the furnace contents is a function of the temperature difference between flame and furnace contents, so that high flame temperatures favor rapid heating. On the other hand, higher flame temperatures are accompanied by greater danger of overheating and destroying the furnace refractories and even sometimes by the danger of locally overheating the furnace charge.

Direct measurement of actual flame temperatures is a complex problem, not often

TABLE 5-2

Theoretical Adiabatic Flame Temperatures for Various Fuels*

Fuel	Analysis, dry, %								Low calorific values Btu/ft^3	Theoretical adiabatic flame temperatures °F	°C
	CO_2	CO	CH_4	C_2H_4	C_2H_6	H_2	O_2	N_2			
Natural gas			84.7	9.4				1.6	935	3780	2080
Coke-oven gas	0.8	6.0	28.2			53.0		12.1	428	3730	2055
Producer gas	9.7	19.0	2.8	0.2		13.5		54.8	128	2930	1610
Blast furnace gas	12.5	25.4				3.5		58.6	92	2600	1425
Water gas	3.5	43.5	0.7			47.3	0.6	4.4	279	4100	2260
	% by weight										
	C	H	O	N	S	Ash			Btu/lb		
Coal	78.86	5.02	4.27	1.86	1.18	7.81			14,080	4020	2215
Coal	70.0	5.0	8.0	2.0	2.0	13.0			12,410	3960	2180
Lignite (dry)	59.9	9.4	18.6	1.2	2.7	13.2			9,897	3740	2060
Fuel oil (No. 6)	86.8	10.2			2.0				17,410	3820	2105

*Adiabatic flame temperatures calculated for complete combustion with theoretical dry air and not corrected for dissociation of products of combustion: reference temperature 80°F (27°C). After W. Trinks, Industrial Furnaces, Vol. II. New York: John Wiley, 1947.

attempted in metallurgical furnaces. However, the adiabatic combustion temperature, calculated as already described, serves as a useful comparative measure of flame temperatures and furnishes a sound basis for estimating the effects of changes in fuel or changes in combustion conditions. Figure 5–5, for example, shows the effect of air adjustment on adiabatic flame temperature for a producer gas, furnishing evidence of the importance of careful air control. Similar calculated data are useful in appraising the effects of various degrees of preheat of fuel or air, effects of enriching the air with oxygen, effects of moisture in fuel and air, effects of changes in fuel composition, and effects of other operating variables.

Critical temperatures of metallurgical processes. Only under unusual conditions do the interior furnace surfaces, the furnace gases, and the solid or molten charge reach a single uniform temperature which could be defined as the process temperature. In fact, as will be seen in the later study of heat flow, substantial temperature differences within the furnace are necessary to obtain the flows of heat on which furnace processes depend. In spite of these variations and differences in temperature, careful study of any given process usually will disclose a rather definite critical or minimum temperature level which must be met or exceeded in the coolest parts of the furnace interior if the process is to operate smoothly and satisfactorily. For example, in the continuous melting of metals or in the various

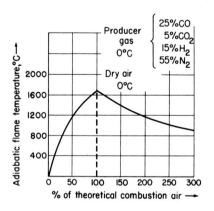

FIG. 5–5. Effect of air adjustment on theoretical adiabatic flame temperature.

processes of refining liquid metals, certainly no large portion of the furnace interior and contents can remain below the melting point of the metal. Actually, in such processes involving liquid metals, the critical temperature might well be the pouring or tapping temperature at which the liquid metal is to be withdrawn from the furnace. In processes characterized by the formation of liquid slags, the critical temperature is generally the so-called free-running temperature of the slag, which is the lowest temperature at which a slag is formed fluid enough to discharge freely from the furnace. In still other processes, the critical temperature may be the temperature required to carry out a certain chemical reaction effectively.

The critical temperature has a simple and direct significance in relation to the utilization of heat from the burning of fuel, that is, the critical temperature is the lowest temperature at which the flue gases, the products of combustion, can leave the working chamber of the furnace. A flue gas temperature lower than the critical temperature could only be obtained if some part of the furnace interior walls or charge were below the critical temperature, which would be inconsistent with the concept of critical temperature outlined in the previous paragraph.

A flue gas temperature well above the critical process temperature shows an unnecessary loss of heat into the flue gases. Such a condition might result from poor furnace design, for example, from failure to provide for good heat transfer from combustion gases to furnace charge. Other causes of excessive flue gas temperatures include driving the furnace over capacity, insufficient cold charge, and too high a fuel rate.

For a multistep batch process, starting with cold charge, the critical temperature will vary from step to step. In appraising fuel utilization for such a process, it may be preferable to estimate an average critical temperature rather than to attempt to take into account the variations with time.

Available heat. The sensible heat in the flue gases at the critical process temperature is not available to the process in the furnace; it is at too low a temperature level to be useful. The higher the process temperature, the larger this sensible-heat item will appear in the heat balance, to the point where it frequently becomes by far the largest item for processes at temperatures over 2000°F. Moreover, this is a loss of heat from the furnace which is inherent in the very nature of heating with fuels and therefore should be placed in a different category from heat losses chargeable only to poor furnace management. Recognition of this feature of fuel utilization leads to the important concept of available heat.

The *gross available heat* may be defined as the total heat input from the combustion of the fuel, less the sensible heat in the products of combustion at the critical process temperature. The total heat input includes sensible heats in fuel and air, as well as heat of combustion. Thus the gross available heat is the heat given up by the combustion products to the furnace and its contents in cooling them to the critical temperature. Available heat may be expressed in various ways according to convenience, for example, Btu/lb of fuel, Btu/hr, % of total heat input, % of calorific power of fuel.

The gross available heat is the heat given up by the combustion products in cooling from the adiabatic flame temperature to the critical temperature. If it is assumed as an approximation that the specific heat of the combustion products does not vary with temperature, the gross available heat is then proportional to $(t_{AFT} - t_{crit})$ and

$$\text{gross available heat, \% of total heat input} = \text{approximately } \frac{t_{AFT} - t_{crit}}{t_{AFT}} \times 100.$$

This rough approximation shows the importance of adiabatic flame temperature in relation to fuel utilization. Thus for a metallurgical process with $t_{crit} = 3000°F$, a combustion process with t_{AFT} of 3000°F or less could not be used, one with $t_{AFT} = 3500°F$ would be only about 14% efficient in supplying available heat, and one with $t_{AFT} = 4500°F$ would be about 33% efficient.

Although the gross available heat truly represents the heat quantity available *from the combustion process* at the critical temperature, it does not represent heat available *to the metallurgical process*. Some high-temperature heat must be used to keep the furnace hot, that is, to supply the heat loss to the furnace surroundings. Once a furnace is designed and built, the rate of heat loss to the surroundings when the furnace is at a given working temperature is relatively fixed and not under the operator's control. Hence, we may define the *net available heat*, available to the metallurgical process itself, as the gross available heat less the heat loss to the surroundings with the furnace at working temperature. The net available heat thus is a criterion for comparing the effectiveness of the system of fuel combustion plus furnace in supplying heat to the process, whereas the gross available heat is a criterion for comparing different fuel-combustion systems.

For some simple smelting and melting processes that require large quantities of heat, the principal factor determining furnace capacity is the rate at which the fuel and furnace can supply heat. For such processes the *smelting power* of the furnace is measured by the *net available heat* which

can be developed in the furnace chamber per unit time. However, as will be brought out in later chapters, other factors like heat transfer rates and chemical reaction rates also limit furnace capacities, so that an increase in rate of production of net available heat does not necessarily lead to an increase in capacity.

Variables affecting heat utilization. The principal objective in combustion control is to supply the requisite process heat as economically as possible. Fuel economy is directly affected by a great many variables, and is quite sensitive to some. Moreover, the effects of these variables frequently cannot be judged properly by visual observation, even by experienced furnace operators. Accordingly, to achieve maximum fuel economy, full use should be made of the various stoichiometric and thermochemical relations which disclose clearly and quantitatively the effects of the important operating variables on fuel utilization. These calculations require only the simplest and most common stoichiometric data from the furnace operation, principally the fuel analysis and fuel rate, the Orsat analysis of the flue gases, and some temperature measurements or estimates. Using the calculations outlined previously in this chapter and discussed more fully below, the combustion variables can be appraised satisfactorily without going into all the complications and difficulties of obtaining a complete heat balance.

The effects of combustion variables can be appraised satisfactorily in most instances by using available heat as a yardstick. Comparisons based on total heat input or on calorific power are often misleading. That is, the function of a fuel-combustion system is to make a required amount of heat available at an appropriate temperature level, not to generate a certain total heat supply, and the criteria of performance should be chosen accordingly.

The choice between gross available heat and net available heat as criteria of heat utilization depends on the kind of problem under consideration. Thus, if the furnace design and the daily process heat needs are fixed, the gross available heat requirement per day is fixed and the usual problem is to find the most economical means of supplying this requirement. Under these conditions,

$$\frac{\text{fuel}}{\text{consumption}} = \frac{\text{required gross available heat per unit time}}{\text{gross available heat per unit of fuel}}.$$

Furthermore, it can be seen from this relation that if both the daily requirement of gross available heat and the calorific power of the fuel are fixed, the fuel consumption will be inversely proportional to the figure giving gross available heat as a percentage of the calorific power. In this case then, in appraising the effects of adjustments in combustion conditions, the percent gross available heat is an appropriate criterion.

If heat supply is a critical factor in determining furnace capacity or tonnage throughput per day, consideration of the effects of combustion variables on net available heat becomes necessary. The following relation then is applicable:

$$\frac{\text{maximum furnace throughput}}{\text{(smelting power)}} = \frac{\text{net available heat generated per unit time}}{\text{required net available heat per unit of throughput}}.$$

This relation might be used, for example, to estimate the increase in smelting rate which might be obtained by better air adjustment, or to estimate the increase in fuel consumption required to take care of a planned increase in smelting rate.

The following list covers some of the variables with important effects on available heat (order is not significant).

(1) Air adjustment, fuel-air ratio, percent excess air, air leakage, furnace draft.
(2) Choice of fuel, changes in fuel analysis.
(3) Preheat temperatures of air and fuel.
(4) Moisture in air and fuel.
(5) Incomplete combustion, unburned fuel.
(6) Oxygen enrichment of air for combustion.
(7) Flue gas temperature, critical temperature.

Air adjustment. Good control of combustion air might well be considered the most important single factor in the efficient use of fuel. At the same time air control may be difficult to achieve. Good burner design and installation will provide dampers or other control of primary and secondary air entering through and around the burner, and in some cases automatic (proportioning) control of the fuel-air ratio at the burner is provided. However, metallurgical furnaces rarely even approach gas-tightness, so that other means of air access must be considered. Charging doors may be open for appreciable intervals, and even when closed leave cracks permitting substantial air leakage. Also air leakage through cracks in the brickwork of the furnace walls can easily amount to an appreciable fraction of theoretical combustion air.

To minimize air leakage through doors, cracks, and other openings into the furnace chamber and to maintain the correct flow of combustion air through the inlet openings at the burner, the furnace operator regulates the furnace draft to keep the pressure in the furnace chamber near, but usually a little below, atmospheric pressure. The furnace draft may be regulated by a damper in the flue, either manually operated or, preferably, automatically controlled to maintain a constant, preset draft at the furnace. A more detailed discussion of air and draft control in terms of the fluid-flow problems involved is given in Chapter 6.

The following example shows how the effect of air adjustment and air leakage can be assessed. A metallurgical furnace is fired with producer gas analyzing 25% CO, 5% CO_2, 15% H_2, and 55% N_2, both producer gas and air entering at 25°C. The process in the furnace is a continuous one, with a critical temperature of 1200°C.

Taking 100 m³ (STP) of producer gas as a basis, the total heat input from combustion ($-\Delta H$ for burning 15 m³ H_2 to H_2O (gas) and for burning 25 m³ CO to CO_2) is found to be 114,200 kg cal.

If exactly 100% of the theoretical air requirement is supplied, and combustion is complete, 175.2 m³ of flue gas (STP) will be formed analyzing 17.1% CO_2, 8.6% H_2O, and 74.3% N_2. The sensible heat in these flue gases at the critical process temperature (1200°C) is found to be 78,000 kg cal.

Subtracting the sensible heat in the flue gases from the total heat input gives the *gross available heat* as 36,200 kg cal/100 m³ of producer gas. This is 31.7% of the low calorific power of the fuel.

Repetition of the stoichiometric calculations with 20% excess air gives 194.3 m³ of flue gas analyzing 15.5% CO_2, 7.7% H_2O, 2.1% O_2, and 74.7% N_2. Heating the excess air to the critical temperature consumes part of the heat input from combustion, reducing the gross available heat to 28,600 kg cal/100 m³ or 25% of the calorific power of the fuel. Thus, if the furnace and furnace process are to be operated at a specified rate, that is, with a fixed gross rate of heat con-

sumption, the fuel consumption will be approximately 27% higher for operation with 20% excess air than for operation with the theoretical air.

Figures 5–6 and 5–7 summarize the results of a series of such calculations for the producer gas of the composition given, showing both the effect of air adjustment and the effect of critical process temperature (or flue gas temperature) on utilization of heat from the fuel. For a given critical temperature, the maximum percentage of available heat and the minimum fuel consumption are theoretically obtained with just the theoretical amount of air. The yield of available heat shows a sharp drop and the fuel consumption a sharp rise as the air is decreased below theoretical, because some of the calorific power is not realized and also because the unburned fuel consumes available heat in being heated to the critical temperature. The decrease in available heat and increase in fuel consumption as the air is raised above theoretical are less sharp because all the calorific power is realized and

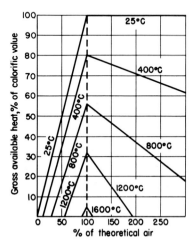

Fig. 5–6. Effect of air adjustment and critical process temperature on gross available heat.

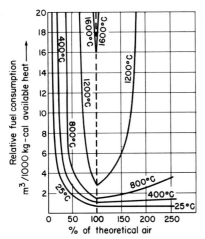

FIG. 5–7. Effect of air adjustment and critical process temperature on fuel consumption.

the loss of available heat is only that required to heat the excess air.

Figures 5–6 and 5–7 also show the major effects of process critical temperature on heat utilization from the producer gas. For low temperature processes with critical temperatures of only a few hundred °C, well over half the calorific power of the fuel will be available heat, and careful air adjustment is worthwhile but not critical. On the other hand, as the critical temperature approaches and exceeds 1000°C the yield of available heat decreases markedly and air adjustment becomes progressively more critical. Even with the most careful control of air, it is clear from these graphs that the producer gas of given analysis, burned without preheating either fuel or air, would not be used for processes requiring temperatures approaching 1600°C and would not be attractive for process temperatures of 1200°C and above.

In actual practice, complete combustion of the fuel and lowest fuel consumption are obtained usually with slightly more than theoretical air, since imperfect mixing of fuel and air and other factors result in some departure from the ideal conditions assumed in the available heat calculations.

Comparison of fuels. The data in the previous section showed low yields of available heat at temperatures above 1200°C. That is, available heat calculations alone prove the unsuitability of producer gas as a fuel for high-temperature furnaces fired directly without preheating fuel or air. On the other hand, similar calculations for natural gas (e.g., CH_4) or water gas ($CO + H_2$) show that these fuels give acceptable yields of available heat up to much higher temperatures. Near the other end of the scale, blast furnace gas is suitable only for relatively low-temperature heating. The approximate relationship given earlier between % available heat, adiabatic flame temperature, and critical temperature affords a preliminary basis of fuel comparison.

In making a quantitative comparison of two fuels for a given process, the fuel cost per unit of available heat might well serve as one of the guiding criteria; it affords a more logical basis of comparison than the cost per unit of calorific power.

Preheating air and fuel. Preheating the combustion air is a practical means of increasing available heat, and is widely applied in metallurgical furnaces requiring temperatures above 1000°C. Some gaseous fuels also may be preheated; this has been common with producer gas in particular. As will be discussed in more detail later, the combustion air or both the air and the gaseous fuel are generally heated with heat recovered from the flue gases, using regenerators or recuperators. The development of several important high-temperature metallurgical processes has been closely tied up with combustion procedures involving preheating and heat recovery from flue gases, since such procedures have afforded in many cases the only economical means of supplying high-temperature heat. The open-hearth furnaces which produce the major tonnage of steel are outstanding examples of this dependence on preheating and will be considered in some detail in the last part of this chapter.

The units of heat added to the heat input by preheating, as sensible heat in air (or in both air and fuel), add directly to the available heat. That is, increasing heat input by preheating involves no increase in the quantity or critical temperature of flue gases, so that the added heat is 100% available. Taking an extreme case as an example, firing with cold air and fuel might furnish available heat equivalent to only 5% of the calorific value of the fuel. By using a fraction of the sensible heat in the flue gases to preheat the air, sensible heat in preheated air might easily be brought to 10 or 15% of the calorific power of the fuel. This addition would multiply the available heat per unit of fuel by a factor of 3 or 4, making possible substantial reduction in fuel consumption or increase in furnace capacity.

The effect of preheating on available heat can be illustrated by further calculations involving the same producer gas used in previous examples. With no preheating of air or gas, 20% excess air, and a process critical temperature of 1200°C, the gross available heat was found to be 28,600 kg cal/100 m³ (STP) of producer gas. If the fuel were burned at the rate of 1000 m³/hr and the heat loss from the furnace were 100,000 kg cal/hr, the net available heat would be 186,000 kg cal/hr under the conditions given. Now suppose the operation is modified by preheating the combustion air to 500°C. The increase in heat input to the furnace per 100 m³ of fuel gas is the sensible heat in 114.2 m³ of air (120% of theoretical) at 500°C, which is 17,600 kg cal. Thus the

gross available heat per 100 m³ of fuel is increased from 28,600 to 46,200 kg cal. This result indicates that the given furnace at given output with preheated air would require only $28,600/46,200 \times 1000 = 619$ m³ of producer gas per hour, a fuel saving of 38.1%. On the other hand, if the fuel rate were left at the original 1000 m³/hr, the net available heat would increase through adoption of air preheating from 186,000 to 362,000 kg cal/hr, representing an increase of 95% in the smelting power of the furnace with no additional fuel.

Heating the combustion air, or both the air and the gaseous fuel, requires additional equipment which must be built, operated, and maintained. Thus, even if the heat is derived from waste heat which would otherwise pass up the stack with the flue gases, the cost of recovering the heat is not by any means negligible. Accordingly, in determining whether or not to use preheating, the savings in fuel must be balanced against the over-all cost of heat recovery. In the large-scale matte smelting of copper ore, fuel consumption is high and might be reduced appreciably by preheating air to combustion. However, in this case the costs for preheating air with heat recovered from flue gases have appeared greater than the possible fuel savings. A serious cost item in heat recovery from copper-smelting furnaces, for example, is the maintenance of the heat recovery equipment with dusty and dirty flue gases, since the furnaces are charged as a rule with fine flotation concentrates.

Moisture in fuel and air. Moisture in fuel and air reduces the available heat yield from a given quantity of fuel, since the water must be heated to leave the furnace at the flue gas temperature or the critical temperature. In addition, if the moisture enters in the liquid state (e.g., as moisture in pulverized coal), the heat of vaporization of the water reduces the available heat per unit of dry fuel. The latter effect is, of course, automatically taken into account when the low calorific power of the moist fuel is used in the available heat calculation.

The effect of varying humidity of the combustion air is generally small, as can be seen by calculations based on the previously discussed producer gas example. It was found that 114.2 m^3 of dry air represented 20% excess over theoretical for the combustion of 100 m^3 producer gas. If, instead of being dry, the combustion air had a relative humidity of 100% on a hot summer day (say 100°F), the moisture in the air would amount to 8 m^3/100 m^3 of producer gas. At a critical process temperature of 1200°C, this quantity of water would reduce the available heat by 4000 kg cal/100 m^3 of producer gas, or by 3.5% of the low calorific power of the fuel. From this calculation it appears that air humidity alone is not ordinarily a serious factor in reducing available heat.

The following calculations will illustrate the effect of liquid moisture in the fuel. A smelting furnace is fired with pulverized coal of the following specifications, on a dry basis:

% C	71.3
% H	5.4
% O	13.4
% N	1.5
% S	1.7
% Ash	6.7
	100.0

Combustion is complete and analysis of the flue gas gives 10.5% CO_2, 81.0% N_2, 6.5% O_2, and 2% SO_2 (dry basis). The operation is satisfactory and smelting proceeds at the desired rate when the flue gas temperature is 1200°C. At the normal rate of operation the coal is dried to 4% moisture before entering the burner and the coal consumption on

a dry basis is 10,000 lb/hr. An increase of 1% in the moisture content of the coal increases the fuel consumption as estimated below, the smelting rate, smelting conditions, and air-fuel ratio remaining constant.

For this estimate the sensible heats in coal and air can be neglected so that the heat input is given by the calorific power of the coal. Using Dulong's formula,

low calorific power (dry basis)
$$= 12{,}230 \text{ Btu/lb,}$$
total heat input
$$= 1.223 \times 10^8 \text{ Btu/hr, normal operation.}$$

The carbon balance gives the volume of dry flue gas:

$$\text{dry flue gas} = 2.03 \times 10^6 \text{ ft}^3 \text{ (STP)/hr.}$$

The hydrogen and moisture in the coal contribute water to the flue gas:

$$H_2O \text{ (gas) in flue gas, normal operation} = 0.105 \times 10^6 \text{ ft}^3 \text{ (STP)/hr.}$$

From data above and heat-content data, the total sensible heat in the moist flue gases at 1200°C is calculated to be 1.032×10^8 Btu/hr. The heat required to evaporate the moisture in the fuel is found to be only 0.004×10^8 Btu/hr. Subtracting these figures from the heat input gives the gross available heat in normal operation as 0.187×10^8 Btu/hr. This corresponds to 1870 Btu/lb of dry coal.

An additional 1% moisture in the coal, or approximately 0.01 lb/lb of dry coal, will require 10.5 Btu for evaporation and 11.1 Btu for heating to 1200°C, reducing the gross available heat to 1848.4 Btu/lb of coal. Dividing this figure into the normal available heat requirement gives a fuel consumption of 10,120 lb/hr (dry basis), representing an increase of 1.2% in fuel consumption.

Calculations similar to those above can be made to compare steam and air as atomizing agents for fuel-oil burners.

Still another effect of moisture becomes important in some circumstances. In the iron blast furnace, for example, heat is generated by burning carbon (coke) with air to carbon monoxide, the heat of reaction at 25°C being about 4400 Btu/lb of carbon, each lb of carbon requiring about 70 cu ft (STP) of air. If the air contains H_2O, the H_2O reacts with C endothermically.

$$C + H_2O \text{ (gas)} \longrightarrow CO + H_2,$$

and the C reacting with water then cannot generate heat by the reaction

$$C + \tfrac{1}{2}O_2 \longrightarrow CO.$$

Accordingly, the loss of available heat caused by moisture is equivalent to the heat absorbed by the reaction

$$H_2O \longrightarrow H_2 + \tfrac{1}{2}O_2$$

for the quantity of water involved. With air at 80°F and a relative humidity of 60%, 70 cu ft of air would contain about 1.4 cu ft (STP) of H_2O (gas), and the dissociation of this amount of moisture would represent a heat consumption or available-heat loss of about 400 Btu, which is 10% of the heat of combustion of the C to CO. When the high critical temperature of the blast furnace (about 1500°C) is considered, it is readily understood that the effect of air humidity on available heat can be serious enough to justify air conditioning or dehumidification of the air supply.

Completeness of combustion. The possible effects of incomplete combustion are readily estimated by methods already presented. If combustion is incomplete, not only is part of the calorific power of the fuel not developed, but also the unreacted fuel and air act as diluents and consume sensible heat in

being raised to flue gas temperature. Both these effects reduce available heat.

Although theoretical calculations lead to maximum available heat when just the theoretical quantity of air is supplied for combustion, in practice complete combustion is not generally obtained without some excess air. A balance is sought between loss of heat input from incomplete combustion and increased sensible heat in flue gas resulting from excess air. The optimum quantity of excess air varies with the fuel; powdered coal, for example, requires a larger excess than gaseous fuels. Burner and port design, size of combustion chamber, and other factors also affect the completeness of combustion and optimum air adjustment for complete combustion.

Oxygen enrichment. The 79% nitrogen in the air normally used for combustion passes through fuel-fired furnaces without reacting and comprises a major part of the volume of the flue gases. Thus, of the sensible heat tied up in the flue gases at the critical temperatures (the nonavailable heat), a large part is that required by the nitrogen. Reduction in the quantity of nitrogen passed through the furnace obviously then increases the yield of available heat, other factors remaining unchanged. This reduction in nitrogen throughput is obtained by oxygen enrichment of combustion air, that is, by substituting straight oxygen for part of the combustion air. Although the maximum effect is obtained by replacing all the air with oxygen, this extreme is rarely justified. Ordinarily oxygen enrichment during a high-demand part of a furnace operating cycle, or enrichment to increase the oxygen content from 21% for air up to 25% or 30%, has proved to be more attractive.

As with other means of increasing available heat per unit of fuel, oxygen enrichment also increases flame temperature. In fact, oxygen-fuel combinations are often used to attain extremely high temperatures on a small scale (oxygen-acetylene, oxygen-hydrogen, oxygen-gas) where the cost of oxygen is not a serious part of the total cost. In metallurgical furnaces, the increase in flame temperature with oxygen enrichment can be used to increase rate of heat transfer from the flame to the charge in the furnace, sometimes greatly speeding up the work of the furnace. On the other hand, the effect on flame temperature may limit the use of oxygen, since the life of the furnace refractories suffers when too high a temperature is reached. This limitation has been observed particularly with furnaces having silica roofs; they fail rapidly at temperatures close to 3000°F.

When a given fuel-air combination is inadequate, either because of poor available heat yield or because of too low a flame temperature, oxygen enrichment is, in theory at least, an alternative to installation of preheating apparatus. Moreover, it is more flexible, will give greater increases in available heat and flame temperature, and is not subject to the often messy maintenance and upkeep problems usually associated with high-temperature heat recovery. Burners using oxygen in place of part of the air can be added to existing furnaces without extensive structural changes. None of these advantages, however, is sufficient to justify the use of oxygen if the cost of the oxygen is greater than the over-all savings in cost resulting from its use.

Figures 5–8 and 5–9 show graphically the calculated effects of oxygen enrichment on gross available heat and fuel consumption when producer gas of the same composition as used in previous examples (25% CO, 5% CO_2, 15% H_2, 55% N_2) is burned with just theoretical oxygen requirements (in air plus oxygen). These curves show that the

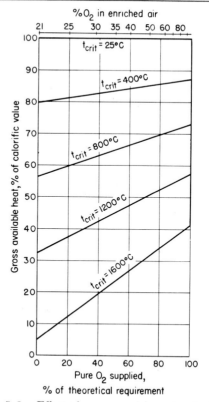

Fig. 5–8. Effect of oxygen enrichment and critical process temperature on gross available heat.

maximum benefit from oxygen enrichment is obtained at high critical temperatures. Also, Fig. 5–9 shows that at high process temperatures, the first units of oxygen used give the most fuel savings. Taking the most extreme case, for a critical temperature of 1600°C, enriching the combustion air from 21% to 25% oxygen (replacing about one-fifth of the combustion air with oxygen) cuts the fuel consumption from about 16 down to 7 units (saving = 9 units), while replacing all the air with oxygen cuts the fuel consumption to about 2 units (saving = 14 units) but requires five times as much oxygen.

Only very recently have processes been developed which will produce oxygen in

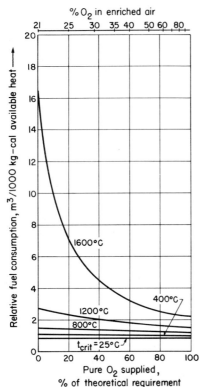

FIG. 5–9. Effect of oxygen enrichment and critical process temperature on fuel consumption.

tonnage quantities at sufficiently low cost to be considered for large-scale metallurgical uses. These processes make oxygen from air by low-temperature fractional distillation and probably will furnish impure oxygen (about 95%) for under $5 a ton, 99% oxygen for somewhat more. Transportation and distribution add considerably to the cost, so for firing large metallurgical furnaces the oxygen-making plant should be in or adjacent to the oxygen-using plant. At the time of this writing only a small number of such installations have been made experimentally, so that too few cost data are available to foretell clearly the extent to which use of oxygen enrichment will spread.

A rough order-of-magnitude cost calcu-

lation will show further the potential importance of cheap oxygen. One ton of oxygen will eliminate 7500 lb of nitrogen from the combustion products, and at a critical temperature of about 1400°C (2550°F) this quantity of nitrogen would require about 5×10^6 Btu of sensible heat. Thus, the cost of securing available heat through enrichment with oxygen at $5 per ton would be about $1.00 per 10^6 Btu. Compared with using coal (12,000 Btu/lb) under conditions giving 40% available heat, the energy cost of using oxygen is equivalent to coal at $10 per ton, which is not expensive energy.

Flue gas temperature. Much of the previous discussion and calculation has been based on the concept of a critical process temperature defined as a minimum flue gas temperature for satisfactory operation. This concept involves a relatively ideal operation in several respects. In the first place, a flue gas temperature near the minimum practicable process temperature implies an efficient furnace design. Second, the critical temperature has been regarded as primarily determined by the more or less static temperature requirements of the process. In actuality, under many operating conditions the controlling requirement is the rate of heat demand. To meet higher than normal heat demands, larger differences in temperature between furnace gases and charge are required to drive the flow of heat as fast as desired. Particularly if the furnace is not designed for the high heat transfer rate, very high flue gas temperatures may be obtained in attempting to reach the desired production rate.

Under conditions of lower heat demand, a fuel rate in excess of process needs will raise the flue gas temperature and presumably also the temperature of the materials in process higher than necessary.

Since the flue gas temperature approaches process critical temperature only under relatively ideal conditions, all the available heat in the flame and combustion products is not necessarily utilized in the furnace and metallurgical process. That is, the available heat criterion applies primarily to the fuel-combustion system and does not take into account efficiency of the furnace in taking the heat from the flame and utilizing it. In many furnaces, then, improvements in furnace design to improve heat transfer and to decrease flue gas temperatures may be more important to fuel economy than changes in the combustion system designed to yield more available heat.

Electric furnaces. When electricity is used for heating, generally no flue gases at all are generated as an inherent feature of the heat supply. Therefore, regardless of the critical temperature of the process, all the electrical heat input might be regarded as gross available heat. This feature of electrical heating is important in making comparisons with fuel heating. For example, the cost per unit of energy should be based not on total energy supply but on energy which can be made available to the process. This type of comparison favors electric heating, especially for high-temperature processes where only a fraction of the calorific power of fuels is available heat. On the other hand, in most localities in the United States electric power costs are so much greater than fuel costs that electric heating becomes competitive only for relatively high-temperature processes (e.g., above 3000°F).

Heat recovery from flue gases; regenerators and recuperators

Since the flue gases from fuel-fired high-temperature furnaces necessarily account for substantial fractions of the heat input from the combustion of the fuel, various means are used to recover waste heat from the gases and return it to the furnace or apply it to other useful purposes. One of several procedures may be followed, depending on the nature of the process, fuel costs, power costs, and other factors. In the first place, the heat recovered may be used to preheat the air for combustion and sometimes the gaseous fuel as well. The effect of this procedure on available heat was discussed previously. Second, through the use of waste-heat boilers, the waste heat may be converted into steam power for use around the plant or even for sale. These two procedures, either alone or combined, account for the major share of heat recovery practice. However, other waste-heat applications, such as using the hot flue gases for drying ore or for preheating air for a different furnace, may be attractive under certain circumstances.

In subsequent sections, typical heat exchangers used to transfer heat from hot flue gases to cold air and gaseous fuels will be described briefly and discussed in terms of heat- and materials-balance criteria. The important design and operating variables relating to rates of heat transfer and rates of gas flow will be discussed in subsequent chapters.

Thermal efficiency of a heat exchanger. The process of heat exchange is schematically represented in Fig. 5–10. Gas leakages and inadvertent mixing of the two flows of

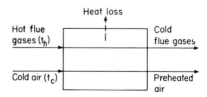

FIG. 5–10. Heat exchange.

gas may be ignored to start with, keeping in mind, however, that their effects can be serious and must be taken into account in many systems, especially those constructed of refractory brick and in a poor state of repair. If the heat exchanger represented in Fig. 5–10 is operated for a period over which there is no net accumulation or diminution in the heat stored in the apparatus itself, the heat balance items are sensible heats in the four streams of incoming and outgoing gases and the heat loss to the surroundings. In defining thermal efficiency, it is convenient to take the temperature of the cool incoming gas (t_c) as the base temperature for the heat balance, so that the cool incoming air has zero sensible heat. On this basis the over-all thermal efficiency is defined simply as the ratio of the heat picked up by the cold gas to the total heat input, or

over-all thermal efficiency (%)

$$= \frac{\text{sensible heat in heated air}}{\text{sensible heat in hot flue gas}} \times 100.$$

Using this figure as the criterion of efficiency, the inefficiencies are the heat loss from the apparatus and the heat left in the flue gases leaving the heat exchanger.

Although the over-all thermal efficiency is a simple and logical measure of efficiency in heat recovery, it does not take into account a fundamental and inherent limitation stemming from the fact that heat flows only from a higher temperature to a lower one and never in the reverse direction. In other words, the flue gas cannot possibly heat the air to any temperature above the temperature of the hot incoming flue gas. In heat exchangers using hot flue gases to preheat the combustion air for the same furnace, it is generally true that an air-preheat temperature equal to the temperature of the incoming hot flue gases would correspond to a thermal efficiency (as defined above) well under 100%, even though it is obvious that no improvement is possible. The reason for this relation is that the heat capacity of the flue gases (quantity flowing × specific heat) is generally greater than the heat capacity of the corresponding combustion air. Accordingly, we may define an efficiency limit as

efficiency limit (%)

$$= \frac{\begin{array}{c}\text{sensible heat in air} \\ \text{at hot flue gas temperature } (t_c \text{ to } t_h)\end{array}}{\begin{array}{c}\text{sensible heat} \\ \text{in hot flue gas } (t_c \text{ to } t_h)\end{array}} \times 100.$$

In these definitions, t_c and t_h are the lower and upper limits of the working temperatures of the heat exchanger, and are, respectively, the temperature of the incoming cold air and the temperature of the incoming hot flue gas.

Since a common purpose in calculating efficiencies is to compare percentage-wise what is being done with the best that can be done, a third figure can be calculated to compare the thermal efficiency with the limiting thermal efficiency:

relative efficiency (%)

$$= \frac{\text{over-all thermal efficiency}}{\text{efficiency limit}} \times 100$$

$$= \frac{\begin{array}{c}\text{sensible heat in air} \\ \text{at preheat temperature}\end{array}}{\begin{array}{c}\text{sensible heat in air} \\ \text{at hot flue gas temperature}\end{array}} \times 100.$$

The relative efficiency is an appropriate criterion for general comparisons of different heat exchangers and of different operating conditions in the same heat exchanger.

Recuperators. A recuperator is a continuous type of heat exchanger with both streams of fluid flowing continuously through the apparatus, the streams being separated by walls across which the heat flows from the hot fluid stream to the cold. This type of heat exchanger, constructed of metal, is very common for heat exchange between liquids in the chemical manufacturing industry. It is relatively uncommon in high-temperature applications around metallurgical furnaces because of the difficulties encountered in sealing and maintaining separate but adjacent gas passages constructed of nonmetallic refractories. Also, since metallurgical flue gases are usually laden with dust, fume, and even slag particles, keeping a recuperator sufficiently clean may be difficult or impossible. Figure 5–11 illustrates one type of recuperator of relatively simple construction. The air to be preheated is passed through silicon carbide tubes which cross the brick duct carrying the hot flue gases. Silicon carbide is a suitable material for this use because it has good mechanical properties, will stand high temperatures, can be made relatively gastight, and, finally, is a much better conductor of heat than other common nonmetallic refractory materials.

Regenerators. A regenerative-type heat exchanger contains heat-storage elements which alternately absorb heat from the hot flue gases and then give up heat to the cold air. Two types of regenerators are in use: (1) continuous gas flow, moving-element regenerators; (2) intermittent gas flow, stationary-element regenerators.

In the first type, each of the two gas streams flows continuously and steadily through its own compartment of the regenerator while the heat-storage elements move back and forth from one chamber to the other. One example of this type is shown in Fig. 5–12. The gas streams are carried in adjacent ducts, and the heat-storage elements are mounted around a vertical shaft whose rotation carries them alternately through one stream and then through the other. With this mechanical arrangement, all-metal construction is almost essential, so that the apparatus is suited primarily for service at relatively low temperature (e.g., boilers). Another example of continuous gas flow, moving-element regenerators is the pebble heater, shown in Fig. 5–13. The hot gases enter at the bottom of the upper chamber and pass up through the chamber countercurrent to the refractory pebbles. The heated pebbles move down by gravity into the lower chamber, where they give up heat to the upward stream of air. The cooled pebbles discharge at the bottom and are elevated outside the

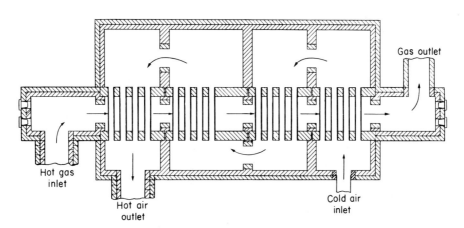

Hot gas inlet

Hot air outlet

Gas outlet

Cold air inlet

Fig. 5–11.   Silicon carbide tube recuperator (courtesy of Fitch Recuperator Company).

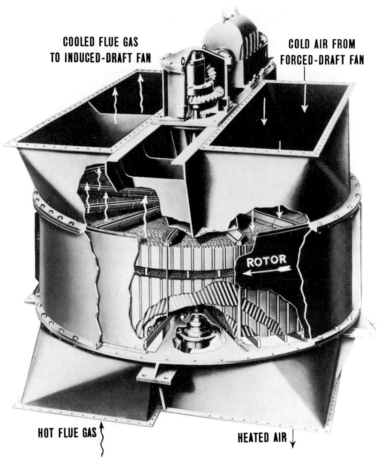

COOLED FLUE GAS TO INDUCED-DRAFT FAN

COLD AIR FROM FORCED-DRAFT FAN

ROTOR

HOT FLUE GAS

HEATED AIR

FIG. 5–12. Ljungstrom regenerative-type air preheater (courtesy of The Air Preheater Corp.).

exchanger and returned at the top to complete the cycle. The pebbles can be fabricated from refractory materials capable of withstanding extremely high temperatures and can be cleaned or freed from dust easily while out of the heat exchanger. Also, high heat transfer rates and high thermal efficiencies are achieved. The principal disadvantage is the pressure required to drive the gases through the pebble beds, since most furnace and flue systems are designed to operate with gas pressures throughout differing very little from the outside atmos-

pheric pressures. Still another exchanger in the continuous gas flow, moving-element class makes use of finely divided solids for the moving elements. In each of the two compartments, the solids in contact with the gases are in a loose bed, kept in a very fluid and highly agitated condition by passing the gases vertically upward through the solids. This fluidized-solids technique has other applications, and is discussed in more detail in Volume II.

Most regenerators for metallurgical furnaces are of the intermittent gas flow, sta-

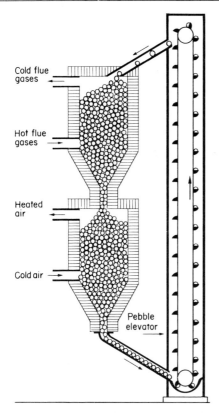

Cold flue gases

Hot flue gases

Heated air

Cold air

Pebble elevator

FIG. 5–13. Pebble heat exchanger.

being heated with flue gases while the other is being cooled with air. Flow reversal may be controlled either manually or automatically (preferably the latter), and the periods between reversals may be as short as 6 or 7 minutes or as long as an hour or more, depending on the size of the regenerator and on operating conditions.

Owing to the cyclic operation of a regenerator, the air preheat temperatures, waste flue gas temperatures, and checker temperatures vary through the operating cycle. Moreover, the furnace process itself may be characterized by its own cycle of varying demand, so that the individual regenerator cycles necessarily vary considerably in duration and character between different stages of the furnace process. As a result of these variations, the regenerator chamber itself is not returned to a given condition (that is, given temperature distribution and heat content) at each reversal, and some storage and release of heat occurs over a long term of hours even when the reversal cycle is only a few minutes.

The above-described characteristics of a regenerator make it difficult to determine any kind of a satisfactory heat balance for an operating system. Other calculations, for example, those relating to heat flow rates, become so complicated for the unsteady conditions in the regenerator that special machines have been built to perform the necessary calculations. In spite of the difficulties, however, thermal calculations based on the best available data, although they do not give a precise accounting, still disclose the important variables in proper perspective and afford a good basis for engineering control. Illustrative data for an open-hearth furnace with regenerators will be given in the next section.

The blast furnace stoves used to preheat the air blown into the iron blast furnace can

tionary-element type. They consist essentially of a chamber filled with a checkerwork of brick or tile laid in a pattern to give a multiplicity of vertical gas passages (Fig. 5–14). The hot flue gases and the air (or other gas being preheated) flow alternately through the same chamber and same passages, so that the operation is cyclic. Generally, the hot gases are introduced at the top and flow down while cold air enters at the bottom and flows up, so that the periodic change from the heating to the cooling part of the cycle is called a reversal. To provide for continuous flow of both flue gases and air, a pair of regenerator chambers is required, so that at any given instant one is

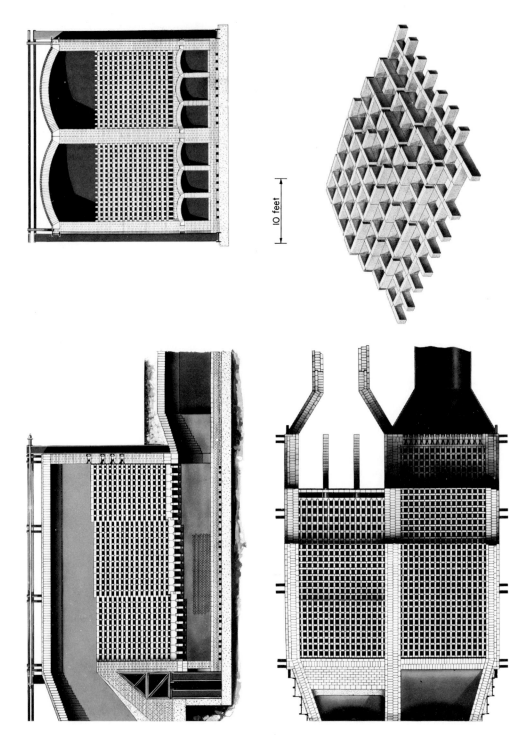

10 feet

FIG. 5–14. Construction of regenerator (courtesy of Harbison-Walker Refractories Company).

be considered as a special case of regenerative heat recovery. The heat is supplied by the gas effluent from the blast furnace, not as sensible heat, which is almost negligible in blast furnace gas, but as heat derived from burning the CO and other constituents. Thus the stoves serve both as combustion chambers and regenerators.

Fuel utilization in the open-hearth furnace

Open-hearth steelmaking furnaces unquestionably represent, in terms of total tonnage production, number of furnaces in operation, and total fuel consumption, the outstanding furnace operation with which metallurgists are concerned. Between 80% and 90% of all the metal fabricated and sold in the United States is steel produced in open-hearth furnaces. Capacity production of open-hearth steel is over 90,000,000 tons per year. The average fuel consumption is roughly equivalent to 4,000,000 Btu (estimate) per ton of steel, or to about 0.15 ton of coal per ton of steel. Combining such figures with current fuel prices shows that many tens of millions of dollars worth of fuels are burned annually in open-hearth furnaces, so that an understanding and careful application of the principles of fuel utilization to open-hearth furnaces is of great economic importance. Accordingly, a general survey of the fuel utilization problem for the open-hearth furnace is here taken as an example of specific application of the principles presented earlier. The point of view and much of the numerical data to be used are based to a considerable extent on the discussion in *Basic Open-Hearth Steelmaking*, (Committee on Physical Chemistry of Steelmaking, AIME), and this reference should be consulted for additional information.

Process characteristics. For this brief outline of the fuel utilization problem in the open-hearth, details of the chemistry, charge materials, products, furnace construction, etc., are not essential; only those process characteristics pertaining directly to fuel utilization will be summarized. The open-hearth process is a complex, multistep batch process, with a typical scale of operation on the order of 100 to 150 tons of steel product per batch and a time of operation of about 10 hr per batch. The process itself, with no allowances for heat losses and other thermal inefficiencies, requires heat from the fuel at the rate of about 1,000,000 Btu per ton of steel, most of the heat consumption being represented by sensible heat in the hot liquid products of the furnace (liquid steel and liquid slag). Working temperatures required are the highest among common fuel-fired metallurgical furnaces, with a mean critical temperature of about 2860°F (1570°C).

A detailed breakdown of the thermal requirements of the steelmaking process shows that the rate of demand for heat and the temperatures of the materials in process vary over wide ranges during the period from start to finish of a heat. Thus a complete thermal accounting would have to include time as a variable and would represent so much numerical data that the really significant and basic thermal relationships might be hidden. For this outline, the open-hearth process will therefore be treated as if it were a continuous, steady state process, and average heat requirements, temperatures, and flow rates will be used as if they prevailed over the whole time of the heat.

Furnace arrangement. Figure 5–15 shows the main features of the open-hearth furnace system. The *furnace chamber* itself, which might be about 50 ft long by 25 ft wide to hold 175 tons of liquid steel, is massively constructed of bricks and other forms of refractory materials bound with steel struc-

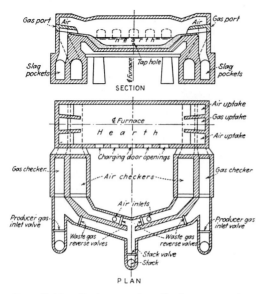

FIG. 5-15. Basic open-hearth furnace system (courtesy of Blaw-Knox Company).

tural members on the outside. Feed materials are charged through a series of doors in one sidewall of the furnace and the liquid products are removed through tap holes in the other sidewall. The furnace is symmetrical longitudinally, so that the ports at each end serve alternately for the entrance of fuel and combustion air and for the exit of flue gases. In fact the whole combustion, regeneration, and gas-flow system is symmetrical, and the direction of gas flow is periodically reversed in accordance with the requirements of the heat regenerators.

The hot flue gases from the furnace chamber pass out through the ports and then down into the *slag pockets*, from which they flow laterally into the top of the *checkers* or *regenerator chambers* (see also Fig. 5–14). Flow of flue gases through the checker brick is vertically downward, to ensure proper flow distribution. From the checkers, the waste gases go into the flue system and then through the stack to the atmosphere.

At the firing end of the system, cold air is brought into the bottom of the checkers and passes up to the furnace air inlet ports. If producer gas is used, it is preheated in one of the regenerator chambers while air is preheated in the other. Provision of two regenerators at each end became standard practice when producer gas was the standard open-hearth fuel, the smaller of the two chambers being used for preheating the fuel. The practice of using two pairs of regenerators has continued even after the widespread replacement of producer gas by other fuels, both chambers now being used for preheating air.

The arrangements of the valves and flues for reversing the flow vary with the method used for supplying and controlling furnace draft. Automatic control of furnace reversal is becoming common, with, for example, a control circuit arranged so that reversal occurs when the waste flue gases leaving the checkers reach a certain preset temperature.

Fuels. The inventions of the open-hearth furnace, regenerator, and the gas producer by the Siemens brothers about 1860 were very closely tied together; in fact, the combination of the three for many years constituted the open-hearth process. For this reason, the design of presently operating furnaces reflects in many respects the requirements of producer gas as the fuel even though other fuels are now used very extensively. In many cases, for example, the open-hearth fuel now is wholly or partly one of the by-products of the rest of the integrated steel plant, such as tar, blast furnace gas, coke-oven gas, etc. Fuels in general use for the open-hearth now include virtually all the important common fuels discussed in the previous chapter, and various mixtures and combinations as well. The only fuels which can be preheated are the relatively low-grade gaseous fuels such as producer gas, blast furnace gas, and mixtures of these with coke-oven gas.

Regardless of which fuel is used, the temperatures required in the open-hearth process are high enough (critical temperatures close to 1600°C) so that regenerative heat recovery and preheating of air are essential to obtain reasonable yields of available heat and economical fuel consumptions. However, from a longer-range point of view there is the possibility that oxygen enrichment of combustion air may afford an attractive alternative to regeneration.

Simplified heat balances. Many of the important thermal relationships of the open-hearth furnace system can be brought out by a few illustrative calculations based on the use of average temperature and thermal data as if the process were a continuous, steady state process. In particular, these calculations show the role of regeneration and regeneration efficiency and the importance of good control of excess air and air leakage. Assuming fuel oil to be the fuel, calculations will be made for the following combustion conditions: (1) 110% of theoretical air, no air leakage, 80% relative regenerator efficiency, (2) 110% of theoretical air, 15% air leakage, 80% relative regenerator efficiency, (3) 125% of theoretical air, 15% air leakage, 80% relative regenerator efficiency, and (4) 125% of theoretical air, 15% air leakage, 60% relative regenerator efficiency.

For the calculations below, the system will be regarded as divided into three zones: the furnace chamber between ports, and the two regenerative zones extending from air inlet (waste gas outlet) to the ports and including downtake and uptake flues and slag pockets as well as the checkers.

(1) *110% of theoretical air, no air leakage, 80% relative regenerator efficiency.* The heat balance may be calculated on the basis of 1 lb of fuel oil, analyzing 90% C and 10% H, with a low calorific value of 17,500 Btu/lb. Theoretical dry air is 171 cu ft (STP); 10% excess gives 188 cu ft (STP). The products of combustion with 10% excess air will amount to 197 cu ft (STP), analyzing 13.7% CO_2, 9.1% H_2O, 75.4% N_2, and 1.8% O_2.

Taking 77°F (25°C) as the base temperature, sensible heat in the flue gases leaving the furnace and entering the regenerative zones at the critical temperature of 2860°F is calculated to be 13,100 Btu/lb of fuel oil. The air preheat temperatures can be estimated, as follows, from regenerator efficiency. First, the efficiency limit of the regenerators is found as the ratio of the sensible heats at 2860°F of air and flue gases:

$$\text{efficiency limit} = \frac{11{,}412}{13{,}100} \times 100 = 87.1\%.$$

If the relative efficiency of the regenerators is 80% (assumed), the over-all thermal efficiency will be 0.80×87.1, or 69.7%. Thus 69.7% of the sensible heat in the flue gases, or 9100 Btu will be recovered in the air. This quantity of heat will raise the temperature of the air from an entering temperature of 77°F (assumed) to 2240°F.

To avoid complication, it may be assumed arbitrarily that the heat loss from the regenerative zones is 10% of the sensible heat input to the regenerators in the hot flue gases. On the basis of this figure and previously calculated data, the heat balance of the regenerative zone can now be set up (reference temperature, 77°F; basis, 1 lb of fuel oil):

Heat input	*Btu*
Sensible heat in entering flue gases	13,100

Heat output	*Btu*
Sensible heat in preheated air	9,100
Sensible heat in waste flue gases	2,700
Heat losses to surroundings	1,300
	13,100

The gross available heat provided for the high-temperature zone of the furnace can now be calculated. The total heat input to the high-temperature zone is the sum of the heat derived from the calorific power of the fuel and the sensible heat of the preheated air, or $17,500 + 9100 = 26,600$ Btu. Subtracting from this the sensible heat in the flue gases leaving the high-temperature zone at the critical temperature gives $26,600 - 13,100 = 13,500$ Btu as the gross available heat released to the high-temperature zone per pound of fuel oil burned. This is $13,500/17,500 \times 100 = 77.2\%$ of the low calorific power of the fuel oil.

The figure of 77.2% for the yield of gross available heat is, as discussed in detail earlier in this chapter, a significant measure of the effectiveness of the whole combustion system, including regenerators, in supplying high-temperature heat to the working chamber and in utilizing fuel. Furthermore, this result is based on the very optimistic assumptions of small percent excess air, no air leakage, high regenerator efficiency, etc., which are unattainable in practice. Therefore, the figure just calculated may be taken as a high standard to be compared with the results of subsequent calculations for less favorable conditions.

(2) *110% of theoretical air, 15% air leakage, 80% relative regenerator efficiency.* The design of the open-hearth working chamber makes some air leakage unavoidable. Air leaking into the working chamber is, of course, cold air and thus consumes large quantities of heat just in being heated to the working temperature. Also, since less air must be brought in through the regenerators, the regenerator efficiency is lowered by air leakage. Accordingly, air leakage has a much more serious effect on fuel utilization than the same quantity of excess air brought in through the regenerator. In fact, as has

been widely recognized in recent years, control and minimization of air leakage is probably the most important single factor in obtaining good fuel economy in open-hearth furnaces.

Repetition of the simplified thermal calculations for the condition of 110% of theoretical air, with 95% of the theoretical quantity of combustion air entering the system through the regenerators and 15% of theoretical air leaking into the high-temperature zone, gives the following data:

Basis:

    1 lb fuel oil, 90% C, 10% H, low calorific power 17,500 Btu/lb

Air quantities:

    theoretical, 171 cu ft (STP)
    actual, 188 cu ft (STP)
    regenerators, 162 cu ft (STP)
    cold air leakage, 26 cu ft (STP)

Flue gases:

    volume, 197 cu ft (STP)
    analysis, 13.7% CO_2, 9.1% H_2O, 75.4% N_2, 1.8% O_2
    sensible heat at 2860°F, 13,100 Btu

Regenerator efficiencies:

    efficiency limit, 75.2%
    relative efficiency (assumed), 80%
    over-all thermal efficiency, 60.2%

Gross available heat to working chamber:

    12,300 Btu, or 70.3% of the low calorific value of the fuel oil.

These operating conditions correspond to about 10% greater fuel consumption than the operation with the same total air supply but no leakage.

(3) *125% of theoretical air, 15% air leakage, 80% relative regenerator efficiency.* If the excess air is increased from 10 to 25%, the chief effect on fuel utilization will be the decrease in available heat resulting

from the increased quantity of sensible heat carried out in the flue gases at critical temperature. However, counteracting this effect, the increased air quantity will afford a net increase in regenerator efficiency because of the increased efficiency limit. Actually, the relative efficiency probably would be decreased a little by the increase in air velocity but, for the calculations below, this figure, which is difficult to determine, is left unchanged at 80%. Repetition of the calculations outlined previously, but with 125% of theoretical air and with 15% air leakage, leads to the following results.

Basis:
  1 lb fuel oil, 90% C, 10% H, low calorific power 17,500 Btu/lb

Air quantities:
  theoretical, 171 cu ft
  actual, 214 cu ft
  regenerators, 188 cu ft
  cold air leakage, 26 cu ft

Flue gases:
  volume, 223 cu ft
  analysis, 12.1% CO_2, 8.1% H_2O, 4.0% O_2, 75.8% N_2
  sensible heat at 2860°F, 14,700 Btu

Regenerator efficiencies:
  efficiency limit, 77.6%
  relative efficiency (assumed), 80%
  over-all thermal efficiency, 62.1%

Heat balance of regenerative zone:

Input	Btu
Sensible heat in entering flue gases	14,700
Output	Btu
Sensible heat in preheated air	9,100
Sensible heat in waste flue gases	4,100
Heat losses	1,500
	14,700

Gross available heat to working chamber: 11,900 Btu or 68.0% of the low calorific value of the fuel oil.

By comparison with the previous calculations, it is seen that excess air passing through the regenerators has a much less serious effect on available heat than excess air leaking directly into the furnace chamber.

(4) *125% of theoretical air, 15% air leakage, 60% relative regenerator efficiency.* In the three previous calculations, the relative efficiency was assumed constant at the rather high value of 80%. Repetition of the latter part of the calculations of (3), but with an assumed relative efficiency of 60%, gives the following results:

Regenerator efficiencies:
  efficiency limit, 77.6%
  relative efficiency (assumed), 60%
  over-all thermal efficiency, 46.6%

Heat balance of regenerative zones:

Input	Btu
Sensible heats in entering flue gases	14,700
Output	Btu
Sensible heat in preheated air	6,900
Sensible heat in waste flue gases	6,300
Heat losses	1,500
	14,700

Gross available heat to working chamber: 9700 Btu or 55.4% of the low calorific value of the fuel oil.

These conditions, the most unfavorable of those considered, represent a fuel consumption about 40% greater than the most ideal conditions considered in (1). Actual open-hearth practice in many instances will show much lower efficiencies of heat utilization than any of the cases considered, especially when careful combustion control is not practiced. When practical limitations on air control, air leakage, regenerator efficiency, heat losses, etc., are considered, a recovery of about 60% of the low calorific value of fuel oil as gross available heat probably is as good as can be attained under the most favorable conditions in the open-hearth system.

SUPPLEMENTARY REFERENCES

AIME, COMMITTEE ON PHYSICAL CHEMISTRY OF STEELMAKING, *Basic Open-Hearth Steelmaking*, revised ed. New York: AIME, 1951.

BUTTS, ALLISON, *Metallurgical Problems*. New York: McGraw-Hill, 1943.

ETHERINGTON H., *Modern Furnace Technology*. London: Charles Griffin, 1944.

HASLAM, ROBERT T., and RUSSELL, ROBERT P., *Fuels and Their Combustion*. New York: McGraw-Hill, 1926.

HOUGEN, O. A., and WATSON, K. M., *Chemical Process Principles. Part I, Material and Energy Balances*. New York: John Wiley, 1943.

LEWIS, WARREN K., and RADASCH, ARTHUR H., *Industrial Stoichiometry*. New York: McGraw-Hill, 1926.

PERRY, JOHN H., *Chemical Engineers' Handbook*. New York: McGraw-Hill, 1949.

TRINKS, W., *Industrial Furnaces*, Vols. I and II. New York: John Wiley, 1950 and 1947.

PROBLEMS

5–1 Producer gas (30% CO, 10% CO_2, 15% H_2, 45% N_2) is burned with dry air, and the Orsat analysis of the flue gas is 15.3% CO_2, 7.5% O_2, 77.2% N_2. Calculate (a) cu ft of dry flue gas per cu ft dry producer gas, both volumes measured at STP, (b) % H_2O in flue gas, by volume, (c) cu ft of air used per cu ft of producer gas (both STP), and (d) % excess air.

5–2 A furnace burns coal of the following ultimate analysis: 72.0% C, 8.0% H, 6.4% O, 1.4% N, 3.2% S, and 9% ash. The moisture content as fired is 3.0%. The barometer is 735 mm Hg, the air temperature is 72°F, and the relative humidity of the air is 45%. The Orsat analysis of the flue gas is 10.5% CO_2, 2.1% CO, 6.0% O_2, and 81.4% N_2. Calculate (a) % excess air, (b) air consumption, actual ft³/lb coal, (c) total volume flue gases, wet basis, ft³/lb coal (STP), and (d) the dew point of the flue gases, °F.

5–3 A melting furnace burns 1500 lb/hr of pulverized coal with 2% moisture and the following ultimate analysis: 76% C, 6% H, 5% O, 1% N, 2% S, and 10% ash. In normal operation, the Orsat analysis of the flue gas is: 13.0% CO_2, 0.5% CO, 6.0% O_2, and 80.5% N_2. However, when the furnace door is opened for charging, with firing rate and burner air adjustment unchanged, the analysis of the flue gas becomes: 11.5% CO_2, 8.2% O_2, and 80.3% N_2. Calculate (a) air leakage through door, ft³/min, and (b) air leakage through door, % of theoretical air for combustion.

5–4 A natural gas analyzing 85% CH_4, 5% C_2H_6, and 10% N_2 is burned with automatic control of combustion air to maintain 2% O_2 (dry basis) in the products of combustion. Assuming complete combustion, calculate (a) flue gas analysis, dry basis, and (b) % excess air.

5–5 Calculate the adiabatic combustion temperature for combustion of a blast furnace gas with 30% excess air. The blast furnace gas is preheated to 500°C and the air is at 25°C before combustion. The blast furnace gas has the following analysis: 12% CO_2, 24% CO, 4% H_2, 60% N_2.

5–6 A fuel oil analyzing 87% C, 11% H, and 2% O_2 is burned in a furnace and the products of combustion analyze as follows: 12.7% CO_2, 1.5% CO, 3.3% O_2, and 82.5% N_2. The furnace operates with a flue gas temperature estimated to be 2300°F. The gross calorific power of the oil is 19,500 Btu/lb. Calculate (a) the gross available heat furnished to the furnace, Btu/lb fuel oil; (b) the gross available heat which would be furnished to the furnace if complete combustion (i.e., no CO in flue gases) were obtained with the same fuel-air ratio, Btu/lb fuel oil, and (c) the % saving in fuel if complete combustion were obtained at the same fuel-air ratio.

5-7 A batch reverberatory furnace is to melt 300 tons of copper and heat it to 2500°F, over a period of 6 hr. The heat loss of the furnace to the surroundings over this period is estimated to be about one-half of that required to heat and melt the copper. Fuel oil analyzing 85% C, 12% H, and 3% O and having a low calorific power of 17,000 Btu/lb is to be burned with dry air, 20% in excess of theoretical combustion requirements. The fuel oil costs $0.05/gal. The flue gases from the furnace enter a waste-heat boiler, where the heat loss is estimated at 15% of the sensible heat of the entering flue gases. The stack gases leave the boiler at 600°F.

Estimate (a) average fuel consumption, lb/hr, (b) fuel cost, dollars/ton Cu melted, (c) air supplied to furnace, ft³/min (STP), (d) heat recovered in boiler, Btu/hr, and (e) fraction of calorific value of oil recovered in boiler.

5-8 A continuous smelting furnace is heated with water gas (50% CO, 50% H_2). Fuel and air enter at 77°F. Excess air is 20% of theoretical air. The operation proceeds satisfactorily and at the desired rate with the temperature of the flue gases leaving the furnace equal to 2200°F. What would be the % increase in fuel consumption if the excess air were increased to 50% of the theoretical air?

5-9 One thousand ft³/hr of natural gas (at 77°F, 1 atm) are burned in a remelting furnace which operates 24 hr a day, 300 days a year. The natural gas analyzes 70% CH_4 and 30% C_2H_6. Dry combustion air (at 77°F, 1 atm) is supplied 20% in excess of theoretical requirements. Flue gas enters the stack at 1800°F. It is proposed to install a preheater for the air

which will cool the flue gases from 1800°F to 1000°F. It may be assumed that 90% of the heat removed from the flue gases will be transferred to the preheated air. Calculate the annual fuel saving in dollars which would result from the preheater installation. The natural gas costs $0.30/1000 cu ft.

5-10 A given metallurgical process and furnace has a gross available heat requirement of 5×10^6 Btu/hr and a critical temperature of 2500°F. The furnace is fired with dry pulverized coal containing 70% C and having a low calorific power of 12,000 Btu/lb. The dry combustion air is preheated with heat recovered from the flue gases in a heat exchanger having a relative efficiency of 50%. The flue gases have the following analysis (% by volume on wet basis): 12% CO_2, 5% H_2O, 7% O_2, 76% N_2.

(a) Calculate the coal consumption in lb/hr if the flue gases leave the furnace at the critical process temperature.

(b) Assume the heat exchanger is not used, but instead the coal is burned with a mixture of cold air (77°F) and oxygen. Also the excess of oxygen (in air and pure oxygen combined) over theoretical for combustion will be kept the same as with the preheater. Calculate the quantity of oxygen required in lb/hr to obtain the same fuel consumption as found in part (a) with the preheater.

5-11 A reverberatory furnace is fired with fuel oil analyzing 85% C, 15% H and having a low calorific power of 17,500 Btu/lb. Combustion air at 77°F is 20% in excess of requirements for complete combustion. Products of combustion leave the furnace at 2500°F. Oxygen enrichment of the combustion air is under consideration. To what % O_2 by volume would it be necessary to enrich the combustion air in order to cut the fuel consumption in half? Assume that total oxygen supply is kept 20% in excess of theoretical combustion requirements and that the flue gases remain at 2500°F. Note any additional assumptions made in your calculations.

5-12 A heat balance for a continuous metallurgical process gives the following data:

Heat input	*% of total*
Combustion of fuel	100.0

Heat output	*% of total*
Process requirements	25.0
Sensible heat in flue gases (combustion products)	50.0
Heat loss	25.0
Total	100.0

The installation of an air preheater is being considered. It is estimated that the proposed preheater would recover one-half of the sensible heat in the flue gases and would return this quantity of heat to the furnace in the combustion air.

(a) If the daily process requirements and daily heat loss are kept the same, what percentage saving in fuel could be achieved by the preheater installation? What would be the new heat balance (in percentages)?

(b) If the daily fuel consumption and daily heat loss are kept the same, what percentage increase could be made in heat furnished to the process as a result of the preheater installation? What would be the new heat balance (in percentages) under these conditions?

5–13 Estimate the gross available heat furnished to the working chamber of a basic open-hearth furnace by a coke-oven gas, in (a) Btu/ft³ (STP) and (b) % of low calorific power. The gas analysis and estimated operating conditions are as follows:

Analysis of coke-oven gas: 45% H_2, 35% CH_4, 5% CO, 10% N_2, 5% CO_2

Total air consumption (dry): 125% of theoretical

Cold air leakage to working chamber: 20% of theoretical combustion air

Relative regenerator efficiency: 60%

Critical temperature: 2900°F.

CHAPTER 6

FLUID FLOW

Each of the unit processes for extracting metals involves at least one gas or liquid phase in large quantity. Air is by far the bulkiest raw material in a number of processes and when air is fed to a process commensurate weights and volumes of hot waste gases are usually produced. For example, the production of one ton of pig iron in the blast furnace requires about 2 tons of iron ore and smaller amounts of other solid raw materials, but requires about 4 tons of air and yields about 6 tons of blast furnace gas. When the process itself does not use air, the combustion of fuels to supply process heat actually requires more air than fuel and gives corresponding quantities of flue gases as a waste product. Even the most casual visitor to a metallurgical plant can hardly fail to notice the conspicuous flues, stacks, blowers, and other facilities for handling tremendous volumes of gases.

The hot molten products of various smelting and refining processes, such as liquid metals and slags, are handled by relatively simple and crude methods in tapping furnaces, pouring from ladles, gravity flow in open channels, etc. However, the principles of fluid flow to be discussed in this chapter apply to these liquids as well as to gases and contribute occasionally to the solution of practical problems in dealing with these materials. Water is used for many incidental purposes around metallurgical plants, and the same general principles of fluid flow are used in making the necessary engineering calculations.

The engineer's chief concerns with fluid flow arise in designing and operating the equipment through which and by which fluids are moved. Fluid movement consumes power, which must be furnished through pumps, fans, compressors, and other devices. Measurements and estimations of flow quantities are often necessary and in gas-flow systems particularly these may present serious difficulties. Control of flow rates is also important, as for example in supplying air for combustion of fuels. For such problems as these, the flow systems are in general too complex for detailed analysis in terms of the flow mechanisms. However, most such problems can be dealt with in terms of simple energy balances obtained by applying the principle of conservation of energy to the flow systems. This approach to fluid-flow problems is often associated with the name of Bernoulli, and various forms of energy balances for flow systems are called Bernoulli equations.

Energy balances

Kinds of energy. Figure 6–1 represents in schematic fashion a steady flow system of any kind with flow in the direction from point 1 to point 2. Since the flow is steady,

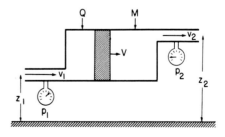

Fig. 6–1. Steady flow system.

the mass rate of flow into the system at point 1 must equal the mass flow rate out of the system at point 2. Also, conditions at any one point in the system are considered not to vary with time. The crosshatched area within the flow system represents a given quantity of fluid, say one pound, during its passage through the system.

Many different kinds of energy and energy changes are to be considered in the flow system shown in Fig. 6–1. In the first place, the entering and leaving fluids possess various kinds of energy, including heat content, energy content, potential energy, kinetic energy, pressure energy, etc. Second, there are such energy exchanges between the flow system and its surroundings as a transfer of heat, or mechanical work added by means of a fan or pump. Third, if the single pound of fluid (crosshatched in the figure) is considered as a system, it may do mechanical work on the fluid surrounding it (by expanding). Finally, within the flow system mechanical energy is converted into heat by friction. For successful use of energy balances in solving engineering problems, a thorough understanding of the various kinds of energy is essential. Care must be taken also to distinguish between energy moving in or out with the fluid, energy exchanges with the surroundings, and energy conversions within the system, since all may appear in the same energy balance.

The customary and convenient unit of energy for flow calculations is the foot-pound, which is defined as the work done by a constant force of one pound on a body when the body moves one foot in the same direction as the force. It is important to keep in mind that this energy unit is based on the use of the pound as the unit of *force*. At the same time, engineers use the pound as the unit of *mass*, and, in fact, in energy balances often take one pound of material

as the basis of calculation so that the terms of the energy balance are in ft-lb *force* per lb *mass*. Accepting the pound as a fundamental unit of *mass*, this practice necessitates the use of a unit conversion factor in such relations as Newton's law, (force) = (mass) × (acceleration), which is then expressed: g_c × (lb *force*) = (lb *mass*) × (ft/sec²). The conversion factor g_c is given by

$$g_c = 32.1740 \frac{\text{lb } mass\text{-ft*}}{\text{lb } force\text{-sec}^2}.$$

Potential energy is energy possessed by the fluid by virtue of its mass, its position, and the effect of gravity; it is energy which can be realized by allowing the fluid to fall toward the earth. Expressed in ft-lb *force*/lb *mass*, the potential energy is numerically equal to height z in feet, above an arbitrary position at which the potential energy is taken to be zero.† Thus in Fig. 6–1, the

---

* The force of gravity on a 1-lb *mass* is exactly 1 lb *force* when the acceleration of gravity g is 32.1740 ft/sec², but this value of g is that measured at sea level and 45° latitude. For other locations, Newton's law gives: lb *force* on a body due to gravity = g/g_c × lb *mass*, in which g takes the value for the location in question but the conversion factor g_c is always 32.1740 lb *mass*-ft/lb *force*-sec². In engineering calculations the small variations of g with altitude and latitude are generally neglected, and both g and g_c are rounded off numerically to 32.2.

If force is expressed in poundals, the factor g_c is not used, so that Newton's law, (force) = (mass) × (acceleration), is expressed: (poundals *force*) = (lb *mass*) × (ft/sec²). In other words, 1 lb *force* = 32.1740 poundals. However, the poundal and the dyne, the corresponding unit in the metric system, are rarely encountered in engineering calculations.

† Strictly speaking, potential energy is z × g/g_c ft-lb *force*/lb *mass*, but since both g and g_c will be given the same numerical value, the ratio g/g_c is omitted in the interest of simplicity.

potential energy of the entering fluid is z_1, that of the leaving fluid is z_2, and the increase in potential energy of the fluid in passing through the system is $z_2 - z_1$.

Pressure energy may be considered as mechanical energy possessed by a fluid because it is under pressure, just as a spring in compression represents a storage of mechanical energy. In setting up an energy balance for the system in Fig. 6–1, the pressure energy of the entering fluid is equal to the work done by the entering fluid on the system at point 1. This work can be considered as the result of a force (from the fluid pressure) acting over a distance (the distance moved by the fluid) and is equal to p_1/ρ_1 per unit mass of fluid, in which p_1 is the pressure and ρ_1 the density of the fluid at point 1. Similarly, the pressure energy of the leaving fluid is p_2/ρ_2 and the increase in pressure energy of the fluid in passing through the system is $(p_2/\rho_2) - (p_1/\rho_1)$. The units of pressure should be lb *force*/ft² and of density lb *mass*/ft³ in order to obtain energy in ft-lb *force*/lb *mass*.

Kinetic energy is energy the fluid has by virtue of its motion. Starting with the familiar relation $KE = \frac{1}{2}mV^2$, it is found that the increase in kinetic energy of a unit mass of fluid in passing through the system from point 1 to point 2 is $(V_2^2 - V_1^2)/2g_c$, in which V_2 and V_1 are the fluid velocities at points 2 and 1, respectively, in feet per second, and g_c is the conversion factor necessary because the pound is used both as a unit of force and as a unit of mass.

Energy content (E) and *heat content* (H) were defined in Chapter 3. The changes in these thermodynamic properties of the fluid in passing through the system, $E_2 - E_1$ and $H_2 - H_1$, should of course be converted into foot-pounds of energy per pound mass of fluid to be dimensionally consistent with the other expressions of flow energy.

Heat absorbed by the fluid-flow system from the surroundings is Q ft-lb *force*/lb *mass* of fluid passed through the system, and has essentially the same significance as Q in the First Law Equation (Chapter 3).

Self-expansion work is the mechanical work which one pound of fluid does on the fluid surrounding it as it expands in passing through the system. This quantity is $\int_1^2 p\,d(1/\rho)$, and is zero if the fluid does not change in density in passing through the system.

Friction within the moving fluid or between the fluid and the apparatus represents a conversion of mechanical energy into heat, and this quantity is designated as F ft-lb *force*/lb *mass* of fluid.

Mechanical energy may be added to the flow from the outside by means of a pump or fan, and this quantity is designated as M ft-lb *force*/lb *mass* of fluid. In some apparatus, for example, a turbine, the system delivers mechanical energy to the surroundings, so that M is negative.

As will be seen, no energy balance will include terms for all the kinds of energy and energy conversions defined above. In fact, for many systems only two or three kinds of energy are significant. Accordingly, the engineer thoroughly familiar with the various kinds of energy is able to set up by inspection the energy balance necessary to the solution of a given problem.

The total energy balance. Application of the principle of conservation of energy to any flow system (Fig. 6–1) leads to the total energy balance as follows:

total energy output = total energy input, (6–1)

$$z_2 + \frac{p_2}{\rho_2} + \frac{V_2^2}{2g_c} + E_2$$
$$= z_1 + \frac{p_1}{\rho_1} + \frac{V_1^2}{2g_c} + E_1 + Q + M. \quad (6\text{–}2)$$

From the definition of heat content, $H = E + p/\rho$,* so that Eq. (6–2) can be placed in the following form, which is more convenient for many purposes:

$$z_2 + \frac{V_2^2}{2g_c} + H_2 = z_1 + \frac{V_1^2}{2g_c} + H_1 + Q + M.$$
$$(6\text{–}3)$$

The total energy balances deal with all forms of energy in and out of the system, both various kinds of mechanical energy and heat, so that terms measuring conversions of one kind of energy to another within the system do not appear in the equations. These total energy balances, Eqs. (6–2) and (6–3), also may be thought of as amplified statements of the First Law of Thermodynamics [compare with Eqs. (3–1) and (3–3) in Chapter 3].

Equations (6–2) and (6–3) are useful for

------

* That is, heat content or enthalpy is the sum of the energy content and the pressure energy.

flow systems in which heat effects and changes in the heat content of the fluid are directly and importantly involved in the flow system. For example, if the flow system is a steam engine, the heat content of the entering steam, acquired in the boiler, is the major input of energy to be balanced against the heat content of the exhaust steam, the mechanical power output of the engine $(-M)$, and the heat loss from the engine $(-Q)$. Other systems for which the total energy balance is useful are the flows of steam and other vapors through pipes and miscellaneous equipment, and various systems with compressed air and other gases, especially those with major pressure or temperature drops through the system. However, for gas flows at substantially atmospheric pressure in furnaces, flues, and ordinary metallurgical equipment, and for liquid-flow systems, the mechanical-energy balance derived in the next section is more useful.

Example 1

Compressed air is withdrawn steadily from a large storage tank through a needle valve and a horizontal $\frac{1}{4}$-inch pipe (inside cross section, 0.00050 sq ft). The air in the tank is at 90°F. In the $\frac{1}{4}$-inch pipe a short distance from the storage tank and needle valve, the air temperature is found to be 70°F, and a gauge at this point shows a pressure of 15 psi (gauge).

Estimate the rate of withdrawal of air, in pounds per minute.

Solution

The total energy balance may be considered from a point 1 within the tank to point 2 in the pipeline, where pressure and temperature are measured. For this system as described above,

$$z_2 - z_1 = 0,$$
$$V_1 = 0,$$
$$M = 0.$$

Since the air in the system is substantially at ambient temperature, and point 2 is close to the tank, it may be assumed that there is no heat exchange with the surroundings, so that

$$Q = 0 \text{ (assumption).}$$

Under these conditions, the total energy balance [from Eq. (6–3)] shows changes only in heat content and kinetic energy:

$$\frac{V_2^2}{2g_c} + H_2 - H_1 = 0.$$

The change in heat content can be estimated from the temperature change and the specific heat of the air (c_p), which is 0.24 Btu/lb per °F.

$$H_2 - H_1 = -(90 - 70)(0.24) = -4.8 \text{ Btu/lb } mass$$

$$= -4.8 \times 778 = -3740 \text{ ft-lb } force/\text{lb } mass.$$

Solving the energy balance,

$$V_2 = \sqrt{2 \times 32.2 \times 3740}$$
$$= 490 \text{ ft/sec.}$$

The density of air at 70°F and a gauge pressure of 15 psi (29.7 psi absolute) is 0.15 lb $mass$/ft^3, so that the required flow rate is found:

$$\text{lb air/min} = 490 \times 60 \times 0.0005 \times 0.15$$
$$= 2.2.$$

Brief consideration of the numerical results of these calculations indicates that the total energy balance will be useful mainly when relatively high velocities and large kinetic energy changes are involved. In this example, the velocity in the pipe was found to be very high (in fact, nearly one-half the velocity of sound), but the temperature change was small enough so that difficulties would be encountered in trying to measure it accurately.

Example 2

An air compressor takes in 100 lb/hr of dry air at 70°F and discharges it at a pressure of 90 psi absolute and at a temperature of 350°F. The compressor is driven by an electric motor, and the metered electrical input to the motor is 5 kw.

Calculate the quantity of heat given off to the surroundings by the compressor and electric motor, in Btu/hr.

Solution

For the system described, it is reasonable to assume that the potential and kinetic energy changes can be ignored, so that the total energy balance becomes

$$H_2 - H_1 - Q - M = 0.$$

The mechanical work input (M) is the electrical energy input, or

$$M = 5 \text{ kw-hr/100 lb of air}$$
$$= 133,000 \text{ ft-lb } force/\text{lb } mass \text{ of air.}$$

The increase in heat content ($H_2 - H_1$) is estimated from the temperature rise and the specific heat: *

---

* The heat content of an ideal gas is independent of pressure, so $H_2 - H_1$ for the change from 70°F, 1 atm to 350°F, 90 psi for an ideal gas is the same as $H_2 - H_1$ for the change from 70°F, 1 atm to 350°F, 1 atm. However, air departs from ideal behavior sufficiently in the range from 14.7 psi to 90 psi so that the calculation of $H_2 - H_1$ from specific heat at constant pressure c_p is only an approximation.

$$H_2 - H_1 = (350 - 70)(0.24)$$
$$= 67.2 \text{ Btu/lb}$$
$$= 52,300 \text{ ft-lb } \textit{force}/\text{lb } \textit{mass}.$$

Solving the energy balance,

$$-Q = 133,000 - 52,300$$
$$= 80,700 \text{ ft-lb } \textit{force}/\text{lb } \textit{mass}$$
$$= 104 \text{ Btu/lb}.$$

For a flow of 100 lb of air/hr, then, the total heat given off by the compressor and motor will be about <u>10,000 Btu/hr.</u>

Example 3

A steam engine receives from a boiler 500 lb/hr of saturated steam at 200 psi absolute and 382°F. The steam is discharged at atmospheric pressure (14.7 psi) and at 212°F, and as discharged is 8% liquid water, 92% gaseous water.

Estimate the horsepower developed by the engine.

Solution

For a steam engine, the changes in potential and kinetic energy from inlet to outlet are generally small, so in Eq. (6–3), $z_2 - z_1 = 0$ and $(V_2^2 - V_1^2)/2g_c = 0$. Also, for an estimate of power developed, we may assume that the heat loss from the engine to the surroundings is small compared with the work output of the engine. Applying these assumptions, the total energy balance is simplified to

$$M = H_2 - H_1.$$

The change in heat content, $H_2 - H_1$ can be obtained accurately from "steam tables" or a "Mollier Chart" which give heat content and other thermodynamic properties of steam over the whole range of conditions of steam applications. From standard steam tables, $H_2 - H_1$ is found to be -125.3 Btu/lb.

In the absence of steam tables, $H_2 - H_1$ can be estimated, assuming the steam behaves as an ideal gas so that heat content is independent of pressure. On this basis $H_2 - H_1$ can be found by adding (1) ΔH for cooling 1 lb of H_2O (gas) from 382°F to 212°F, and (2) ΔH for condensing 0.08 lb H_2O (gas) to H_2O (liq) at 212°F. This calculation with heat-content data from the Appendix (Table A–3) gives $H_2 - H_1 = -156$ Btu/lb. This result is about 25% higher than the enthalpy change found from the steam tables. The difference is due to the assumption in the calculation that H_2O (gas) behaves as an ideal gas, that is, it was assumed that its heat content at 382°F and 200 psi was the same as at 382°F and 1 atm.

Solving the energy balance for M (using the value of $H_2 - H_1$ from the steam tables),

$$-M = 125.3 \text{ Btu/lb}$$
$$= 97,500 \text{ ft-lb } \textit{force}/\text{lb } \textit{mass}.$$

With steam supplied at the rate of 500 lb/hr,

$$\text{estimated horsepower} = \frac{500 \times 97{,}500}{60 \times 33{,}000}$$
$$= \underline{24.6}.$$

The mechanical energy balance. The input and output of mechanical energy alone cannot be equated for a flow system because of the conversions of heat into mechanical energy and other energy conversions occurring within the system. However, the very useful mechanical energy balance can be derived by allowing for these conversions, as follows:

mechanical energy output + mechanical energy converted to heat = mechanical energy input + other energy converted to mechanical energy.　　(6–4)

The mechanical energy turned into heat is friction (F), and the other forms of energy are turned into mechanical energy by expansion of the fluid, furnishing the self-expansion work. Accordingly, the complete mechanical energy balance becomes

$$z_2 + \frac{p_2}{\rho_2} + \frac{V_2^2}{2g_c} + F = z_1 + \frac{p_1}{\rho_1}$$
$$+ \frac{V_1^2}{2g_c} + M + \int_1^2 p\, d\!\left(\frac{1}{\rho}\right), \quad (6–5)$$

or

$$z_2 - z_1 + \frac{p_2}{\rho_2} - \frac{p_1}{\rho_1} + \frac{V_2^2 - V_1^2}{2g_c} -$$
$$\int_1^2 p\, d\!\left(\frac{1}{\rho}\right) + F - M = 0. \quad (6–6)$$

Since

$$\frac{p_2}{\rho_2} - \frac{p_1}{\rho_1} = \int_1^2 p\, d\!\left(\frac{1}{\rho}\right) + \int_1^2 \left(\frac{1}{\rho}\right) dp,$$

pressure energy and self-expansion energy can be combined in one term to obtain:

$$z_2 - z_1 + \frac{V_2^2 - V_1^2}{2g_c} +$$
$$\int_1^2 \left(\frac{1}{\rho}\right) dp + F - M = 0. \quad (6–7)$$

For liquids (that is, incompressible fluids), the self-expansion energy is negligible and the density is constant, so that Eq. (6–6) becomes

$$z_2 - z_1 + \frac{p_2 - p_1}{\rho} + \frac{V_2^2 - V_1^2}{2g_c}$$
$$+ F - M = 0. \quad (6–8)$$

For gas flows at constant temperature and substantially at atmospheric pressure (for example, in air ducts), the density is practically constant and Eq. (6–8) is again applicable. However, when the gas density varies through the flow system it becomes necessary to evaluate the integral in Eq. (6–7). For a rigorous solution, quantitative knowledge of the relation between density and pressure is necessary. In the absence of this information, the integral can be estimated by

$$\int_1^2 \left(\frac{1}{\rho}\right) dp = (p_2 - p_1)\left(\frac{1}{\rho}\right)_{\text{average}},$$

so that numerical solution requires a choice of the right kind of *average reciprocal density*.* For example, if a flue gas passes through a long duct at substantially atmospheric pressure and cools during passage so that the pressure and specific volume both vary linearly with distance (and therefore linearly with each other), it can be shown that the arithmetic average $(1/\rho_2 + 1/\rho_1)/2$ should be used. This arithmetic average of the reciprocal densities is a good approximation for many gas-flow systems and there-

---

* Note that $1/\rho$ is the specific volume in cubic feet per pound, so that $(1/\rho)_{\text{average}}$ is the average specific volume.

fore is commonly used, especially when the density change is not large.* Accordingly, a generally used form of the mechanical energy balance for gas-flow systems is

$$z_2 - z_1 + (p_2 - p_1)\left(\frac{1}{\rho}\right)_{\text{average}}$$

$$+ \frac{V_2^2 - V_1^2}{2g_c} + F - M = 0. \quad (6\text{-}9)$$

For many types of flow systems, one or more of the terms in Eq. (6-9) can be neglected. Thus where the inlet and outlet are at the same height, the potential energy change $(z_2 - z_1)$ is zero. The mechanical energy input M enters the balance only when a fan, turbine, or other similar item of mechanical equipment is present. Also, as will be seen, friction (F) can be neglected in some cases without serious errors. In other cases the friction term is omitted and an empirical correction factor is applied to the result calculated on the assumption of a frictionless system.

The mechanical energy balances as stated in Eqs. (6-5) to (6-9) are easily derived and are in simple mathematical form. However, application of these relations to actual engineering situations in order to obtain useful quantitative results or estimates requires skill and judgment in translating general information as to how the fluid is moving into specific terms of the various kinds of energy involved. Accordingly, the succeeding parts of this chapter consist mainly of detailed developments of the applications of the energy equations just derived to flow systems common in metallurgical plants. Although a number of formulas are derived by which a

variety of problems can be solved by simple substitution and "turning the crank," it should be emphasized that a thorough understanding of the kind of thinking used in deriving the formulas is much more important than knowledge of the formulas (which are more concisely presented in engineering handbooks, anyway) in the solution of new problems.

As the student goes through the subsequent sections dealing with specific flow systems, he will find it helpful to refer back frequently to the complete mechanical energy balances, Eqs. (6-6) to (6-8), as check lists of the forms of energy which have to be considered. Thus, a logical way of examining a flow system involves asking and answering the following questions: First, what is the system (sketch essential features)? Is this a steady system, with the same mass flow rates at 1 and 2? Second, checking off the energy terms one by one, is there a change in z from 1 to 2 and thus a change in potential energy? Is there a change in pressure energy ("yes" for most systems) and is it known or is it to be calculated? Are V_1 and V_2 equal or is there a significant kinetic energy change? And so on for each term in Eq. (6-6). If the student does not learn a mental procedure of this kind, it can almost be guaranteed that fluid-flow problems will be difficult just because there are a half-dozen or so different kinds of energy to be considered. Moreover, this mental procedure is taken for granted in the abbreviated derivations of energy-balance equations for specific systems in subsequent sections, starting with Eq. (6-28).

Friction

In deriving the mechanical energy balances, friction was defined simply as the quantity of mechanical energy converted

---

* Other methods of evaluating the integral, $\int_1^2 \frac{1}{\rho}\, dp$, are discussed later in the section on *Compressed Air and Blast.*

into heat within the flow system. Actually this definition is equivalent to defining the term F in the mechanical energy balance as the quantity to be added to the equation to make it balance. The mechanisms by which mechanical energy is dissipated are very complex, so that theoretical calculations of friction are almost impossible except for the most ideal systems. Accordingly, friction is evaluated for engineering purposes by empirical and semi-empirical methods; the primary purpose of this section is to present these methods of evaluating the friction term F in the mechanical energy balance.

Viscosity; internal friction. Viscosity is a property of a fluid which can be defined in terms of the system shown in Fig. 6-2. The fluid is contained between two parallel plates, the bottom one stationary and the top one moving with a velocity u. Under the ideal conditions of streamline or laminar flow, the fluid flow is in the nature of shearing along planes parallel to the two confining plates, and the fluid velocity varies linearly with distance between the plates from 0 at the bottom to u at the top. A force f must be applied to the top plate in the direction of movement and an equal force to the bottom plate in the opposite direction to maintain a steady flow as described. The work done by this force is the work necessary to overcome the internal friction of the fluid. For a given fluid, regardless of the dimensions of the system, the force required is found to be proportional to the velocity gradient, so that the

viscosity (μ) can be defined as

$$\mu = \frac{f/A}{du/dx} g_c, *\qquad (6\text{--}10)$$

in which f/A is the shear force per unit area and du/dx is the velocity gradient at right angles to the direction of flow. The cgs unit of viscosity is the poise, for which f is in dynes, A in sq cm, u in cm/sec, and x in cm. The more convenient unit for engineering purposes has the dimensions of lb *mass*/ft-sec (1 poise = 1 g/cm-sec = 0.0672 lb *mass*/ft-sec).

For the flow system in Fig. 6-2, it can be shown that the rate at which mechanical energy is dissipated through friction, per unit mass of fluid, is

$$\frac{F}{\theta} = \frac{\mu}{\rho g_c}\left(\frac{du}{dx}\right)^2, \qquad (6\text{--}11)$$

in which F is the frictional work per unit mass of fluid as defined previously (ft-lb *force*/lb *mass*), θ is time (sec), μ is viscosity (lb *mass*/ft-sec), ρ is the fluid density (lb *mass*/ft³), (du/dx) is the velocity gradient (ft/sec-ft), and g_c is the conversion factor, 32.2 lb *mass*-ft/lb *force*-sec.² The ratio μ/ρ is known as the kinematic viscosity ν. Equation (6-11) thus shows that the rate of dissipation of mechanical energy by friction in simple laminar flow is proportional to the kinematic viscosity and to the square of the rate of shear of the fluid.

When flow systems more complex than the one in Fig. 6-2 are considered, it becomes difficult to derive on theoretical grounds any relation between friction energy, viscosity, and flow characteristics such as the one in Eq. (6-11). However, it is clear that the property of fluid viscosity will, in general, play an important role in

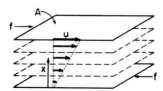

FIG. 6-2. Simple laminar flow.

---

* If cgs units are used, the unit conversion g_c is omitted.

evaluating frictional losses in common engineering flow systems.

The kinetic theory of gases sheds some light on the nature of internal friction or viscosity (see, for example, Sears, *Mechanics, Heat, and Sound*, Chapter 25, Addison-Wesley Press). According to the kinetic theory, under ideal conditions the viscosity of a gas should be independent of pressure and should be proportional to the square root of the absolute temperature. Actual gases deviate considerably from this behavior, especially at very low and very high pressures, but for gases at substantially 1 atm the following empirical relation is useful:

$$\frac{\mu}{\mu_0} = \left(\frac{T}{273}\right)^n. \qquad (6\text{-}12)$$

In this relation μ is the viscosity at a temperature of $T°$K, μ_0 is the viscosity at 0°C (273°K), and n is an empirical constant. For air, $\mu_0 = 1.15 \times 10^{-5}$ lb *mass*/ft-sec and $n = 0.768$. Viscosities of the common gases are not far different from each other, so that for engineering calculations in the absence of precise data, Eq. (6–12) may be simplified to give a good approximation of the viscosity of air, flue gases, and other common gases at about 1 atm as follows:

$$\mu = 1.3 \times 10^{-7} \times T^{0.8} \text{ lb } mass/\text{ft-sec}. \qquad (6\text{-}13)$$

Dimensional analysis. Frictional work in a flow system is complexly dependent on fluid viscosity, density, velocity, velocity gradients, and several other characteristics of the system. At first glance it might seem that friction is a function of so many variables that each system would require separate experimental study. However, application of the methods of dimensional analysis results in a considerable simplification and in effect reduces the number of variables to be related experimentally to a small

number, only two in many cases. These methods are useful not only in dealing with fluid flow but also in dealing with heat flow and other complex phenomena. The usefulness of dimensional analysis is an expression of the fact that any one system can be considered as a model of many other systems, and dimensional analysis indicates simple relationships between the behavior of models. Thus, for example, it becomes possible to use the results of laboratory experiments on water flow in 1-inch pipes to predict the behavior of hot flue gases in large ducts.

All the variables in physical systems, such as mass, length, time, area, volume, density, viscosity, velocity, energy, etc., are measured quantitatively in units such as pounds, feet, seconds, square feet, etc. Each of these units in turn can be expressed in terms of a small number of primary units. For mechanical systems the dimensions of these primary units can be taken as mass (M), length (L), and time (θ); when heat quantities are involved, a fourth dimension is necessary and is usually taken as temperature (T). Table 6–1 gives the dimensions of common variables in flow systems.

Quantitative relationships and laws relating physical variables are generally written as algebraic equations in which the various quantities appear with various exponents. A fundamental requirement of such equations is that they must be dimensionally homogeneous; that is, each term must have

TABLE 6-1

Dimensions of Quantities in Fluid-Flow Systems

(Mass, length, and time as fundamental dimensions)

Mass	M	Energy	$ML^2\theta^{-2}$
Length	L	Density (ρ)	ML^{-3}
Time	θ	Pressure	$ML^{-1}\theta^{-2}$
Velocity	$L\theta^{-1}$	Mass velocity	$ML^{-2}\theta^{-1}$
Acceleration	$L\theta^{-2}$	Viscosity	$ML^{-1}\theta^{-1}$
Force	$ML\theta^{-2}$	Energy per unit mass	$L^2\theta^{-2}$

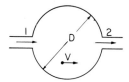

FIG. 6–3. Flow system with one characteristic dimension D.

the same dimensions. It is the application of this principle of dimensional homogeneity that effects the simplifications in correlating data on flow systems and other systems without which many engineering calculations could not be made.

Dimensional analysis for a steady flow. Figure 6–3 represents a fluid flow of fixed geometry or shape, so that only one characteristic linear dimension (D) need be given to specify the system. With the geometry fixed, a suitable measure of the flow through the system is the velocity at some point or the average velocity at a given section (V). It is reasonable to assume that the frictional energy loss per unit mass of fluid through the system (F') will depend on D, V, and on the properties of the fluid, density (ρ) and viscosity (μ). Since the form of the relationship is unknown and probably very complex, it can be expressed functionally

$$F' = \psi_1(D, V, \rho, \mu). \qquad (6\text{--}14)$$

This functional equation can be expressed in the form of an infinite series

$$F' = \alpha D^a V^b \rho^c \mu^d + \beta D^{a'} V^{b'} \rho^{c'} \mu^{d'}$$
$$+ \gamma D^{a''} V^{b''} \rho^{c''} \mu^{d''} \ldots, \qquad (6\text{--}15)$$

in which α, β, and γ are dimensionless coefficients and a, a', a'', b, b', b'', etc., are dimensionless exponents. By proper choice of the values for these dimensionless numbers, it is clear that an infinite series can be set up to represent quantitatively the actual relation between F' and the four properties of the system.

The principle of dimensional homogeneity can now be applied to Eq. (6–15). To start with, only one term in the series need be considered. Assigning the proper dimensions to each of the five variables, we obtain

$$(L^2\theta^{-2}) = (L)^a(L\theta^{-1})^b(ML^{-3})^c(ML^{-1}\theta^{-1})^d. \qquad (6\text{--}16)$$

Dimensional homogeneity requires that the combined exponents of M, L, and θ each be the same in both terms of the equation. Accordingly, equating the exponents for each of the three dimensions,

$$M: \ 0 = c + d,$$
$$L: \ 2 = a + b - 3c - d, \qquad (6\text{--}17)$$
$$\theta: -2 = -b - d.$$

Three simultaneous equations in four unknowns cannot be solved numerically, but can be solved to express all the unknowns in terms of one. Solving Eq. (6–17) in this way, and expressing all the exponents in terms of c,

$$a = c; \quad b = 2 + c; \quad d = -c. \qquad (6\text{--}18)$$

Similar relations are found for the other terms of the infinite series in Eq. (6–15), so that this equation can now be rewritten:

$$F' = \alpha D^c V^{2+c} \rho^c \mu^{-c} + \beta D^{c'} V^{2+c'} \rho^{c'} \mu^{-c'}$$
$$+ \gamma D^{c''} V^{2+c''} \rho^{c''} \mu^{-c''} + \cdots, \qquad (6\text{--}19)$$

or

$$\frac{F'}{V^2} = \alpha \left(\frac{DV\rho}{\mu}\right)^c + \beta \left(\frac{DV\rho}{\mu}\right)^{c'}$$
$$+ \gamma \left(\frac{DV\rho}{\mu}\right)^{c''} + \cdots \qquad (6\text{--}20)$$

Returning to the functional form of equation,

$$\frac{F'}{V^2} = \psi_2 \left(\frac{DV\rho}{\mu}\right). \qquad (6\text{--}21)$$

A reconsideration of the dimensions of the terms in Eq. (6–21) shows that the term (F'/V^2) and the term ($DV\rho/\mu$) are both dimensionless, so that the final result of applying dimensional analysis to this system is a general relation between two dimensionless variables. That is, instead of having the nearly impossible problem of measuring and tabulating the relationships between five dimensional variables (F', D, V, ρ, and μ) in all their permutations and combinations, we find it necessary only to measure a relationship between two dimensionless variables, (F'/V^2) and $(DV\rho/\mu)$.

In Table 6–1 and in the foregoing analysis, F' was given the absolute units of energy per unit mass to maintain consistency. If, however, the pound is used both as a unit of mass and as a unit of force, the unit conversion factor g_c is necessary, and Eq. (6–21) can now be rewritten:

$$F = \phi \frac{V^2}{2g_c},\qquad (6\text{--}22)$$

in which $\phi/2$ represents the function ψ_2 $(DV\rho/\mu)$ of Eq. (6–21). In this equation F is in ft-lb *force*/lb *mass*, V is in ft/sec, g_c is in lb *mass*-ft/lb *force*-sec^2, and ϕ, of course, is dimensionless. This equation suggests a rather direct relationship between frictional loss and kinetic energy of the flow, so that it is reasonable in some systems to regard friction as dissipation of the kinetic energy of the flow and to consider the dimensionless factor ϕ as the fractional kinetic-energy loss through friction.

The dimensionless number, $DV\rho/\mu$, for a flow system is known as the Reynolds Number (N_{Re}), and is a very important quantity in fluid-flow problems. The previous derivations have shown that *for a given geometry of flow system* there must exist a single relationship between the two dimensionless variables, ϕ and N_{Re}. This relationship can be determined experimentally for a few fluids and a few values of size of system (D) and fluid velocity (V) and then applied with

some confidence to predict the behavior of still other systems of the same geometry.

Pipes and ducts. The analysis in the previous section applied to a system of fixed geometry with one characteristic dimension (D). When friction in a pipe is considered, obviously two dimensions must be considered (length and diameter). Moreover it is clear from simple considerations of the nature of the frictional loss in long pipes that F should be directly proportional to the length (l) of the pipe. When l is introduced as an additional variable, dimensional analysis leads to the following relation, corresponding to Eq. (6–22) for a single size parameter:

$$F = f \frac{V^2}{2g_c} \frac{l}{D}. \qquad (6\text{--}23)$$

In this equation, f represents a function of the Reynolds number, $\psi_3(DV\rho/\mu)$. For flow in circular pipes, the dimensionless number f is known as the *friction factor*, and is defined [*] by Eq. (6–23), so that

$$f = \frac{2g_c}{V^2} \frac{D}{l} F. \qquad (6\text{--}24)$$

For flows of noncircular cross section, the size parameter usually employed is the *hydraulic radius R_H*, defined as the ratio of cross-sectional area (at right angles to the flow) to the wetted perimeter of the flow.[†] Using this parameter,

$$f \text{ (friction factor)} = \frac{8g_c}{V^2} \frac{R_H}{l} F \qquad (6\text{--}25)$$

---

[*] Care must be taken to avoid confusion here, because in some engineering fields it is customary to use a friction factor f' which is numerically equal to $f/4$. For this definition of friction factor, Eq. (6–23) would be written:

$$F = 4f' \frac{V^2}{2g_c} \frac{l}{D}.$$

[†] For a circular pipe, $R_H = D/4$.

and

$$N_{\mathrm{Re}} \text{ (Reynolds number)} = \frac{4R_H V \rho}{\mu}. \quad (6\text{--}26)$$

Equations (6–24) and (6–25), rearranged, give the well-known Fanning equations:

$$\frac{F}{l} = f \frac{V^2}{2g_c} \frac{1}{D} = \frac{f}{4} \frac{V^2}{2g_c} \frac{1}{R_H}, \quad (6\text{--}27)$$

or

$$F = f \frac{V^2}{2g_c} \frac{l}{D} = \frac{f}{4} \frac{V^2}{2g_c} \frac{l}{R_H}.$$

For a *fluid of constant density flowing in a long horizontal, pipe of constant cross section,* a mechanical energy balance [from Eq. (6–8)] is

$$\frac{\Delta p}{\rho} + F = 0, \quad (6\text{--}28)$$

which, combined with Eq. (6–27), gives the pressure drop per unit length $-\Delta p/l$ due to friction:

$$\frac{-\Delta p}{l} = f \frac{V^2}{2g_c} \frac{\rho}{D} = \frac{f}{4} \frac{V^2}{2g_c} \frac{\rho}{R_H}. \quad (6\text{--}29)$$

In the above relations for flow in pipes, the velocity V is the average fluid velocity over the whole flow cross section, which can be found by dividing the total volume rate of flow (ft³/sec) by the cross-sectional area of the flow (sq ft). The product $V\rho$ thus is the mass velocity G (lb fluid per sec/sq ft of cross section). Substitution of G for $V\rho$, to give $N_{\mathrm{Re}} = DG/\mu$, often simplifies calculations. For example, with gas flowing in a duct of constant cross section, both V and ρ vary with temperature of the gas but G does not.

Figure 6–4 shows graphically the well-established empirical relationship between friction factor (f) and Reynolds number (N_{Re}) for straight ducts. This relation is based on hundreds of measurements by

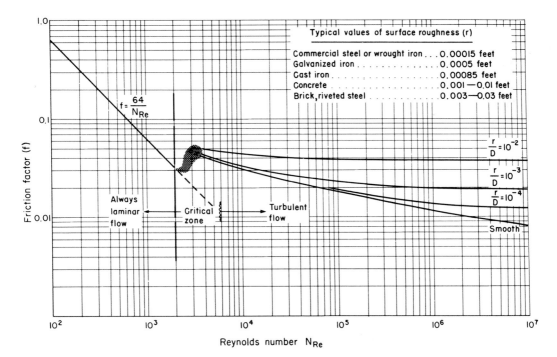

FIG. 6–4. Friction factor vs. Reynolds number in long, straight ducts (after Moody).

independent investigators, using gases and liquids of widely different densities and viscosities and covering wide ranges of fluid velocities and pipe sizes, so that it can be used with confidence for all problems of flow in long straight ducts. This chart is based on data for ducts of circular cross section, but it serves satisfactorily for engineering calculations of friction in noncircular ducts if the size is expressed in terms of the hydraulic radius R_H, as indicated in Eqs. (6–25) to (6–29). It will be noted that curves are given for different degrees of pipe roughness r/D over part of the range. The roughness of a pipe surface can be measured by the average size r (ft) of the surface irregularities. When the ratio of r to D (pipe diameter) is used as a dimensionless parameter, the simple correlation in Fig. 6–4 results. The example below illustrates the method of using Fig. 6–4 to estimate frictional losses.

EXAMPLE

30,000 ft³/min (STP) of flue gases at about 600°F pass through a horizontal flue of circular cross section, 175 ft long and 8 ft in diameter. Estimate the following: (a) friction loss F, in ft-lb *force*/lb *mass* of flue gas, (b) equivalent power consumption by friction, in hp, and (c) pressure drop, in inches of water.

In the absence of detailed data, it may be assumed that the temperature change of the flue gases is small, so that average density and viscosity can be calculated at 600°F.

The first step is calculation of the Reynolds number DG/μ:

$$D = 8 \text{ ft},$$

$$G = \frac{30,000}{60} \times \frac{30^*}{359} \times \frac{4}{\pi(8)^2} = 0.83 \text{ lb } mass/\text{sec-ft}^2,$$

$$600°F = 590°K,$$

$$\mu = 1.3 \times 10^{-7} \times (590)^{0.8} = 2.15 \times 10^{-5} \text{ lb } mass/\text{ft-sec},$$

$$N_{\text{Re}} = \frac{8 \times 0.83}{2.15 \times 10^{-5}} = 3.1 \times 10^5.$$

Now the friction factor f for this value of Reynolds number can be found from Fig. 6–4 (curve for smooth pipes):

$$f = 0.0145.$$

The friction loss F can now be found by using the Fanning equation (6–27):

$$l = 175, \quad D = 8,$$

$$V = \frac{30,000}{60} \times \frac{4}{\pi(8)^2} \times \frac{590}{273} = 21.4 \text{ ft/sec}.$$

(a) $F = 0.0145 \times \dfrac{(21.4)^2}{64.4} \times \dfrac{175}{8} = \underline{2.26 \text{ ft-lb } force/\text{lb } mass}.$

---

* Assumed average molecular weight of flue gases = 30.

The total rate of energy loss can now be found:

$$2.26 \times 30{,}000 \times \frac{30}{359} = 5{,}670 \text{ ft-lb/min.}$$

(b) $\dfrac{5670}{33{,}000} = \underline{0.17 \text{ hp.}}$

From the mechanical energy balance, Eq. (6–28), the pressure drop $(-\Delta p)$ equals ρF,

$$-\Delta p = \frac{30}{359} \times \frac{492}{1060} \times 2.26 = 0.09 \text{ lb } force/\text{ft}^2.$$

(c) Taking the weight of water as 62.4 lb $force$/ft³, the pressure drop is converted to inches water:

$$0.09 \times \frac{12}{62.4} = \underline{0.017 \text{ inch water.}}$$

Types of flow; critical Reynolds numbers. Two distinct types of flow are encountered in fluid-flow systems, *streamline* and *turbulent*. Under conditions of streamline flow in a pipe, the fluid particles move in straight lines parallel to the pipe axis. This type of flow is also known as laminar flow or as viscous flow. For a pipe of circular cross section, the pressure drop due to internal friction in streamline flow can be calculated rigorously, starting with the definition of viscosity, and gives the equation of Poiseuille:

$$\frac{-\Delta p}{l} = \frac{32\mu V}{g_c D^2}. \qquad (6\text{–}30)$$

By combining this equation with Eq. (6–29), it is found that for streamline flow in pipes,

$$f = \frac{64}{N_{\text{Re}}}, \qquad (6\text{–}31)$$

which represents the straight line of slope -1 to the left on the log-log plot in Fig. 6–4. Experiments have established that streamline flow is always obtained in circular pipes when the Reynolds number is less than about 2100, and this value is known as the lower critical Reynolds number. Small Reynolds numbers and streamline flow are favored by high values of μ and low values of D, V, and ρ, so that they are most easily obtained in the laboratory in capillary tubes. On the other hand, streamline flow in industrial ducts and pipes is very uncommon.

Turbulent flow is characterized by irregular eddies and velocity fluctuations, that is, by turbulence. Fluid flow in circular pipes is generally turbulent for Reynolds numbers above about 3000, but by careful laboratory manipulation the onset of turbulence sometimes has been prevented with Reynolds numbers as high as 13,000 in smooth pipes. Surface roughness causes the onset of turbulence at lower Reynolds numbers. A useful empirical equation for long smooth pipes, with Reynolds numbers between 10^4 and 10^6, is

$$f = 0.185 \, (N_{\text{Re}})^{-0.2}. \qquad (6\text{–}32)$$

Reference to Fig. 6–4, or comparison of Eqs. (6–31) and (6–32), shows that f is relatively insensitive to changes in N_{Re} for turbulent flow as compared with streamline flow, especially in rough pipes.

Between $N_{\text{Re}} = 2100$ and $N_{\text{Re}} =$ about

4000 for commercial pipes is the transition region in which the flow is sometimes stream-line, sometimes turbulent. No simple and clear-cut relation between f and N_{Re} can be given for this flow region.

Pipe fittings and bends. A fitting, bend, or other fixture in a long pipe creates additional flow disturbance and increases the friction loss. If necessary another f vs. N_{Re} correlation could be set up and measured for each kind of such fixture. In the first place, however, the Reynolds number can be given the same numerical value as that for the pipe by taking the pipe diameter D as the characteristic dimension and the average velocity V in the pipe as the characteristic velocity. When this procedure is followed, it has been found further that for turbulent flow the Fanning friction factor f given by Fig. 6–4 and the Fanning equation (6–27) can be used to calculate friction, simply by considering the friction as being due to an additional length of pipe expressed in pipe diameters. Thus a standard 45° elbow for a given pipe diameter adds a frictional loss equivalent to that in a length of pipe equal to 15 times the diameter. Calculation of friction losses in this way amounts to the definition of $(l/D)_{equiv}$ by comparing Eqs. (6–22) and (6–27):

$$\phi = f\left(\frac{l}{D}\right)_{equiv}. \qquad (6\text{–}33)$$

In this relation ϕ is the fractional kinetic energy loss in the fitting through friction, f is the friction factor for the flow in the pipe itself, and $(l/D)_{equiv}$ is a dimensionless factor characteristic of the fitting. Values of $(l/D)_{equiv}$ for various fittings are given in Table 6–2. When friction is calculated for a complete pipe system, the most convenient procedure is to add (l/D) for the whole run of pipe to $\Sigma(l/D)_{equiv}$ for the fittings and then substitute the sum for the (l/D) term in

TABLE 6-2

Friction Losses in Screwed Fittings and Valves*

(Turbulent flow only)

Type of fitting or valve	Equivalent length in pipe diameters: $\left(\frac{l}{D}\right)_{equiv}$
45° elbows	15
90° elbows, standard radius	32
90° elbows, medium radius	26
90° elbows, long sweep	20
90° square elbow	75
180° close return bends	75
180° medium radius return bends	50
Tee (used as elbow, entering run)	60
Tee (used as elbow, entering branch)	90
Couplings	Negligible
Unions	Negligible
Gate valves, open	7
Gate valves, one-fourth closed	40
Gate valves, one-half closed	200
Gate valves, three-fourths closed	800
Globe valves, open	300
Angle valves, open	170
Water meters, disk	400
Water meters, piston	600
Water meters, impulse wheel	300

*After J. H. Perry, Chemical Engineers' Handbook. New York: McGraw-Hill, 1949.

solving the Fanning equation (6–27) for F.

As the curves in Fig. 6–4 indicate, the friction factor f for fluid flow in pipes does not vary greatly for relatively large changes in Reynolds number in the turbulent-flow region. Similarly, as long as turbulent flow is considered, the factor ϕ in Eq. (6–22) does not vary greatly with Reynolds number. Accordingly, for many systems the friction loss can be estimated satisfactorily in terms of a constant value of ϕ, representing the fractional loss of kinetic energy through friction. Figure 6–5 gives values of ϕ for various types of bends in ducts. Values of ϕ for various kinds of entrance and exit openings and for changes in flow cross section are discussed later and are shown in Figs. 6–18

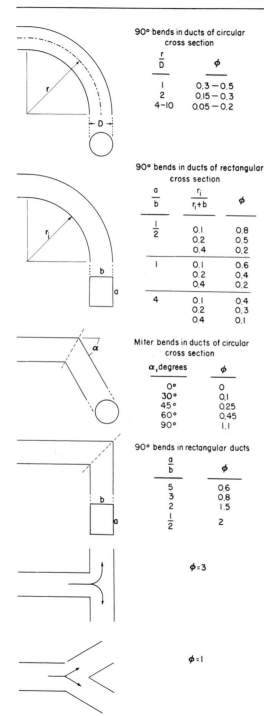

90° bends in ducts of circular cross section

$\dfrac{r}{D}$	ϕ
1	0.3 − 0.5
2	0.15 − 0.3
4–10	0.05 − 0.2

90° bends in ducts of rectangular cross section

$\dfrac{a}{b}$	$\dfrac{r_i}{r_i + b}$	ϕ
$\dfrac{1}{2}$	0.1	0.8
	0.2	0.5
	0.4	0.2
1	0.1	0.6
	0.2	0.4
	0.4	0.2
4	0.1	0.4
	0.2	0.3
	0.4	0.1

Miter bends in ducts of circular cross section

α, degrees	ϕ
0°	0
30°	0.1
45°	0.25
60°	0.45
90°	1.1

90° bends in rectangular ducts

$\dfrac{a}{b}$	ϕ
5	0.6
3	0.8
2	1.5
$\dfrac{1}{2}$	2

$\phi = 3$

$\phi = 1$

and 6–19. The energy losses calculated using these factors are, of course, to be added to the losses calculated for the friction in straight duct of the same running length.

Sizes of pipes and ducts. A common problem is to choose the best size of pipe or duct to handle a given flow of gas or liquid. Generally speaking, the larger the diameter, the more expensive is the installation, and the maintenance, depreciation, and related costs are roughly proportional to the first cost. On the other hand, the larger the diameter, the lower is the pressure drop and the lower is the power consumed in passing the fluid through the pipe or duct. Hence, for a given quantity of flow, the determination of the best size of duct is a matter for cost calculations, balancing the cost of the duct against pumping cost to obtain minimum over-all costs. Common engineering procedure in this situation is to estimate operating conditions and costs for a series of tentative designs, so that the general relation between costs and design variables becomes apparent. In a final choice, of course, various other factors of judgment may be important, such as allowances for future expansion, peak loads, availability of material, etc.

When friction calculations are made for a series of tentative duct sizes, it is found that the friction, pressure drop, and power requirements change very steeply with small changes in diameter. If a given total rate of flow (total lb/min) is considered, the mass velocity G varies inversely as the area or inversely as the square of the diameter D. Thus, the Reynolds number varies inversely as the diameter, with the result that the friction factor f increases slowly as the diameter

Fig. 6–5. Fractional losses (ϕ) in kinetic energy for turbulent flow through bends.

is increased.* However, in the Fanning equation V^2 is inversely proportional to the fourth power of the diameter and D already

* Approximately as the 0.2 power of the diameter, for turbulent flow [see Eq. (6–32)].

appears in the denominator, so that without considering the relatively small change in f, the frictional losses are seen to be inversely proportional to the 5th power of the duct size.

EXAMPLE

Calculate the pressure drop, friction losses, and equivalent power consumption for the system described in the example on page 156, but with duct diameters of 2, 4, 6, and 10 ft. The results are tabulated below together with those previously calculated for an 8-ft duct:

Diameter of duct (ft)	F (ft-lb/lb)	Δp (inches water)	Hp
2	1790	13.3	135
4	63.3	0.47	4.8
6	9.1	0.07	0.7
8	2.3	0.02	0.2
10	0.8	0.01	0.1

Beds of solid particles. A fluid-flow system which appears in a variety of forms and apparatus in metallurgical processes is a bed of solids with a gas or a liquid flowing through the pore spaces between particles (Fig. 6–6). One very important example is the shaft of a blast furnace, which is a vertical, cylindrical column filled from the top with lump iron ore, coke, and limestone, and carrying a high upward flow of hot gases which heat and react with the solid materials. Other examples are the beds of solids on filters, in percolation leaching vats, and in pebble heaters (Fig. 5–13).

Flow through a bed of solids has much in common with flow through pipes, and in fact might be considered as flow through a set of many more or less parallel, but tortuous, channels. Accordingly, dimensional analysis leads to the same general result as for pipes, namely a functional relation between a friction factor and a Reynolds number, both of the same form as for pipes. However, dimensional analysis alone cannot indicate

the manner in which the volume fraction of voids, ϵ, must enter the calculations, and also dimensional analysis does not show the particular value of velocity or characteristic dimension of the bed which should be used. Accordingly, several slightly different forms of friction coefficients and Reynolds numbers have been proposed and used, of which the following appear particularly useful:

(modified Reynolds number):

$$N'_{Re} = \frac{V\rho}{\mu S} = \frac{G}{\mu S},\qquad (6\text{–}34)$$

FIG. 6–6.   Flow through beds of solids.

(modified friction factor):

$$f' = -\frac{Fg_c\epsilon^3}{lSV^2},\qquad(6\text{--}35)$$

(modified Fanning equation):

$$F = 2f'\frac{V^2}{2g_c}\frac{lS}{\epsilon^3}.\qquad(6\text{--}36)$$

In these equations ρ, μ, F, g_c, and l are defined in the same way as for flow in pipes, V is the superficial velocity obtained by dividing the volumetric rate of flow by the total cross-sectional area of the bed, G is the superficial mass velocity, ϵ is the volume fraction of pore space available to fluid flow, and S is the specific surface of the bed expressed as fluid-solid interface area per unit volume of bed. Comparison with the corresponding relations for pipes shows that the principal modifications for solid beds are, first, the use of $1/S$ as a characteristic dimension* instead of D and, second, the insertion of the porosity term (ϵ^3) in the modified friction factor. Figure 6–7 gives the correlation compiled by Carman on the basis of these definitions. This graph is suitable for estimating purposes, but does not have as broad an experimental grounding as the one given previously for pipes and therefore should be used cautiously. It can be seen that the shape of the curve is quite similar to that for pipes. At the left is a straight line of slope -1 for streamline flow. The flow starts to become turbulent at values of N'_{Re} around 2, and at

---

* The reciprocal of the specific surface can in fact be considered as a measure of average pore size in the bed. This relation is analogous to the more familiar relation for loose solids: specific surface, S_o (area per unit volume of solids) = ψ/average size. For spheres or cubes, it is easily shown that $\psi = 6$. For solids packed in a bed, $S = S_o(1 - \epsilon)$ so that $1/S =$ approximately $x/6(1 - \epsilon)$, in which x is the particle size in a bed of uniformly sized particles or is an average particle size in a nonuniform bed.

higher Reynolds numbers in the turbulent flow region flattens out to indicate a small rate of decrease of friction factor with increasing Reynolds number of about the same form as the relation for straight pipes.

Other systems. From the foregoing considerations of a few specific types of flow systems, it should be evident that the method of correlating dimensionless variables will be an important engineering tool in evaluating friction in all types of flow systems. Another system with which many metallurgists become familiar is that of a single particle moving in a body of fluid, and the friction factor vs. Reynolds number correlation for this system has been well established experimentally. Still another system for which correlations have been made is a fluidized bed of solid particles, in which a thick suspension is maintained in an agitated and semifluid condition by an upward flow of gas.

When a new system is being studied, a system of geometric configuration for which no previous data are available, the first experimental studies of operating characteristics should be planned to establish a general correlation of dimensionless variables. If, instead, measurements are made to find the effect of one variable at a time, such as the effect of fluid velocity on pressure drop, it will be found that the resulting information has rather limited usefulness because it is true only for the specific experimental conditions considered.

Flow measurement

Efficient handling, utilization, and disposal of fluids in engineering processes usually require information as to quantities of fluids flowing. Often this information can be obtained indirectly, by stoichiometric calculations and the like. In some cases, however, direct and accurate metering of

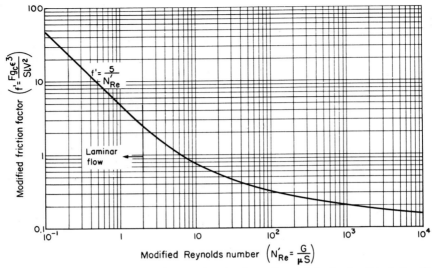

FIG. 6–7.   Friction factor vs. Reynolds number for flows through beds of solids (after Carman).

flow rates becomes essential to efficient operation. Most of the flow-measuring apparatus used for large-scale engineering purposes can be best understood by setting up and examining the mechanical energy balances for the apparatus. For details of design, construction, and installation, reference should be made to engineering handbooks and to special publications on flow-measuring devices, such as the report *Fluid Meters* (Special Committee on Fluid Meters, ASME, New York, 1937).

Venturis. Figure 6–8 shows the general form of a Venturi meter. This device involves a gradual reduction in flow cross sec-

tion to the throat, which is a short straight section, then a very gradual return of the flow to its original cross section. Usual design involves an angle of convergence of about 25° on the upstream side, a throat diameter between one-quarter and one-half the upstream diameter, and a downstream angle of divergence of 7° or less. Pressure taps are installed upstream and at the throat, and the difference in pressure at these two points is a measure of the total rate of fluid flow through the Venturi.

A Venturi tube is nearly frictionless under most operating conditions and also the change in fluid density from tap 1 to tap 2

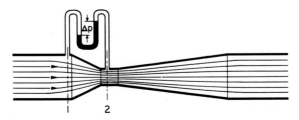

FIG. 6–8.   Venturi meter.

is usually small. Accordingly, the mechanical energy balance for the system between the two pressure taps (asssuming $z_2 - z_1 = 0$; $\int_1^2 p\, d\left(\frac{1}{\rho}\right) = 0$; $F = 0$; $M = 0$) becomes simply:

$$\frac{p_2 - p_1}{\rho} + \frac{V_2^2 - V_1^2}{2g_c} = 0. \quad (6\text{--}37)$$

That is, as the fluid section is reduced from point 1 to point 2, the velocity necessarily must increase in inverse ratio to the area, with the result that pressure energy must be converted into kinetic energy. If β^2 is the ratio of the cross-sectional area at tap 2 to that at tap 1 ($\beta = D_2/D_1$ for circular sections), $V_1 = \beta^2 V_2$ and the energy balance can be solved to obtain the velocity V_2 in terms of the difference in pressure between the two pressure taps:

$$V_2 = \sqrt{\frac{2g_c(p_1 - p_2)}{\rho(1 - \beta^4)}}. \quad (6\text{--}38)$$

This equation can be modified to a more general form, giving the total quantity of fluid passing through the Venturi:

$$w = CYA_2\sqrt{\frac{2g_c\rho_1(p_1 - p_2)}{1 - \beta^4}}, \quad (6\text{--}39)$$

in which w = total rate of flow, lb of fluid/sec; C is the coefficient of discharge; Y is the expansion factor; A_2 is the cross-sectional area of the flow at the throat; and the other symbols are as already defined. The dimensionless coefficients C and Y are empirical coefficients which are arbitrarily inserted to allow for deviations from the assumptions originally made in writing the simple mechanical energy balance in Eq. (6–37). The *coefficient of discharge C* allows for the friction and the *expansion factor Y* allows for the self-expansion of the fluid. When turbulent flow prevails (that is, N_{Re} in duct up or downstream > 2100), the coefficient of discharge C is substantially unity,* but with streamline flow, C drops rapidly with decreasing N_{Re}. The expansion factor Y is unity for liquids and also is unity for gases when the pressure difference is small compared with the total pressure. For gas-flow conditions under which an appreciable expansion and pressure change occurs through the Venturi, values of Y may be found in handbooks, or in some cases may be calculated on a theoretical basis (for example, assuming adiabatic expansion) as discussed later in the section on *Compressed Air and Blast.*

If another pressure tap is placed downstream from the Venturi and the pressure is compared with that upstream (for the same cross-sectional area of flow), the fluid pressure is found to have recovered substantially its upstream value, since an energy balance will show no difference in kinetic energy between the two points. The small pressure drop across the Venturi as a whole will correspond to the friction. Thus the advantage of the nearly frictionless design of the Venturi over cruder devices is that it permits flow measurement without appreciable loss of fluid pressure.

Orifices and nozzles. An orifice meter is of very simple construction, consisting of a thin plate with a hole in it placed across the flow (Fig. 6–9). The most common design in pipes of circular section has a centrally located, square-edged, circular hole. Burrs are carefully removed from the hole after drilling. The main path of flow through the orifice is similar to that through a Venturi, but the orifice design results in considerable disturbance of the flow, so that flow energy is dissipated in eddy currents in the vicinity of the orifice. As indicated in the figure, the flow continues to contract for a short dis-

---

* For values of N_{Re} above 10,000, it is commonly assumed that $C = 0.98$.

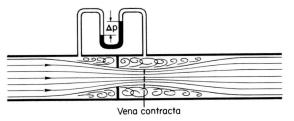

Vena contracta

Fig. 6–9.   Orifice meter

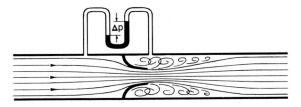

Fig. 6–10.   Nozzle.

tance downstream from the orifice so that the smallest flow section, known as the *vena contracta*, where the kinetic energy is highest, is downstream from the orifice. The two pressure taps may be placed in different ways, and different methods of placing the taps do not give exactly the same readings. *Corner taps* are drilled upstream and downstream with openings as close as possible to the orifice plate. *Radius taps* are located, respectively, 1 pipe diameter upstream and $\frac{1}{2}$ pipe diameter downstream, so that the pressure difference corresponds fairly well to that between the flow in the full area of pipe and that in the *vena contracta*. Various other methods of locating the pressure taps are also used.

Nozzles are similar to orifices in general and may be similarly installed in pipe lines, but are designed so that the discharge opening is preceded by some kind of a smooth contracting passage (Fig. 6–10). As a result, the nozzle has less eddying upstream from the discharge, so that the measured pressure drop corresponds approximately to that for a Venturi rather than to that for an orifice. Downstream from the nozzle discharge, however, the flow behavior is more similar to that downstream from an orifice, so that there is a relatively large loss of flow energy in friction across the nozzle installation as a whole.

In both orifices and nozzles, the mechanical energy balances involve the same principles as for the Venturi; that is, the primary effect is the conversion of pressure energy into increased kinetic energy in the contracted section of the flow. Accordingly, Eq. (6–39), already derived for Venturis, is applicable to orifices and nozzles as well. For nozzles the coefficient of discharge C to be used in Eq. (6–39) is, like that for a Venturi, substantially unity in the turbulent-flow region. However, because of friction, discharge coefficients for sharp-edged orifices are in general well below unity, as shown in Fig. 6–11. The data given in the figure were obtained on circular square-edged ori-

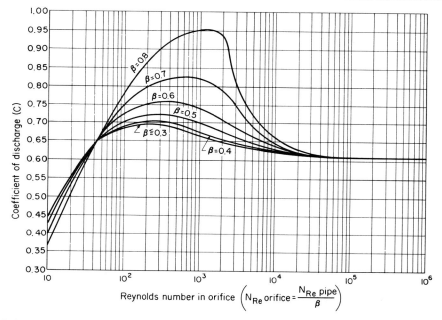

FIG. 6–11. Discharge coefficients for sharp-edged orifices (after Tuve and Sprenkle).

fices with corner taps, but also represent orifice behavior with radius taps to within 1 or 2%. For Reynolds numbers above 10,000 a constant value of $C = 0.61$ is used. However, in metering flows with all these devices, it should be kept in mind that the results are affected by surface roughness and by small differences in details of construction. Accordingly, for accurate work a direct experimental calibration should be made instead of relying on published values of the coefficients, which are accurate only for very carefully standardized designs.

Pitot tubes. A pitot-tube assembly (Fig. 6–12) measures the difference between the *static* pressure and the *impact* pressure at a given point in the flow. The impact opening faces directly upstream to receive the impact of the flow, while the static opening or openings are parallel to the direction of flow. To find the quantitative relation between pressure difference and velocity, the mechanical

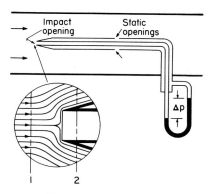

FIG. 6–12. Pitot tube.

energy balance can be set up for the system from point 1 directly upstream from the impact opening to point 2 just inside the impact opening. Ignoring friction and other secondary effects, the principal phenomenon is a conversion of the kinetic energy at point 1 to pressure energy at point 2. At point 1 the velocity is the unknown velocity to be measured and the pressure is that detected by the static openings. At point 2, the velocity is zero and the pressure is that detected by the impact opening. Thus the basic relation for the pitot tube is found to be

$$u = C' \sqrt{\frac{2g_c \, \Delta p}{\rho}}, \qquad (6\text{-}40)$$

where the coefficient C' is used to allow for friction or other failure of the design to meet the assumption of a simple conversion of kinetic energy to pressure energy. For well-designed pitot tubes, the theoretical formula can be used with accuracy (that is, it can be assumed that $C' = 1$). However, if the instrument is not calibrated, particular care should be taken with the location and form of the static openings, so that they give a true measure of the actual static pressure along the same flow line as that for which impact pressure is measured. Burrs or nonparallel location of these openings may introduce errors, or difficulties may be encountered because of variations in static pressure across the stream.

The behavior of the pitot tube illustrates the importance of proper design of openings for pressure measurements in general, especially when small pressure differences are significant. That is, the pressure value usually desired is that of the static pressure, so that the effect of impact must be avoided.

Unlike Venturis, orifice meters, and the like, which measure total flow, the pitot tube measures fluid velocity u along a given flow line and thus does not give directly the total flow or the average velocity V across the whole cross section of flow. In long circular pipes, a given flow reaches a steady state in which the variation of velocity across the section is readily predictable, with maximum velocity at the center and no velocity at the walls. Under these conditions, data such as shown in Fig. 6–13, giving the relation between maximum velocity on the pipe axis and average velocity for various values of Reynolds number, can be used to calculate total flow from pitot-tube readings taken on the pipe axis.

Unfortunately disturbances from bends, fittings, deposits in the ducts, and other causes make it unsafe to assume a regular pattern of velocity distribution across the flow section. Instead, it becomes necessary

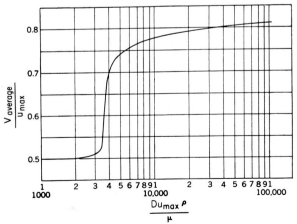

FIG. 6–13.   Ratios of average velocity to maximum velocity for flow in pipes (after Stanton and Pannel).

to measure the velocity with the pitot tube at enough points to establish accurately the velocity distribution and hence the average velocity across the whole section. The common procedure of exploring the flow section with a pitot tube is called a traverse, or an N-point traverse, in which N represents the number of readings. For an N-point traverse, the cross-sectional area of the flow is divided systematically into N equal areas, so that the average velocity can be taken simply as the arithmetic average of the velocities measured at the centers of each of the small areas. In ducts of circular cross section, it is customary to divide the area with concentric circles so that readings are taken from top to bottom at distances from the axis equal to $100\sqrt{2n - 1/N}\%$ of the pipe radius, in which n varies from 1 to $N/2$. In measuring flue gases, frequently a similar traverse is made to obtain fume samples and samples for chemical analysis, the analysis of each individual sample being weighted in proportion to the gas velocity measured at the point of sampling. Such a traverse is of little value for obtaining dust samples, since the dust particles do not necessarily move at the gas velocities which are measured by the pitot tube.

Capillary flow meters. For measurements of small flows in the laboratory, capillary flow meters are easily set up and calibrated, and consist usually of a short section of glass capillary tubing with a water manometer installed to measure the pressure drop across the capillary. Generally the relation between flow rate and manometer reading is established by experimental calibration. However, it should be pointed out that in spite of the superficial similarity to an orifice meter, the capillary flow meter operates on an entirely different basis. Specifically, the pressure connections are made at low-velocity, large-area sections, respectively up and downstream from the capillary, so that kinetic energy changes are generally not important in the energy balance. The important terms in the mechanical energy balance are the pressure energy change and the friction in the capillary. Often an effort is made to design the capillary so that streamline flow is obtained, since this gives a simple linear relation between pressure drop and flow rate [see Poiseuille's equation (6–30), or combine the Fanning equation (6–27) with Eq. (6–31) for streamline flow].

Other flow meters. A number of different designs measure and record the total quantity of fluid passing through the meter in a period of time. Many of these meters operate on the principle of filling and emptying a given volume, with a revolution counter or similar device to record the number of times the given volume is filled and emptied. Fuel-gas meters and water meters operate in this way.

Another kind of device which has proved attractive for measuring relatively small flows of both liquids and gases is the rotameter (Fig. 6–14), which consists of a vertical tapered tube in which a float moves up and down with small clearance to take up a position which measures the rate of flow up through the tube. The chief flow-energy effects are: first, the weight of the float must be balanced by the difference between the impact pressure of the flow acting up on the bottom of the float and the static pressure acting down on the top of the float; second, the annulus between the float and the tube acts as an orifice of variable degree of contraction, depending on the flow rate. However, instead of depending on the relationship derived on this theoretical basis, rotameters are usually calibrated experimentally.

In dealing with water flow in open channels it is often convenient to estimate the flow rate from measurements of the flow over

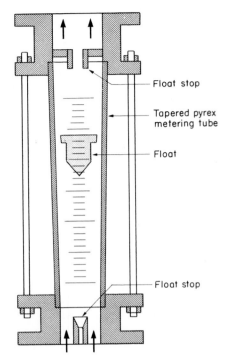

Float stop

Tapered pyrex
metering tube

Float

Float stop

Fig. 6–14.   Rotameter.

a weir. Details of the measurements and calculations are available in texts on hydraulics and in handbooks.

Dilution methods. A method of flow measurement which sometimes proves useful under unusual conditions is dilution metering. This method consists of adding to the flow at a known rate a foreign substance which mixes with or dissolves entirely in the stream. At a point sufficiently downstream to allow thorough mixing to take place, the flow is sampled and analyzed for the foreign substance. The flow rate in the main stream is readily calculated from the known rate of admixture of the foreign substance and the final percentage of the foreign substance in the main stream.

Similarly, heat may be added at a known rate to a fluid stream and the rate of flow calculated from the measured temperature increase and the specific heat of the fluid. Obviously, this procedure can have only very specialized applications.

Draft

Problems relating to the flow of gases in furnaces, flues, stacks, and other equipment operating substantially at atmospheric pressure generally are more conveniently handled by expressing fluid pressure in terms of draft rather than in terms of absolute pressure (p). In fact, in such systems direct measurement of p_2 and p_1 for use in the previously derived energy balances is not practical, especially since the measurement of p_2 and p_1 would have to be made with great precision in order that the small difference $p_2 - p_1$ could be known with sufficient accuracy. The *draft* at any point in a gas-flow system is the difference between the absolute pressure in the system and the absolute pressure of the surrounding atmosphere at the same level, and is conveniently defined to be positive in sign when the pressure in the system is less than atmospheric pressure.   Thus,

$$d_1 = \frac{p_1^\circ - p_1}{5.2}; \quad d_2 = \frac{p_2^\circ - p_2}{5.2}, \quad (6\text{–}41)$$

in which d_1 and d_2 represent the drafts in inches water, p_1 and p_2 the absolute pressures in lb *force*/ft^2, and p_1° and p_2° the atmospheric pressures at points 1 and 2 in any gas-flow system. The conversion factor 5.2 is the specific weight of water (62.4 lb/ft^3) divided by the number of inches per foot (12). The simplest arrangement for measuring draft is an ordinary water manometer, with one leg connected to the flow system and the other leg open to the atmosphere.   Greater sensitivity can be obtained by using inclined manometers.   However, special instruments for measuring furnace draft are now commonly used, and are designed to give pressure readings on an instrument at a point remote from the flow system if necessary. Such instruments are available with sensitivities as high as 0.001 inch water, and may be arranged to actuate automatic control devices of various kinds.

Differences in draft of small fractions of an inch of water are generally quite important in furnace systems, and often the total draft in a complex system may be no more than an inch or two of water. To illustrate the magnitudes involved, a pressure difference of 1 inch of water in a flow of air at standard conditions corresponds to a difference in pressure energy (p/ρ) of 64.5 ft-lb *force*/lb *mass*. This difference in pressure energy in a water-flow system corresponds to a difference in head of 64.5 feet of water or a pressure difference of about 28 psi. That is, the effect of a pressure difference of 1 inch water in a gas-flow system is broadly equivalent to the effect of a 28 psi pressure drop in a water-flow system.

Since small pressure differences are controlling factors in furnace systems, variations of pressures with height in a gas column are especially important. In a static column of fluid at constant temperature, the pressure decreases with height in accordance with $dp = -\rho\, dz$, because of the effect of gravity.* Over the limited range in z involved in any metallurgical gas-flow system, the variation of ρ with z can be neglected, so that for static systems,

$$p = p_0 - \rho z, \qquad (6\text{--}42)$$

in which p is the pressure at height z, p_0 is the pressure at the arbitrary datum plane of

zero height, and ρ is the average gas density. A similar relation expresses the variation of atmospheric pressure with height.

Natural draft. Figure 6–15 shows a vertical column of hot gases surrounded by air at ordinary temperature and pressure. Manometers at heights z_1 and z_2 measure the respective drafts d_1 and d_2 at the two levels. If the average density of the hot gas is ρ_{hot} and that of the surrounding air is ρ_{air}, Eqs. (6–41) and (6–42) show that the difference in draft at the two levels is given by

$$d_2 - d_1 = -\frac{(z_2 - z_1)(\rho_{air} - \rho_{hot})}{5.2}, \quad (6\text{--}43)$$

provided the gases are static. If the column is open to the atmosphere at the top (at height z_2), $d_2 = 0$ so that the draft or suction produced at the bottom is

$$d_1 = \frac{(z_2 - z_1)(\rho_{air} - \rho_{hot})}{5.2}. \quad (6\text{--}44)$$

Equation (6–44) gives the so-called *static draft* of a stack or chimney of total height $(z_2 - z_1)$, which is the draft with no flow in the stack. When gases are flowing, the draft is reduced by frictional losses and other effects in the stack, so that the static draft

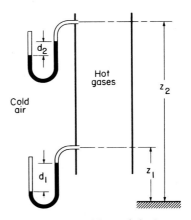

Fig. 6–15. Natural draft.

---

* More vigorously, $dp = -\rho\, dz(g/g_c)$. For the sake of simplicity, the ratio g/g_c is omitted from this relation, from Eq. (6–42), and from subsequent equations in which z is a variable (see footnote on page 144).

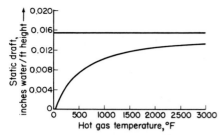

FIG. 6–16.   Static draft vs. gas temperature.

measures the maximum draft which the stack can produce under limiting conditions. Making the approximation that furnace gases have about the same density as air and assuming that outside air is at standard conditions (0°C, 1 atm), Fig. 6–16 shows graphically the variation of static draft per foot height with hot-gas temperature. From these data it can be seen that a 500-ft smelter stack carrying gases at about 600°F would produce a static draft or suction at the bottom of about 4 inches of water. Also these data show that the variation in draft which might be observed over a height of 10 ft within a furnace chamber at high temperature (say 3000°F) would be a little greater than 0.1 inch of water. Thus natural-draft effects are important within furnaces as well as in bona fide chimneys.

Using Eqs. (6–41) and (6–42), the mechanical energy balance for a gas-flow system in Eq. (6–9) can be expressed in terms of draft instead of absolute pressure, as follows:

$$(z_2 - z_1)\left[1 - \rho_{air}\left(\frac{1}{\rho}\right)_{av}\right]$$
$$- 5.2(d_2 - d_1)\left(\frac{1}{\rho}\right)_{av}$$
$$+ \frac{V_2^2 - V_1^2}{2g_c} + F - M = 0. \qquad (6\text{–}45)$$

This form of the mechanical energy balance, expressed in terms of draft, is applicable to a variety of furnace, flue, and stack systems.

Openings and ports. Access to furnaces for charging, observation, and other purposes requires various kinds of openings, doors, and peepholes which connect the furnace interior with the atmosphere. In addition to these deliberately provided openings, high-temperature furnaces of refractory-brick construction have cracks and expansion joints of various kinds which act in the same way. A major objective in draft control is the control of furnace pressure to minimize flow through these furnace openings, or at least to maintain such flow in a desired direction and magnitude.

Essentially the flow through furnace openings, ports, dampers, and the like is the same problem as flow through an orifice, which was discussed previously under *Flow Measurement.* A more general treatment is required because furnace openings are not standardized in design as are orifices for flow measurement. Flows of all kinds through orifices and constricted openings involve conversion of upstream pressure energy into kinetic energy in the orifice and also a loss of energy through friction which depends on the design of the orifice and on the flow conditions up and downstream from the orifice.

In many cases, the velocities of the main bodies of gases on each side of the orifice can be neglected in comparison to the velocity through the opening, so that flow is through the opening between two still bodies of gas at different pressures. Such a system is indicated schematically in Fig. 6–17, in which the gas at the left is under higher pressure, so that the flow is from point 1 through point 2 in the opening, to point 3 in the interior of the right-hand body of gas. From point 1 to point 2 pressure energy is converted into kinetic energy, and some energy is lost as friction about the entrance. From point 2 to point 3 the kinetic energy is all lost as friction through eddy currents

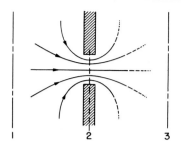

FIG. 6–17.   Flow through furnace openings.

and agitation in the body of the gas. That is, there is no pressure recovery from kinetic energy downstream from the opening. The mechanical energy balance for the flow from 1 to 3 then is

$$\frac{\Delta p}{\rho} + F_{1-3} = -\frac{5.2\,\Delta d}{\rho} + F_{1-3} = 0, \quad (6\text{–}46)$$

in which Δp is the difference in pressure across the opening. From dimensional analysis, F_{1-3} may be considered as $\phi_{1-3}\,(V_2^2/2g_c)$, in which ϕ_{1-3} is a function of the Reynolds number for the flow in the opening. Accordingly,

$$\frac{\Delta p}{\rho} + \phi_{1-3}\left(\frac{V_2^2}{2g_c}\right) = 0. \quad (6\text{–}47)$$

Solving for the velocity in the opening and then for the pounds of gas w flowing per second through an opening of area A_2,

$$w = A_2\sqrt{\frac{1}{\phi_{1-3}}}\,\sqrt{-2g_c\rho\,\Delta p}$$

$$= A_2\sqrt{\frac{1}{\phi_{1-3}}}\,\sqrt{10.4g_c\rho\,\Delta d}$$

$$= CA_2\sqrt{10.4g_c\rho\,\Delta d}, \quad (6\text{–}48)$$

in which C is equal to $\sqrt{1/\phi_{1-3}}$ and is the *coefficient of discharge* of the opening. The coefficient of discharge is primarily a function of the design of the entrance to the opening, and has a maximum value of unity if the entrance is smoothed or "streamlined" to make friction losses negligible. Under these

conditions $\phi_{1-3} = 1$, so that the total frictional loss through the opening, F_{1-3}, equals $V_2^2/2g_c$ and is the loss downstream from the opening. That is, the frictional energy loss results from the fact that all the kinetic energy of the stream is dissipated by eddying in the right-hand body of gas.

For the system in Fig. 6–17, $F_{1-3} = F_{1-2} + V_2^2/2g_c$, and $\phi_{1-3} = \phi_{1-2} + 1$. Also, then, $C = \sqrt{1/(1 + \phi_{1-2})}$. The quantities F_{1-2} and ϕ_{1-2} refer to the *entrance losses*, and data are frequently given in this form. If the opening is long in comparison with its diameter, an additional term must be included to allow for friction within the passage, so that $\phi_{1-3} = \phi_{\text{entrance}} + f\dfrac{l}{D} + 1$, in which f is the friction factor, l is the length, and D is the diameter as evaluated for the Fanning equation.

The friction factors (ϕ) for furnace openings vary with Reynolds number, and therefore corresponding variations occur in discharge coefficients (C). However, as pointed out previously, such factors do not vary much for a given flow system *under turbulent-flow conditions*. Accordingly, in the absence of detailed information, it is customary to use a single value of ϕ or C for all turbulent-flow conditions in a given type of opening. Figure 6–18 gives representative values of the various dimensionless constants for a number of different kinds of furnace openings.

The previous discussion dealt with openings of various kinds between bodies of fluid which could be considered as stationary. Other types of openings in which upstream and downstream velocities cannot be ignored may be handled as enlargements or contractions of flow. Figure 6–19 indicates the friction calculations for several such systems. The formulas given are empirical and are based to a large extent on experimental data

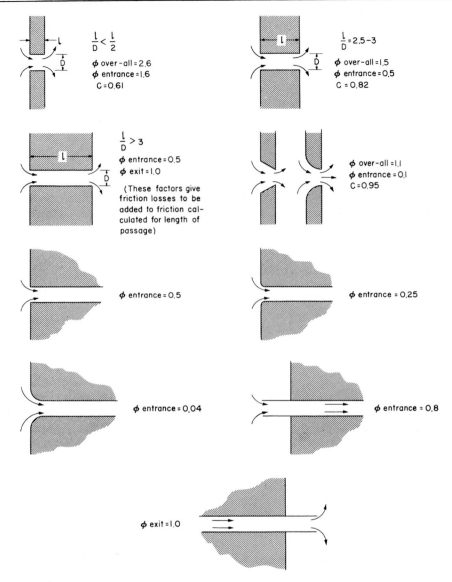

FIG. 6–18. Friction losses in openings of various types.

secured with water as the fluid. It is to be noted that only very gradual enlargement of the flow section is effective in reducing friction, whereas moderate tapering is quite effective in reducing friction when the flow is contracting. This difference is reflected in the design of a Venturi meter (Fig. 6–8).

Furnace pressure control. Simple calculations based on the relations just discussed show the necessity of maintaining furnace

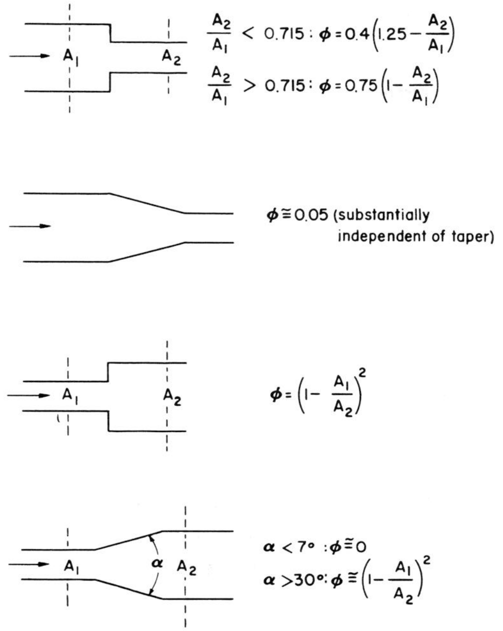

$$\frac{A_2}{A_1} < 0.715 : \phi = 0.4\left(1.25 - \frac{A_2}{A_1}\right)$$

$$\frac{A_2}{A_1} > 0.715 : \phi = 0.75\left(1 - \frac{A_2}{A_1}\right)$$

$\phi \cong 0.05$ (substantially independent of taper)

$$\phi = \left(1 - \frac{A_1}{A_2}\right)^2$$

$\alpha < 7° : \phi \cong 0$

$\alpha > 30° : \phi \cong \left(1 - \frac{A_1}{A_2}\right)^2$

FIG. 6–19. Friction losses accompanying changes in cross section.

pressures very close to atmospheric pressure, that is, with positive or negative drafts of a small fraction of an inch of water at the most. For example, using Eq. (6–48), it is found that a positive draft of only 0.05 inch water across a sharp-edged opening through a furnace wall (for example, an open door) will cause a flow of cold air into the furnace of about 0.7 lb/sec, or about 2500 lb/hr, for each square foot of area of the opening. If the furnace process is a high-temperature process, such a rate of cold air leakage into the furnace can consume a large amount of available heat. Thus, the 2500 lb of air/hr estimated above for a 1 sq ft opening would consume over 2×10^6 Btu/hr of available heat for a process with a critical temperature of about 3000°F. If the fuel were powdered coal having a calorific value of around 12,000 Btu/lb, of which 50% was gross available

heat at 3000°F, this air leakage would cause an additional coal consumption of over 4 tons per day!

Just as a slight positive furnace draft causes air flow into the furnace chamber, a negative draft (furnace pressure above atmospheric) causes outward flow of hot gases. The most desirable furnace draft condition is different for different furnaces and different processes. For example, in heating and annealing steel, air leakage causes oxidation and scaling, so that it is desirable to keep the furnace under pressure or slight negative draft, say −0.01 inch water near the hearth level. On the other hand, in other metallurgical furnaces a positive draft is desired to bring in part of the air for combustion and for process needs, or to prevent any leakage of fumes and gases into the furnace surroundings. Whatever the optimum draft condition, it is clearly important that provisions be made to maintain the conditions desired, since small changes in furnace pressure can wreak havoc in various ways with the furnace process.

The action of the furnace chamber itself as a chimney makes it impossible to maintain the draft constant at all levels. Thus, in a chamber 10 ft high at 3000°F the natural static draft is about 0.1 inch water, so that if the draft were zero at the hearth it would be −0.1 inch water at the roof, or if the draft were zero 5 ft above the hearth, the furnace would draw air in with a positive draft of 0.05 inch at the hearth and blow furnace gases out at the roof with a negative draft of 0.05 inch.

Close control of draft under changing process conditions, changing fuel rate, etc., is difficult to achieve manually. Accordingly, automatic draft control is now widely used for metallurgical furnaces. Figure 6–20 shows schematically the essential features of one typical arrangement for automatic draft

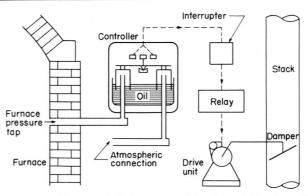

Fig. 6–20. Automatic draft control (courtesy of Leeds and Northrup Company).

control. The draft is measured by a bell-type instrument, in which two oil-sealed bells are mounted at the ends of a beam, one connected to the furnace pressure and the other to the atmosphere, so that the beam moves between electrical contacts in response to small changes in draft (as small as 0.001 inch water). The contacts are connected through appropriate relays and other electrical devices so that a damper in the flue or stack is moved to maintain constant furnace pressure at all times.

Instead of installing a damper in the flue or stack to regulate the flow, a cold-air inlet with damper may be placed near the stack. As more cold air is admitted by opening this damper, the stack temperature is decreased and the stack flow rate is simultaneously increased by the addition of air, both effects serving to decrease the draft in the flue upstream from the cold-air inlet. An arrangement of this kind is easier to maintain than a damper placed in the flue and exposed to the erosive and corrosive effects of hot flue gases.

In old coal furnaces with grate firing of coarse coal, the furnace design provided for use of natural draft to draw the combustion air through the bed of coal. For such systems draft control often represented the chief

control of the quantity of air supplied for combustion. In modern furnaces, however, control of fuel rate and combustion air is usually provided separately through burner design or forced air supply. Thus, the function of the flue and stack system is primarily that of removing gases from the furnace chamber in such a way as to maintain the desired pressure in the furnace.

An excellent general discussion of the control of draft and furnace pressures in regenerative, open-hearth furnaces is given in *Basic Open-Hearth Steelmaking* (AIME, New York, 1951).

Dampers. A damper is a flow-regulating device which introduces a controllable amount of friction into the flow system. Generally the friction is introduced as a constricted opening which acts like an orifice placed in the duct but which has somewhat more complex geometry than a simple orifice. The operating characteristics of dampers, louvres, and gas valves of various kinds can accordingly be considered in terms of relations already discussed for openings and orifices. That is, for a given setting of the damper,

$$F_d = \phi_d \frac{V_2^2}{2g_c} = \phi_d \frac{V_1^2}{2g_c}\left(\frac{A_1}{A_2}\right)^2, \quad (6\text{--}49)$$

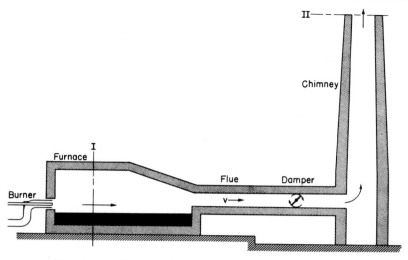

Fig. 6–21. Furnace, flue, and stack system with damper.

in which F_d is the loss of mechanical energy by friction through the damper (ft-lb/lb of fluid), V_1 is the average velocity of flow in the duct up or downstream from the damper, V_2 is the average velocity through the constricted opening in the damper, A_1/A_2 is the ratio of the duct area to the constricted area in the damper, and ϕ_d is a friction factor depending on the Reynolds number for the flow through the damper. In the absence of good data, ϕ_d may be assumed constant for all turbulent-flow conditions at a given damper setting. However, ϕ_d will vary with damper setting and can in some cases be estimated by comparison with data for openings of various kinds, such as the data given in Figs. 6–18 and 6–19.

The behavior of the damper also may be described in terms of quantity of gas flowing (w), loss of draft (Δd), and a discharge coefficient ($C_d = \sqrt{1/\phi_d}$):

$$w = C_d A_2 \sqrt{10.4 g_c \rho \, \Delta d}, \qquad (6\text{--}50)$$

in which A_2 is the constricted area through the damper.

The effect of damper adjustment on a complete furnace and flue system is complex and not easily quantified in a rigorous way. However, to show the main principles of damper control, a simplified analysis can be made of the complete furnace, flue, and stack system shown in Fig. 6–21.

First, the mechanical energy balance between point I in the furnace chamber and point II at the top of the chimney can be set up. Since $d_{II} = 0$ and $V_I = 0$,* Eq. (6–45) becomes (with damper open or omitted from system):

$$(z_{II} - z_I)\left[1 - \rho_{air}\left(\frac{1}{\rho}\right)_{av}\right] + 5.2 \, d_I \left(\frac{1}{\rho}\right)_{av}$$
$$+ \frac{V_{II}^2}{2g_c} + F = 0. \quad (6\text{--}51)$$

To simplify the subsequent equations, it will be convenient to deal with the average velocity V in the flue just up or downstream

---

*It is assumed for simplicity that the velocity in the furnace chamber is small compared with that in flue and stack.

from the damper. The friction loss F for the entire system can then be expressed as $\phi(V^2/2g_c)$ when the coefficient ϕ is applied to the system as a whole. Since both $V_{II}^2/2g_c$ and F in Eq. (6–51) are proportional to $V^2/2g_c$, the kinetic energy and friction terms can be combined in a single term with a coefficient ϕ' to obtain the following relation for the system without a damper:

$$(z_{II} - z_I)\left[1 - \rho_{air}\left(\frac{1}{\rho}\right)_{av}\right] + 5.2\,d_I\left(\frac{1}{\rho}\right)_{av}$$
$$+ \phi'\left(\frac{V^2}{2g_c}\right) = 0. \quad (6\text{–}52)$$

The effect of a damper can now be added to the energy balance as another term in V^2 [Eq. (6–49)]:

$$(z_{II} - z_I)\left[1 - \rho_{air}\left(\frac{1}{\rho}\right)_{av}\right] + 5.2\,d_I\left(\frac{1}{\rho}\right)_{av}$$
$$+ \phi'\left(\frac{V^2}{2g_c}\right) + \phi_d\left(\frac{V^2}{2g_c}\right)\left(\frac{A_{flue}}{A_d}\right)^2 = 0. \quad (6\text{–}53)$$

Since the total flow rate $w = VA_{flue}\rho$, in which V, A_{flue}, and ρ represent average velocity, area, and gas density in the duct section containing the damper, Eq. (6–53) may be modified further:

$$(z_{II} - z_I)\left[1 - \rho_{air}\left(\frac{1}{\rho}\right)_{av}\right] + 5.2\,d_I\left(\frac{1}{\rho}\right)_{av}$$
$$+ \frac{\phi'}{2g_c A_{flue}^2\rho^2}\,w^2 + \frac{\phi_d}{A_d^2 2g_c\rho^2}\,w^2 = 0. \quad (6\text{–}54)$$

For a given system, z_{II}, z_I, ρ_{air}, A_{flue}, and ϕ' are fixed. Also, as an approximation and simplifying assumption, it may be assumed that $(1/\rho)_{av}$ for the system as a whole and ρ for the gas at the damper do not vary with damper setting. Accordingly, Eq. (6–54) gives a relation between the three variables of most interest in damper adjustment: damper setting as measured by ϕ_d/A_d^2, furnace draft d_I, and total flow of gas per second w.

One relation of obvious interest in applying dampers to furnace systems is the one showing how damper setting must be varied to take care of a varying flow (w) of flue gases and to maintain *constant furnace draft* (constant d_I). Such a relation, derived from Eq. (6–54), is given by

$$\phi_d\left(\frac{A_{flue}}{A_d}\right)^2 = \phi'\left[\left(\frac{w_0}{w}\right)^2 - 1\right], \quad (6\text{–}55)$$

in which w_0 is the mass flow rate for the system without damper friction ($\phi_d = 0$) and from Eq. (6–54) is given by

$$w_0 = A_{flue}\rho\sqrt{\frac{2g_c}{\phi'}} \times$$
$$\sqrt{- (z_{II} - z_I)\left[1 - \rho_{air}\left(\frac{1}{\rho}\right)_{av}\right] + 5.2\,d_1\left(\frac{1}{\rho}\right)_{av}}. \quad (6\text{–}56)$$

Another relation which might be of interest in damper operation is that between damper setting and furnace draft under conditions of *constant gas flow*. That is, if the quantity of flue gases is fixed independently by the supply of fuel and forced air to combustion, this relation will show how the furnace draft changes as the damper is adjusted. For these conditions, from Eq. (6–54),

$$d_I = d_0 - \beta\phi_d\left(\frac{A_{flue}}{A_d}\right)^2, \quad (6\text{–}57)$$

where d_0 is the furnace draft when the system is operated without damper friction and is given by

$$d_0 = -\frac{(z_{II} - z_I)\left[1 - \rho_{air}\left(\frac{1}{\rho}\right)_{av}\right] - \phi'\left(\frac{V^2}{2g_c}\right)}{5.2\left(\frac{1}{\rho}\right)_{av}}, \quad (6\text{–}58)$$

and β is a constant for the conditions considered:

$$\beta = \frac{V^2}{2g_c} \cdot \frac{1}{5.2\left(\frac{1}{\rho}\right)_{av}} = \frac{w^2}{2g_c A_{flue}^2 \rho^2} \cdot \frac{1}{5.2\left(\frac{1}{\rho}\right)_{av}}.$$

$$(6\text{--}59)$$

The difference in draft measured across the damper is

$$\Delta d_d = \frac{\rho}{5.2}\,\phi_d\left(\frac{A_{flue}}{A_d}\right)^2\frac{V^2}{2g_c} = \beta\phi_d\left(\frac{A_{flue}}{A_d}\right)^2\frac{\left(\frac{1}{\rho}\right)_{av}}{\left(\frac{1}{\rho}\right)},$$

$$(6\text{--}60)$$

so that Eq. (6–57) can be written

$$d_{\mathrm{I}} = d_0 - \Delta d_d\,\frac{\left(\frac{1}{\rho}\right)}{\left(\frac{1}{\rho}\right)_{av}}.\qquad (6\text{--}61)$$

Equation (6–61) shows that the effect of the damper on furnace draft (at constant w) is proportional to but not equal to the pressure drop across the damper, the proportionality constant being the ratio of the reciprocal density of the gas at the damper to the average reciprocal density for the system as a whole.

Forced draft. The natural draft derived from the buoyancy of hot gases furnishes the chief motive power for moving gases through many metallurgical systems. A stack or chimney built to supply draft is a simple device with no moving parts; it also serves the purpose of dispersing waste gases in the atmosphere in such a way that they will not harm people or vegetation. In fact, the design of smelter stacks is in many instances fixed more by the requirements of dispersing waste gases than by draft requirements. Another condition which often dictates the use of natural draft rather than forced draft is that the gases handled are hot, dirty, and corrosive, so that fans will not stand up for an appreciable length of time. However, in some circumstances the use of fans to produce gas movement is advantageous. In

particular, forced draft may afford better and more positive control of gas quantity, is capable of giving greater drafts, and gives greater design flexibility. Probably the most common use of fans around metallurgical plants is for supplying air to combustion in fuel-fired furnaces, since the fans can be placed in a part of the flow system where they handle only cold and relatively clean air. When a fan is used for the air supply it may be designed to supply sufficient draft in excess of that required by the flow system to permit addition of an orifice, Venturi, or other flow-measuring device, and these devices in turn may furnish the basis for better manual or automatic control of the quantity of air supplied.

Figure 6–22 shows schematically the Morgan system for obtaining forced draft of hot flue gases using a cold-air fan. Under ideal conditions, the cold air and hot gases mix

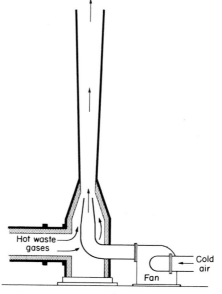

FIG. 6–22. Morgan ejector (courtesy of Morgan Construction Company).

Fig. 6–23.   Centrifugal fan.

thoroughly in the Venturi so that the kinetic energy of the added stream of cold air becomes part of the pressure energy of the whole mixture beyond the Venturi. Draft control is obtained by regulating the speed of the cold-air fan.

Centrifugal fans are generally used for low pressures, up to about 5 inches of water. Figure 6–23 shows the general design of a common type of centrifugal fan. For greater pressures centrifugal compressors (turbocompressors), rotary positive-displacement blowers, piston compressors, or other devices are used, the choice depending on the nature of the service.

The contribution of a fan or pump of any kind to the mechanical energy of a flow system is represented by the term M, which is foot-pounds of energy per pound of fluid and is sometimes called simply the *total head* in feet contributed to the flow by the fan or pump. The theoretical power consumption can be calculated directly from the *total head* M and the rate of flow of the fluid. The efficiency is the theoretical horsepower divided by the actual brake horsepower input to the shaft of the pump. The performance characteristics of a fan or pump are com-

monly given by a set of curves such as those shown in Fig. 6–24 for a centrifugal fan, in which total head, power input, and efficiency are plotted against capacity or discharge rate. The curves shown are for one fan speed. The commercial capacity rating is usually the capacity at maximum efficiency.

Compressed air and blast

The gas-flow systems considered previously in this chapter have involved rather small pressure differences, and in fact have dealt with gases under absolute pressures for the most part within 1% or so of the ambient atmospheric pressure. Somewhat different techniques are used in compressing and handling gases at substantial gauge pressures, and these are of particular importance to the metallurgist in supplying large quantities of air under pressure to unit processes such as the blast furnace process and the converting process. One iron blast furnace, for example, will consume from 50,000 to 100,000 ft³/min of air at a gauge pressure of an atmosphere or so (about 15 psi gauge), and a Bessemer converter may consume something like 40,000 ft³/min of air at about

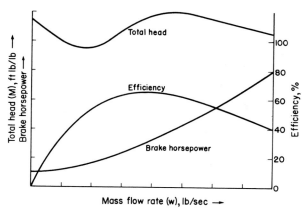

Fig. 6–24. Performance curves of a centrifugal fan.

30 psi. Such quantities of blast require very large blowing engines which, as will be seen, use up large amounts of power.

Whereas centrifugal fans are efficient in generating drafts up to a few inches of water, quite different equipment is used to obtain air under substantial pressures. The principal types of blowers for higher pressures include (1) turbocompressors, (2) reciprocating, piston-type engines, and (3) rotary, positive-displacement blowers. A detailed discussion of the principles of operation of these is beyond the scope of this book, but the general features of each are shown schematically in Fig. 6–25. In recent years there has been a definite trend toward use of turbocompressors in large, new installations because of their high efficiencies and because their operating characteristics (for example, the relation between quantity of air supplied and pressure) match better the requirements of the processes using the air.

Although the design and specification of high-pressure equipment are matters which should concern mainly the specialist in mechanical engineering, the metallurgist should be familiar with a few guiding principles which apply to all the common types

of blowers. In particular, the relations between air quantities, pressures, temperatures, power consumption, and blower efficiencies are of direct concern to the user of air under pressure, and therefore will be considered briefly in the following paragraphs.

Theoretical mechanical energy for compression. The theoretical mechanical energy required to obtain a given pressure increase in a blower is found by applying a mechanical energy balance [Eq. (6–7)] across the blower and assuming that friction (F), potential energy change ($z_2 - z_1$), and kinetic energy change $[(V_2^2 - V_1^2)/2g_c]$ are negligible. These assumptions give

$$M = \int_1^2 \frac{1}{\rho}\, dp. \qquad (6\text{–}62)$$

Thus evaluation of the theoretical work input requires knowledge of the path of the compression process, or of the way in which ρ varies with p as the gas passes through the blower. Without this kind of data, the integral in Eq. (6–62) cannot be solved.

For small pressure changes, such as are produced by the ordinary centrifugal fans considered in the previous section, the

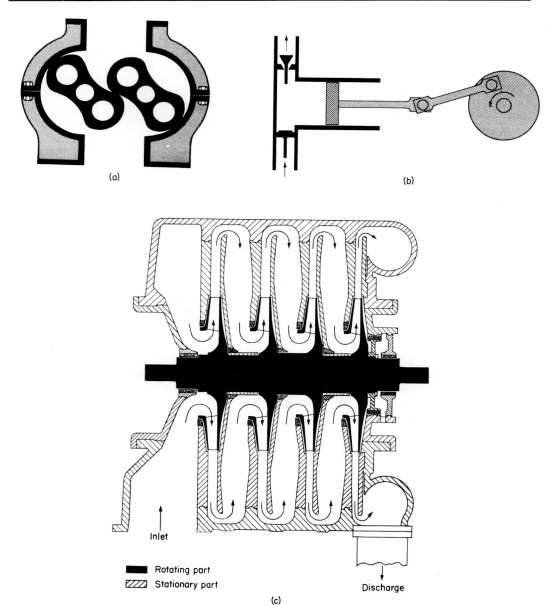

FIG. 6–25. High-pressure blowers.

density can be assumed constant, and Eq. (6–62) then becomes

$$M = \frac{1}{\rho}\,(p_2 - p_1). \qquad (6\text{–}63)$$

The path of the compression process, where appreciable pressure changes are produced, can be considered conveniently in terms of a pressure-volume diagram or a p vs. $1/\rho$ diagram such as the one shown in Fig. 6–26. Three different paths are shown as examples, all starting at the same point $(p_1, 1/\rho_1)$ and all being carried to the same final pressure p_2. However, the final densities and gas temperatures are all different. Path I is an isothermal path, II is an adiabatic path, and III is an intermediate path which might represent better the behavior of the gas in passing through an actual compressor or blower. For any one of these paths the value of the integral in Eq. (6–62) is the area between the curve and the p axis. Thus, if actual measurements are made of the way p varies with $1/\rho$ as the gas is compressed, the theoretical mechanical energy consumption is found by plotting the measurements as in Fig. 6–26 and measuring the area between the curve and the p axis. This graphical procedure is particularly applicable to the study of reciprocating-piston blowers, as the necessary data are readily obtained by making an "indicator-card" record of the compression cycle.

If it is assumed that the gas is ideal, theoretical mechanical energy consumptions are readily calculated for isothermal compression (path I) and adiabatic compression (path II). For these calculations the ideal-gas equation may be written in the form

$$\frac{p}{\rho T} = c_1, \qquad (6\text{–}64)$$

in which c_1 is a constant for a given gas.

For *isothermal* compression T is constant, so solving Eq. (6–64) for $1/\rho$ and substituting in Eq. (6–62) gives

$$M = c_1 T \int_1^2 \frac{dp}{p} = c_1 T \ln_e \frac{p_2}{p_1}. \qquad (6\text{–}65)$$

Since $p_1/\rho_1 T = c_1$, the theoretical mechanical energy requirement for *reversible (frictionless) isothermal compression of an ideal gas* can be expressed in terms of initial and final pres-

sures and densities as follows:

$$M = \frac{p_1}{\rho_1} \ln_e \frac{p_2}{p_1} = 2.303 \frac{p_1}{\rho_1} \log_{10} \frac{p_2}{p_1}. \qquad (6\text{–}66)$$

For *adiabatic* compression of an ideal gas, p and $1/\rho$ follow the relation

$$p\left(\frac{1}{\rho}\right)^\kappa = c_2 = p_1\left(\frac{1}{\rho_1}\right)^{\kappa}{}^*, \qquad (6\text{–}67)$$

in which κ is the ratio of the heat capacity at constant pressure

$$\left[C_p = \left(\frac{\partial H}{\partial T}\right)_p\right],$$

---

* Proof of this relation is as follows: The First Law can be stated for an infinitesimal process by $dH = dQ - dW + d(pV)$, in which dH, dQ, and V (volume) refer to one mol [see Eq. (3–3), Chapter 3]. For an adiabatic process $dQ = 0$, and if only expansion work is involved, $dW = p\,dV$. Accordingly, $dH = -p\,dV + p\,dV + V\,dp = V\,dp$. But for an ideal gas H varies only with T and not with p or V. Hence $dH = C_p\,dT$. By differentiating the ideal gas equation $pV = RT$, it is found that $dT = (1/R)[p\,dV + V\,dp]$. Combining these relations,

$$dH = C_p\,dT = \frac{C_p}{R}(p\,dV + V\,dp) = V\,dp.$$

Rearranging the two right-hand members of this equation and integrating gives

$$pV^{\frac{C_p}{(C_p - R)}} = \text{constant}.$$

The definition of heat content, $H = E + pV$, for a mol of ideal gas can be written $H = E + RT$. For an ideal gas H and E depend only on the temperature, so differentiation with respect to T gives

$$\frac{dH}{dT} = \frac{dE}{dT} + R.$$

From the heat capacity definitions, then

$$C_p = C_v + R,$$

so that the p-V relationship for adiabatic compression of an ideal gas can be written

$$pV^{\frac{C_p}{C_v}} = \text{constant}.$$

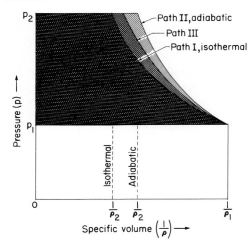

FIG. 6-26. p vs. $1/\rho$ for compression of a gas.

to the heat capacity at constant volume

$$\left[C_v = \left(\frac{\partial E}{\partial T}\right)_v\right].$$

Strictly speaking, κ for any given gas varies with temperature and pressure, but this variation is small enough so that it can be ignored in most technical calculations. Thus, κ for air at standard conditions is 1.40 and if the temperature is raised to 600°F, κ becomes 1.37. The value of κ is closely related to the kind of gas molecules, so that the following approximate values can be used:

monatomic gases (He, A, etc.):

$$\kappa = 1.67,$$

diatomic gases (H_2, CO, N_2, O_2, air):

$$\kappa = 1.40,$$

triatomic, tetra-atomic gases (CO_2, H_2O):

$$\kappa = 1.30.$$

Solving Eq. (6-67) for $1/\rho$ and substituting in Eq. (6-62) gives

$$M = \frac{p_1^{1/\kappa}}{\rho_1} \int_1^2 p^{\frac{-1}{\kappa}}\, dp. \qquad (6\text{-}68)$$

Integrating, the theoretical mechanical en-

ergy requirement for *reversible (frictionless), adiabatic compression of an ideal gas* is found to be

$$M = \frac{\kappa}{\kappa - 1}\frac{p_1}{\rho_1}\left[\left(\frac{p_2}{p_1}\right)^{\frac{\kappa-1}{\kappa}} - 1\right]. \qquad (6\text{-}69)$$

Comparison of Eqs. (6-69) and (6-66) shows that isothermal compression requires less mechanical work than adiabatic compression for a given degree of compression. This relation is also indicated in Fig. 6-26, where it can be seen that the area between the adiabatic curve (I) and the p axis is somewhat greater than that between the isothermal curve (II) and the p axis. For example, to compress 1 lb of air at standard conditions (0°C, 1 atm) to a gauge pressure of 1 atm theoretically requires 18,100 ft-lb by the isothermal process, whereas it requires 20,100 ft-lb by the adiabatic process. Also, the air compressed by the isothermal process will be at 0°C and at a density just double that of the original air, whereas the air compressed adiabatically will have a density equal to 1.64 times that of the original air and will be at a temperature of 60°C. To give a further idea of the magnitudes of the energy quantities involved, the adiabatic compression of 10,000 ft³/min (STP) of air, as just described, theoretically would require close to 500 hp.

Heat effects attending compression. The total energy balance for the flow through a blower or compressor, under the conditions of no changes in potential energy or kinetic energy, is simply [from Eq. (6-3)]:

$$H_2 - H_1^\bullet = Q + M. \qquad (6\text{-}70)$$

For the *isothermal compression* of an ideal gas, $H_2 - H_1 = 0$, so that $Q = -M$, and thus, from Eq. (6-66),

$$Q = -2.303\frac{p_1}{\rho_1}\log_{10}\frac{p_2}{p_1}. \qquad (6\text{-}71)$$

That is, in isothermal compression of an

ideal gas, a quantity of heat exactly equal to the input of mechanical energy must be removed from the gas during compression. Thus, although the isothermal process theoretically requires less power than an adiabatic process, isothermal compression is not practical because it would require an efficient and elaborate cooling system as part of the blowing engine.

In *adiabatic compression* $Q = 0$, so that the total energy balance under the conditions of no changes in potential or kinetic energy becomes

$$H_2 - H_1 = M. \qquad (6\text{–}72)$$

From Eq. (6–69), which gives M for frictionless, adiabatic compression of an ideal gas, the increase in enthalpy is thus given by

$$H_2 - H_1 = \frac{\kappa}{\kappa - 1} \frac{p_1}{\rho_1}\left[\left(\frac{p_2}{p_1}\right)^{\frac{\kappa-1}{\kappa}} - 1\right]. \qquad (6\text{–}73)$$

Thus, in adiabatic compression a quantity of energy equal to the work input is added to the heat content of the gas. If desired, this energy can be removed from the gas after compression by using a cooler.

Air compressors are generally designed with some kind of integral cooling arrangement, either water jackets or exterior fins for air cooling. Thus the compression process falls somewhere between an isothermal and an adiabatic process. When very high pressures are required, e.g., over 100 psi, so that compression is accomplished in two or more stages, an *intercooler* may be used to cool the gas between stages about to its initial temperature. An intercooler is simply a water-cooled heat exchanger.

Compressor and blower efficiencies. The process in an actual compressor or blower differs in a number of respects from the ideal reversible processes considered above. Friction, changes in kinetic energy, and changes in potential energy were neglected in calculating the theoretical energy consumption.

Also, it was assumed that the fluid behaved as an ideal gas. It has already been pointed out that the actual p vs. $1/\rho$ path of compression may differ considerably from both isothermal and adiabatic paths. No consideration was given to energy losses in bearings and in the engine or motor used to drive the blower. The combined effects of all these factors can be expressed in terms of the overall efficiency as follows:

$$\text{over-all efficiency} = 100 \times \frac{\text{theoretical work}}{\text{work input to prime mover}}.$$

In reporting over-all efficiencies, it is customary to calculate the theoretical work as that required for *adiabatic, reversible compression of an ideal gas*, as given by Eq. (6–69). Taking the work input as the electrical work input to the driving motor, probably an over-all efficiency of 75% may be taken as an average for estimating purposes in the absence of more specific information.

Gas flow with large pressure drops. In applying mechanical energy balances to gas-flow systems considered earlier in this chapter, it has been possible to treat the gases essentially as incompressible fluids (that is, the same as liquids). In particular, the density changes have been small enough so that they could be ignored, making it possible to evaluate the term $\int_1^2 \frac{1}{\rho}\,dp$ in Eq. (6–7) by the simple expression $(p_2 - p_1)/\rho$. Where changes in fluid density have been small, the use of the arithmetic average reciprocal density $[(1/\rho_1 + 1/\rho_2)/2]$ has been suggested, to give the approximation

$$\int_1^2 \frac{1}{\rho}\,dp = (p_2 - p_1)\left(\frac{1}{\rho}\right)_{av}.$$

These assumptions introduce no appreciable errors when the pressure drop and change in density are small compared with the pressure and density of the gas entering the system, and even serve fairly well for estimating purposes when substantial changes in pressure and density are encountered. However, in some calculations relating to flow of air under pressure and involving pressure drops of several pounds per square inch, a more rigorous approach becomes desirable. Strictly speaking, evaluation of the integral $\int_1^2 \frac{1}{\rho} \, dp$ for many systems would require actual measurement of the quantitative relation between $1/\rho$ and p through the flow system. Such data are rarely available, so recourse is usually had to the ideal $1/\rho$ vs. p relationships which have been found to represent the gas behavior satisfactorily for most practical needs.

The most useful ideal process is that of *reversible* or *frictionless adiabatic* flow, for which the important quantitative relations have already been derived in connection with the discussion of compression. Actually, it is reasonable to assume adiabatic flow ($Q = 0$) for many systems such as orifices, Venturis, nozzles, etc., because the gas passes through so quickly that no real opportunity is allowed for exchange of heat with the surroundings. The further assumption that the flow is frictionless is, however, often not valid, so that allowance for friction must be made through direct friction calculations or through use of correction factors such as discharge coefficients.

The mechanical energy balance [Eq. (6–7)] for a gas-flow system with negligible change in potential energy ($z_2 - z_1 = 0$) and with no exchange of mechanical energy with the surroundings through a fan, turbine, or the like ($M = 0$) becomes

$$\frac{V_2^2 - V_1^2}{2g_c} + \int_1^2 \frac{1}{\rho} \, dp + F = 0. \quad (6\text{--}74)$$

If it is further assumed that the flow is frictionless, $F = 0$, and that the flow is adiabatic and that the gas is ideal, the value of $\int_1^2 \frac{1}{\rho} \, dp$ is given by Eq. (6–69). With these assumptions, the mechanical energy balance for the *ideal* case becomes

$$\frac{V_2^2 - V_1^2}{2g_c} + \frac{\kappa}{\kappa - 1} \frac{p_1}{\rho_1} \left[\left(\frac{p_2}{p_1} \right)^{\frac{\kappa-1}{\kappa}} - 1 \right] = 0. \quad (6\text{--}75)$$

In using this equation care should be taken not to overlook the various assumptions made in its derivation. This equation is to be compared with Eq. (6–37), which was based on the assumption of constant density instead of on the assumption of adiabatic behavior of an ideal gas.

Equation (6–75) can be used to obtain a theoretical equation for the flow of gases through Venturis, nozzles, and orifices of various kinds. If w is the total flow in pounds per second, A_1 and A_2 are the cross-sectional areas at sections 1 and 2, and the other symbols are as already specified, then

$$\frac{w^2}{2g_c} \left[\frac{1}{A_2^2 \rho_2^2} - \frac{1}{A_1^2 \rho_1^2} \right]$$
$$+ \frac{\kappa}{\kappa - 1} \frac{p_1}{\rho_1} \left[\left(\frac{p_2}{p_1} \right)^{\frac{\kappa-1}{\kappa}} - 1 \right] = 0. \quad (6\text{--}76)$$

Using the relation $p_1 \left(\frac{1}{\rho_1} \right)^{\kappa} = p_2 \left(\frac{1}{\rho_2} \right)^{\kappa}$ for adiabatic expansion and simplifying,

$$\frac{w^2}{2g_c} \left[\frac{1}{A_2^2} \left(\frac{p_1}{p_2} \right)^{\frac{2}{\kappa}} - \frac{1}{A_1^2} \right]$$
$$+ \frac{\kappa}{\kappa - 1} p_1 \rho_1 \left[\left(\frac{p_2}{p_1} \right)^{\frac{\kappa-1}{\kappa}} - 1 \right] = 0. \quad (6\text{--}77)$$

Letting $\beta^2 = A_2/A_1$ ($\beta = D_2/D_1$ for circular sections) and solving for w,

$$w = C_a A_2 \sqrt{\dfrac{2g_c\kappa p_1\rho_1\left[1 - \left(\dfrac{p_2}{p_1}\right)^{\frac{\kappa-1}{\kappa}}\right]}{(\kappa - 1)\left[\left(\dfrac{p_2}{p_1}\right)^{\frac{-2}{\kappa}} - \beta^4\right]}}, \quad (6\text{-}78)$$

or

$$w = C_a A_2 \sqrt{\dfrac{2g_c\kappa p_1\rho_1(1 - r^{\frac{\kappa-1}{\kappa}})}{(\kappa - 1)(r^{\frac{-2}{\kappa}} - \beta^4)}}, \quad (6\text{-}79)$$

where $r = p_2/p_1$ and is known as the *pressure ratio*. The discharge coefficient C_a, which is called the *adiabatic-discharge coefficient*, has been inserted arbitrarily in these equations to allow for the deviations obtained in actual flow systems from the ideal assumptions made in deriving the equation. This equation is to be compared with Eq. (6-39), which was derived on the assumption of constant fluid density. Comparison shows that these two equations give the same result for w if

$$C \text{ [in Eq. (6-39)]} = C_a,$$

and

$$Y \text{ [in Eq. (6-39)]} =$$

$$\sqrt{\dfrac{\kappa}{\kappa - 1}\dfrac{(1 - r)^{\frac{\kappa-1}{\kappa}}(1 - \beta^4)}{(1 - r)(r^{\frac{-2}{\kappa}} - \beta^4)}}.$$

These two relations appear to be valid for nozzles and Venturis, but not for orifices of all kinds. In other words, either Eq. (6-39) or Eq. (6-79) may be used satisfactorily for flow calculations with nozzles and Venturis. If Eq. (6-39) is used, the expansion factor Y corrects the result from the assumption of constant density to the assumption of ideal adiabatic expansion.

Characteristics of nozzles. Air is blown into the working chambers of various metallurgical furnaces (for example, blast furnaces, converters) through relatively crude nozzles or tuyères. Although the conditions in and around tuyères are likely to be complex and

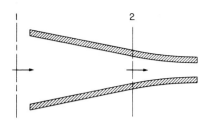

FIG. 6-27. Convergent nozzle.

difficult to define quantitatively, a knowledge of some of the special characteristics of ideal flow through nozzles and some acquaintance with nozzle design may be helpful in understanding the behavior of tuyères.

Figure 6-27 shows in cross section a typical convergent nozzle, with the flow passage shaped to fit the flow and to afford minimum friction. The gas-flow system may be taken from point 1 where the velocity is small or negligible to any point 2 within the nozzle. Under these conditons, Eq. (6-79) is applicable with $\beta = 0$, and experience shows that with good nozzle design C_a may be close to unity. If, to start with, point 2 of the flow system is taken at the throat, where the area is smallest,

$$w = C_a A_{\text{throat}} \sqrt{\dfrac{2g_c\kappa p_1\rho_1\left(1 - r_{\text{throat}}^{(\kappa-1)/\kappa}\right)}{(\kappa - 1)\left(r_{\text{throat}}^{-2/\kappa}\right)}}.$$
$$(6\text{-}80)$$

Now let it be assumed that the initial conditions of the gas (p_1, ρ_1, κ) as well as the area of the throat (A_{throat}) are fixed. Under these conditions, the total flow w will vary with pressure ratio r, which is the ratio of the absolute pressure in the throat p_2 to the inlet pressure. The nature of this variation can be found by solving Eq. (6-80) for w at various values of r. Results of one such set of calculations are plotted in Fig. 6-28. The important and striking feature of nozzle flow brought out by these calculations is that w

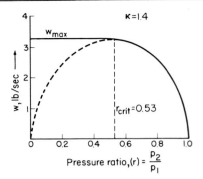

FIG. 6–28. Nozzle flow rate vs. pressure ratio.

is a maximum for a given value of r. This value is known as the *critical pressure ratio* and can be shown to be a function of κ, as follows:

$$r_{\text{crit}} = \left(\frac{2}{\kappa + 1}\right)^{\frac{\kappa}{\kappa - 1}}. \qquad (6\text{–}81)$$

Substituting $\kappa = 1.4$, a value of $r_{\text{crit}} = 0.53$ is obtained for air. The maximum value of w for the critical pressure ratio is given by

$$w_{\max} = C_a A_2 \sqrt{g_c \kappa p_1 \rho_1 \left(\frac{2}{\kappa + 1}\right)^{\frac{\kappa + 1}{\kappa - 1}}}. \qquad (6\text{–}82)$$

For air, this becomes

$$w_{\max} = 3.9 C_a A_2 \sqrt{p_1 \rho_1}. \qquad (6\text{–}83)$$

Thus the effect of changing discharge pressure, p_2, on the flow through the convergent nozzle may be described as follows: For given conditions of gas supply to the nozzle, that is, with pressure p_1 and density ρ_1 fixed, the flow w in pounds per second through the nozzle will increase as the discharge pressure is decreased until $p_2 = r_{\text{crit}} p_1$. For discharge pressures above the critical pressure, the flow rate is given by Eq. (6–80). However, if the discharge pressure is decreased further, so that $p_2 < r_{\text{crit}} p_1$, no further increase in flow rate can occur, and the flow rate through the nozzle is then given by Eq. (6–82). Thus, if the nozzle discharges air into a space where the absolute pressure is less than 53%

of the initial pressure, the quantity of discharge is independent of the actual value of pressure and density in the discharge space.

From the above considerations of flow through nozzles, it can be seen that a nozzle fed with gas under high pressure, so that $p_2/p_1 < r_{\text{crit}}$, is a convenient and simple device for obtaining a known flow rate for regulating or calibrating purposes. A particular advantage of using a nozzle in this way is that the quantity flowing is constant, regardless of variations in downstream conditions, as long as the ratio of discharge to inlet pressure remains below critical. With careful streamlining to eliminate friction, the coefficient of discharge can be made close to unity so that the flow rate of air, for example, is found from the throat area A_2, and the upstream pressure p_1 and ρ_1 by Eq. (6–83).

The approach outlined above is applicable in a general way to the behavior of tuyères through which air is blown into a blast furnace or converter. However, allowance must be made for the fact that tuyères are poorly designed nozzles, often nothing more than short sections of straight pipe, so that the discharge coefficient C_a will be considerably less than unity, perhaps on the order of 0.7. By assuming a value for C_a in Eqs. (6–80) and (6–83), engineering calculations can readily be made, for example, of the cross-sectional tuyère area required to handle a required quantity of blast under any given conditions of supply and discharge pressure.

A nozzle converts pressure and internal energy of the gas into kinetic energy. For nozzle applications where efficient conversion is desired, the convergent-divergent nozzle (Fig. 6–29) is used. From the intake up to the throat, the flow system is the same as in the previously discussed simple convergent nozzle. Moreover, the convergent-divergent nozzle is designed to operate with maxi-

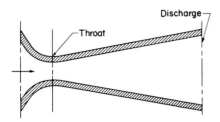

FIG. 6–29.   Convergent-divergent nozzle.

mum flow in the throat, as given by Eq. (6–82), and therefore with the pressure in the throat at just the critical point.   The divergent section after the throat permits the efficient expansion of the gas from critical pressure to discharge pressure, thereby obtaining in the discharged stream the maximum kinetic energy.   The proper amount of divergence, as measured by the ratio of the discharge area to the throat area, is readily found as follows.   Since the nozzle is designed for full capacity flow in the throat, the flow rate will be given by Eq. (6–82), in which A_2 is the throat area.   The flow

through the discharge section will satisfy Eq. (6–80) ($\beta = 0$ because A_1 has been assumed large).   Therefore,

$$w = C_a A_{\text{throat}} \sqrt{g_c \kappa p_1 \rho_1 \left(\frac{2}{\kappa + 1}\right)^{\frac{\kappa+1}{\kappa-1}}}$$

$$= C_a A_{\text{dis}} \sqrt{\frac{2 g_c \kappa p_1 \rho_1 \left(1 - r_{\text{dis}}^{(\kappa-1)/\kappa}\right)}{(\kappa - 1)\left(r_{\text{dis}}^{-2/\kappa}\right)}}.$$

$$(6\text{–}84)$$

Solving,

$$\text{divergence} = \frac{A_{\text{dis}}}{A_{\text{throat}}} = \sqrt{\frac{\left(\dfrac{2}{\kappa+1}\right)^{\frac{\kappa+1}{\kappa-1}}(\kappa-1)\left(r_{\text{dis}}^{-2/\kappa}\right)}{2\left(1 - r_{\text{dis}}^{(\kappa-1)/\kappa}\right)}}.$$

Thus if the pressure ratio from supply to discharge for the nozzle as a whole (r_{dis}) is fixed, this equation gives the theoretical nozzle design which will result in maximum flow and in critical pressure conditions in the throat or smallest section of the nozzle. Moreover, this nozzle design theoretically affords maximum conversion of pressure and internal energy into kinetic energy at the discharge of the nozzle.

Supplementary References

AIME, Committee on Physical Chemistry of Steelmaking, *Basic Open-Hearth Steelmaking*, revised ed.   New York: AIME, 1951.

ASME, Special Committee on Fluid Meters, *Fluid Meters*.   New York: ASME, 1937.

Binder, R. C., *Fluid Mechanics*.   New York: Prentice-Hall, 1949.

Bradley, M. J., "Open-Hearth Furnace Control." *Iron and Steel Engineer*, **18,** No. VI, 35–43, June, 1941.

Bridgman, P. W., *Dimensional Analysis*.   New Haven: Yale University Press, 1931.

Brown, George Granger, et al., *Unit Operations*.   New York: John Wiley, 1950.

Brownell, L. E., and Katz, D. L., "Flow of Fluids through Porous Media." *Chem. Eng. Prog.*, **43,** 537–548, 1947.

Carman, P. C., "Fluid Flow through Granular Beds." *Trans. Inst. Chem. Engrs.* (London), **15,** 150–166, 1937.

Etherington, H., *Modern Furnace Technology*. London: Charles Griffin, 1944.

Everett, H. A., *Thermodynamics*.   New York: Van Nostrand, 1941.

Fulton, J. S., "Analysis of the Generation and Delivery of the Blast to the Metal in a Bessemer Converter." *Trans. AIME*, **145,** 175–193, 1941.

Furnas, C. C., *Flow of Gases through Beds of Broken Solids*.   U. S. Bureau of Mines, Bull. 307, 1929.

Gooding, E. J., and Thring, M. W., "The Flow of Gases in Natural Draft Furnaces." *Trans. Soc. Glass Tech.*, **25,** 21–85, 1941.

GROUME-GRJIMAILO, W. E., *Flow of Gases in Furnaces,* Translation. New York: John Wiley, 1923.

HUNSAKER, J. C., and RICHTMIRE, B. G., *Engineering Applications of Fluid Mechanics.* New York: McGraw-Hill, 1947.

LEVA, MAX, " Fluid Flow through Packed Beds." *Chem. Eng.,* **56,** No. 5, 115–117, May, 1949.

LEWIS, W. K., GILLILAND, E. R., and BAUER, W. C., "Characteristics of Fluidized Particles." *Ind. Eng. Chem.,* **41,** 1104–1117, 1949.

MCADAMS, W. H., *Heat Transmission.* New York: McGraw-Hill, 1942.

MOODY, L. F., "Friction Factors for Pipe Flow." *Trans. ASME,* **66,** 671–684, 1944.

MURPHY, NELSON F., *Dimensional Analysis.* Virginia Polytechnic Institute, Eng. Expt. Station, Bull. 73, 1949.

PERRY, JOHN H., *Chemical Engineers' Handbook.* New York: McGraw-Hill, 1949.

ROUSE, HUNTER, *Elementary Mechanics of Fluids.* New York: John Wiley, 1946.

STEWART, F. C., and DOOLITTLE, J. S., "Fluid Flow Measurement by Head Type Metering Elements." *Instruments,* **12,** 175–193, 1939.

TRINKS, W., *Industrial Furnaces,* Vol. I. New York: John Wiley, 1950.

WALKER, W. H., LEWIS, W. K., MCADAMS, W. H., and GILLILAND, E. R., *Principles of Chemical Engineering.* New York: McGraw-Hill, 1937.

PROBLEMS

6–1 Combustion air for a gas burner is passed through a Venturi and the flow rate is automatically regulated to maintain a fixed pressure drop in the Venturi corresponding to 120% of theoretical air for combustion when the air temperature just upstream from the Venturi is 70°F. If the fuel rate and the pressure drop in the air Venturi are kept the same, but the air temperature rises to 90°F, what will be the new flow rate of air, in % of theoretical for combustion?

6–2 A pitot tube is installed with its impact opening along the center line of a long pipe 1 ft in diameter. The pipe carries air at 150°F and 12 psi gauge. Atmospheric pressure is 745 mm. The pressure difference measured by the pitot tube is 0.42 inch water. Calculate the total flow rate in the pipe, lb air/hr.

6–3 A flowmeter is needed in the laboratory to measure gas flow rates in the range from 200 to 500 ml/min. This flowmeter will be made using a straight length of standard glass capillary tubing and a water manometer. The manometer should not read less than 1 inch pressure difference for the lowest flow rate or more than 15 inches for the highest flow rate. Also, the calibration should be linear, that is, with pressure drop directly proportional to flow rate. Capillary tubing is available with bores of $\frac{1}{2}$, 1, $1\frac{1}{2}$, and 2 mm. (a) Design and sketch a flowmeter to meet these requirements. (b) Would you consider using a mercury manometer instead of a water manometer?

6–4 An air duct of circular cross section $1\frac{1}{2}$ ft in diameter carries up to a maximum of 2500 ft³/min (STP) of air at 80°F and 1 atm. An orifice meter with corner taps is to be installed to meter the air flow, and the pressure difference will be measured with an instrument reading up to 2 inches water, full scale. (a) What diameter of sharp-edged orifice should be used to obtain full-scale reading with the maximum flow rate of air? (Assume a discharge coefficient of 0.61 and then check this assumption against Fig. 6–11 after diameter of orifice is calculated.) (b) With the proposed orifice meter, over what range of flow rates and pressure readings would you expect close proportionality between flow rate and $(\Delta p)^{1/2}$? (c) Estimate the total pressure drop added to the duct system by friction in the orifice installation at the maximum flow rate. (d) If the air is supplied by a blower operating at 50% over-all efficiency, what power consumption would be added by the orifice installation? (e) How would you decide whether or not to add a Venturi instead of an orifice?

6–5 A duct 50 ft in length and circular in cross section is to carry 500 ft³/min of air at substantially standard conditions with a pressure drop of no more than 0.2 inch water. (a) What is the smallest diameter of duct that can be used? (b) What horsepower is consumed in maintaining air flow through the duct specified in (a)?

6–6 (a) Estimate the horsepower required for a pump installation to deliver 100 gal of water/min in the following system: Lift, 50 ft; length of pipe, 200 ft; pipe size, 3-inch standard (3 inches inside diameter); fittings, four 90° elbows and one open gate valve; free discharge; submerged intake; mechanical efficiency of pump, 70%. The viscosity of water at the operating temperature is 0.01 poise. (b) If the gate valve is three-quarters closed, what will be the power consumption and rate of discharge of water? Assume pump is operated to give same pressure difference across pump as in (a).

6–7 (a) Estimate the air velocity (ft/min) and flow rate (ft³/min, STP) in a 6-inch square hole through an 18-inch wall into a furnace chamber at a level where the draft is 0.08 inch water. (b) Calculate the Btu/day required to heat this air to 2500°F and the corresponding coal consumption (tons/day) if the gross available heat yielded by the coal at 2500°F is 4000 Btu/lb.

6–8 A brick chimney 10 ft inside diameter (round) and 150 ft high is to handle waste gases (average molecular weight = 30) at 600°F. The barometric pressure is 735 mm and the outside air is at 80°F. It may be assumed that the gases do not cool appreciably in passing through the chimney. Make the necessary calculations and prepare a graph with two curves showing, respectively, (a) draft (inches water) at bottom of stack as *ordinate* and flow rate of waste gases (ft³/min, STP) as *abscissa* and (b) horsepower equivalent of the flow energy available for furnace draft at the bottom of the stack as *ordinate* and flow rate of waste gases as *abscissa*. The graph should cover the complete range of stack flow rates from 0 to the flow rate at which the available draft at the bottom of the stack is nil.

6–9 A brick flue must be designed to conduct 15,000 ft³/min (STP) of flue gases from a furnace to a stack. This flue will be horizontal, with a total length of 300 ft, and with four sharp 90° bends. Also the flue will be of rectangular cross section with a 2:1 ratio of height to width inside. The flue gases will analyze approximately 10% CO_2, 10% H_2O, 5% O_2, and 75% N_2, and will be at an average temperature of 600°F.

(a) Calculate the pressure drop in inches water to be expected if this flue were 4 ft high by 2 ft wide, inside. (b) Calculate the energy consumed by friction in the 4 by 2 ft flue, in horsepower. (c) What would be the minimum cross-sectional dimensions of the flue if the pressure drop were limited to 0.1 inch of water?

6–10 1200 ft³/min (STP) of air at 546°C flows through a duct 200 ft long and 1 sq ft in cross-sectional area which delivers it at a positive gauge pressure of 1 inch water to a fuel burner. The duct drops 50 ft in height in the 200-ft run to the burner. The duct is well insulated, so it can be assumed that the air does not cool appreciably as it passes through the duct. Outside air is at 25°C. What will be the reading of a water manometer, which has one leg open to the atmosphere and the other connected to a pressure tap at the beginning of the 200-ft run?

6–11 A smelter stack is 200 ft high by 15 ft in diameter and receives 300,000 ft³/min (STP) of flue gases at 200°C. The gas enters the bottom of the stack from a 12-ft square flue, and loses all its kinetic energy as friction on entering the stack. The flue and furnace system require a suction of 2 inches water in the 12-ft flue at a point near the stack. What horsepower fan would be required in the 12-ft flue to obtain the necessary draft? Assume 50% mechanical efficiency.

6–12 A Bessemer converter is a cylindrical vessel in which air is blown through liquid iron to oxidize the carbon, silicon, and other elements present in the iron. In a given 25-ton converter, the air is introduced into the liquid iron through 200 tuyères, each tuyère consisting of a tubular passage 30 inches long and $\frac{5}{8}$ inch in diameter. Each tuyère may be considered as a crude adiabatic nozzle with a discharge coefficient of 0.7 compared to an ideal, frictionless, adiabatic nozzle. The tuyères discharge at a pressure of 19 psi (atmospheric pressure + head of liquid iron). The air is supplied to the tuyères at 200°F and 40 psi absolute.

Calculate: (a) The quantity of air consumed by the converter in ft³/min (STP), allowing 25% extra for leakage. (b) The horsepower required to produce this quantity of air at 30 psi gauge (at the blower), assuming an over-all blower efficiency of 75%. The air entering the blower is at 14 psi absolute and 70°F. (c) The temperature of the air leaving the blower at 30 psi gauge if the compression process were frictionless and adiabatic. (d) Express the total kinetic energy of the air moving through the tuyères (assume temperature is 200°F) in horsepower. How is the bulk of this mechanical energy expended?

CHAPTER 7

STEADY HEAT FLOW

Previous chapters (3, 4, and 5) have dealt with the heat requirements of metallurgical processes, with the heat supplied by burning fuels, and with the determination of the various quantities of heat produced and used in metallurgical processes. The relationships developed in the previous chapters were based mainly on the principle of conservation of energy, and on balances of energy input and output, which do not give information about the mechanisms of heat production and utilization. However, an understanding of the mechanisms by which heat energy is transferred from the points of generation to the points of utilization or loss is necessary to the engineer because these mechanisms determine rates of heat flow. That is, accomplishing the flow of heat from point of supply to point of use is just as essential as supplying the right quantity of heat. Frequently, it is the rate at which heat is transferred during some part of a process that determines the practical rate at which the process as a whole can be conducted, or determines the rate of production and capacity of a given piece of equipment. A typical case is the industrial calcining of lump limestone at high temperatures by the endothermic reaction $CaCO_3 \longrightarrow CaO + CO_2$, in which the rate of production of CaO is fixed mainly by the rate at which heat is conducted from the surface to the interior of a lump of calcining limestone. In many melting and smelting processes in which cold solid materials are received and hot molten products discharged, furnace rates are determined by the rates at which heat can be transferred from the gaseous combustion products to the charge. Heat exchangers represent still another kind of system in which rates of heat flow are directly and obviously related to equipment design and equipment capacity.

While some directions of heat flow are essential to a given process, others such as the flow of heat to the surroundings may represent loss. Accordingly, a common problem in metallurgical systems is to evaluate heat losses and then to control them by proper design of equipment and by choice of materials.

Another class of heat flow problem important to metallurgists deals with heating and cooling of solid bodies; that is, with unsteady heat flows. When metal is cast, not only does the rate of heat transfer to the surroundings determine the rate of solidification, but also the nature and pattern of heat flow during solidification have major effects on the structure and properties of the casting. Heat treatment of metals also involves close control over heating and cooling rates of metal bodies.

Heat moves by three different mechanisms: *conduction*, *convection*, and *radiation*. These mechanisms are described and the principles governing rate of heat flow by each mechanism are developed in the next three sections of this chapter. Then methods of applying these principles to a few more complex systems of steady heat flow are considered. Chapter 8 deals with the special methods of handling problems in heating and cooling of solids, or unsteady heat flow.

Conduction

Conduction is the flow of heat through a body occurring without displacement of the particles which make up the body. Although most engineering problems in conduction of heat deal with conduction through solids, this mechanism also accounts in part for flow of heat through liquids and gases. The transmission of heat energy by conduction consists in the transmission of kinetic energy from particle to particle within the substance. This kinetic energy is primarily the kinetic energy of vibration and other motions of the molecules and atoms, and is transferred from molecule to molecule by collisions and other interactions. In metals, which generally are much better conductors than nonmetals, the thermal energies and motions of the free electrons play a major role in heat conduction.

The basic principle of heat conduction, usually associated with the name of Fourier, may be stated simply as follows: *the rate of heat flow across a unit of surface is proportional to the temperature gradient perpendicular to the surface.* A convenient mathematical form of this principle is given in Eq. (7–1), referring to the system sketched in Fig. 7–1.

Fourier's Law

$$q = - kA \frac{dt}{ds}. \qquad (7\text{–}1)$$

steady heat flow

In this equation, q is the rate of heat flow (heat units per unit time) along any path

like that shown in Fig. 7–1, A is a cross-sectional area perpendicular to the path of flow, and dt/ds is the temperature gradient or rate of change of temperature with distance in the direction of the flow at plane A. The proportionality constant k is the *thermal conductivity* of the body through which heat is flowing, measured at the plane A and at the temperature t in this plane. Strictly speaking, Eq. (7–1) applies only to a path of *steady heat flow*, in which q remains steady and the temperature t at any position s along the path does not vary with time.

If *unsteady heat flow* along the path of Fig. 7–1 is considered, the relation between heat flow and temperature gradient should be written

unsteady heat flow

$$q = - kA \left(\frac{\partial t}{\partial s} \right)_\theta, \qquad (7\text{–}2)$$

in which q is the rate of heat flow at time θ, and $(\partial t/\partial s)_\theta$, the partial derivative of t with respect to s at constant θ, is the value of the temperature gradient at time θ. That is, for a system in the unsteady state, temperature varies with time as well as with position, $t = \psi(s, \theta)$. The rate of change of temperature with time at a given position then is the partial derivative, $(\partial t/\partial \theta)_s$.

Now the effect of unsteady heat flow can be considered with respect to a thin slice of the flow path between two planes of area A and area $A + dA$ and having a thickness δs. The volume of this slice will be $A\,\delta s$ and the weight will be $\rho A\,\delta s$, ρ being the density. If the rate of heat flow at a given time into this slice, q_s, is greater than the rate of heat flow out of this slice, $q_{s+\delta s}$, the difference in these two flows $-(\delta q)_\theta$ will serve to heat and raise the temperature of the material. That is,

$$-(\delta q)_\theta = (q_s - q_{s+\delta s})_\theta = c_p \rho A\,\delta s \left(\frac{\partial t}{\partial \theta} \right)_s. \qquad (7\text{–}3)$$

Dividing by δs,

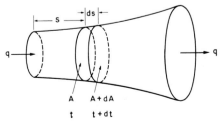

FIG. 7–1.   A heat-conduction path.

$$\left(\frac{\partial q}{\partial s}\right)_\theta = -c_p \rho A \left(\frac{\partial t}{\partial \theta}\right)_s. \qquad (7\text{-}4)$$

Differentiating Eq. (7–2) with respect to s at constant θ, and assuming k is constant,

$$\left(\frac{\partial q}{\partial s}\right)_\theta = -kA \left(\frac{\partial^2 t}{\partial s^2}\right)_\theta - k \left(\frac{dA}{ds}\right)\left(\frac{\partial t}{\partial s}\right)_\theta. \qquad (7\text{-}5)$$

Eliminating $(\partial q/\partial s)_\theta$ between Eqs. (7–4) and (7–5) and rearranging terms leads to a general equation for unsteady heat flow along a fixed path:

$$\left(\frac{\partial t}{\partial \theta}\right)_s = \frac{k}{c_p \rho}\left[\left(\frac{\partial^2 t}{\partial s^2}\right)_\theta + \left(\frac{d \ln A}{ds}\right)\left(\frac{\partial t}{\partial s}\right)_\theta\right]. \qquad (7\text{-}6)$$

The quantity $k/c_p\rho$ is a relationship among three properties of the material; and it can therefore itself be considered a property, and is defined as the *thermal diffusivity*, α. In dealing with unsteady systems it is convenient to deal directly with this property rather than with conductivity, specific heat, and density because these will always appear in the same combination $k/c_p\rho$ in any solution of Eq. (7–6).

Practical systems of heat conduction require integrations and solutions of Eq. (7–1) for steady systems, and of Eq. (7–6) for unsteady systems. The nature of the solution will depend in general on the geometrical configuration of the system and on the properties of the materials through which heat is conducted, specifically the thermal conductivity k and the thermal diffusivity α. As might be expected, Eq. (7–1) for steady systems is readily solved in most instances, and the most common solutions are taken up in the remainder of this section. On the other hand, solutions of the partial differential equation (7–6) for unsteady heat flows pose great mathematical difficulties even for systems of simple geometry. A number of the most useful special solutions of Eq.

(7–6) are considered in Chapter 8 which deals with heating and cooling of solids.

Units. Unfortunately, the various quantities in Eqs. (7–1) to (7–6) can be expressed in a great many different combinations of units, and in addition some special units have been invented for heat flow calculations. Accordingly, to avoid confusion, only one consistent set of units will be used in this chapter, with the understanding that conversions to other units can be made where necessary. The system chosen is probably the one most widely used in engineering work, and involves the following units:

Rate of heat flow (q)	Btu/hr
Area (A)	sq ft
Temperature (t)	°F
Absolute temperature (T)	°R (= °F + 460)
Distance (s)	ft*
Thermal conductivity (k)	Btu-ft/hr-ft²-°F
Thermal diffusivity (α)	ft²/hr
Heat transfer coefficients (h, U)	Btu/hr-ft²-°F

Thermal conductivities of solids. Solids exhibit about a 10,000-fold range of conductivities, from over 200 Btu-ft/hr-ft²-°F for the metals silver and copper to 0.025 Btu-ft/hr-ft²-°F for light, porous, nonmetallic materials such as mineral wool, cotton wool, cork, balsa, etc. Table A–6 in the Appendix gives values of k for various common materials.

Metals as a group are the best conductors, principally because of the large contribution their free electrons make to heat conductivity. Accordingly, it is not surprising to

---

* A slightly different set of units also quite common in American engineering practice involves the unit inches for s and Btu-inch/hr-ft²-°F for k, so that care must be taken to avoid confusion.

find for metals a close relationship between heat conductivity and electrical conductivity, since the latter property of metals is also largely determined by the behavior of the free electrons characteristic of the metallic state. As a first approximation, the ratio of thermal to electrical conductivity at a given temperature, known as the Wiedemann-Franz ratio, is the same for all metals. Also, both the thermal and electrical conductivities for metals are in many cases quite sensitive to the addition of small quantities of alloying elements or fractional percentages of impurities.

Thermal-conductivity data for nonmetallic solids are relatively scanty, probably because of the great difficulties involved in making satisfactory experimental measurements. However, fairly adequate conductivity figures are available for many of the common refractory materials and other materials of construction used in metallurgical engineering systems. The k values for most of the common heavy furnace refractories fall in the general range of 0.5 to 3 Btu-ft/hr-ft²-°F. On the other hand, the unusual materials graphite and silicon carbide, both of which are fairly good electrical conductors, have heat conductivities of the order of 50 to 100 and 6 to 15, respectively, depending on the temperature and form of the material.

The low thermal conductivities of insulating materials generally are the result of porous structure. Thus for a given solid substance, k will vary considerably with the over-all bulk density and will also depend on whether the material is in the form of loose grains or is in a sintered type of structure. Strictly speaking, heat flow through a porous solid system involves mechanisms other than simple heat conduction, such as convection and radiation within the pore spaces. However, some of the lowest apparent conductivities of insulating materials are obtained with light, loosely-packed powders or fibers. Such systems may approach closely the conductivity of still air, which is about 0.02 Btu-ft/hr-ft²-°F. Moreover, in such systems the conductivity of the solid itself has only a secondary effect on that of the aggregate. A striking illustration of this is the behavior of graphite; solid, dense graphite is a good conductor ($k > 50$) while a minus 100-mesh powder of the same material is a good insulator ($k < 0.1$).

The heat conductivity of a given solid varies with temperature, and this variation is large enough so that it should be taken into account in engineering calculations whenever the necessary data are available. Manufacturers of furnace refractories and insulating materials often supply plotted curves for their products showing how k varies with t over the temperature range in which the material is used. Often the relation between k and t is linear, or can be considered linear over a moderate temperature range, and therefore can be expressed as

$$k = k_0(1 + at), \qquad (7-7)$$

in which k_0 is the conductivity at 0°F and a is the temperature coefficient of the thermal conductivity with dimensions of $(°F)^{-1}$. The temperature coefficient can be either positive or negative.

Thermal conductivities of liquids and gases. Knowledge of the conductivities of liquids and gases is scantier than for solids, since the measurements are more difficult to make. The particular difficulty is that in most fluid systems the flow of heat by convection or fluid movement predominates over the much smaller effect of conduction. By the same token, of course, for engineering purposes there is somewhat less need for accurate conductivity measurements in such systems. However, as will be seen in the subsequent discussion of convection, the rate of heat flow by convection itself is related to the conductivity of the fluid.

Liquid metals are relatively good conductors of heat, although not as good as solid metals. Thus k for most liquid metals falls between $\frac{1}{3}$ and $\frac{2}{3}$ the value of k for the solid metal, both being measured at the melting point of the metal. Water and water solutions are better conductors than other nonmetallic liquids; water has a conductivity of about 0.35 at ordinary temperatures, while most organic liquids have conductivities in the range 0.07 to 0.04 Btu-ft/hr-ft²-°F.

The concept of heat conduction by transfer of kinetic energy from molecule to molecule by collision can be applied to gases in a relatively straightforward manner. The kinetic theory of gases, in its simplest form, accounts for such properties of gases as conductivity, viscosity, heat capacity, diffusivity, etc., by considering the gas as composed of elastic spheres that move at high velocities through the space occupied by the gas and constantly collide with each other and with the walls of the vessel. The heat energy contained in the gas is the kinetic energy of motion of these spheres. By applying the ordinary physical laws of motion to these spheres, important relations among the properties of gases can be deduced. The kinetic theory leads to the conclusion that the thermal conductivity should be proportional to the product of the specific heat and the viscosity. This relationship can be expressed in the form

$$\frac{c_p\mu}{k} = N_{\text{Pr.}} \qquad (7\text{--}8)$$

The expression $c_p\mu/k$ is a dimensionless number * known as the Prandtl number ($N_{\text{Pr.}}$)

---

* Note that μ should be in lb *mass*/hr-ft (and not in lb *mass*/sec-ft as in Chapter 6) if k is in Btu-ft/hr-ft²-°F and c_p is in Btu/lb-°F.

and is useful in dealing with convection heat flow. This number is substantially independent of temperature and pressure under ordinary conditions, and is 0.74 for air and other diatomic gases, 0.67 for monatomic gases, and 0.78 to 0.80 for triatomic and more complex gases. Using these values, k can be calculated from c_p and μ. (Note: Methods of evaluating μ were discussed in Chapter 6, page 152.)

It should be emphasized that conduction of heat through bodies of gas is usually negligible in quantity compared with the heat flows by convection and radiation. Accordingly, engineering calculations of heat conduction through gases are rare. The chief uses of gaseous conductivities are their indirect uses in estimating convection, to be discussed later. Another indirect application of interest to the metallurgist is automatic gas analysis based on the variation of thermal conductivity with composition.

Simple heat conduction systems. For steady heat conduction along a given path (Fig. 7–1), q is constant and does not vary either with position (s) along the path or with time (θ). Accordingly, Eq. (7–1) is readily placed in a convenient form for integration along the path of heat flow from position 1 to position 2:

$$q \int_{s_1}^{s_2} \frac{ds}{A} = - \int_{t_1}^{t_2} k\, dt. \qquad (7\text{--}9)$$

In this equation k is grouped under the integral sign with dt because k varies with t and the relationship of k to t is frequently known. The area A is under the integral sign with ds because in the general case the cross-sectional area of the path varies with position along the path and for simple paths can be expressed as an algebraic function of s. Since q is a constant for steady flow on a given path, it is not under either integral sign. The usual calculation of rate of heat

conduction then can be expressed as

$$q = \frac{-\int_{t_1}^{t_2} k\, dt}{\int_{s_1}^{s_2} \dfrac{ds}{A}}. \qquad (7\text{–}10)$$

Evaluation of the integral $\int_{t_1}^{t_2} k\, dt$ requires information as to how k varies with t over the the temperature range t_1 to t_2, and with such data a standard graphical integration can be made. For many materials, only a single value of k at one temperature is available, so it is necessary to assume that k is constant, and the integral $\int_{t_1}^{t_2} k\, dt$ is taken simply as $k(t_2 - t_1)$. However, if k is linearly related to t or is assumed so [Eq. (7–7)], it is easily shown that

$$\int_{t_1}^{t_2} k\, dt = \frac{k_1 + k_2}{2}(t_2 - t_1) = k_{av}(t_2 - t_1).$$
$$(7\text{–}11)$$

Following this equation, then, heat conduction calculations for a path with temperatures t_1 and t_2 at the two ends of the path make use of the arithmetic average k_{av} of the conductivities k_1 and k_2 at the two ends of the path. This simple average is commonly used in engineering calculations, since the assumption of a linear relation between k and t usually affords an adequate representation of the available conductivity data.

The integral $\int_{s_1}^{s_2} (ds/A)$ depends on the geometric configuration of the heat flow path. For many common systems, A is constant along the path, so that

$$\int_{s_1}^{s_2} \frac{ds}{A} = \frac{s_2 - s_1}{A} = \frac{\Delta s}{A}.$$

This simple result applies to flow through a flat wall, the flow being perpendicular to the wall surfaces. Also, the same relation is

used for flow along a well-insulated rod or wire of uniform cross section. For these systems, Eq. (7–10) is often used in the form

$$q = -\,k_{av}A\left(\frac{t_2 - t_1}{s_2 - s_1}\right) = -\,k_{av}A\,\frac{\Delta t}{\Delta s}, \quad (7\text{–}12)$$

which will be recognized as being very similar to Eq. (7–1) except that Δt and Δs are finite increments representing, respectively, the total temperature drop and the total length of the path of heat flow. Although Eq. (7–12) applies rigorously only to a flat wall with thickness small in comparison with other dimensions (Fig. 7–2), it serves satisfactorily for estimations on other, more complex systems in which the assumption of constant area is only a first approximation. For many estimations, good accuracy is often attained by using an average area for A in Eq. (7–12).

In applying Eq. (7–12) and other integrated heat conduction relations, care must be taken in evaluating the temperature drop $(t_2 - t_1,\text{ or } \Delta t)$. These relations are valid only for temperatures and temperature differences *within* the conducting body. When conduction across a body is considered, t_2 and t_1 are the *surface temperatures*, but must be temperatures *just within the surface of the conducting body*. As will be explained later,

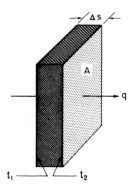

FIG. 7–2.   Heat flow through a flat wall.

large temperature drops generally occur over short distances at the interfaces between two bodies or two phases. For example, at a gas-solid interface, the major part of the temperature difference between the solid surface (just within the solid) and the body of the gas may be found across a thin film of stagnant gas adjacent to the solid. Also, when two solid bodies (for example, two bricks) are placed in contact, the contact region generally affords additional resistance to heat flow, so that there may be a considerable diffe in surface temperatures of the two adjacent brick surfaces. Temperature differences such as these clearly should be omitted from the temperature differences involved in calculation of conduction. Experimental measurement of temperatures just inside a body requires special techniques and often proves troublesome, but it is essential for accurate evaluation of conduction.

EXAMPLE

The inside and outside surface temperatures in a 9-inch silica-brick furnace wall are 2400°F and 650°F, respectively. The thermal conductivity of the brick is 0.90 Btu-ft/hr-ft²-°F at 1000°F and 1.20 Btu-ft/hr-ft²-°F at 2000°F. The wall area is 200 sq ft and the furnace is heated with a fuel gas having a low calorific value of 550 Btu/ft³. Under normal operating conditions, the gross available heat at furnace temperature is 55% of the low calorific value of the fuel gas.

(a) What is the rate of heat loss through this wall per sq ft of wall area?

(b) What is the temperature exactly halfway through the wall?

(c) What part of the fuel consumption can be attributed to the heat loss through the wall?

SOLUTIONS

(a) From Eq. (7–12),

$$\frac{q}{A} = -k_{av}\frac{\Delta t}{\Delta s}.$$

Since k is given at only two temperatures, it is necessary to assume that k varies linearly with temperature. Therefore, k_{av} is equal to the value of k at the mean wall temperature of 1525°F and is found by interpolation between the values given for 1000°F and 2000°F:

$$k_{av} = 1.06 \text{ Btu-ft/hr-ft}^2\text{-°F}.$$

Substituting in the equation given previously, and converting wall thickness into feet,

$$\frac{q}{A} = 1.06\frac{(2400 - 650)}{9/12} = 2480 \text{ Btu/hr-ft}^2.$$

(b) As a rough first approximation only, it might be assumed that the temperature halfway through the wall is the average of the two surface temperatures, which is 1525°F. Using this approximation, a tentative value of k_{av} (2400 to 1525°F) for the inside half of the wall is found:

$$k_{av} = k_{1960} = 1.19.$$

Using this tentative value and q/A found in (a), Eq. (7–12) can be solved for Δt across the inside half of the wall:

$$\Delta t = -\frac{q\,\Delta s}{k_{av}A}$$
$$= -\frac{4.5/12}{1.19} \times 2480$$
$$= -780°\text{F.}$$

Accordingly, a second approximation of the midway temperature is found:

$$t = 2400 - 780 = 1620°\text{F.}$$

Using this value, a second tentative figure for k_{av} for the inside half of the wall is found to be 1.20 Btu-ft/hr-ft²-°F and recalculation of Δt gives $-775°\text{F}$. The third approximation of the midway temperature is then

$$t = 2400 - 775 = \underline{1625°\text{F, midway through the wall.}}$$

Further repetition of the successive approximations will give no significant change, so this answer can be considered as correct for the data given. It is to be noted that the finally calculated midway temperature is 100°F higher than estimated for a linear temperature gradient across the wall. This difference is, of course, due to the variation of k with temperature.

(c) The total heat loss for 200 sq ft of wall is

$$200 \times 2480 = 496{,}000 \text{ Btu/hr.}$$

This quantity of heat leaves the high-temperature zone of the furnace, and thus is a loss of gross available heat. The fuel gas furnishes gross available heat at the rate of

$$550 \times 0.55 = 302.5 \text{ Btu/ft}^3.$$

Accordingly, the portion of the fuel consumption which should be attributed to heat loss through the silica brick wall is

$$\frac{496{,}000}{302.5} = \underline{1640 \text{ ft}^3/\text{hr.}}$$

Concentric cylinders. Figure 7–3 illustrates another kind of heat conduction often encountered, radial heat flow between two concentric cylinders. Typical cases are flow through the walls of circular pipes and through pipe insulation, and flow of heat out through the insulation in cylinder-shaped laboratory tube furnaces.

The general path integral $\int_{s_1}^{s_2} ds/A$ is readily solved for flow between concentric cylinders by expressing s and A in terms of

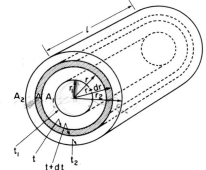

Fig. 7–3. Radial heat flow between concentric cylinders.

the appropriate dimensions of the cylinders. If r represents the radius from the axis to any cylindrical area A of the heat flow path and l is the length of the concentric cylinders, $ds = dr$ and $A = 2\pi rl$; then

$$\int_{s_1}^{s_2} \frac{ds}{A} = \int_{r_1}^{r_2} \frac{dr}{2\pi rl} = \frac{1}{2\pi l} \ln \frac{r_2}{r_1} = \frac{2.3}{2\pi l} \log \frac{r_2}{r_1}, \quad (7\text{-}13)$$

and Eq. (7–10) can be rewritten

Conc. cylinders (thin walled)

$$q = -k_{av} \frac{2\pi l\, \Delta t}{2.3 \log \frac{r_2}{r_1}}. \quad (7\text{-}14)$$

For some purposes it is helpful to deal with the length of the path of heat flow, $\Delta s = r_2 - r_1 = \Delta r$ (for example, Δr might be the thickness of the pipe wall or of the insulation) and with the inside and outside areas, $A_1 = 2\pi r_1 l$ and $A_2 = 2\pi r_2 l$. In terms of these variables, Eq. (7–14) becomes

thick-walled
$A_2/A_1 > 2$

$$q = -k_{av} \left(\frac{A_2 - A_1}{2.3 \log \frac{A_2}{A_1}} \right) \frac{\Delta t}{\Delta r}. \quad (7\text{-}15)$$

The expression $(A_2 - A_1)/[2.3 \log (A_2/A_1)]$ is known as the *logarithmic mean area*. Thus Eq. (7–15), like Eq. (7–12), shows q as

the product of k_{av}, a characteristic or average area, and the ratio of total temperature drop to length of path. If the radial distance from inner to outer cylinder is less than the diameter of the inner cylinder (that is, a thin-walled pipe or a duct with wall thickness less than inside diameter), A_2/A_1 is less than 2 and the logarithmic mean area is substantially equal to the arithmetic mean area $(A_2 + A_1)/2$. However, if A_2/A_1 is greater than 2 (thick-walled pipes, thick insulation on small pipes, tube furnaces), the arithmetic mean is inaccurate and the log mean area should be calculated as shown in Eq. (7–15).

The relations just derived were for radial heat flow at right angles to the cylinder axes. If the cylinder is short in comparison to its diameter, end effects become important and the direction of heat flow may no longer be perpendicular to the axis, especially near the ends. Under these conditions, Eqs. (7–14) and (7–15) may lead to large errors. In intermediate cases, the end effects may be allowed for by arbitrarily considering the end effects as equivalent to an added length, the percentage addition being based on experience.

thick = $r_o - r_i$

EXAMPLE

A $\frac{3}{4}$-inch hot-water pipe (1.05-inch outside diameter; 0.824-inch inside diameter) is covered with a 2-inch layer of insulation having a thermal conductivity of 0.05 Btu-ft/hr-ft²-°F. The inside and outside surface temperatures in the insulation are 175°F and 90°F, respectively, and the water averages 180°F. The normal flow rate of the water is 5 gal/min.

(a) Estimate the rate of cooling of the water, in °F per 100 ft of pipe, with hot water flowing at the normal rate.

(b) Estimate the initial cooling rate of the water at 180°F with no flow in the pipe.

SOLUTIONS

(a) To find the rate of cooling of water, it is first necessary to calculate the rate of heat loss. Then the rate of cooling of the flowing water is found by a simple heat balance.

For a 100-ft length of pipe, the inner and outer areas of the insulation are found as follows:

$$A_1 = 2\pi r_1 l = \pi \times \frac{1.05}{12} \times 100 = 28 \text{ sq ft,}$$

$$A_2 = 2\pi r_2 l = \pi \times \frac{5.05}{12} \times 100 = 132 \text{ sq ft.}$$

Since $A_2/A_1 > 2$, average area must be calculated as the logarithmic mean:

$$A_{av} = \frac{(132 - 28)}{2.3 \log (132/28)} = 66 \text{ sq ft.}$$

(Note that the arithmetic mean area for this system would be 80 sq ft, which is 20% too large.)   Substituting in Eq. (7–15),

$$q = 0.05 \times 66 \times \frac{(175 - 90)}{2/12}$$
$$= 1680 \text{ Btu/hr for 100 ft of pipe.}$$

The flow rate of hot water (density 8.34 lb/gal) is

$$5 \times 8.34 \times 60 = 2500 \text{ lb/hr.}$$

Taking the specific heat of water as unity,

$$\text{rate of cooling} = \frac{1680}{2500} = 0.7°\text{F per 100 ft of pipe.}$$

(b) If the water stops flowing, the heat is lost by the static contents of the pipe.   The volume in 100 ft of pipe is

$$100 \times \pi \times \frac{(0.824)^2}{4 \times 144} = 0.37 \text{ cu ft.}$$

Taking the density of water as 62.4 lb/ft³, the initial cooling rate of the hot water in the pipe just after flow is stopped is therefore

$$\text{initial cooling rate (no flow)} = \frac{1680}{62.4 \times 0.37} = 73°\text{F/hr.}$$

As increasing thicknesses of insulation are added to a pipe or other long cylindrical object, the outside area (A_2) increases in direct proportion to the increase in thickness (Δr).   As a result, the outermost layer of the insulation, because of its greater area, offers less resistance to heat flow and has less insulating value than inner layers of the same thickness.   Accordingly, for practical purposes, a point of diminishing returns is soon reached where the cost of additional insulation becomes greater than the heat savings.   These characteristics are shown in a general way by the curves in Fig. 7–4, based on Eq. (7–14).   For these curves, the relative heat loss is measured by solving Eq. (7–14) for $q/(l\,\Delta t)$, which is the loss of heat in Btu/hr for a 1-ft length of pipe per °F temperature drop across the insulation.   The graphs show clearly that the first layers of insulation cause the largest decreases in heat loss and require less volume of insulating material.   Furthermore, comparisons of the curves for different values of k show that the thermal conductivity of the insulating material is the major factor determining the effectiveness of an insulating cover, regardless of the thickness.   That is, contrary to the situation in insulating a flat wall, it is not feasible to use a heavy coating of a poor insulator to

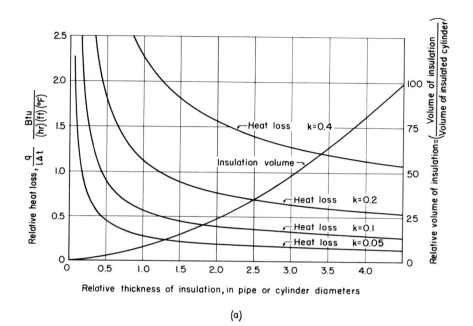

(a)

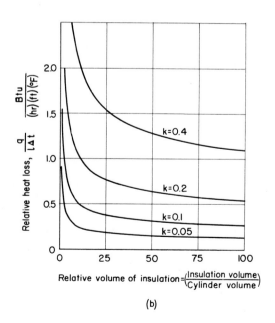

(b)

FIG. 7–4. Effects of insulation on heat loss from pipes and long cylinders.

obtain the same result as a moderate coating of a good insulator.

Concentric spheres. Radial heat conduction between two concentric spherical surfaces involves a heat flow path along which the cross-sectional area (A) varies as the square of the distance along the path (r^2). One practical system involving this type of path is a solid particle which is being chemically decomposed by heating. The outer shell of the particle is decomposed first, and the inside kernel is decomposed as heat flows in radially through this shell. For steady conduction between two concentric spheres of radii r_1 and r_2, it is readily shown that the integrations indicated in Eq. (7–10) lead to

$$q = -\ k_{av} \frac{4\pi r_1 r_2}{r_2 - r_1} \Delta t = -\ k_{av} \sqrt{A_1 A_2} \frac{\Delta t}{\Delta r}.$$
$$(7\text{--}16)$$

Thus, for two concentric spherical surfaces the proper average of the two areas, to be multiplied by the conductivity and the gross temperature gradient $(\Delta t / \Delta r)$, is the *geometric mean area*, $\sqrt{A_1 A_2}$.

Equation 7–16 leads to some interesting conclusions when applied as an approximation to some systems which are only roughly spherical. For example, we may consider the problem of insulating a small object or a small furnace cavity by enclosure in a large volume of insulating material. As the thickness of insulation becomes large compared with the size of the object,

$$\frac{r_2}{r_2 - r_1} \to 1 \text{ and } q \to -\ k_{av}(4\pi r_1)\ \Delta t. \quad (7\text{--}17)$$

Equation (7–17) gives the lowest rate of heat loss attainable by insulation for a given size of object (r_1), given temperature drop (Δt), and given insulation conductivity (k). This limit is reached when the insulation thickness becomes large compared with the size of the object. Reference to Eq. (7–16), for example, shows that when the insulation thick-

ness is only just greater than twice the diameter of the object, q has already been reduced to within 25% of the limiting value given by Eq. (7–17). Thus, the point of diminishing returns is reached more rapidly in insulating spheres or small, approximately equidimensional bodies than in insulating cylinders (see Fig. 7–4). Also, Eq. (7–17) shows that the conductivity of the insulating material determines the lowest rate of heat loss for a given hot body, so that it becomes impossible to use a larger volume of a poor insulating material to obtain the same results as with a lesser volume of a good insulator.

The design of small laboratory furnaces for high temperatures often involves a small working chamber with relatively thick insulation. The problem of minimizing heat loss from the furnace chamber is likely to be somewhat intermediate between the problem of insulating a cylinder (cross-sectional area of heat flow path proportional to distance from furnace axis) and that of insulating a sphere (cross-sectional area of heat flow path proportional to square of distance from center point of furnace). Regardless of where the furnace behavior falls between these two ideal behaviors, it should be clear from the foregoing discussion that little is gained by very thick insulation but that much is gained by using an insulating material of the lowest possible thermal conductivity.

Other geometric configurations. The three integrated heat conduction equations considered previously were based on three simple heat flow paths along which A was constant (flat wall), A was proportional to s (concentric cylinders), and A was proportional to s^2 (concentric spheres), respectively. When the geometric shape of the path does not follow one of these simple relations, derivation of a rigorous equation may involve considerable mathematical difficulties and is rarely attempted in connection with metallurgical systems. However, if necessity demands, graphical methods described in the literature on the subject serve reasonably well.

Estimations of heat flows through the walls of metallurgical furnaces are made in a variety of ways. Usually it is necessary to add up separate estimates for different parts of the furnace enclosure, where the wall thicknesses, refractory materials, or temperatures are different. Direct conduction calculations require measurements of surface temperatures inside and outside. However, methods of combining conduction, convection, and radiation calculations to obtain estimates of surface temperatures will be discussed later. When the necessary temperature and conductivity data are available, the calculations for large furnaces (inside dimensions large compared with wall thickness) are generally based on the simple flat-wall equation, Eq. (7–12). However, for small furnaces the area is not a constant and it becomes necessary to use some kind of average of the inside wall area A_1 and the outside area A_2. As long as A_2/A_1 does not exceed 2, the arithmetic, geometric, and logarithmic means of A_1 and A_2 are nearly the same, so that it is most convenient to use the arithmetic mean, $(A_1 + A_2)/2$.

When A_2/A_1 is greater than 2, the arithmetic mean area is too large and the geometric mean area $\sqrt{A_1 A_2}$ appears most suitable. Use of the geometric mean as an approximation corresponds to considering the furnace walls as bounded by concentric spherical surfaces with heat flow at right angles to the surfaces. Where the furnace walls are bounded inside and outside by rectangular parallelepipeds, the heat flow, especially in the corners, cannot be perpendicular to the outer bounding surfaces and the simple geometric mean is too large. For this configuration, with $A_2/A_1 > 2$, it is appropriate to use a modified geometric mean, 0.725 $\sqrt{A_1 A_2}$, in which 0.725 $(= \sqrt{\pi/6})$ is a semi-empirical factor which for a cube surface gives in effect the mean area as the geometric mean of A_1 and $\pi A_2/6$, the latter term representing the area of a sphere inscribed in the outer cube. If the furnace is unusually long in one dimension, the logarithmic mean $\dfrac{A_2 - A_1}{2.3 \log \left(\dfrac{A_2}{A_1} \right)}$ should be applied. Attempts to apply more rigorous conduction calculations to small furnaces do not seem justified, since data on surface conditions, temperatures, and the accuracy of available k data do not justify precise calculations.

EXAMPLE

The working chamber of a laboratory furnace is $6 \times 8 \times 15$ inches, and the walls, 6 inches thick on all sides, are made of a refractory brick with a thermal conductivity of 0.2 Btu-ft/hr-ft²-°F. When the furnace chamber is held at 1800°F, the average temperature of the outside surface is estimated to be 300°F.

Estimate the electrical power consumption of the furnace.

SOLUTION

When the furnace reaches a steady state, the power consumption will equal the heat loss, if both are expressed in the same units.

$$A_1 = \frac{2(6 \times 8 + 6 \times 15 + 8 \times 15)}{144} = 3.6 \text{ sq ft.}$$

The outer dimensions of the furnace will be $18 \times 20 \times 27$ inches.

$$A_2 = \frac{2(18 \times 20 + 18 \times 27 + 20 \times 27)}{144} = 19.2 \text{ sq ft.}$$

The modified geometric mean area can now be found:

$$A_{av} = 0.725\sqrt{A_1 A_2} = 0.725\sqrt{3.6 \times 19.2} = 6.0 \text{ sq ft.}$$

(Note: For comparison, the arithmetic mean area is 11.4 sq ft and the logarithmic mean is 9.3 sq ft. The calculated geometric mean of 6.0 sq ft is close to the result of 6.5 sq ft obtained by applying a more complicated formula often recommended for this type of system.)

Using the estimated mean area and other given data in Eq. (7–12),

$$q = 0.2 \times 6 \times \frac{1800 - 300}{6/12}$$

$$= 3600 \text{ Btu/hr} = \frac{3600}{3413} \text{ kw}$$

$$= \underline{1.05 \text{ kw, estimated power consumption.}}$$

Figure 7–5 illustrates the conduction of heat under the hearth of a furnace resting directly on the ground (a nonventilated hearth) or on a solid massive foundation, as worked out by Keller. The solid curves represent isothermal surfaces at various values of relative temperature * and the dotted lines show the directions of heat flow. This diagram shows one principle of heat conduction which is often useful in analyzing complex systems; that is, the heat flow paths always intersect isothermal surfaces at right angles. Once the pattern of isotherms and heat flow paths is worked out, reliable estimates can be made of the rate of heat flow. For example, application of graphical methods to estimate heat loss from furnace hearths such as the one shown in Fig. 7–5 leads to the following relation (J. D. Keller, "The Flow of Heat through Furnace Hearths," *Trans. ASME*, Vol. 50, p. 111, 1928):

---

* Relative temperature =

$$\frac{\text{actual temp.} - \text{ambient temp.}}{\text{furnace temp.} - \text{ambient temp.}}$$

$$q = SkA\frac{\Delta t}{D}, \qquad (7\text{–}18)$$

in which k is the conductivity of the hearth material, A is the hearth area, D is the smallest width of the hearth, and Δt is the temperature difference between the furnace interior and the furnace surroundings. The symbol S represents a dimensionless hearth shape factor varying from 3.75 for long rectangular hearths to 4.4 for square hearths.

Thermal resistivity and thermal resistance; series heat flow. The close analogies between conduction of heat and conduction of electricity are helpful in clarifying many aspects of heat flow. The underlying principles of both are the same; that is, the rate of conduction is proportional to the area of the path and to the rate of change of a driving force with distance in the direction of flow. Thus, starting with Eq. (7–1), all the quantitative heat-conduction equations considered so far are valid for conduction of electricity if current in amperes (coulombs per second) is substituted for q, if electrical conductivity is substituted for k, and if electromotive force

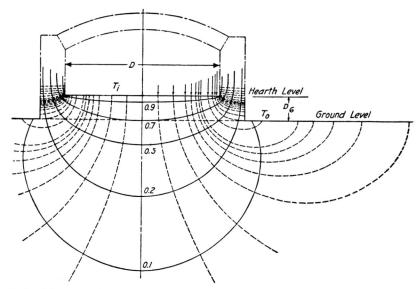

FIG. 7-5. Heat flow in nonventilated furnace hearths (after Keller, *Trans. ASME*).

in volts is substituted for t. As one result of this close relationship, complex heat-flow systems are sometimes studied by means of electrical models because electrical measurements are easier than heat measurements. It should also be pointed out that diffusion, another mechanism important to metallurgists, is governed by the same general principle and the same forms of mathematical relations. The rate of diffusion is proportional to area and to concentration gradient, and the proportionality constant is the diffusivity.

Just as the quantity of electrical resistance is more convenient to work with than electrical conductance in studying many electrical circuits, so in many thermal circuits it is advantageous to think in terms of thermal resistance rather than conductance. For this purpose, heat flow equations take the form analogous to Ohm's Law ($I = E/R$), or

$$q = \frac{\Delta t}{R}, \qquad (7\text{-}19)$$

in which R is the thermal resistance. Thus R for a given body might be defined as the temperature difference (°F) required to cause a heat flow of 1 Btu/hr. Comparison of Eq. (7-19) with Eqs. (7-12), (7-15), and (7-16) shows that the thermal resistances for conducting bodies of various shapes can be expressed as

$$R = \frac{\Delta s}{k_{av}A_{av}}, \qquad (7\text{-}20)$$

in which A_{av} is the value of average area appropriate to the geometric configuration of the heat flow path. For a cube of unit dimensions, R is the thermal resistivity and is equal to the reciprocal of the thermal conductivity.

The concept of thermal resistance is usefully applied to heat flow by mechanisms other than conduction. Thus, in the usual multilayered brick wall, the surfaces of contact between adjacent bricks are imperfect joints, so that each joint has a contact resistance ($R_{contact}$) in addition to the resistance to heat flow through the bricks by conduction. Furthermore, heat flows by convection and radiation also can be analyzed in terms of resistances, using Eq. (7–19) as the definition of resistance. However, for these mechanisms, Eq. (7–20) is not valid, and other ways of calculating R are used. Calculations of thermal resistances to heat flow by other mechanisms and by various combinations are taken up later in this chapter.

The value of the thermal-resistance concept is that it affords a simple approach to heat flow systems consisting of various kinds of resistances in series. Most practical systems are of this type. Again in analogy to electrical circuits, the total resistance along a given heat flow path is found simply by adding the resistances for all the parts of the path in series. Thus, for a given series path of steady heat flow,

$$q = \frac{(\Delta t)_1}{R_1} = \frac{(\Delta t)_2}{R_2} = \frac{(\Delta t)_3}{R_3} \cdots,$$

$$q = \frac{(\Delta t)_1 + (\Delta t)_2 + (\Delta t)_3 + \cdots}{R_1 + R_2 + R_3 + \cdots}$$

$$= \frac{\Sigma \, \Delta t}{\Sigma R}. \tag{7–21}$$

Example

A plane furnace wall consists of two layers, 9 inches of firebrick ($k = 1.2$) and $4\frac{1}{2}$ inches of insulating brick ($k = 0.15$). The interior surface of the firebrick is at 2400°F and the outer surface of the insulating brick is at 350°F.

(a) Estimate the rate of heat loss per sq ft of wall.

(b) Estimate the maximum temperature in the insulating brick.

Solutions

(a) In the absence of data, contact resistance at the joint between the firebrick and the insulating brick will be ignored. Applying Eqs. (7–21) and (7–20),

$$\frac{q}{A} = \frac{2400 - 350}{\dfrac{9/12}{1.2} + \dfrac{4.5/12}{0.15}} = \frac{2050}{0.625 + 2.5} = \underline{656 \text{ Btu/hr-ft}^2}.$$

It is noted that the insulating brick, while representing only $\frac{1}{3}$ of the wall thickness, accounts for $\frac{4}{5}$ of the thermal resistance.

(b) The temperature drop in the insulating brick is, applying Eq. 7–19,

$$\Delta t = qR = 656 \times 2.5 = 1640°F.$$

Thus the temperature at the inside surface of the insulating brick is $1640 + 350 = \underline{1990°F}$.

The temperature relationships in this example illustrate some important consequences of good insulation. In the first place, since most of the temperature drop is in the insulation, the full thickness of refractory brick is subjected to temperatures not far from the inside furnace temperature, and this may have serious effects on strength, service life, and properties. Second, the inside surface of the insulation is also subjected to a temperature close to that of the furnace and the material must be selected with this in mind.

Convection

When fluids are in motion, whether it be unidirectional flow, mixing, eddying, or gravitational currents, the fluid particles carry heat energy and their movements result in a flow of heat. Convection thus may be defined as heat flow associated with fluid motion. The most familiar everyday examples of convection are systems of *natural convection* or *free convection,* in which the fluid movement itself is caused by a difference in temperature. Thus, air is heated by a steam radiator and then rises to displace colder air at the top of a room. Similar currents are produced in heating a liquid, the fluid motion resulting from the fact that fluid density decreases with increase in temperature. When the heat flow is associated with systematic fluid movement produced by means other than natural convection (usually produced mechanically by pumps, blowers, agitators, and the like), it is known as *forced convection.*

For engineering calculations, it is not often necessary to deal directly with convection heat flow within a mass of gas or liquid. In most systems involving convection, the heat flow of practical interest is that between a body of gas or liquid and its containing surfaces. Thus, in a reverberatory furnace, the transfer of heat from gaseous products of combustion to the furnace walls and furnace charge is generally evaluated without detailed consideration of the convection from one part of the gases to another.

Figure 7–6 illustrates several characteristics of the process of heat flow from a hot fluid to a cold surface. The over-all heat flow occurs as the result of the difference Δt between the bulk temperature of the fluid and the surface temperature of the solid. Experimental measurements on many different kinds of convection systems have shown that most of the temperature drop from the surface to the interior of the fluid occurs in a very thin layer of fluid adjacent to the surface, as indicated by the temperature vs. distance curve in Fig. 7–6. In other words, most of the resistance to heat transfer between a fluid body and a surface is concentrated in a thin fluid film just at the surface. Within the body of fluid, the fluid movement, whether natural or forced convection, usually is rapid enough so that temperature differences are small. However, as the confining surface is approached, the fluid velocities decrease because of friction and, in fact, become zero just at the surface. For analytical purposes, it is helpful to consider the characteristics of the system in terms of a simplified model with all resistance to heat flow embodied in a *stagnant layer* of equivalent thickness Δs and having zero velocity with respect to the surface. Heat flow across the stagnant layer is mainly conduction, so that Eq. (7–12) can be applied,

$$q = - kA \frac{\Delta t}{\Delta s}, \qquad (7\text{–}12)$$

in which k, A, and Δs refer to the thermal conductivity of the fluid, the area of the sur-

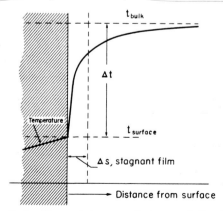

FIG. 7–6. Temperature vs. distance relationship for convection heat flow from a fluid to a solid surface.

face, and the equivalent thickness of the stagnant layer, respectively, and Δt is the difference between the surface temperature and the bulk fluid temperature.

The relation given above [Eq. (7–12)] is not especially useful for ordinary convection problems because the equivalent film thickness Δs is difficult to define and determine. Instead, it is more convenient to use the following empirical relation for heat flow between fluids and their confining surfaces:

$$q = h_c A \, \Delta t. \qquad (7\text{–}22)$$

This equation is consistent with the stagnant-layer concept of the heat flow mechanism in that the rate of heat flow q is shown to be directly proportional to surface area A and to the difference Δt between the surface temperature and the bulk fluid temperature. The proportionality constant h_c is known as the *convection heat flow coefficient* and has the dimensions of Btu/hr-ft²-°F in the English system. When the heat transfer between the fluid and surface is part of a series heat path, as is often the case, Eqs. (7–19) and (7–21) may be used, expressing the thermal resistance R by

$$R = \frac{1}{h_c A}. \qquad (7\text{–}23)$$

In terms of the stagnant-layer concept, $h_c = k/\Delta s$, as can be seen by comparing Eqs. (7–22) and (7–12). Since the coefficient h_c is inversely proportional to equivalent film thickness Δs, it is of interest to consider briefly the kinds of variables which are likely to affect Δs. To start with, it is clear that the bulk velocities of the fluid with respect to the surface should affect the thickness of the stagnant layer, higher velocities naturally resulting in thinner stagnant layers (smaller Δs, larger h_c). Also, the stagnant-film thickness should be greater for fluids of higher viscosity. Thus, the coefficient h_c

will depend on such characteristics of the fluid motion as the velocity, viscosity, density, etc. As is the case with other aspects of fluid flow, the relations between h_c and the variables describing the fluid motion are so complex that a theoretical derivation is difficult even for the simplest kinds of fluid flow.

Since convection heat flow is not readily evaluated on theoretical grounds, engineering calculations are usually based on Eq. (7–22) and on empirical data for the coefficient h_c. Sufficient work has been done in this field so that values of h_c are readily found or estimated for many of the systems usually encountered. Methods of evaluating h_c for a few of the most common types of natural and forced convection are discussed in subsequent parts of this section, and more detailed information is available in such books as *Heat Transmission* by W. H. McAdams, and in current engineering periodicals.

Natural convection within a fluid body. The nature of free convection within a body of fluid is shown by the simple system in Fig. 7–7. Fluid circulates between the side tube and the main fluid reservoir because the heated fluid in the side tube has a lower density than the cooler fluid in the main reservoir. For simplicity we may assume that the system reaches a steady state in

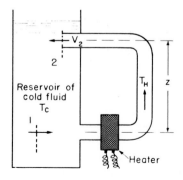

Fig. 7–7. Simple natural-convection system.

which the main reservoir dissipates heat to the surroundings at the same rate as heat is supplied through the heater. The density difference over the height z produces a difference in pressure which drives the flow. A mechanical energy balance for the fluid flow from section 1 in the main reservoir to section 2 at the discharge of the side tube gives the following relation:

$$\left(\frac{\rho_1 - \rho_2}{\rho_2}\right) z = \frac{V_2^2}{2g_c} + F = F' = \phi\left(\frac{V_2^2}{2g_c}\right),$$

$$(7\text{--}24)$$

in which ρ_2 and ρ_1 are the hot and cold fluid densities, respectively, z is the height of the side tube, and F is the friction loss in the side tube. Since the kinetic energy at section 2 is also dissipated as friction after the fluid enters the main reservoir, Eq. (7–24) signifies essentially that in a steady natural-convection system, the flow-energy input resulting from a height z of two adjacent fluid masses of different densities is $z(\rho_1 - \rho_2)/\rho_2$ ft-lb *force*/lb *mass* of fluid, and this energy is consumed as friction (F'), which by dimensional analysis (Chapter 6) can be expressed as the product of a dimensionless factor ϕ and the kinetic energy at some point in the system $V_2^2/2g_c$. The factor ϕ will depend, of course, on the geometry of the system and on the nature of the flow (as measured, for example, by the Reynolds number). For gases, the ideal-gas equation gives as energy input

$$\left(\frac{\rho_1 - \rho_2}{\rho_2}\right) z = \left(\frac{T_H - T_C}{T_C}\right) z; \quad (7\text{--}25)$$

while for liquids,

$$\left(\frac{\rho_1 - \rho_2}{\rho_2}\right) z = \beta(T_H - T_C)z, \quad (7\text{--}26)$$

in which T_H and T_C are the temperatures (°K) of the fluid in the side tube and in the reservoir, respectively, and β is the volume coefficient of expansion of the liquid. Also, the friction term in Eq. (7–24) can be expressed in terms of the total mass flow w in lb/sec or in terms of the rate of heat transfer q into the reservoir in Btu/sec;

$$\phi\left(\frac{V_2^2}{2g_c}\right) = \frac{\phi}{2g_c} \frac{w^2}{\rho_2^2 A_2^2}$$

$$= \frac{\phi}{2g_c}\left(\frac{q}{\rho_2 A_2 c_p(T_H - T_C)}\right)^2. \quad (7\text{--}27)$$

By using Eq. (7–27) with Eq. (7–26) or (7–25), it is seen that velocity V_2 and flow rate w generated by natural convection are proportional to the square root of the temperature difference $T_H - T_C$, while heat transfer rate q is proportional to the $\frac{3}{2}$-power of the temperature difference, for a given system and a given value of ϕ.

In most natural-convection systems, the flow of the fluid does not follow a path so definitely constrained as that in Fig. 7–7. In general, a hot surface in a fluid body causes a more or less systematic circulation of the fluid, with the largest velocities and temperature variations occurring not far from the hot surface. Some workers have measured the variations of fluid velocity and temperature near vertical plane surfaces, hot cylinders, and other geometrical surfaces and have in some cases correlated their measurements successfully on a theoretical basis. For example, such measurements in the vicinity of a heated vertical plate exposed to air have shown the upward air velocity increasing sharply from zero at the surface to a maximum a few millimeters away and then decreasing to a low value a few centimeters out from the surface. Temperature measurements for the same system have shown the temperature of the upward moving fluid to drop from a maximum at the surface to nearly the bulk air temperature a centimeter or so from the surface.

Effects of natural convection on furnace design. The tendency of a hot fluid to rise and displace colder fluid must be taken into account in designing for the flow of gases through furnaces and other equipment. The simplest illustration is gas flow through a vertical chamber, of which several examples are shown in Fig. 7–8. To start with, consider the hot and cold gases as two separate fluids of slightly different density, like kerosene and water. Systems (a) and (b) are analogous to many furnace operations in that hot gases are supplied to the working chamber while cold gases are withdrawn from the chamber. If the hot gases are supplied at the top and cold gases are removed at the bottom, as in (a), good displacement of cold gases by hot is attained because the hotter, less dense gas remains above the colder gas. On the other hand, if hot gas is supplied at the bottom as in (b), displacement of the cooler gases is poor because the lighter hot fluid tends to rise through the cold gas and pass directly out the exit port. Although for continuously operating furnaces the concept of a gas at one temperature displacing a gas at another temperature is a considerable oversimplification, nevertheless the same principles are applicable. That is, the exit port for cooled waste gases should be at or near the hearth level, or below if possible, to avoid the short-circuiting of hot gases directly to the flue without coming in contact with the charge. When the exit port must be high because of other design requirements, the tendency for the flame and hot combustion products to float over colder gases on the hearth is overcome by jetting the flame toward the hearth with just sufficient kinetic energy to overcome the buoyant effect.

Similar considerations apply to systems in which a cold gas is to displace a hot gas

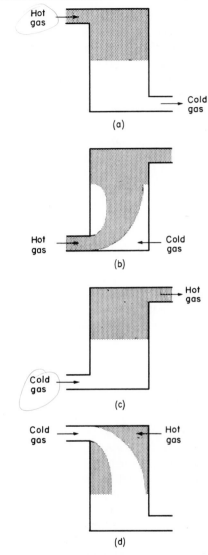

Fig. 7–8. Gas displacement in vertical chambers.

and to systems in which a cold gas is being heated. Thus, whereas a regenerator chamber is operated with downward flow of cooling gases during the flue-gas part of the cycle, the flow of gases must be upward during the air-heating part of the cycle.

Similarly, if a vertical laboratory tube furnace is to be flushed with cold gas from a tank, the cold gas should be introduced at the bottom to obtain the most effective displacement. These effects are illustrated in Fig. 7–8(c) and (d).

It will be recognized that the problems of attaining rational gas flow in furnaces of various shapes can be very complex. The specialist in furnace design may find it necessary to deal with these problems in some detail and to modify in various ways the general principle that cooling gases should flow downward and heating gases upward.

Heat transfer by natural convection. As indicated previously, the heat flow quantity which is of the greatest engineering importance in convection systems is the rate of transfer of heat between the bulk of the fluid and a surface in contact with the fluid. This quantity is generally calculated by Eq. (7–22), and the calculation therefore requires an evaluation of the convection coefficient of heat transfer, h_c. From the preceding discussions of various aspects of natural convection, it should be evident that the heat flow mechanisms are too complex to hope for a simple theoretical relation between h_c and the many variables on which it depends. This situation is similar, then, to the problem of estimating friction in fluid flow and can be handled in much the same way. That is, dimensional analysis can be applied to show that the behavior of the system can be characterized by two or more dimensionless variables, each of which is expressed in terms of the physical variables giving the properties of the fluid and of the system.

The three characteristic dimensionless numbers for natural convection are shown in the following functional equation for natural or free convection:

$$\left(\frac{h_c D}{k}\right) = \psi\left[\left(\frac{c_p \mu}{k}\right), \left(\frac{D^3 \rho^2 \beta g (\Delta t)}{\mu^2}\right)\right]. \quad (7\text{--}28)$$

The three dimensionless numbers are named as follows:

$$\frac{h_c D}{k} = N_{\text{Nu}}, \text{ Nusselt number;}$$

$$\frac{c_p \mu}{k} = N_{\text{Pr}}, \text{ Prandtl number;}$$

$$\frac{D^3 \rho^2 \beta g (\Delta t)}{\mu^2} = N_{\text{Gr}}, \text{ Grashof number.}$$

Equation (7–28) thus shows that the convection coefficient h_c for natural convection depends on the following physical variables: D, a characteristic linear dimension of the system; k, the thermal conductivity of the fluid; μ, viscosity; ρ, density; β, coefficient of expansion; c_p, specific heat; g, the acceleration of gravity; and Δt, the difference between the fluid temperature and the surface temperature. As is the case for the relationships between Reynolds number and the friction factor (Chapter 6), the relationship among Nusselt number, Prandtl number, and Grashof number will be different for systems of different geometrical configuration. For any given geometry, the numerical relationship found experimentally may be represented by an empirical equation or by graphs.

Figure 7–9 shows the experimentally measured relationship of dimensionless numbers for heat transfer by natural convection from horizontal cylinders to fluids. This correlation is based on experiments of many workers using both liquids and gases over a considerable range of cylinder diameters, from fine wires to large pipes. The curve drawn from the data is nearly asymptotic to the line representing the equation

$$N_{\text{Nu}} = \alpha (N_{\text{Pr}} N_{\text{Gr}})^{0.25},$$

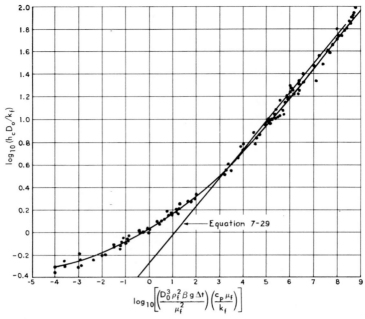

FIG. 7-9.   Correlation of dimensionless numbers for natural convection from horizontal cylinders to fluids (after McAdams).

or

for short vertical planes

$$\frac{h_c D}{k_f} = \alpha \left[\left(\frac{c_p \mu_f}{k_f} \right) \left(\frac{D^3 \rho_f^2 \beta g \, \Delta t}{\mu_f^2} \right) \right]^{0.25}, * \quad (7-29)$$

in which $\alpha = 0.53$, so that the equation itself can be used for estimations over a considerable range of conditions. In fact, equations in the form of Eq. (7-29) but with different values of α, have been used for other natural-convection systems. Data for short vertical planes, for example, have been correlated by Eq. (7-29) with $\alpha = 0.59$ and D taken as the height of the plane.

Both the Nusselt number $h_c D/k$ and the Prandtl number $c_p \mu/k$ are involved in forced convection. The Grashof number, however,

is significant just with respect to natural convection. Attention was called earlier in this chapter to the Prandtl number [see Eq. (7-8)] and it was noted that $N_{Pr} = 0.74$ for air and other diatomic gases, substantially independent of temperature and pressure, and is not much different for other gases. Accordingly, Eq. (7-29) is often reduced to simplified equations for air and other gases of similar properties (for example, flue gases) by substituting this value of the Prandtl number and other properties of air into Eq. (7-29). These substitutions lead to relations such as

$$h_c = C \left(\frac{\Delta t}{D} \right)^{0.25}, \quad (7-30)$$

with $C = 0.28$ for vertical plates less than 1 ft high (D is height of plate) and with $C = 0.27$ for horizontal pipes and for vertical pipes more than 1 ft high (D taken as

---

* The subscripts on k_f, μ_f, and ρ_f signify that these three fluid properties are measured at the film temperature, the arithmetic mean of the fluid temperature and the surface temperature.

pipe diameter). A similar simplification is used for large plane surfaces,

$$h_c = C'(\Delta t)^{0.25}, \qquad (7\text{-}31)$$

in which $C' = 0.38$ for horizontal plates facing upward, $C' = 0.2$ for horizontal plates facing downward, $C' = 0.27$ for vertical plates. The values of C and C' given above for use in Eqs. (7–30) and (7–31) apply only for English engineering units; that is, with h_c in Btu/hr-ft²-°F, Δt in °F, and D in ft. Equations (7–30) and (7–31) suffice in estimating h_c for most of the natural-convection problems encountered in connection with metallurgical furnaces.

Forced-convection systems. The metallurgist works with a wide variety of systems in which heat flow occurs between moving fluids and solid surfaces. Some of the most common systems are shown in Fig. 7–10. Heat exchangers constitute one important group, but in addition, forced convection is a vital mechanism in several unit processes, as in the blast furnace where large quantities of heat are transferred from the upward-moving gases to the downward-moving lumps of ore, coke, and limestone. As in the case of natural convection, the engineer is usually concerned with rate of heat transfer from the bulk of the fluid to the solid surfaces confining the flowing fluid and does not often have to consider in detail the heat flow within the fluid from one part of the stream to another. Just as in natural-convection systems, it is useful to assume that the resistance to heat flow is concentrated in a thin stagnant layer of fluid at the solid surface. However, a qualitative visualization of the convection mechanism within the flowing fluid is often helpful in understanding the engineering calculations.

The flow of heat across a turbulent stream is readily visualized since the convection heat flow results from the mixing and the

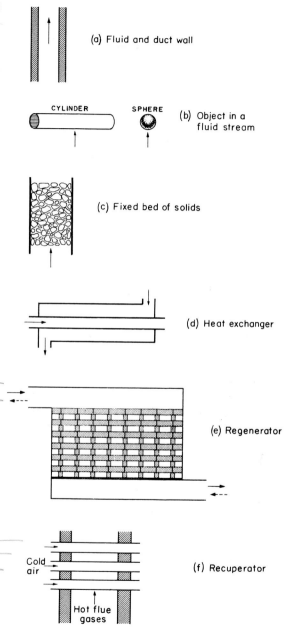

Fig. 7–10. Forced-convection systems.

random eddying of fluid particles which are mechanisms characteristic of turbulent flow. Fluid particles acquiring heat at a hot surface in contact with the flowing fluid are rapidly and completely mixed with the bulk of the colder fluid. For example, Fig. 7–11 based on actual temperature measurements by Pannell across a turbulent air stream in a vertical heated brass pipe shows the effectiveness of the turbulence in minimizing temperature differences across the body of the flowing field. These data also show directly the validity of the assumption that the temperature drop between fluid and walls is largely concentrated in a thin layer adjacent to the wall. It is to be noted that measurements of fluid velocities across a pipe with turbulent flow give a relationship very similar to Fig. 7–11, and use is sometimes made of the close analogy between heat transfer and friction.

Streamline flow involves a somewhat different convection mechanism, since there is no random eddying and mixing. In ideal streamline flow in the absence of temperature gradients, the fluid moves in parallel laminas and the only movement of mass at right angles to the laminas is the molecular diffusion which occurs even if the fluid is stationary. However, if the fluid is heated at a confining surface, its properties (especially viscosity and density) necessarily change so that the velocities must change along the flow path. These longitudinal velocity changes, from simple materials balance considerations, must modify the laminar flow as compared to the isothermal laminar flow, and this modification must result in some fluid movement at right angles to the normal flow direction. Another factor contributing to convection in streamline flow, and often in turbulent flow as well, is the effect of natural convection generated by differences in density and the effect of gravity. Thus, different convection rates may be obtained in identical systems if the flow is vertical in one case and horizontal in another. Also, it is readily seen that convection rates for a given overall temperature difference in a given system may be different depending on whether the heat flow is to or from the fluid.

Steady heat transfer by forced convection. Heat transfer rates by forced convection are estimated by Eq. (7–22), using empirical methods of estimating the value of the convection coefficient h_c. From the discussion in the preceding section it is evident that for a given system h_c will depend on all the variables considered in fluid-flow and fluid-friction problems (Chapter 6); that is, on the size of the system as measured by a characteristic dimension (D), on velocity (V), on fluid density (ρ), and on viscosity (μ). Also, it is logical to expect that h_c will be related to the Reynolds number ($DV\rho/\mu$ or DG/μ, in which G is the mass velocity). In addition to these fluid-flow variables, convection depends on thermal properties of the fluid, such as the specific heat (c_p) and the thermal conductivity (k). Clearly, then, h_c is related complexly to so many physical variables that recourse is again profitably made to the simplifications of

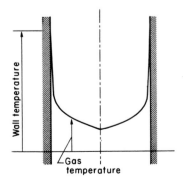

FIG. 7–11. Temperature variation across a turbulent flow in a heated pipe.

dimensional analysis. The assumption that $h_c = \psi(D, V, \rho, \mu, c_p, k)$ leads to the conclusion that three dimensionless numbers are involved in forced-convection systems: the Nusselt number $(h_c D/k)$, the Reynolds number $(DV\rho/\mu)$, and the Prandtl number $(c_p\mu/k)$.[*] All three of these numbers have been considered previously, the Nusselt and Prandtl numbers in connection with natural convection and the Reynolds number in connection with fluid friction (Chapter 6). Accordingly, for *forced convection*,

$$N_{\mathrm{Nu}}{}^{\bullet} = \psi(N_{\mathrm{Re}}, N_{\mathrm{Pr}}),$$

or

Forced Convection

$$\frac{h_c D}{k} = \psi\left[\left(\frac{DV\rho}{\mu}\right), \left(\frac{c_p\mu}{k}\right)\right]. \qquad (7\text{-}32)$$

The application of this relationship follows the same pattern as for fluid-friction calculations. For a flow system of given geometrical configuration, a unique relation should exist between the three dimensionless variables, and this relation should be valid regardless of the size of the system, of the kind of fluid, and of other dimensional properties. Accordingly, known experimental data are used to establish the quantitative relationship in the form of graphs or

---

[*]Another number sometimes used is the Stanton number, $h_c/(c_p V\rho)$. However, this is not a fourth independent parameter because it is equal to

$$\frac{[\text{Nusselt no.}]}{[\text{Reynolds no.}][\text{Prandtl no.}]}.$$

empirical equations, and these graphs or equations are used to predict the behavior of new systems of the same geometry.

As discussed previously in Chapter 6, dimensional analysis indicates relations between systems which serve as models of each other, so that quantitative data on one system can be used to predict the behavior of other models. Thus systems with the same numerical values of the Nusselt number $(h_c D/k)$ are models of each other as far as forced convection is concerned. In terms of the stagnant-layer concept, $h_c = k/\Delta s$, in which Δs is the equivalent thickness of the stagnant layer. Accordingly, the Nusselt number is equal to $D/\Delta s$ and can be visualized as the ratio of the size of the system to the thickness of the stagnant layer.

Many of the correlations commonly used for forced convection are empirical equations of the form

$$\frac{h_c D}{k} = a\left(\frac{DG}{\mu}\right)^m\left(\frac{c_p\mu}{k}\right)^n, \qquad (7\text{-}33)$$

in which the exponent m is usually between 0.5 and 0.8, and the exponent n is between 0.3 and 0.4. For example, the following equation is useful for *turbulent flow through pipes*, in calculating the heat transfer between the fluid and the internal surface of the pipe:

$$\frac{h_c D}{k} = 0.023\left(\frac{DG}{\mu}\right)^{0.8}\left(\frac{c_p\mu}{k}\right)^{0.4}. \qquad (7\text{-}34)$$

In applying this equation, the fluid properties are evaluated at the bulk temperature of the fluid.

turbulent flow in pipes

EXAMPLE

A round duct 2 ft in diameter carries 3000 ft³/min (STP) of preheated air at about 400°F. The inside surface temperature of the duct is estimated at 340°F. (a) Estimate the heat loss per running ft. (b) Estimate the drop in air temperature per 100 running ft.

(a) The heat loss q in Btu/hr is estimated by applying Eq. (7–22) after estimating h_c from Eq. (7–34).

First, the Reynolds number is calculated:

$$D = 2 \text{ ft},$$

$$G = \frac{3000 \times 29 \times 4}{60 \times 359 \times \pi \times 4} = 1.29 \text{ lb/ft}^2\text{-sec},$$

$$\mu = 1.8 \times 10^{-5} \text{ lb/ft-sec at } 400°F \text{ (Eq. 6–13)},$$

$$Re = \frac{DG}{\mu} = 143{,}000.$$

The Prandtl number for air and diatomic gases is 0.74, substantially independent of temperature.

The Nusselt number is found by Eq. (7–34):

$$\frac{h_c D}{k} = 0.023(143{,}000)^{0.8}(0.74)^{0.4} = 272.$$

Taking the specific heat (c_p) of air as 0.25 Btu/lb-°F, μ as 1.8×10^{-5} lb/ft-sec, and the Prandtl number as 0.74, k is estimated as follows:

$$k = \frac{c_p\mu}{0.74} = \frac{0.25 \times 1.8 \times 10^{-5} \times 3600}{0.74} = 0.0219 \text{ Btu-ft/hr-ft}^2\text{-°F}.$$

Substituting D and k into the value of the Nusselt number estimated above,

$$h_c = \frac{272 \times 0.0219}{2} = 2.98 \text{ Btu/hr-ft}^2\text{-°F}.$$

Applying Eq. (7–22),

$$q = 2.98 \times \pi \times 2 \times (400 - 340)$$
$$= 1120 \text{ Btu/hr per running ft}.$$

(b) The heat loss of 1120 Btu/hr corresponds to a loss of sensible heat from the air. A heat balance on the 1-ft length for a 1-hr interval gives the temperature drop of the air.

$$\text{lb air/hr} = 3000 \times 60 \times \frac{29}{359} = 14{,}500,$$

$$c_p = 0.25,$$

$$\text{temperature drop} = \frac{1120}{0.25 \times 14{,}500} = 0.31°F/\text{ft, or } 31°F/100 \text{ running ft}.$$

Since the Prandtl number is approximately the same for all gases and is substantially independent of temperature and pressure, Eq. (7–34) can be simplified considerably for estimating forced convection heat transfer between gases and the surfaces of ducts in which the gases are flowing. Taking an average value of 0.78 for $c_p\mu/k$,

Eq. (7–34) becomes

$$h = 0.027 \frac{c_p G^{0.8} \mu^{0.2}}{D^{0.2}}. \qquad (7\text{–}35)$$

Since this is not a dimensionless equation, consistent English units must be used.* For ducts of noncircular cross section, D may be replaced in these equations by $4R_H$, in which R_H is the hydraulic radius defined as the ratio of the cross-sectional area to the perimeter.

For *fluids flowing at right angles to single cylinders under turbulent-flow conditions,* Eq. (7–33) is used with $a = 0.26$, $m = 0.6$, and $n = 0.3$. The fluid properties in this case are evaluated at the film temperature, which can be taken as the arithmetic mean of the surface temperature and the bulk fluid temperature. For heat transfer between *single spheres and gases,* estimates can be made by the following relation (after McAdams):

$$\frac{h_c D}{k_f} = 0.33 \left(\frac{DG}{\mu_f}\right)^{0.6}, \qquad (7\text{–}36)$$

k_f and μ_f being measured at the film temperature.

A system important to metallurgists, but unfortunately one for which few data are available, is that of a fluid flowing through a porous bed of solids. The outstanding example is the transfer of heat from gases to solid lumps in the shaft of an iron blast furnace. Another example is the heat transfer in the pebble heater described in Chapter 5. In spite of the fact that quantitative correlations of the dimensionless numbers for this type of system are not well established, some general features can be pointed out. Recent work by Denton, Robinson, and

Tibbs on beds of uniform-sized spheres is well represented by Eq. (7–33), with $a = 0.58$, $m = 0.7$, and $n = 0.3$. Bed porosity was not varied significantly in these tests, but almost certainly will be an important variable in packed beds of lump solids of nonuniform size. Accordingly, assuming a constant Prandtl number for gases, heat transfer between gases and the solid surfaces in porous beds should be governed by a relation analogous to Eq. (7–35):

$$h_c = a \frac{c_p G^{0.7} \mu^{0.3}}{D^{0.3}} \times f(\epsilon), \qquad (7\text{–}37)$$

in which a is a constant, G is the superficial mass velocity based on the total cross-sectional area of the bed, D represents average particle size, and $f(\epsilon)$ is a function of the fraction of voids in the bed. In dealing with fixed beds, it is usually more convenient to refer the heat transfer coefficient to a unit volume of bed rather than to a unit surface of gas-solid contact. Since the surface area per unit volume of bed is inversely proportional to the particle size D, a heat transfer coefficient per unit volume corresponding to Eq. (7–37) would be given by

$$h_c' = a' \frac{c_p G^{0.7} \mu^{0.3}}{D^{1.3}} f'(\epsilon), \qquad (7\text{–}38)$$

in which h_c' is in Btu/hr-ft³-°F. This relation has no direct experimental justification. However, indirect determinations of h_c' by Furnas indicated that h_c' is proportional to the 0.7-power of G and inversely proportional to the 0.9-power of D for beds of ore, and inversely proportional to the 1.3-power of D for beds of coke. Thus it is reasonable to use Eq. (7–38) as the basis, at least, of predicting the way in which h_c' is affected by the important variables in this type of system. For example, if pebble size in the pebble heater were to be doubled, it would be reasonable to expect that the heat transfer

---

* To obtain h_c in Btu/ft²-hr-°F, c_p must be in Btu/lb-°F, G in lb/hr-ft², μ in lb/ft-hr, and D in ft.

rate would be cut about in half, principally because the gas-pebble surface area is halved for a given total apparatus volume.

Heat transfer by forced convection is a very important step in chemical plants of all kinds, so that data on new systems, especially new geometrical arrangements, appear fairly regularly in chemical engineering literature. These data are generally presented in terms of correlations of the dimensionless variables of the form of Eq. (7–33) or in simplifications of Eq. (7–33) similar to Eq. (7–35), for example. Likewise, when the metallurgical engineer encounters a new system for which he can find no data usable for design purposes, he may find it necessary to make some experimental measurements. Such experimental work should be planned, if possible, to yield a correlation along the lines of the correlations just considered. The work should not be planned on the philosophy of varying the dimensional physical variables one at a time to determine the effect on a single dimensional variable of practical interest, although this procedure might seem logical if we did not have the benefit of the information available from dimensional analysis.

Radiation

Energy is continually radiated or emitted from the surfaces of all bodies in the form of electromagnetic waves of various wavelengths. These waves include visible light waves, and they move through space with the velocity of light. The radiation most important in heat flow involves wavelengths just beyond the visible range, often called infrared. Since radiant heat energy is essentially the same as light, the differences being of degree rather than of kind, the general behavior of light can serve as a guide to the characteristics of radiant heat energy. For example, the waves move in straight lines, and thus one body can receive radiant heat from another only if it can "see" the radiating body. As with visible light, the phenomena of absorption, reflection, emission, etc., are important. Most solids and also the liquid phases common in metallurgical systems (liquid metals, slags, etc.) are opaque to heat radiation. On the other hand, gases are relatively transparent. Accordingly, in considering heat flow by radiation, we are concerned primarily with flow of heat through gas-filled spaces, and specifically with exchange of heat between solid and liquid surfaces separated by gases. In some systems, the emission and absorption of heat radiation within a body of gas are also important.

The rate of radiation of heat from a surface increases with the fourth power of the absolute temperature, so that the relative importance of radiation compared with other heat flow mechanisms becomes greater and greater as higher temperatures are considered. Within the working chambers of high-temperature metallurgical furnaces, radiation is often the chief heat transfer mechanism. However, the temperature range is not the only guiding factor; for example, a fine, electrically heated filament loses heat more rapidly by convection than by radiation.

Radiation properties of a blackbody. For quantitative information as to emission and absorption of radiant heat energy at surfaces, it is convenient to refer to the behavior of a blackbody, which is an ideal or perfect absorber and emitter of radiation. A black surface absorbs all radiant energy which strikes it and reflects none. Also the rate of emission of energy from a unit area of black surface at a given temperature is greater than that from any other surface of unit area at the same temperature. The characteristics of a blackbody can be derived from fundamental principles of thermodynamics and theoretical physics. Properties of particular interest in radiant heat flow are (1) *the total rate of emission of energy at a given temperature*, (2) *the distribution of the energy according to wavelength at a given temperature*, (3) *the shifting of wavelengths of the radiant energy with changing temperature*, and (4) *the distribution of radiated energy according to direction from the surface.*

The *emissive power E* of a radiating surface is the amount of energy emitted per unit time per unit area (Btu/hr-ft^2). At a given temperature T, in degrees Rankine (°F + 460), the Stefan-Boltzmann Law gives the emissive power of a blackbody:

$$E_B = 0.173 \times 10^{-8}T^4 = 0.173\left(\frac{T}{100}\right)^4. \tag{7-39}$$

This relation gives the total amount of heat radiated in electromagnetic waves of all wavelengths leaving the surface in all directions. The distribution of radiant energy according to wavelength at any given temperature can be represented graphically as shown in Fig. 7–12, in which the *monochromatic emissive power* is plotted against wavelength for a blackbody. The monochromatic emissive power at a given wavelength λ is simply $(\partial E/\partial\lambda)_T$, so that $(\partial E/\partial\lambda)_T \, d\lambda$ represents the amount of energy emitted per unit area per unit time in the wavelength range from λ to $\lambda + d\lambda$. Thus it can be seen that the energy radiated from the black surface in a given range of wavelengths is equal to the area under the distribution curve of Fig. 7–12 over the given wavelength range. The shape of the distribution curve is given by Planck's Law as

$$\left(\frac{\partial E_B}{\partial\lambda}\right)_T = \frac{c_1\lambda^{-5}}{e^{\frac{c_2}{\lambda T}} - 1}, \tag{7-40}$$

in which c_1 is 1.19×10^{-8} and c_2 is 2.59 when E_B is in Btu/hr-ft^2, T is in °R, and λ is in cm.

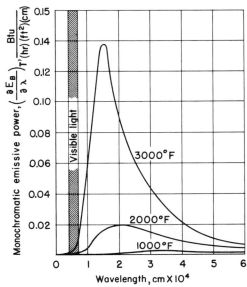

FIG. 7–12. Wavelength distribution of blackbody radiation.

Planck's equation, first set forth around 1900, has played a large role in the development of modern physics because of its close association with the origin and development of the quantum theory by Planck. Integration of Planck's equation leads to the Stefan-Boltzmann relation already presented as Eq. (7–39). Returning to Fig. 7–12, it can be seen that as the temperature is increased, the wavelength corresponding to maximum monochromatic emissive power decreases. That is, at higher and higher temperatures the energy is emitted as waves of shorter and shorter wavelengths, and larger and larger proportions of the radiant energy are in the visible range. These effects account for the fact that, starting at about 500°C, a body becomes dull red, and then progressively red, orange, yellow, and white as the temperature is increased.

Gray bodies and selectively radiating bodies. Although surfaces closely approaching ideal blackbody behavior have been prepared in the laboratory, most common surfaces exhibit considerable deviations. In the first place, the total rate at which energy is radiated, the emissive power E, is always less than that of a blackbody. The ratio E/E_B of the emissive power to that of a blackbody at the same temperature is the *emissivity* ϵ. Combining this definition with the Stefan-Boltzmann Law [Eq. (7-39)],

$$E = 0.173\epsilon \left(\frac{T}{100}\right)^4 \qquad (7\text{-}41)$$

for the emissive power of any body. Although at first glance this equation indicates E varying with the fourth power of absolute temperature, actually for most common substances the emissivity ϵ is not a constant but is found to increase slowly with temperature.

Nonblack surfaces do not absorb all the radiant energy incident on them. The fraction absorbed is the absorptivity α, which like ϵ has a maximum value of 1 for a blackbody. For a given body a close relationship exists between absorptivity and emissivity; in fact, under some circumstances $\alpha = \epsilon$, and under other circumstances α and ϵ may be taken as equal without serious error. However, care must be exercised, because in some systems α and ϵ will not be equal and the assumption of equality can lead to error.

Figure 7-13 represents any isothermal enclosure at temperature T, and A represents a small area of radiating surface forming part of the enclosure. The enclosure is filled with a medium which neither emits nor absorbs radiation; that is, a vacuum or a gas transparent to radiation. If the small portion of the surface A is black, it is emitting energy to the rest of the enclosure at the rate $E_B A$. Since the surface A is at the same temperature as the rest of the enclosure, it must be at thermal equilibrium with the rest of the enclosure. Therefore, it is absorbing radiant energy at the same rate as

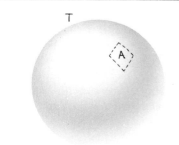

Fig. 7-13. Isothermal enclosure.

it is emitting, or $E_B A$. The black surface absorbs all incident radiation. Hence, we may conclude that the intensity of radiant energy incident on A, expressed as energy per unit area, must equal E_B. This conclusion is reached without considering the nature of the rest of the surface of the enclosure. Since A has been made a small fraction of the total surface, the nature of A cannot affect the intensity of radiant energy incident on itself. Thus we can make the important generalization that *the intensity of radiant energy incident on any surface in an isothermal enclosure at temperature T is equal to the emissive power E_B of a blackbody at this temperature.* This generalization incidentally points to a simple method for obtaining experimentally the equivalent of a blackbody. If an isothermal enclosure is observed from the outside through a hole too small to have a significant effect on the intensity of radiant energy within, the emissive power of the hole to the observer will be the same as that of a blackbody, regardless of the true nature of the internal surface of the enclosure [Fig. 7-14(a)]. If the hole is made too large, of course, the enclosure is no longer isothermal because part of the enclosing surface (the hole) is at the temperature of the surroundings. A more detailed analysis shows, however, that equivalent blackbodies are approached very closely by

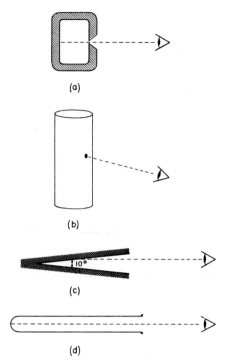

(a)

(b)

(c)

(d)

Fig. 7–14. Experimental arrangements equivalent to blackbodies.

various other geometrical arrangements such as those shown in Fig. 7–14(b) and (c). Such systems as these are of interest in radiation pyrometry and also in relation to some problems of heat transfer in furnace enclosures.

The behavior of a nonblack surface of emissivity ϵ and absorptivity α in an isothermal enclosure can now be examined. The rate of emission of energy per unit area is ϵE_B. The intensity of incident radiation is E_B, of which αE_B is absorbed and $(1 - \alpha)E_B$ is reflected. If the surface is at thermal equilibrium (i.e., at the same temperature) with the enclosure, the rate of emission must equal the rate of absorption: $\epsilon E_B = \alpha E_B$. Accordingly, we may generalize that $\alpha = \epsilon$ *for any body at the same temperature as its*

surroundings, and further that *the emissivity ϵ of any body at a given temperature T is equal to its absorptivity α for blackbody radiation from a source at the same temperature.*

A *gray body* may be defined as a body which has the same absorptivity for all wavelengths of incident radiation. Such a body will also have the same emissivity at all wavelengths. Now if absorptivity is independent of wavelength, it follows that total absorptivity will be independent of the nature of the incident radiation and therefore will be independent of the temperature of the source of the incident radiation. Thus, for gray bodies, the generalization of the previous paragraph can be extended: *the emissivity ϵ of a gray body is always equal to its absorptivity α.*

Most real bodies are neither black nor gray, but are selective absorbers and emitters of radiant energy. For common solids, the monochromatic emissivity and monochromatic absorptivity increase as wavelength decreases. Since at high temperatures larger proportions of the total radiant energy are in shorter wavelengths, the general tendency is for total emissivities and absorptivities of common solids to increase with increasing temperature. For the gases which are not transparent to thermal radiation, absorption and emission occur predominantly in certain wavelength bands.

As might be imagined from the preceding discussion, calculations of radiant heat exchange become complex and unwieldy if variations of emissivity and absorptivity with wavelength and temperature are rigorously taken into account. Moreover, the complete data necessary for rigorous calculations are rarely available for practical systems and the accuracy of other needed information usually does not justify attempts at fine distinctions. Emissivity and absorptivity are sensitive to surface conditions such as roughness, oxidation, dirt, etc., which are not readily measured or controlled except under special laboratory conditions. Accordingly, for engineering purposes it is common to assume gray-body behavior, which means using a single estimated or averaged figure for emissivity and absorptivity of a given surface, without attempting to allow directly for the detailed interactions of different selective radiators at different temperatures.

Emissivities of common substances (emissivities of various common materials are given in Table A–7 in the Appendix). A fair correlation exists between the behaviors of surfaces with respect to visible light and their behaviors with respect to thermal radiation. Surfaces with values of emissivity near unity are good absorbers and hence poor reflectors of incident radiation. Surface changes which increase the reflectivity of the surface decrease the emissivity. Thus most highly polished, unoxidized metal surfaces are recognized visually as good reflectors, reflecting perhaps more than 0.9 of the incident light. The same surfaces then are found to be good reflectors of thermal radiation, with total emissivities less than 0.1. The same metals when roughened or oxidized become to the eye poorer reflectors and are found also to have correspondingly higher emissivities for thermal radiation. Thus, various forms of iron exhibit nearly a 20-fold range in ϵ, from 0.05 for highly polished electrolytic iron to 0.95 or so for rough oxidized steel plate. When heat flow calculations involving emission and absorption at metal surfaces are made, a visual appraisal of surface condition is generally the basis for choosing the most appropriate emissivity value from laboratory-measured data.

Nonmetallic surfaces generally have emissivities and absorptivities above 0.8. For furnace refractories and bricks at high temperatures, ϵ averages about 0.85, but may be reduced to 0.6 or 0.7 by slagging. The expression *blackbody* can be misleading in judging emissivity, because it might imply that black-colored surfaces should have much higher emissivities and absorptivities than white-colored surfaces. Actually, published figures for black lacquered surfaces and white lacquered surfaces at 100–200°F fall into the same range, $\epsilon = 0.80$–0.85. On the other hand, aluminum paints on various surfaces give ϵ from 0.3 to 0.7.

When engineering calculations of heat exchange are made, it is, of course, the total emissivity and absorptivity for radiation of all wavelengths that must be used. The metallurgist, however, also makes use of other kinds of emissivity in temperature measurement by optical and radiation pyrometry, and the data in Table A–7 are not appropriate for these purposes. For example, an optical pyrometer measures the brightness of an object in a certain relatively narrow wavelength range in the visible region, so that calculation of the temperature of the object requires knowledge of the *spectral emissivity* or *monochromatic emissivity* for the wavelength or wavelength region used. This application is readily understood by reference to Planck's Law, Eq. (7–40). For the relatively short wavelengths utilized in optical pyrometry, Planck's equation for monochromatic emissive power of a blackbody [Eq. (7–40)] can be simplified to the following equation because $e^{\frac{c_2}{\lambda T}}$ is large compared with 1:

$$\left(\frac{\partial E_B}{\partial \lambda}\right)_T = c_1 e^{\frac{-c_2}{\lambda T}} \lambda^{-5}. \qquad (7\text{–}42)$$

The monochromatic emissive power of a nonblack body for the same narrow wavelength band depends on the monochromatic emissivity ϵ_λ:

$$\left(\frac{\partial E}{\partial \lambda}\right)_T = c_1 \epsilon_\lambda e^{\frac{-c_2}{\lambda T}} \lambda^{-5}. \qquad (7\text{-}43)$$

Now the optical pyrometer under favorable conditions measures the monochromatic emissive power $(\partial E/\partial \lambda)_T$ in Eq. (7-43), but the scale is usually calibrated under black-body conditions so that the pyrometer reading is T_B, which would be obtained as a solution of Eq. (7-42). Thus, T_B is known as the equivalent blackbody temperature of the object and is always lower than the true temperature T. The necessary correction is found by equating the two expressions for the measured emissive power, Eqs. (7-42) and (7-43), to obtain after cancellation

$$e^{\frac{-c_2}{\lambda T_B}} = \epsilon_\lambda e^{\frac{-c_2}{\lambda T}}, \qquad (7\text{-}44)$$

from which can be shown

$$\frac{1}{T_B} - \frac{1}{T} = -\frac{\lambda}{c_2} \ln \epsilon_\lambda = k. \qquad (7\text{-}45)$$

If the wavelength is known, measurement of T_B with the optical pyrometer and simultaneous measurement of true temperature T with a thermocouple permit calculation of ϵ_λ. For many purposes, it is convenient to calibrate a system by using the measurements of T_B and T to determine the correction term k as a whole, and this correction often embodies the effects of experimental variables other than ϵ_λ and λ. Thus, an optical pyrometer focussed on a hot, non-black object sees and measures not only the radiation emitted from the object but also reflected radiation whose magnitude depends on temperatures, emissivities, and geometric arrangement of the surroundings of the object.

Direct heat exchange between two black surfaces. Thermal energy is emitted from a surface in all directions, and the quantitative relations so far considered have dealt with the total energy regardless of direction.

When heat exchange between two surfaces is considered, obviously direction must be taken into account because an exchange of energy occurs only as the result of energy leaving one surface and striking the other. Two small surfaces which are a large distance apart will not exchange heat very rapidly because each intercepts only a small fraction of the radiation from the other. In taking direction of radiation into account, reference again is made to the characteristics of a black surface as the standard. Since a black surface always has the same brightness from whatever angle it is viewed, the intensity of radiant flux in a given direction from the black surface is proportional to the cosine of the angle the direction vector makes with a normal to the black surface. Also, of course, the intensity is inversely proportional to the square of the distance from the radiating surface. Most real surfaces encountered in engineering systems are also *diffuse radiators;* that is, the energy distribution according to direction is like that of a blackbody and follows the so-called cosine law.

Considering now the rate of heat exchange between any two surfaces A_1 and A_2, let us express the rate of heat exchange from the point of view of one surface only, say A_1. The surface A_1 is emitting at a total rate proportional to $(T_1/100)^4$, and a certain fraction of this radiation is eventually intercepted and absorbed by A_2. Similarly A_1 intercepts and absorbs a certain fraction of the total energy emitted from A_2, and this rate of emission is proportional to $(T_2/100)^4$. The net heat exchange between A_1 and A_2 can be expressed generally as the net rate of heat loss of A_1 to A_2 by the relation

$$q = 0.173 F A \left[\left(\frac{T_1}{100}\right)^4 - \left(\frac{T_2}{100}\right)^4 \right]. \qquad (7\text{-}46)$$

In this equation, F is a dimensionless composite factor for a given system which takes into account (1) the geometric relationship of the two surfaces or the fraction of the total radiation from one surface which is intercepted directly by the other, (2) the geometric relationships of the two surfaces to other reflecting or refractory surfaces in the system which add indirect paths of heat flow between the two surfaces, and (3) the emissivities and absorptivities of the two surfaces. Thus, the calculation of radiant heat exchange for any two surfaces becomes primarily a problem in evaluating F for the system. To avoid undue complications, F for a given pair of surfaces is evaluated by a stepwise procedure in which each of the three characteristics of the system listed above is considered in order.* First, a factor F_B is evaluated which takes into account only the direct geometric relationship of the two surfaces. F_B is the value of F for calculating q by Eq. (7–46) if the two surfaces A_1 and A_2 are both black and if only the direct radiation between the two surfaces is considered. The evaluation of F_B is discussed in the two following paragraphs. Second, after F_B is found, the effect of reflecting and refractory surfaces on the rate of heat flow is considered, and the factor F_{BR} is found which takes into account both direct radiation and reflections, but gives the rate of heat flow which would be obtained if A_1 and A_2 were both black surfaces. Third, the value of F_{BR} can be combined with data on the emissivities of A_1 and A_2 to obtain a final value of F which with Eq. (7–46) gives the net heat-flow rate from A_1 to A_2, taking into proper account the surface emissivities of A_1 and A_2 and both direct and indirect paths of heat flow between A_1 and A_2.

* This procedure is taken from Hottel, H. C., "Radiant Heat Transmission," Chapter III in McAdams, *Heat Transmission*.

If the two surfaces exchanging heat are both black and only direct heat exchange is considered, the dimensionless factor in Eq. (7–46), which we have designated as F_B for this case, can be evaluated by relatively simple geometrical calculations based on the cosine law and the inverse-square law. The calculations are simple because all radiation incident on either surface is absorbed by that surface and there are no reflections to be considered. Then, since the total energy emitted from A_1 in all directions is $0.173A_1(T_1/100)^4$, and the portion absorbed by A_2 is the first term of the right-hand member of Eq. (7–46), or $0.173F_BA_1(T_1/100)^4$, it is evident that F_B is simply the fraction of the radiation leaving A_1 which is intercepted by A_2. Similarly, by examining the second term of the right-hand member of Eq. (7–46), it is seen that the fraction of the radiation leaving A_2 which is intercepted by A_1 is given by F_BA_1/A_2.† That is, as might be expected, the larger surface intercepts a proportionally larger fraction of the energy from the smaller surface than the smaller surface does from the larger one. Either of these fractions, F_B or F_BA_1/A_2, can be calculated rigorously for any given geometrical arrangement of A_1 and A_2 by applying the cosine law of directional distribution of radiation from a black surface. Figure 7–15 summarizes the results of such calculations of F_B for systems of two parallel plane surfaces. These curves rise to approach $F_B = 1$

† This can be shown geometrically, but it also follows from the fact that the coefficients of T_1^4 and T_2^4 in Eq. (7–46) must be equal so that $q = 0$ when both surfaces are at the same temperature ($T_1 = T_2$). From the last term of Eq. (7–46), then, the radiation leaving A_2 and absorbed by A_1 is seen to be $0.173F_BA_1(T_2/100)^4$. The total radiation leaving A_2 is $0.173A_2(T_2/100)^4$. Dividing, we find the fraction of the radiation from A_2 intercepted by A_1 is F_BA_1/A_2.

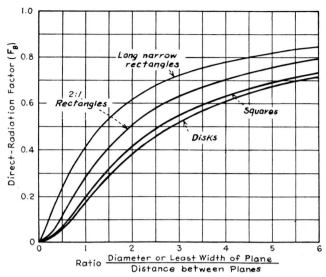

FIG. 7–15. Geometric factors (F_B) for direct radiation between equal, parallel planes (after Hottel and Keller).

for two parallel surfaces where the distance of separation is small compared with the dimensions of the surfaces. Similar data can be found for various other arrangements of two surfaces. Often a satisfactory estimate can be made by inspection.

A geometrical arrangement which occurs frequently in practical problems, and even more frequently serves as a good engineering approximation, involves a surface A_1 completely surrounded by the surface A_2, each surface itself having a substantially uniform temperature. For this geometry, $F_B = 1$ because surface A_2 intercepts all radiation from A_1. Perhaps the most common example is an object or a surface (A_1) surrounded by a room and contents (A_2), all at ambient temperature. Often, too, a value of unity for F_B accurately represents the system of an object placed in a furnace in which all the surrounding surfaces are at substantially the same high temperature.

The role of refractory surfaces. The refractory surfaces on the interiors of furnaces play an important role in radiation heat exchange which may be described loosely as "keeping the heat in the furnace" or "reflecting heat." The over-all effect of refractory surfaces is, of course, to increase the rate of heat exchange between two surfaces.

Before attempting to appraise the effect of refractory surfaces on the heat exchange between two nonrefractory surfaces, it is well to consider the behavior of a refractory wall under simplified conditions. The total incident heat flux on the refractory surface can be thought of as equivalent to a blackbody temperature T_B of the furnace interior. If all the interior furnace surfaces are at the same temperature, T_B will be the same as the true temperature; on the other hand, if the temperatures in sight of the refractory surface vary, T_B will be an average furnace temperature as observed from the refractory surface. On this basis, the intensity of radiation striking the refractory surface is $0.173 (T_B/100)^4$ Btu/hr-ft². Of this incident radiation, $\epsilon_R[0.173(T_B/100)^4]$ is absorbed and

$(1 - \epsilon_R)[0.173(T_B/100)^4]$ is reflected, ϵ_R representing the emissivity or absorptivity of the refractory surface. If the surface temperature of the refractory is T_R, the surface is emitting energy at the rate of $0.173\epsilon_R$ $(T_R/100)^4$ Btu/hr-ft². Assuming that the refractory wall is in a steady state and ignoring convection from furnace gases to the wall, the difference between the rate of absorption of heat from the furnace and the rate of emission back into the furnace must be equal to the flow of heat out through the wall to the surroundings. This heat balance can be written as

$$\text{heat loss rate/unit area} = \frac{q}{A}$$
$$= 0.173\epsilon_R\left[\left(\frac{T_B}{100}\right)^4 - \left(\frac{T_R}{100}\right)^4\right]. \quad (7\text{--}47)$$

The heat loss through the wall depends on the inside surface temperature T_R, the temperature of the furnace surroundings, and the thermal resistance of the heat flow path from inner refractory surface to surroundings. To take a numerical example, q/A is approximately 1800 Btu/hr-ft² for a 9-inch firebrick wall ($k = 1.0$ Btu-ft/hr-ft²-°F) with inner surface at 1840°F ($T_R = 2300$°R) and surroundings at 70°F. Using these values of q/A and T_R and letting $\epsilon_R = 0.8$, Eq. (7-47) gives $T_B = 2326$°R, only 26°F greater than T_R. With thicker walls, better insulated walls, or higher furnace temperatures, T_B will be even closer to T_R. Also, proper allowance for convection from furnace gases to the wall will further reduce the calculated difference between T_B and T_R. From this example, it is apparent that for most furnace conditions the inner refractory surface is (well within engineering accuracy) at the average furnace temperature, especially if this average is defined in terms of an equivalent blackbody temperature T_B, as in the preceding analysis.

For the firebrick wall just discussed, the total incident heat flux is 0.173 $(T_B/100)^4$ or 50,660 Btu/hr-ft². Of this total, 10,130 is reflected back into the furnace, 1800 flows out through the wall and is lost, and the balance, 38,730, is absorbed and re-radiated by the refractory back into the furnace. Thus, in effect, the refractory surface does serve as an efficient reflector (96.5%), but the mechanism of the reflection is predominantly absorption and re-radiation. Furthermore, it is apparent that the emissivity of the hot refractory surface has no practical effect, but merely determines the relative extents to which direct reflection and reflection by absorption and re-emission occur.

The effect of refractory surfaces on the heat flow between two black surfaces A_1 and A_2 can now be considered. To avoid complex and cumbersome relations not justified by the usually available engineering data, the analysis can be limited to a relatively simple system, shown schematically in Fig. 7-16. In this system, the three surfaces A_1, A_2, and A_R form an enclosure from which no radiant energy escapes. Each surface is at a uniform temperature, so that the system is characterized by the three temperatures T_1, T_2, and T_R. Surfaces A_1 and A_2 are shaped so that A_1 can see no part of itself and A_2 can see no part of itself. Also it is assumed that the refractories A_R reflect or re-radiate all incident energy, with negligible heat loss. The *direct* heat exchange between A_1 and A_2 is given by Eq. (7-46), with appropriate changes in symbols:

$$q_{1\to2} = 0.173F_BA_1\left[\left(\frac{T_1}{100}\right)^4 - \left(\frac{T_2}{100}\right)^4\right].$$
$$(7\text{--}48)$$

Of the total radiant energy leaving A_1, the fraction F_B is absorbed by A_2 and the remaining fraction $(1 - F_B)$ is intercepted by A_R. Of the total radiant energy leaving

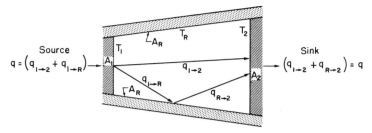

FIG. 7–16.   Combined direct and indirect radiation heat exchange.

A_2, the fraction $F_B A_1 / A_2$ is absorbed by A_1 and the remaining fraction $(1 - F_B A_1 / A_2)$ is intercepted by A_R. Accordingly, the direct heat exchanges between A_1 and A_R and between A_2 and A_R are given by

$$q_{1 \to R} = 0.173(1 - F_B)A_1 \left[\left(\frac{T_1}{100} \right)^4 - \left(\frac{T_R}{100} \right)^4 \right] \tag{7–49}$$

and

$$g_{2 \to R} = 0.173 \left(1 - \frac{F_B A_1}{A_2} \right) A_2$$

$$\left[\left(\frac{T_2}{100} \right)^4 - \left(\frac{T_R}{100} \right)^4 \right]. \tag{7–50}$$

Since there is no net heat flow to the refractory,

$$q_{1 \to R} + q_{2 \to R} = 0. \tag{7–51}$$

Equations (7–49), (7–50), and (7–51) are readily solved to find

$$T_R = \sqrt[4]{\frac{(A_1 - F_B A_1)T_1^4 + (A_2 - F_B A_1)T_2^4}{A_1 + A_2 - 2F_B A_1}} \tag{7–52}$$

and

$$q_{1 \to R} = -q_{2 \to R} = 0.173 \frac{(A_1 - F_B A_1)(A_2 - F_B A_1)}{A_1 + A_2 - 2F_B A_1}$$

$$\left[\left(\frac{T_1}{100} \right)^4 - \left(\frac{T_2}{100} \right)^4 \right]. \tag{7–53}$$

Equation (7–53) thus gives the additional *indirect* heat exchange between A_1 and A_2 obtained by reflection and re-radiation from A_R. The total heat flow q from A_1 to A_2 is

then found by adding Eqs. (7–48) and (7–53):

$$q = q_{1 \to 2} + q_{1 \to R}$$

$$= 0.173 F_{BR} A_1 \left[\left(\frac{T_1}{100} \right)^4 - \left(\frac{T_2}{100} \right)^4 \right], \tag{7–54}$$

in which F_{BR} is a composite geometrical factor for the closed system of two black heat-exchanging surfaces A_1 and A_2 plus refractory surfaces and is given by

$$F_{BR} = \frac{\frac{A_2}{A_1} - F_B^2}{1 + \frac{A_2}{A_1} - 2F_B}. \tag{7–55}$$

Figure 7–17 illustrates the application of Eq. (7–55) to several important limiting cases. A careful study of these cases will show the important role of refractory surfaces in accomplishing heat transfer from source of heat to charge in furnaces of various kinds. In many of the cases shown, the heat exchange by direct radiation is small $(F_B \to 0)$ and heat flow is accomplished almost entirely by reflection.

The various relationships derived above apply rigorously only to systems which are rather idealized in some respects. In particular, they involve division of furnace surfaces into three zones, a source, a sink, and a refractory zone, each at a uniform temperature. In actual operating furnaces, the division into the three zones is not always

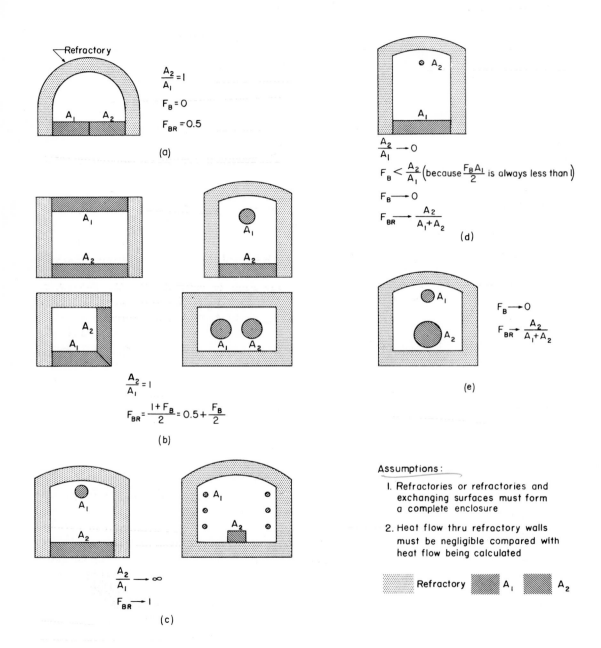

FIG. 7-17.   Effects of refractory surfaces.

clear-cut and generally the temperatures in any one zone are far from uniform. Thus it becomes necessary to use an average or equivalent temperature for each zone and to exercise careful engineering judgment. If circumstances warrant, it is possible to derive more complex equations to take care of more temperature zones, to allow for geometries in which part of the radiation from A_1 is absorbed by A_1 (for example, concave surfaces), etc. However, the metallurgist has little need for high precision in calculations of this kind; and even if he had the need, he probably would not find it possible to secure temperature and other data of sufficient precision to justify more elaborate analyses. Therefore, to obtain a treatment sufficiently general for most practical calculations, it is necessary now only to show how allowance is made for emissivities and absorptivities of the source and sink surfaces.

Heat transfer between gray surfaces. If A_1 and A_2 in the idealized system of Fig. 7–16 are not black, but are gray surfaces with emissivities of ϵ_1 and ϵ_2, the calculations of total heat flow from A_1 to A_2 become complicated by the multiple reflections which occur. Again, Eq. (7–46) may be taken as the basis for calculation:

$$q = 0.173FA_1\left[\left(\frac{T_1}{100}\right)^4 - \left(\frac{T_2}{100}\right)^4\right]. \quad (7\text{–}56)$$

The dimensionless factor F can be calculated from F_{BR} by the following equation:*

$$F = \frac{1}{\dfrac{1}{F_{BR}} + \left(\dfrac{1}{\epsilon_1} - 1\right) + \dfrac{A_1}{A_2}\left(\dfrac{1}{\epsilon_2} - 1\right)}. \quad (7\text{–}57)$$

Solving this equation for two black surfaces ($\epsilon_1 = \epsilon_2 = 1$), $F = F_{BR}$, consistent with the

* After H. C. Hottel in McAdams, *Heat Transmission.*

definitions of both factors. If A_1 is small compared with A_2, F and therefore q are not affected by the value of ϵ_2. A rather common system, at least to a good engineering approximation, is that of a small surface A_1 entirely surrounded by a large surface A_2 or by a large surface A_2 and by refractory surfaces. In this case, $A_1/A_2 = \sim 0$ and $F_B = 1$ or $F_{BR} = 1$, so that Eq. (7–57) gives $F = \epsilon_1$.

The general procedure for calculating total rate of heat transfer between two surfaces within a furnace can now be summarized:

(1) Represent furnace conditions and geometry as closely as possible by a 2-zone (A_1 and A_2) or 3-zone (A_1, A_2, and A_R) model and determine or estimate T_1, T_2, A_1, A_2, ϵ_1, and ϵ_2.

(2) Find or estimate F_B, the geometric factor expressing the fraction of radiant energy from A_1 intercepted directly by A_2.

(3) Calculate F_{BR} from F_B, A_1, and A_2, to allow for the effect of the refractory surfaces [Eq. (7–55)].

(4) Calculate F, using Eq. (7–57) to allow for nonblack conditions.

(5) Calculate q by Eq. (7–46).

Radiation coefficients; thermal resistance. All the equations for radiation heat transfer so far considered have been in the form of Eq. (7–46):

$$q = 0.173FA_1\left[\left(\frac{T_1}{100}\right)^4 - \left(\frac{T_2}{100}\right)^4\right], \quad (7\text{–}46)$$

in which the factor F, evaluated in various ways for different systems, allows for the geometrical arrangement of the system and for the radiation properties of the bodies which are exchanging heat. For a number of purposes, especially in dealing with combinations of radiation with other heat flow mechanisms, it is convenient to deal with a simplified empirical equation similar in mathematical form to the basic convection equation [Eq. (7–22)], as follows:

$$q_r = h_r A_1(t_1 - t_2) = h_r A_1(T_1 - T_2). \quad (7\text{–}58)$$

This relation is useful even though the radiation coefficient h_r varies with temperature and is not as much of a constant as conductivity k or the convection coefficient h_c. Comparing Eqs. (7–46) and (7–58), it is found that

$$h_r = 0.173 \times 10^{-8} F(T_1 + T_2)(T_1^2 + T_2^2). \quad (7\text{–}59)$$

If, as is frequently the case, the temperature difference is relatively small, so that T_1 and T_2 are nearly equal, Eq. (7–59) simplifies to

$$h_r = 0.692 \times 10^{-8} F T^3 = 0.00692 F \left(\frac{T}{100}\right)^3. \quad (7\text{–}60)$$

If T_1/T_2 is between 0.7 and 1.4, and T is taken as the arithmetic average of T_1 and T_2, Eq. (7–60) gives h_r to within 3%, which is entirely adequate for most engineering calculations.

For the rather common situation in which heat flow to a surface involves both convection and radiation in parallel,

$$q_{\text{total}} = q_{\text{convection}} + q_{\text{radiation}} = (h_c + h_r) A \, \Delta t. \quad (7\text{–}61)$$

The use of a radiation coefficient also facilitates calculations for series heat-flow paths in which thermal resistances are additive. The thermal resistance R_r of a radiation path is found by combining Eq. (7–58) with the definition of thermal resistance in Eq. (7–19):

$$R_r = \frac{1}{h_r A}. \quad (7\text{–}62)$$

For example, if the heat flow path involves the circuit of conduction to a surface and then radiation and convection from the surface, with an over-all temperature difference Δt,

$$q = \frac{\Delta t}{\dfrac{\Delta s}{k A_{\text{av}}} + \dfrac{1}{(h_c + h_r) A}}. \quad (7\text{–}63)$$

EXAMPLE 1

The outside steel shell of a furnace has a surface temperature of 400°F and the surroundings are at 80°F. Calculate the rate at which the furnace is losing heat to the surroundings by radiation per square foot of surface and compare this rate with the heat loss by convection.

SOLUTION

The outer surface of the furnace (A_1) is a gray surface at 400°F (860°R) entirely surrounded by a much larger gray surface (A_2) at 80°F (540°R). Since all radiation from A_1 is intercepted by A_2, $F_B = 1$. Taking $A_1/A_2 = 0$, Eq. (7–57) gives $F = \epsilon_1$. For the steel shell, we can assume $\epsilon_1 = 0.8$.

$$\frac{q_{\text{rad}}}{A_1} = 0.173 \times 0.8 [(8.6)^4 - (5.4)^4]$$

$$= 640 \text{ Btu/hr-ft}^2.$$

The radiation coefficient h_r can be found:

$$q_{\text{rad}} = h_r A \, \Delta t$$

$$h_r = \frac{q_{\text{rad}}}{A \Delta t} = \frac{640}{400 - 80}$$

$$= 2.0 \text{ Btu/hr-ft}^2\text{-°F}.$$

[Use of Eq. (7–60) as an approximation with $F = 0.8$ and $T = \dfrac{860 + 540}{2} = 700°R$ gives $h_r = 1.9$.]

$h_r = 0.00692(F)\left(\dfrac{T}{100}\right)^3$

The heat transfer coefficient for natural convection at a vertical surface is given by Eq. (7–31):

$$h_c = 0.27(\Delta t)^{0.25} = 0.27(320)^{0.25}$$
$$= 1.14 \text{ Btu/hr-ft}^2\text{-}°F.$$

Accordingly, the heat loss by convection is

$$\frac{q_{conv}}{A} = h_c \,\Delta t = 1.14 \times 320$$
$$= 365 \text{ Btu/hr-ft}^2,$$

and the total heat loss to the surroundings from the vertical surface is

$$q = 640 + 365 = 1005 \text{ Btu/hr-ft}^2,$$

with radiation accounting for almost $\frac{2}{3}$ of the total.

EXAMPLE 2

The laboratory furnace shown in Fig. 7–18 is 16 inches square by 12 inches inside dimensions and is heated by eight Globar (silicon carbide) resistors, each resistor rod with $\frac{1}{2}$-inch diameter and 16-inch heating length. The furnace is operated steadily with 60 amp ac at 150 v power input. The inside surface temperature of the brick is found to be 1250°C. Assuming convection heat transfer is negligible, (a) estimate the operating temperature of the Globar rods, and (b) estimate the temperature of the Globar rods at the start of heating up a cold furnace (70°F brick temperature) with the same power input.

SOLUTIONS

(a) When the steady state is attained, the flow of heat is from the Globar elements (A_1) to the inside furnace wall (A_2) and then out through the wall to the surroundings. To solve this problem, only the radiation path from A_1 to A_2 need be considered.

$q = I \times V = W$

$$q = 60 \times 150 = 9000 \text{ watts}$$
$$= 9000 \times 3.413 = 30{,}720 \text{ Btu/hr},$$

$$A_1 = 8 \times \frac{\pi \times 0.5 \times 16}{144} = 1.40 \text{ sq ft},$$

$$A_2 = \frac{2 \times 16 \times 16 + 4 \times 12 \times 16}{144} = 8.89 \text{ sq ft},$$

$$\frac{A_1}{A_2} = \frac{1.40}{8.89} = 0.157,$$

$$T_2 = 1250°C = 2280°F = 2740°R.$$

Surface A_1 at first glance appears to be entirely surrounded by A_2, but careful examination shows that part of the total emitted radiation from A_1 is intercepted by another Globar or another part of A_1. For the dimensions given, however, this fraction is clearly quite small, so it is reasonable to assume $F_B = 1$. An average value of 0.8 may be estimated for both ϵ_1 and ϵ_2. Applying Eq. (7–57),

$$F = \frac{1}{1 + \frac{1}{0.8} - 1 + 0.157\left(\frac{1}{0.8} - 1\right)} = 0.78.$$

(For this system, with $A_1/A_2 = 0.157$, the assumption that $A_1/A_2 = 0$, which is equivalent to ignoring the effect of ϵ_2, gives $F = \epsilon_1 = 0.8$.)

Now we can solve Eq. (7–46) to find T_1, the Globar temperature:

$$T_1 = 100 \sqrt[4]{\frac{q}{0.173FA_1} + \left(\frac{T_2}{100}\right)^4}$$

$$= 100 \sqrt[4]{\frac{30,720}{0.173 \times 0.78 \times 1.4} + (27.4)^4}$$

$$= 2920°R = 2460°F = 1350°C.$$

(b) An estimate of Globar temperature for a cold furnace with the same power input involves the same calculations as in (a), but with $T_2 = 530°R$. The estimated Globar temperature is 1550°F, or 845°C.

Estimates of this type are useful in electric-furnace design, especially in selecting the proper size of resistor element. Too great a loading in Btu/hr-ft^2 at a given furnace temperature results in too high a resistor temperature, leading to early failure of the heating element. Loading is commonly expressed in watts/inch^2, which in this example is about 45 watts/inch^2. The manufacturer's rating at the furnace temperature of part (a) (2280°F) is 72 watts/inch^2, so it may be concluded that the calculated Globar temperature of 2460°F is well under the safe operating temperature.

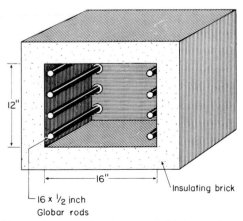

FIG. 7–18. Sectional view of Globar resistance furnace.

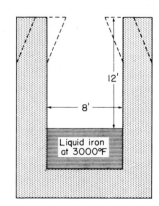

FIG. 7–19. Open refractory vessel containing liquid iron.

EXAMPLE 3

25 tons of liquid iron at 3000°F is held in the open vertical cylindrical vessel shown in cross section in Fig. 7–19. The walls of the vessel are of massive refractory construction to withstand erosion and to prevent excessive heat loss by conduction. (a) Estimate the rate of heat loss by radiation from the liquid metal to the surroundings through the mouth of the vessel, and (b) estimate the rate of temperature drop of the liquid iron due to this radiation loss, in °F/min.

SOLUTIONS

(a) Viewed from any point within the vessel, the imaginary surface across the mouth has substantially all the characteristics of a black surface at the temperature of the surroundings (say 80°F or 540°R). Accordingly, an estimate of radiation loss from the liquid-iron surface through the mouth can be made by calculating the rate of heat flow between two parallel disks 8 ft in diameter and 12 ft apart, with the enclosure completed by a refractory surface.

From Fig. 7–15, for two parallel disks with a diameter/distance ratio of 0.67, we find

$$F_B = 0.1.$$

That is, about 10% of the total radiation leaving the iron passes directly out the mouth of the vessel to the surroundings. Now, applying Eq. (7–55) for $A_2/A_1 = 1$ and $F_B = 0.1$,

$$F_{BR} = \frac{1 - (0.1)^2}{1 + 1 - 0.2} = 0.55.$$

Let us diverge from the problem at hand to consider other possible geometrical arrangements. In the above calculation, F_{BR} is determined mainly by the ratio A_2/A_1, and a large percentage error in F_B has little effect on F_{BR}. In fact, assuming $F_B = 0$ gives $F_{BR} = 0.50$, only 10% lower than the value based on $F_B = 0.1$. Now if the mouth of the vessel were only half the diameter of the liquid-metal surface and thus one-fourth the area, F_B would be a little more than one-quarter of the original value. Assuming $F_B = 0.0$ to 0.05 and taking $A_2/A_1 = 0.25$, we obtain $F_{BR} = 0.20$ to 0.205.

Returning to the original geometry and using $\epsilon_1 = 0.3$ for liquid iron and $\epsilon_2 = 1$ for the opening of the vessel, Eq. (7–57) gives

$$F = \frac{1}{\dfrac{1}{0.55} + \dfrac{1}{0.3} - 1} = 0.24,$$

$$A_1 = A_2 = \frac{\pi}{4} \times 8^2 = 50.4 \text{ sq ft},$$

$$T_1 = 3000°F = 3460°R,$$

$$T_2 = 80°R = 540°R.$$

Accordingly, $\quad q = 0.173 \times 0.24 \times 50.4[(34.6)^4 - (5.4)^4]$

$$= \underline{3 \times 10^6 \text{ Btu/hr}} \text{ radiated through the mouth of the vessel.}$$

If the vessel mouth were only 4 ft in diameter, as already discussed, repetition of the calculations gives $F = 0.136$ and $q = 1.7 \times 10^6$ Btu/hr.

(b) Taking the specific heat of liquid iron to be 0.15, the rate of cooling due to the radiant heat loss just calculated will be

$$\frac{3 \times 10^6}{60 \times 50,000 \times 0.15} = \underline{6.7°F/min.}$$

Gas radiation. Many common gases are transparent to thermal radiation and neither emit nor absorb appreciable quantities of radiant energy at ordinary furnace temperatures. These transparent gases include the common diatomic gases such as N_2, O_2, H_2, etc. These gases, then, are heated or cooled only by conduction and convection. On the other hand, heteropolar gases with somewhat unsymmetrical molecules, such as CO_2, H_2O, SO_2, and even CO, absorb and emit radiation to a sufficient degree to be of considerable practical importance in heat exchange between gases and other bodies. However, the radiation characteristics of a gas body containing one or more of these heteropolar gases are quite different from those of a black or gray solid body of the same geometry. In the first place, gases are very selective radiators, emitting and absorbing only in relatively narrow wavelength bands. Second, emission and absorption occur at molecules throughout the gas body rather than just at the bounding surface of the gas body. The basic physical data are available and the phenomena are well understood, so that accurate calculations of gas radiation are possible for a variety of practical systems. Many such calculations are so long and complex that they are well left to the specialists. However, by making a few approximations and simplifications, the important quantitative principles can be brought out and a satisfactory basis for engineering estimates of gas radiation can be developed.

First the radiant heat exchange between a body of gas at T_G and a black enclosing surface at T_S may be considered (Fig. 7–20). Let A_G represent the area of the enclosing surface and L a characteristic linear dimension of the body of gas. In order to make best use of published data, it is convenient to define the size L of the gas body as $3.5 \times$ volume/area. Also, let p represent the partial pressure in atmospheres of the absorbing gas. Now the behavior of a single narrow beam of radiation passing through the body of the gas can be considered. Since the energy is absorbed by the individual molecules in the body of gas, it is reasonable to expect that the proportion of the radiant energy absorbed will depend on the number of molecules in the path of the beam. At a given temperature, this number is proportional to p and also to L. Also, since the gas body is a selective absorber, the proportion absorbed will depend on the wavelength distribution of the beam and therefore on the temperature of the source T_S. In functional notation, then, we may write for the absorptivity of the gas

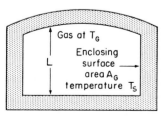

FIG. 7–20.   Gas body and enclosing surface.

$$\alpha_G = f_\alpha(pL, T_S). \qquad (7\text{–}64)$$

By similar reasoning, the emissivity of the gas body can be written

$$\epsilon_G = f_\epsilon(pL, T_G.) \qquad (7\text{–}65)$$

It is to be noted that the emissivity and absorptivity of the gas body are not equal, except when $T_S = T_G$, as a result of the selectively radiating character of the gas. The rate of heat flow from the gas to the black enclosing surface is the difference between emitted and absorbed radiation:

$$q = 0.173 A_G \left[\epsilon_G \left(\frac{T_G}{100} \right)^4 - \alpha_G \left(\frac{T_S}{100} \right)^4 \right]. \qquad (7\text{–}66)$$

To simplify subsequent relationships, the rate of heat flow can be given by

$$q = 0.173 A_G \epsilon'_G \left[\left(\frac{T_G}{100} \right)^4 - \left(\frac{T_S}{100} \right)^4 \right], \qquad (7\text{–}67)$$

in which ϵ'_G is the equivalent gray-body emissivity or absorptivity of the body of gas and can be expressed in terms of ϵ_G, α_G, T_G, and T_S, if desired, by combining Eqs. (7–66) and (7–67). However, a good approximation is made by assuming $\epsilon'_G = \epsilon_G$ if the gas is hotter than the enclosing surface ($T_G > T_S$) or assuming $\epsilon'_G = \alpha_G$ if the enclosing surface is hotter than the gas ($T_G < T_S$). In other words, we will measure gas emissivity at the temperature of the source of heat and assume

the absorptivity has the same numerical value, regardless of the temperature of the sink.

The principal absorbing gases in flames, flue gases, and metallurgical systems in general are CO_2 and H_2O. Emissivities of these gases are given in Fig. 7–21 and 7–22 as a function of pL in ft-atm and t in °F. Flue gases commonly contain both CO_2 and H_2O, so that the total emissivity is found by adding the emissivities of the two constituents. The emissivity figure obtained by addition is a little high, but addition is justified because the emission and absorption by the two kinds of gas molecules are nearly independent of each other since they occur in different wavelength bands with only slight overlapping.

Continuing to regard the gas as an equivalent gray body of emissivity ϵ'_G, the rate of heat flow between a gas body at T_G and a gray enclosing surface of emissivity ϵ_S and temperature T_S is found by applying Eq. (7–57), taking $F_{BR} = 1$ because all energy emitted from the gas body is incident on the enclosing gray surface, and taking $A_1/A_2 = 1$ because the outer surface of the gas body and the gray enclosing surface are the same:

$$q = 0.173 \frac{1}{\dfrac{1}{\epsilon'_G} + \dfrac{1}{\epsilon_S} - 1} A_G \left[\left(\frac{T_G}{100} \right)^4 - \left(\frac{T_S}{100} \right)^4 \right].$$
$$(7\text{–}68)$$

EXAMPLE

A 3-ft square flue carries 3000 ft³/min (STP) of hot gases, and the inside surface temperature of the flue is 100°F cooler than the gases. Calculate the rate of heat flow from gases to the flue wall per foot of length under the following conditions:

(a) The gas is dry air at 2500°F.
(b) The gas is moist air at 2500°F, containing 2% H_2O.
(c) The gas is a flue gas at 2500°F, analyzing 12% CO_2, 8% H_2O, 4% O_2, and 76% N_2.
(d) The gas is dry air at 1000°F.
(e) The gas is a flue gas of the same analysis as in part (c), but is at 1000°F.

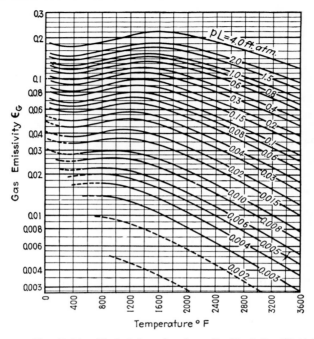

FIG. 7–21.   Emissivity of carbon dioxide (after Hottel).

SOLUTIONS

(a) For dry air, heat transfer is brought about solely by forced convection.   Accordingly, the convection coefficient h_c can be estimated by Eq. (7–35):

$$h_c = 0.027 \frac{c_p G^{0.8} \mu^{0.2}}{D^{0.2}}.$$

This relation is for a duct of circular cross section but can also be used for a square duct, which has the same hydraulic radius as a circular duct of the same diameter.

$$D = 3 \text{ ft,}$$

$$c_p = 0.29 \text{ Btu/lb-°F (air at 2500°F),}$$

$$G = \frac{3000 \times 29 \times 60}{359 \times 9} = 1620 \text{ lb/hr-ft}^2,$$

$$\mu = 4.9 \times 10^{-5} \text{ lb/ft-sec [from Eq. (6–13)]}$$

$$= 0.177 \text{ lb/ft-hr.}$$

Substituting these values in the equation given above, we find

$$h_c = 1.65 \text{ Btu/hr-ft}^2.$$

Then the convection heat flow from the gas to the brick can be found:

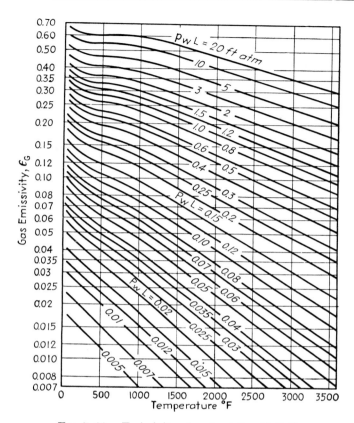

FIG. 7-22. Emissivity of water (after Hottel).

$$q_c = h_c A \, \Delta t = 1.65 \times 12 \times 100$$
$$= \underline{1980 \text{ Btu/hr per running ft of duct.}}$$

(b) The small percentage of moisture (2%) in the air does not affect the convection calculations appreciably, but the gas radiation from the moisture must be considered.

The size of the gas body (L) is $3.5 \times$ volume/area, or

$$L = \frac{3.5 \times 9}{12} = 2.6 \text{ ft,}$$
$$pL = 0.02 \times 2.6 = 0.052 \text{ ft-atm.}$$

From Fig. 7–22, for $pL = 0.05$ and $t_G = 2500°F$, we find $\epsilon_G = 0.016$. Assuming $\epsilon_S = 0.8$, Eq. (7–68) gives

$$q_{\text{rad}} = \frac{0.173 \times 12}{\dfrac{1}{0.016} + \dfrac{1}{0.8} - 1} \left[(29.6)^4 - (28.6)^4 \right]$$

$$= 3260 \text{ Btu/hr per running ft of duct.}$$

The total heat flow from the moist air by convection and gas radiation is therefore 1980 + 3260 or 5240 Btu/hr per running ft. Thus it can be seen that the small moisture content of the air is extremely important in heat transfer to or from the air, accounting for more transfer than is obtained by convection.

(c) For the CO_2 in the flue gas at 2500°F,

$$pL = 0.12 \times 2.6 = 0.31 \text{ ft-atm.}$$

From Fig. 7–21, $\epsilon_{CO_2} = 0.075$.

For the H_2O,

$$pL = 0.08 \times 2.6 = 0.21 \text{ ft-atm.}$$

From Fig. 7–22, $\epsilon_{H_2O} = 0.053$.

Assuming additivity of the gas emissivities,

$$\epsilon_G = 0.075 + 0.053 = 0.128.$$

Applying Eq. (7–68),

$$q_{\text{rad}} = \frac{0.173 \times 12}{\dfrac{1}{0.128} + \dfrac{1}{0.8} - 1} [(29.6)^4 - (28.6)^4]$$

$$= 25{,}400 \text{ Btu/hr per running ft.}$$

The heat transfer by convection will be a little greater than that calculated for air in (a), but the difference is not significant in view of the fact that convection is small compared with gas radiation. Thus taking q_c as 2200 Btu/hr per running ft,

$$q_{\text{total}} = 25{,}400 + 2200 = \underline{27{,}600 \text{ Btu/hr per running ft.}}$$

(d) Repetition of the calculations of part (a), but with a gas temperature of 1000°F, gives

$$q_c = \underline{1650 \text{ Btu/hr per running ft.}}$$

Thus, at the lower temperature the rate of heat loss from dry air for a 100°F temperature difference is nearly the same as at the higher temperature.

(e) Repetition of the calculations of gas radiation of part (c), but with a flue-gas temperature of 1000°F instead of 2500°F, gives

$$q_r = 5070 \text{ Btu/hr per running ft.}$$

The convection heat flow in this case will also be a little greater than for air and is estimated as

$$q_c = 1800 \text{ Btu/hr per running ft.}$$

The total is then

$$q = \underline{6870 \text{ Btu/hr per running ft.}}$$

This result shows that even at 1000°F the predominant heat flow mechanism for the flue gases is still gas radiation, even though the rate of gas radiation is but a fraction of that at 2500°F.

Comparison of the results of the various calculations above shows the great relative importance of gas radiation in high-temperature furnaces, heat exchangers, and related equipment. Also, these calculations show the large differences in heat exchange rates between nonradiating gases such as air and strongly radiating gases like flue gases. These differences are reflected, for example, in the behavior of a regenerator which gives up heat to air largely by convection in one part of the cycle, and then absorbs heat from flue gases predominantly by gas radiation in the other part of its cycle.

Radiation from flames. Gas radiation is important in reverberatory furnaces as a principal mechanism of heat transfer from flame to charge. To evaluate this heat transfer rate, the system can be divided into three zones, source, sink, and refractory, as was done previously for other systems. Taking the surface area of the charge (the sink) as A_C, and the area of the flame surface (the source) as A_F, Eq. (7–57) may be used to write the following approximation of the total rate of heat flow from flame to charge, regarding the flame as an equivalent gray body of emissivity ϵ_G':

$$q = 0.173 A_C \frac{1}{\dfrac{1}{\epsilon_C} + \dfrac{A_C}{A_F}\left(\dfrac{1}{\epsilon_G'} - 1\right)} \times$$
$$\left[\left(\frac{T_G}{100}\right)^4 - \left(\frac{T_C}{100}\right)^4\right]. \quad (7\text{–}69)$$

Gas radiation, together with convection, accounts for heat flow from nonluminous flames such as the pale blue or visually transparent flames obtained from burning H_2 and CO and from burning other gaseous fuels with adequate primary air well mixed with the fuel. The luminous flames of oil and pulverized coal, as well as those of many gaseous fuels, contain clouds of fine particles, especially carbon and ash. Not only do these clouds of particles make the flame yellow and opaque to the eye, but also they add greatly to its emissivity. Accordingly, to obtain the optimum heat transfer rate, the furnace operator strives to produce a flame of maximum luminosity and with a high degree of luminosity over as much of the furnace length as is consistent with other process requirements. For large pulverized-coal flames it is likely that the emissivity is close to unity, so that the rate of heat flow from the flame to a completely blanketed charge of area A_C and emissivity ϵ_C approaches the maximum given by

$$q_{\max} = 0.173 A_C \epsilon_C \left[\left(\frac{T_F}{100}\right)^4 - \left(\frac{T_C}{100}\right)^4\right]. \quad (7\text{–}70)$$

Unfortunately, the usefulness of equations for heat transfer by flame radiation is limited by the lack of data on flame temperatures in actual furnaces and by the further lack of simple means of even estimating flame temperatures. However, the equations do show the important variables in flame radiation in their proper relationships and therefore can serve as guides in predicting magnitudes of heat flow attainable by changes in the combustion and furnace system.

Steady heat flow in some metallurgical systems

The first, and in many respects the most important, step in applying heat flow principles to metallurgical processes is the recognition of the part heat flow plays in the process. No difficulty arises in recognizing furnace heat losses as resulting from heat flow, but it is easy to overlook the very great role of heat flow in ordinary temperature measurement. Careful study and analysis of various metallurgical processes show that heat transfer is very often the rate-limiting factor for the process as a whole. After the function of heat flow in the process is recognized, it is generally possible at least to make useful engineering estimates of rates, and under favorable conditions accurate design calculations become possible.

A second essential and often difficult step in applying heat flow principles to metallurgical systems is that of defining the heat flow path in a sufficiently quantitative fashion to justify calculations. In defining the heat flow path, care must be taken not to overlook such effects as contact resistances between adjacent bricks, changes in heat flow properties during service, effects of thin surface deposits, etc. On the other hand, judgment of a different kind is necessary to avoid such a complex specification of the path that the calculations can be carried out with difficulty only by a graduate mathematician. In particular, the geometric shapes and arrangements of actual systems are often so complex that simplifying assumptions have to be made to obtain manageable calculations. If these simplifications are made intelligently, the errors introduced are often smaller than some of the uncertainties in the data used for the calculations.

Heat flow has so many functions in metallurgical processes that no attempt will be made in this book to consider or even to list all the important applications. The nature and quantitative principles of each of the three mechanisms of heat flow have already been presented in detail in this chapter, and a few important cases of heat flow by combinations of these mechanisms have been considered. It remains now to give some further consideration to methods used for complex systems and to examine a few of the systems most important in metallurgical work.

Complex heat flow paths. Formulation of complex paths of heat flow is understood most readily by reference to analogous electrical circuits consisting of various combinations of resistances in series and parallel connections. For this purpose, the thermal conductance of a given component of the thermal circuit is the product of the heat transfer coefficient and the area, or hA, if convection or radiation is the mechanism, and $kA_{av}/\Delta s$ if conduction is the mechanism. The units of thermal conductance thus are Btu/hr-°F. As in electrical circuits, thermal resistance is the reciprocal of thermal conductance and thus is $1/(hA)$ or $\Delta s/(kA_{av})$, depending on the mechanism. When two or more components of a thermal circuit are in parallel, their conductances are additive, for example, when a surface is losing heat simultaneously by convection and radiation. When two or more components of the circuit are in series, their resistances are additive. That is, in a series circuit the total conductance is the reciprocal of the sum of the resistances, as has already been illustrated in Eqs. (7–21) and (7–63).

Another complexity often met is a continuous variation in conditions in directions at right angles to the direction of heat flow. For example, the outside surface temperature of a furnace wall may vary continuously from one place to another. Another example is a cooling fluid flowing in a pipe, where the temperature decreases continuously in the direction of flow. For some systems of this kind a differential equation is readily set up and integrated over the area of the heat flow path. For other practical systems the best approach is to divide the system into sections and then to calculate each section separately, using averaged data to represent conditions in each section.

Heat losses from furnaces. The heat balance of any high-temperature process always has the item "heat loss to the surroundings" in the heat output, and usually this item accounts for a substantial fraction of the total heat input. More often than not, the rate of heat loss of a system is calculated by difference between the total heat input and the heat output accountable as sensible heats, heats of endothermic reactions, etc. The calculation by difference is common because for practical systems it is usually more difficult to obtain the data necessary for a reliable direct calculation of heat loss than it is to find the other heat balance items. On the other hand, the difference calculation is never entirely satisfactory because all the errors in the other heat balance items are accumulated and reflected in the heat loss figure. Thus a direct evaluation of heat loss is helpful as an over-all check on the heat balance. The direct evaluation and its correlation with the heat balance serve to bring out the relative importance of the heat loss, in terms of equivalent fuel consumption, for example, and further disclose in quantitative fashion the controllable and variable factors on which the heat loss depends. Predictions can be made of the savings and other effects which might be obtained by adding insulation, changing wall thickness, water cooling, etc. Similar predictions are involved in designing new furnaces.

Several kinds of heat losses from a furnace chamber may be considered: (1) loss by conduction through walls, and convection and radiation from the outside wall surface to the surroundings, (2) loss by radiation from the furnace interior through openings to the surroundings, (3) loss to cooling water, and (4) loss into foundations (see page 204). In a loose sense, an outward leakage of hot gases constitutes a loss of heat, but this type of loss is ordinarily classified in a heat balance not as a heat loss but rather as a sensible heat in part of the gases leaving the system.

The rate of heat flow through a simple wall of known thickness and thermal con-ductivity is readily estimated by Eq. (7–12) if the inside and outside surface temperatures are known. Accordingly, one method of estimating losses through furnace walls involves measurements of surface temperatures over a sufficiently representative portion of the wall to obtain a reliable result. Special techniques are required to measure surface temperature, so that this method calls for expenditure of considerable experimental effort at the furnace.

The more common method of estimating heat losses through walls involves consideration of the whole path from furnace interior to surroundings (Fig. 7–23). The inside surface temperature, if not measured, can usually be estimated as equal to furnace temperature or process temperature because, as discussed earlier, radiation intensity within a high-temperature furnace is large compared with heat loss and leads to refractory surface temperatures close to the equivalent blackbody temperature of the furnace enclosure. The temperature of the furnace surroundings is always readily obtained. The unknown outside surface temperature of the furnace wall may be designated as t_x. For this path, assuming a steady state with no heat accumulating in the wall, the heat loss q can be written

$$q = q_{cond} = q_{conv} + q_{rad}, \qquad (7\text{--}71)$$

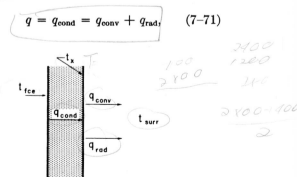

FIG. 7–23. Heat flow from furnace interior to surroundings.

in which q_{cond} is the rate of heat flow by conduction from inner to outer surface, q_{conv} is the rate of heat flow from outer surface to surrounding air by convection, and q_{rad} is the rate of heat radiation from outer surface to surroundings. Substituting Eqs. (7–12), (7–22), and (7–46) (with $F = \epsilon$) in Eq. (7–71),

$$q = k_{av}A_{av}\frac{(t_{fce} - t_x)}{\Delta s} = h_c A (t_x - t_{surr})$$
$$+ 0.173\epsilon A \left[\left(\frac{t_x + 460}{100}\right)^4 - \left(\frac{t_{surr} + 460}{100}\right)^4\right]. \quad (7\text{–}72)$$

For the usual case of natural convection to air at the outer surface, h_c is given by Eq. (7–31), so that

$$q = k_{av}A_{av}\frac{(t_{fce} - t_x)}{\Delta s} = C'A (t_x - t_{surr})^{1.25}$$
$$+ 0.173\epsilon A \left[\left(\frac{t_x + 460}{100}\right)^4 - \left(\frac{t_{surr} + 460}{100}\right)^4\right], \quad (7\text{–}73)$$

in which C' has the value of 0.27 for vertical surfaces. In the two right-hand members of Eq. (7–73), the surface temperature t_x should be the only unknown, and therefore t_x can be calculated numerically. The assortment of powers of t_x in the equation makes necessary either a trial-and-error or a graphi-

cal solution. After t_x is found, the heat loss q is readily calculated.

For a composite wall consisting of two or more conductors in series (for example, firebrick plus insulating brick) the same general procedure is applicable, but the trial-and-error calculation would be very long if all the surface temperatures were treated as unknowns at the start. In such cases it is preferable to solve for the outer surface temperature t_x first as before, substituting for $k_{av}A_{av}/\Delta s$ in Eq. (7–73) the equivalent conductance of the composite wall, calculated from the dimensions and conductivities of the separate layers and including, if necessary, estimated contact resistances (R):

$$\frac{k_{av}A_{av}}{\Delta s} =$$
$$\frac{1}{\frac{\Delta s_1}{k_1 A_1} + \frac{\Delta s_2}{k_2 A_2} + \frac{\Delta s_3}{k_3 A_3} + \cdots + R_{1,2} + R_{2,3} + \cdots} \quad (7\text{–}74)$$

After t_x and q are found, the other unknown temperatures are calculated by using the simple conduction equation for each layer of the composite wall. These temperature estimates may lead to revisions of the k values used in Eq. (7–74) and a repetition of the calculations with the revised figures.

EXAMPLE

The roof of a reverberatory furnace is 22 ft wide by 50 ft long and consists of 12 inches of magnesite brick having a thermal conductivity of 2.5 at 400°F and 1.5 Btu-ft/hr-ft²-°F at 2400°F. The furnace temperature is 2800°F. Estimate the heat loss through the roof.

SOLUTION

Applying Eq. (7–73),

$$t_{fce} = 2800°F,$$
$$t_{surr} = 70°F,$$
$$A = 50 \times 22 = 1100 \text{ sq ft},$$
$$C' = 0.38 \text{ for horizontal surface facing up},$$
$$\epsilon = 0.85 \text{ (estimated for outer brick surface)},$$
$$\Delta s = 1 \text{ ft}.$$

Therefore,

$$q = k_{av}(1100)\frac{2800 - t_x}{1} = 0.38(1100)(t_x - 70)^{1.25} + 0.173(0.85)(1100)\left[\left(\frac{t_x + 460}{100}\right)^4 - 780\right].$$

Assuming a linear relation between k and t, the given values of $k_{400°F} = 2.5$ and $k_{2400°F} = 1.5$ lead to the relation

$$k_{av} = 2.7 - 5 \times 10^{-4} t_{av},$$

in which t_{av} is the mean wall temperature, or $(2800 + t_x)/2$. Now a trial-and-error solution can be made by assuming different values of t_x until a value is found which satisfies the equations given above. If an assumed value of t_x gives $q_{cond} > (q_{conv} + q_{rad})$, a higher value of t_x is assumed for the next trial, and vice versa. This solution is conveniently represented in tabular form:

Assumed value of t_x	t_{av}	k_{av}	Calculated heat flows $\times 10^{-6}$, Btu/hr			
			q_{conv}	q_{rad}	$q_{conv} + q_{rad}$	q_{cond}
1000°F	1900°F	1.75	2.14	7.35	9.49	3.46
500	1650	1.88	0.815	1.245	2.06	4.75
600	1700	1.85	1.07	1.92	2.99	4.48
700	1750	1.825	1.32	2.80	4.12	4.22
710	1755	1.82	1.35	2.90	4.25	4.19

Interpolating between the last two calculations, it can be seen that t_x will be about 705°F and q, the heat loss, will be substantially 4.2×10^6 Btu/hr.

As pointed out previously, the radiant heat flux incident on a square foot of refractory surface inside a high-temperature furnace is normally quite large compared with the rate of heat flow through the wall by conduction. Accordingly, wall openings such as peepholes, open furnace doors, cracks in the brickwork, and the like, can account for large heat losses by direct radiation. For the limiting case of large openings in a thin wall, the radiation to the surroundings through the opening will be the same as the radiation from a black surface having the same area as the opening and the same temperature as the furnace enclosure. However, when the dimensions of the opening are comparable to the wall thickness, part of the direct radiation from the furnace interior is intercepted within the opening and re-radiated back into the furnace to reduce the loss considerably below the rate for a blackbody of area equal to the cross-sectional area of the opening. Table 7–1, based on calculations by Hottel and Keller, shows this behavior for various kinds of furnace openings. The tabulated factors, together with Eq. (7–46), are useful in estimating radiation losses through openings.

TABLE 7-1

Radiation through Openings in Furnace Walls

(Sides of openings perpendicular to wall)

Shape of opening	Ratio: least dimension of opening / wall thickness						
	0.01	0.1	0.2	0.5	1	2	4
	Values of F for Eq. (7–46)						
Circular	0.02	0.10	0.18	0.35	0.52	0.67	0.80
Square	0.02	0.11	0.20	0.36	0.53	0.69	0.82
Rectangular, 2:1	0.03	0.13	0.24	0.43	0.60	0.75	0.86
Long slot	0.05	0.22	0.34	0.54	0.68	0.81	0.89

EXAMPLE

A furnace is fired with pulverized coal of low calorific power, 13,000 Btu/lb, of which about 50% is gross available heat at the furnace temperature of 2500°F.   (a) What is the heat loss through a round peephole 6 inches in diameter through the wall at a point where the wall is 12 inches thick?   (b) Estimate the fuel saving in lb/24 hr which might be obtained by closing up the opening.

SOLUTIONS

(a) From Table 7–1, the radiation loss corresponds to 0.35 of the loss from a black-body of the same area and temperature.   Accordingly, applying Eq. (7–46),

$$q = 0.173(0.35) \left(\frac{\pi \times 0.5^2}{4} \right) [(29.6)^4 - (5.3)^4]$$

$$= 9140 \text{ Btu/hr.}$$

This is roughly 30 times the heat loss which would be obtained through the same area and thickness of a solid firebrick wall.

(b) Elimination of the radiation heat loss reduces the requirement of gross available heat by 9140 Btu/hr or 219,000 Btu/24 hr.   The gross available heat furnished by 1 lb of coal is 13,000 × 0.50, or 6500 Btu.   Accordingly, the fuel saving to be expected by closing the hole is 219,000/6500 or about 34 lb/day.   By reference to calculations in Chapter 6 it is seen that this part of the fuel saving resulting from closing the hole is small compared to the probable fuel saving through reduction of air leakage.

In designing a new furnace or planning alterations to existing furnaces, heat-loss estimations are useful as part of the job of estimating the fuel consumption or electrical power consumption of the new system. Also, it may be of interest to estimate the energy input (either from fuel or electrical power) required to maintain the empty furnace at a given steady temperature. This energy input is simply equal to the total rate of heat loss.   For laboratory furnaces and various small industrial furnaces, the heat loss, or the heat required just to keep the system hot, accounts for substantially all the heat input, so that the estimation of heat loss is likely to be a major part of the design calculations.

A principal objective in designing furnaces and high-temperature equipment in general is to find the design which will give the least over-all costs per unit of material processed.   Thus, the decision whether or not to add insulation to a furnace exterior may be reached by balancing the costs of the insulation against the expected savings in fuel costs.   In economic calculations of this kind, care must be taken to anticipate all possible direct and indirect costs and savings. One mistake sometimes made is the addition of insulation to save fuel in circumstances where the increased operating temperature of the furnace refractories results in de-creased service life and increased refractory costs which more than outweigh the fuel savings.

A system interesting from the heat flow standpoint is that of a hot liquid contained in a cooled shell of frozen liquid.   This ar-rangement is always obtained in one shape or another during solidification of liquid metals and other liquids as well.   Also, in some furnaces the walls are water-cooled steel plates which freeze onto themselves a thick layer of slag which serves as the container for liquid slag. A typical system of this type is shown schematically in Fig. 7–24, together with a plot of the relative temperature along the heat flow path. The key to a quantitative understanding of systems of this kind is the assumption that the inside surface temperature of the frozen liquid, immediately adjacent to the unfrozen liquid, is the melting point. Accordingly, the rate of heat flow q_1 from the hot liquid to the frozen surface, occurring by convection, is given by

$$\frac{q_1}{A} = h_{L,S}(t_L - t_{mpt}), \qquad (7\text{–}75)$$

in which $h_{L,S}$ is the convection coefficient (liquid to solid), t_L is the bulk temperature of the liquid, and t_{mpt} is the melting point of the solid. The rate of heat flow q_2 from the inside of the frozen surface to the surroundings is given by

$$\frac{q_2}{A} = \frac{t_{mpt} - t_{surr}}{\dfrac{\Delta s_S}{k_S} + \dfrac{\Delta s_{Fe}}{k_{Fe}} + \dfrac{1}{h_{Fe,surr}}}, \qquad (7\text{–}76)$$

in which t_{surr} is the temperature of the surroundings or cooling medium, Δs_S and k_S are the thickness and thermal conductivity of the frozen layer, Δs_{Fe} and k_{Fe} are the thickness and conductivity of the steel shell, and $h_{Fe,surr}$ is the coefficient of heat transfer from the steel to the surroundings or cooling medium (by convection or convection plus radiation).* A steady state in the system shown in Fig. 7–24 is obtained if the heat flow to the solid surface q_1 is equal to the heat flow away from the surface q_2. On the other hand, if $q_1 > q_2$, the difference $(q_1 - q_2)$ will

---

* If contact between frozen liquid and steel shell is poor, as may often be the case, an additional contact-resistance term should be added to the denominator in Eq. (7–76).

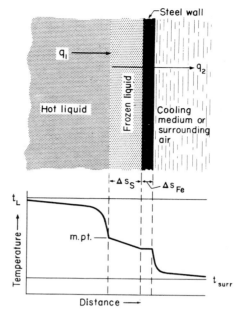

FIG. 7–24. Heat flow in freezing or melting.

melt the solid so that the thickness of solid will decrease with time. If $q_1 < q_2$, more liquid will freeze and the solid thickness will increase with time. The steady state condition, for which the thickness of frozen liquid remains constant with time, is found by combining Eqs. (7–75) and (7–76):

$$h_{L,S}(t_L - t_{mpt}) = \frac{t_{mpt} - t_{surr}}{\dfrac{\Delta s_S}{k_S} + \dfrac{\Delta s_{Fe}}{k_{Fe}} + \dfrac{1}{h_{Fe,surr}}}. \qquad (7\text{–}77)$$

This equation can be solved to find the stable thickness Δs_S as a function of the temperatures and heat flow coefficients:

$$\Delta s_S = \frac{(t_{mpt} - t_{surr})k_S}{(t_L - t_{mpt})h_{L,S}} - \Delta s_{Fe}\frac{k_S}{k_{Fe}} - \frac{k_S}{h_{Fe,surr}}. \qquad (7\text{–}78)$$

This analysis affords clues to the understanding of the mechanism of solidification of ingots and other castings, but does not afford a complete basis for studying these processes. In particular, Eq. (7–76) can apply strictly only to steady heat flow for which the temperatures and temperature gradients at all points along the heat flow path do not vary with time.

Radiation shields. Figure 7–25 shows a simple case of radiation shielding in which a source of heat H at T_H is separated from cold surroundings C at T_C by a series of parallel and closely spaced shields. If the system is evacuated and the shields are very thin, the only mechanism of heat flow from H to C which need be considered is radiation. For the steady state, the heat flow from H to C through the series thermal circuit, in Btu/hr-ft², will be, for a system of N shields,

$$\frac{q}{A} = 0.173 F_{H,Sh}\left[\left(\frac{T_H}{100}\right)^4 - \left(\frac{T_I}{100}\right)^4\right]$$

$$= 0.173 F_{Sh}\left[\left(\frac{T_I}{100}\right)^4 - \left(\frac{T_{II}}{100}\right)^4\right]$$

$$= 0.173 F_{Sh}\left[\left(\frac{T_{II}}{100}\right)^4 - \left(\frac{T_{III}}{100}\right)^4\right]$$

$$= \cdots 0.173 F_{Sh}\left[\left(\frac{T_{N-1}}{100}\right)^4 - \left(\frac{T_N}{100}\right)^4\right]$$

$$= 0.173 F_{Sh,C}\left[\left(\frac{T_N}{100}\right)^4 - \left(\frac{T_C}{100}\right)^4\right] \quad (7\text{–}78)$$

in which

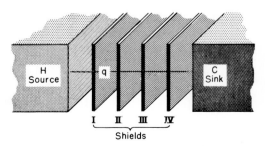

FIG. 7–25. Radiation shields.

$$F_{H,Sh} = \frac{1}{\dfrac{1}{\epsilon_H} + \dfrac{1}{\epsilon_{Sh}} - 1}; \quad F_{Sh} = \frac{1}{\dfrac{2}{\epsilon_{Sh}} - 1};$$

$$F_{Sh,C} = \frac{1}{\dfrac{1}{\epsilon_C} + \dfrac{1}{\epsilon_{Sh}} - 1}.$$

From Eq. (7–78), we can derive

$$\frac{q}{A} = \frac{0.173\left[\left(\dfrac{T_H}{100}\right)^4 - \left(\dfrac{T_C}{100}\right)^4\right]}{\dfrac{1}{F_{H,Sh}} + \dfrac{1}{F_{Sh,C}} + \dfrac{N-1}{F_{Sh}}}$$

$$= \frac{0.173\left[\left(\dfrac{T_H}{100}\right)^4 - \left(\dfrac{T_C}{100}\right)^4\right]}{\dfrac{1}{\epsilon_H} + \dfrac{1}{\epsilon_C} + \dfrac{2N}{\epsilon_{Sh}} - N - 1}. \quad (7\text{–}79)$$

This relation shows how an evacuated space with a multiplicity of thin shields of a material of low emissivity can be a very effective barrier to heat flow. As long as the sheets are not in effective direct contact and are oriented at right angles to the direction of heat flow, they can be made of a good conductor, such as a metal, just as satisfactorily as from a poor conductor. If the system is not evacuated but has a gas present between shields, convection can be substantially eliminated by close spacing. In this way the radiation shields can be used to give a barrier with a thermal conductance approaching that which might be calculated from the thermal conductivity of the gas and the dimensions of the system. These effects are illustrated by the common use of aluminum foil "insulation" and are also quite relevant to the use of fine powders of good conductors as "insulating" layers.

If the shields and heat-exchanging surfaces are blackbodies, as is almost the case in many high-temperature systems, Eq. (7–79) shows that the heat loss is inversely proportional to $N + 1$. Thus the first shield cuts q to $\frac{1}{2}$, the second from $\frac{1}{2}$ to $\frac{1}{3}$, the fifth from $\frac{1}{4}$ to $\frac{1}{5}$, etc., of the rate of heat loss with no shields. Since the first few shields have the greatest effect, common practice in specialized high-temperature furnaces is to use only 1, 2, or 3 shields.

Recuperators. Heat balances. Before considering the heat flow calculations for heat exchangers, consideration should be given to certain other fundamentals of continuous heat exchange processes. These fundamentals can be examined most easily for a continuous, countercurrent, recuperative-type heat exchanger, shown schematically in Fig. 7–26. The two fluids pass through the unit in parallel passages but in opposite directions, and heat flows from the hot to the cold fluid through a separating wall. Conditions vary through the exchanger, so it is necessary to consider first the process in an infinitesimal element of length dx. The heat balance for such an element can be written (assuming a steady state):

$$w_I c_I \, dt_I - w_{II} c_{II} \, dt_{II} + dQ = 0, \quad (7\text{–}80)$$

in which w_I and c_I, respectively, are the mass flow rate and specific heat of the cold fluid, w_{II} and c_{II} are the corresponding quantities for the hot fluid, dt_I and dt_{II} are the temperature increases from position x to $x + dx$, and dQ is the rate of heat loss from the element dx to the surroundings.

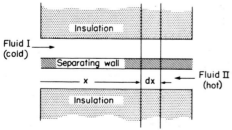

FIG. 7–26. Schematic diagram of continuous, countercurrent recuperator.

The temperature-position relationships for several conditions of operating the heat exchanger of Fig. 7–26 are shown in Fig. 7–27. Part (a) shows the most ideal kind of operation, adiabatic, reversible, and 100% efficient in all respects. The process is adiabatic in that $dQ = 0$, and is reversible because only an infinitesimal temperature difference is maintained between the two fluids at all points.* According to Eq. (7–80), if $dt_I = dt_{II}$ and $dQ = 0$, then $w_I c_I = w_{II} c_{II}$. Thus, from heat-balance considerations alone, it can be shown that ideal, reversible, adiabatic operation of a heat exchanger requires that the two fluids must be equivalent in total heat capacity (that is, flow rate times specific heat). When the heat transfer mechanism is considered, the infinitesimal temperature difference between the two fluids at any point implies either zero resistance to heat flow from fluid II to fluid I across the separating wall or substantially zero flow rates. That is, heat flow at any finite rate against thermal resistance is inherently an irreversible process.

Part (b) of Fig. 7–27 illustrates an adiabatic, irreversible heat exchange between two fluid streams of the same total heat capacities. By Eq. (7–80), if $dQ = 0$ and $w_I c_I = w_{II} c_{II}$, then $dt_I = dt_{II}$. Accordingly, $t_{II} - t_I$, or Δt, the temperature difference driving the heat flow, is constant over the length of the heat exchanger. The practical result of irreversibility is that heat exchange is incomplete, each fluid falling short of attaining the original temperature level of the other. For a given system of this kind, of course, the more rapid the rate

---

* The movement of heat along the heat exchanger by conduction is ignored in comparison to the flows of heat as sensible heats in the moving fluids.

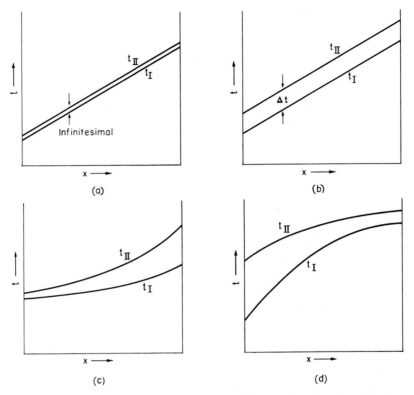

Fig. 7-27. Theoretical temperatures-vs.-position relationships in *adiabatic,*
countercurrent recuperators.

of heat transfer required, the greater the temperature difference.

Parts (c) and (d) illustrate the effects of differences in total heat capacities on fluid temperatures during the heat exchange process. If the entering hot fluid has greater total heat capacity than the cold fluid, it may be possible to heat the cold fluid nearly to the entering temperature of the hot fluid, but the temperature to which the hot fluid can be cooled is limited by the relationship of the two total heat capacities.

Figure 7-28 gives curves showing Δt or $t_{II} - t_{I}$, the difference in bulk temperatures between the two fluids, corresponding to the conditions discussed above. For an adia-

batic process $(dQ = 0)$, the rate of heat flow dq from fluid II to fluid I in the element dx is given by

$$dq = w_{I} c_{I} \, dt_{I} = w_{II} c_{II} \, dt_{II}. \quad (7\text{–}81)$$

Since $\Delta t = t_{II} - t_{I}$, $d(\Delta t) = dt_{II} - dt_{I}$, Eq. (7–81) can be modified to express dq in terms of Δt and the total heat capacities of the two fluids:

$$dq = \frac{d(\Delta t)}{\dfrac{1}{w_{II} c_{II}} - \dfrac{1}{w_{I} c_{I}}}. \quad (7\text{–}82)$$

It should be kept in mind that this relation between q and Δt is based on heat balance considerations alone, assuming no heat loss.

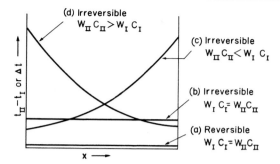

Fig. 7–28. Theoretical Δt-vs.-x relationships in *adiabatic*, countercurrent recuperators.

Integration of Eq. (7–82) to find the quantity of heat $q_{1,2}$ exchanged between two points at which the temperature differences are Δt_1 and Δt_2 gives

$$q_{1,2} = \frac{\Delta t_2 - \Delta t_1}{\dfrac{1}{w_{II}c_{II}} - \dfrac{1}{w_{I}c_{I}}}. \quad (7\text{–}83)$$

Recuperators. Over-all heat transfer coefficients. Another relation between q and Δt for the heat exchanger of Fig. 7–26 can be written on the basis of heat flow principles. That is, the rate of heat flow dq from fluid II to fluid I in the element dx should be proportional to the temperature difference Δt and to the cross-sectional area dA of the heat flow path:

$$dq = U(\Delta t)dA. \quad (7\text{–}84)$$

The constant of proportionality U is known as the *local over-all heat transfer coefficient*, and has the same dimensions (Btu/hr-ft²-°F) as other heat transfer coefficients used previously and designated by h. However, U is distinguished from h in that U is a coefficient for a heat flow path consisting of several thermal resistances in series, whereas h refers to only one thermal resistance at a time. Applying the resistance concept, U can be calculated from data on the individual parts of the heat flow path, as follows:

$$U\,dA = \frac{1}{\dfrac{1}{h_I\,dA_I} + \dfrac{\Delta s}{k\,dA} + \dfrac{1}{h_{II}\,dA_{II}}}, \quad (7\text{–}85)$$

in which h_I and h_{II} are the heat exchange coefficients between the fluids and the two surfaces of the separating wall, Δs and k are, respectively, the thickness and conductivity of the separating wall, dA is the average area of the heat flow path through the separating wall, and dA_I and dA_{II} are the surface areas in contact with the two fluids. For thin-walled pipes and flat walls, $dA = dA_I = dA_{II}$, so Eq. (7–85) simplifies to

$$U = \frac{1}{\dfrac{1}{h_I} + \dfrac{\Delta s}{k} + \dfrac{1}{h_{II}}}. \quad (7\text{–}86)$$

Additional resistance terms may be added to the denominators of Eqs. (7–85) and (7–86) to allow for the thermal resistances of soot or scale deposits, if necessary.

Unless close estimates are being attempted, it can be assumed that U is constant over the length of the heat exchanger. That is, h_I, h_{II}, s, k, and the area relationships which by Eq. (7–85) determine U should not vary greatly from one section of a simple heat exchanger to another. On the basis of this assumption, the variation of Δt with position in the heat exchanger can now be examined. Combining the heat balance equation (7–82) with the heat flow equation (7–84),

$$\frac{d(\Delta t)}{(\Delta t)} = U\left(\frac{1}{w_{II}c_{II}} - \frac{1}{w_{I}c_{I}}\right)dA. \quad (7\text{–}87)$$

Letting $dA = \bar{A}\,dx$, in which $\bar{A}$ is the heat exchange area per unit length, and integrating Eq. (7–87) over the portion from x_1 to x_2,

$$2.303 \log\left(\frac{\Delta t_2}{\Delta t_1}\right)$$

$$= U\left(\frac{1}{w_{II}c_{II}} - \frac{1}{w_{I}c_{I}}\right)\bar{A}(x_2 - x_1). \quad (7\text{–}88)$$

This equation shows that a graph of log (Δt) vs. x should be a straight line or, in other words, that the temperature difference Δt between the two fluids is an exponential function of the position x measured along the fluid flow path through the exchanger. The curves in Fig. 7–28 were drawn in accordance with this finding.

The total-heat-capacity terms ($w_\text{I}c_\text{I}$ and $w_\text{II}c_\text{II}$) can be eliminated between Eqs. (7–83) and (7–88) to obtain the important relation

$$q_{1,2} = U \left(\frac{\Delta t_2 - \Delta t_1}{2.3 \log \left(\frac{\Delta t_2}{\Delta t_1} \right)} \right) A_{1,2}. \quad (7\text{–}89)$$

The expression in parentheses is the logarithmic mean of Δt_2 and Δt_1. Hence, U can now be regarded as the *over-all coefficient* for the heat exchanger as a whole, which can be multiplied by the log mean Δt and the total heat transfer area to obtain the total rate of heat flow from fluid II to fluid I.

In the derivations leading up to Eqs. (7–88) and (7–89), a rather simple, ideal heat exchanger has been considered, with countercurrent flow of the fluids, no heat loss, constant local over-all coefficient, etc. Without these assumptions, much more complicated equations would be necessary and these would be of interest mainly to specialists in heat exchanger design. However, if now as a simple engineering approach we take the ideal exchanger as a standard, we can consider Eq. (7–90) as a definition of the over-all coefficient U' for any type of heat exchanger and can compare different types of apparatus and different operating conditions on this basis:

$$U' = \frac{q}{(A)(\log \text{ mean } \Delta t)}. \quad (7\text{–}90)$$

Regenerators. Regenerators are systems of unsteady heat flow to which the steady state heat-flow relationships so far considered are not strictly applicable. However, the difficulties in a complete quantitative representation of unsteady heat flow in a system such as a regenerative checkerwork are so great that some compromise is necessary to obtain relationships which can be used for practical engineering purposes. One type of analysis, to be outlined below, consists of treating the regenerator much the same as a recuperator. Although this method is not suited for precise calculations and sheds no light on certain important aspects of regenerator design and operation, it does afford a basis for rough estimates of regenerator size for a given heat transfer job and also shows quite well how several of the important variables affect design and operation.

Regenerators were described briefly in Chapter 5, and the general construction of a typical regenerative chamber was shown in Fig. 5–14. The major path of heat flow is

$$
\text{heating cycle} \begin{cases} \text{hot flue gas} \\ \downarrow \\ \text{brick surface} \\ \downarrow \\ \text{storage in} \end{cases}
$$

$$
\text{cooling cycle} \begin{cases} \text{brick interior} \\ \downarrow \\ \text{brick surface} \\ \downarrow \\ \text{cold air} \end{cases}
$$

This is an unsteady heat flow in that the gas temperatures, surface and interior brick temperatures, and heat flow rates vary continuously over relatively wide ranges both with time and position. In fact, the heating and cooling portions of the heat flow path occur at different times, as much as an hour or more apart under some operating conditions. In spite of these conditions, however, it is profitable to regard the heat flow path as analogous to that in a continuous recuperator, as

hot flue gas → brick surface → conduction
  through brick → brick surface → cold air,

and to deal with averaged temperatures to eliminate time as an independent variable. By this analogy, the fact that the brick surfaces in a regenerator are at a higher average temperature during the flue-gas cycle than during the cold-air cycle corresponds to the temperature difference across the separating wall in a recuperator. Also, the resistance to heat flow in and out of a checker brick is equivalent to resistance to heat flow across the separating wall in a recuperator. Applying this analogy, the over-all coefficient of heat transfer for a regenerator can be expressed by an equation similar to Eq. (7–86):

$$U' = \frac{1}{\dfrac{1}{h_{\text{flue gas}}} + \left(\dfrac{\Delta s}{k}\right)_{\text{eq}} + \dfrac{1}{h_{\text{air}}}}, \quad (7\text{–}91)$$

in which $h_{\text{flue gas}}$ and h_{air} are the heat transfer coefficients from gas to brick surface for the flue-gas and air cycles, and $(\Delta s/k)_{\text{eq}}$ represents the resistance, or thickness to conductivity ratio, for a recuperator wall equivalent to the impedance afforded by the checker bricks themselves to heat transfer. Estimation of the equivalent resistance of the brick, or $(\Delta s/k)_{\text{eq}}$, is a complex problem in unsteady heat flow. The equivalent resistance varies particularly with the thickness of the heat storage element and with the time duration of the cycle. Studies by several German workers have shown $(\Delta s/k)_{\text{eq}}$ to be of the order of 0.1 (Btu/hr-ft^2-°F)$^{-1}$, accounting for something like 15 to 30% of the total resistance to heat flow from flue gas to air.

With the over-all heat transfer coefficient U' for the regenerator defined in terms of an equivalent countercurrent recuperator, the average heat-transfer rate is given by a relation analogous to Eq. (7–89):

$$q = U'\left[\frac{\Delta t_2 - \Delta t_1}{2.3 \log\left(\dfrac{\Delta t_2}{\Delta t_1}\right)}\right] A$$

$$= U' \,(\text{log mean } \Delta t)\, A. \quad (7\text{–}92)$$

This relation gives the average heat-flow rate from flue gas to brick in the heating cycle and from brick to air in the cooling cycle. Thus the over-all rate of heat transfer from flue gas to air in the complete cycle in one regenerative chamber should be $q/2$ Btu/hr.

Calculation procedure for heat exchangers. The following procedure, suggested as the basis for heat exchanger calculations, affords a calculation of U', A, or the product $U'A$, if quantities and entering temperatures for the fluids are known and if a value is assumed for heat exchange efficiency. If U' and A are both given and an estimation of the quantity of heat transferred is wanted, this procedure can be used on a trial-and-error basis.

(1) Assume a value for the relative heat-exchange efficiency desired (see Chapter 5).

(2) From the entering temperatures and quantities of the two fluids and standard heat-content data calculate the *efficiency limit* for the heat exchange process.

(3) Combine the relative efficiency and efficiency limit to calculate *over-all thermal efficiency*. Then calculate q, the rate of heat exchange in Btu/hr, as the product of the over-all efficiency and the sensible heat of entering hot fluid per hour above the temperature of the entering cold fluid.

(4) Determine the exit temperature of the hot fluid, assuming it has lost q Btu/hr and allowing also for the heat loss of the heat exchanger if necessary. Determine the exit temperature of the cooler fluid, assuming it has gained q Btu/hr over the entering state. With these temperatures and the given entering temperatures, Δt_2, Δt_1, and

then log mean (Δt) are calculated for use in Eq. (7–89) or (7–92).

(5) Substitute q and log mean (Δt) in Eq. (7–89) or (7–92) to calculate $U'A$. If U' is evaluated by Eq. (7–86) or (7–91), the required area A of heat-exchanging surface is determined.

Pyrometry. In measuring temperatures with a mercury thermometer, thermocouple, resistance thermometer, or other nonoptical device, it must be kept in mind that the indicated temperature is that of the measuring element itself and does not necessarily represent the true temperature of the medium in which the measuring element is placed. Heat flow calculations are occasionally useful in estimating magnitudes of errors in temperature measurement, and heat flow thinking is essential to all temperature measurements.

When a thermocouple in a protection tube is inserted into a gas or liquid, the outer surface receives heat from the fluid by convection or by convection plus radiation. At the same time, owing to the fact that the part of the protection tube projecting out of the fluid is at a different temperature from the immersed part, conduction heat flow will take place along the protecting tube and measuring element. When the period of immersion is long enough for the system to reach a steady state, as evidenced by a constancy of the temperature reading, the rate of heat flow to the temperature-sensitive thermocouple junction by convection and/or radiation to the outside of the protection tube, conduction through the tube wall, etc., must be equal to the rate of flow from the junction to the surroundings by conduction up the thermocouple wires. These steady heat flows are driven by temperature differences, so that if the thermocouple is immersed from cold sur-

roundings, its reading will be low by a finite amount even after a steady state is reached. This error is minimized by immersing the protection tube and couple over a length several times the outside diameter.

Measurement of gas temperature under conditions where the gas and enclosing wall are at different temperatures introduces errors which are more difficult to minimize. Several possible arrangements for measuring gas temperatures with thermocouples are sketched in Fig. 7–29. In all cases shown it is assumed that the couples are sufficiently immersed in the gas so that the effect of conduction along the wires from the junction can be ignored. Then the two principal heat flows involved are (1) the flow between gas and couple by convection or convection and gas radiation, which may be represented by a heat transfer coefficient h_{gas}, and (2) the

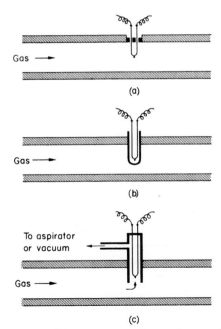

FIG. 7–29. Measurement of gas temperature with a thermocouple.

radiation between the confining wall and the couple, represented by a coefficient h_{wall}. The steady state temperatures of the gas, wall, and couple may be designated as t_{gas}, t_{wall}, and t_{couple}, respectively. When the steady state is reached, the net heat flow to the junction must be nil, or

$$h_{gas}(t_{gas} - t_{couple}) = h_{wall}(t_{couple} - t_{wall}), \quad (7\text{--}93)$$

so that the error in measuring the gas temperature, or $(t_{couple} - t_{gas})$, is given by $(h_{wall}/h_{gas})(t_{wall} - t_{couple})$. Both h_{wall} and h_{gas} are readily calculated by methods already discussed, so that estimates of the errors in measuring gas temperature for the arrangements (a) and (b) in Fig. 7–29 are readily made. The heat flow relation shows that the error in reading the gas temperature is reduced by increasing h_{gas}, by decreasing h_{wall}, or by measuring under conditions where the wall and gas temperatures are close together. This suggests various means for improving accuracy, such as increasing the gas velocity past the couple, using fine wire and omitting the protection tube, silvering the couple to reduce its emissivity, providing radiation shields, etc. Figure 7–29(c) shows a shielded, high-velocity thermocouple suitable for measuring gas temperatures with good engineering accuracy.

Electrical conductors. Earlier in this chapter a calculation was made of the operating temperature of an electrical resistance unit used for heating a furnace. The basis of this calculation was the equating of the electrical work input (EI or I^2R) to the rate of flow of heat from the surface of the resistor to the surroundings. It should now be pointed out that the same energy relation is involved in determining the temperature rise of wires and conductors in electrical circuits in general. In this case, of course,

the temperature rise is undesirable and must be limited by design to avoid overheating of the conductor or its electrical insulation. This is often a primary consideration in choosing the best size of conductor to carry a given current. The general relationship is

$$q = 3.413 I^2 R = UA(t_{cond} - t_{surr}), \quad (7\text{--}94)$$

in which 3.413 is the conversion factor from watts to Btu/hr, U is the over-all coefficient of heat flow from the conductor surface to the surroundings, and the other symbols are self-explanatory. This relation can be applied to a wire of circular cross section, $R = K(4l/\pi D^2)$, in which K is the resistivity in ohm-ft, l is the length of the conductor in ft, and D is the diameter in ft. The surface area A is $\pi l D$. Combining these relations with Eq. (7–94) and solving for D,

$$D = 1.14 \sqrt[3]{\frac{I^2 K}{U(\Delta t)}}. \quad (7\text{--}95)$$

This relationship gives the minimum diameter (ft) of conductor of specific resistivity K (ohm-ft) which can be used to carry I amperes without increasing the conductor temperature by more than $\Delta t°F$ above the surroundings. Other factors being equal, the diameter required increases with the $\frac{3}{2}$-power of the current to be carried by the conductor. When electrical insulation is added to the conductor, the added layers also act as thermal insulation, decreasing U and thereby increasing the size of conductor required. Also, the properties of the electrical insulation may under some conditions be the principal factors affecting the allowable temperature rise Δt.

Heat flow as a rate- and capacity-limiting factor in metallurgical processes. The discussion of heat balances in Chapter 3 brought out the fact that high-temperature chemical reactions and changes in state of aggregation carried out on a large scale as metallurgical unit processes are almost invariably accompanied by substantial heat effects, either heat absorptions or heat evolutions. When the individual unit processes are studied one by one, as will be done in Volume II of this book, it is found that a principal limitation to the rates of carrying out many processes on a large scale is a limitation to the rate at which heat is supplied to or removed from the materials in process. In other words, the capacity of a given apparatus in throughput per unit time and the required retention times for the materials treated are often determined directly by heat flow rates and, in fact, can often be calculated by combining heat flow principles covered in this chapter with information about process heat requirements. These relationships are quite important because of the obvious further relationships between treatment rates and treatment costs. This role of heat flow will be considered further in Volume II in connection with those unit processes in which it is important. At this point it will suffice to consider briefly and qualitatively a few examples of the role of heat flow in specific unit processes.

Melting and *simple smelting* are unit processes generally conducted in reverberatory-type furnaces. Cold solid materials make up the feed, and hot liquids are delivered as products, the major heat consumptions being sensible heats and heats of fusion. Since a material fuses immediately when it absorbs the requisite quantity of heat, ordinary melting is a rather clear-cut case of a process whose rate depends primarily on the rate of heat supply to the material. When melting must be followed by refining reactions, or when a variety of relatively infusible substances must combine to form a fusible solution or slag, other factors combine with heat flow in determining the rate of treatment.

Casting and *solidification* processes are the opposite of melting processes in that rates are determined by heat flow from the liquid to the surroundings, the heat flow path generally being from liquid to the liquid-solid interface, through the solidified metal, and then through a mold wall to the surroundings. Control of solidification rate and mechanism is achieved through control of heat flow by mold design, auxiliary water cooling, auxiliary insulation, and related means, and such control is of great importance in determining the structure, properties, and homogeneity of castings. In the recent development work on continuous casting of metals, the key problems have been to a large extent problems of obtaining satisfactory rates and directions of heat flow.

Gas-solid unit processes present a variety of heat transfer problems. Rates of drying of ores and other granular solids in rotary kilns, for example, are equivalent to rates of supply of the heat of vaporization of water. In the endothermic process of decomposing $CaCO_3$ to give CaO and CO_2, the time required to calcine large lumps depends primarily on conduction of heat from the outside to the inside of the lump. On the other hand, in strongly exothermic roasting processes, such as the oxidation of metal sulfides to form oxides and sulfur dioxide, the process may be deliberately slowed down to avoid overheating and resultant sintering.

Blast furnaces are characterized by a vertical shaft in which lump solids move slowly down, countercurrent to a stream of hot gases. One of the principal functions of the shaft is to act as a heat exchanger, utilizing the sensible heat in the gases coming up from the smelting zone to heat the downward moving solids to smelting temperatures.

Supplementary References

AIME, Committee on Physical Chemistry of Steelmaking, *Basic Open-Hearth Steelmaking*, revised edition. New York: AIME, 1951.

Austin, J. B., *Heat Flow in Metals*. Cleveland: American Society for Metals, 1942.

Brown, George Granger, et al., *Unit Operations*. New York: John Wiley, 1950.

Denton, W. H., Robinson, C. H., and Tibbs, R. S., *The Heat Transfer and Pressure Loss in Fluid Flow through Randomly Packed Spheres*. Oak Ridge, Tenn.: U. S. Atomic Energy Commission, Technical Information Division, HPC–35, June 28, 1949.

Etherington, H., *Modern Furnace Technology*. London: Charles Griffin, 1944.

Furnas, C. C., *Heat Transfer from a Gas to a Bed of Broken Solids*. U. S. Bureau of Mines, Bull. 361, 1932.

Keller, J. D., "The Flow of Heat through Furnace Hearths." *Trans. ASME*, **50,** 111, 1928.

Kern, Donald Q., *Process Heat Transfer*. New York: McGraw-Hill, 1950.

McAdams, W. H., *Heat Transmission*. New York: McGraw-Hill, 1942.

Perry, John H., *Chemical Engineers' Handbook*. New York: McGraw-Hill, 1949.

Saunders, O. A., and Ford, H., "Heat Transfer in the Flow of Gas through a Bed of Solids." *J. Iron and Steel Inst.* (London), **141,** 291, 1940.

Schack, A., *Industrial Heat Transfer* (translated from the German by H. Goldschmidt and E. P. Partridge). New York: John Wiley, 1933.

Trinks, W., *Industrial Furnaces*, Vol. I. New York: John Wiley, 1950.

Vilbrandt, F. C., et al., *Heat Transfer Bibliography*. Virginia Polytechnic Institute, Eng. Expt. Station, Bull. 53, 1943.

Walker, W. H., Lewis, W. K., McAdams, W. H., and Gilliland, E. R., *Principles of Chemical Engineering*. New York: McGraw-Hill, 1937.

Problems

7–1 A furnace wall consists of $4\frac{1}{2}$ inches of firebrick on the inside, then $4\frac{1}{2}$ inches of insulating brick, and finally $4\frac{1}{2}$ inches of red brick on the outside. The furnace temperature is 2500°F. A thermocouple peened into the outside surface of the red brick gives a temperature of 500°F. The thermal conductivities of the firebrick, insulating brick, and red brick are, respectively, 1.0, 0.1, and 0.5 (Btu-ft/hr-ft²-°F). (a) What is the highest temperature to which the insulating brick will be exposed? (b) What is the effect on your calculation in part (a) of neglecting contact resistance between the courses of brick?

7–2 A flat furnace wall consists of a 9-inch course of firebrick ($k = 0.6$ at 400°F, 1.0 at 1800°F) and a $4\frac{1}{2}$-inch layer of insulation ($k = 0.08$ at 500°F, 0.12 at 1500°F). The inside surface temperature of the firebrick is 2050°F and the outside surface temperature of the insulation is 350°F. Calculate (a) heat flow through wall, Btu/ft²-hr, (b) temperature at joint of firebrick and insulation, (c) temperature 2 inches back from hot surface of firebrick, and (d) temperature at midpoint of insulation.

7–3 A long, electrically heated cylinder 1 inch in diameter is covered with 3 inches of insulation for which $k = 0.05$. The temperature in the insulation varies from 800°F at the inside surface to 240°F at the outside surface. (a) What is the heat loss, in watts/running ft? (b) Make the necessary calculations and prepare a graph of insulation temperature vs. distance from cylinder axis.

7–4 (a) Derive Eq. (7–30) for horizontal cylinders and evaluate C for dry air at 1 atm and at an assumed film temperature of 150°F. Base the derivation on Eq. (7–29), with $\alpha = 0.53$. (b) Repeat the calculations of part (a) to find C for dry H_2 under the same film conditions. (c) Look up the "thermal conductivity" method of gas analysis and outline the operating principles which quantitatively relate the ultimate electrical measurements to the gas composition.

7–5 A nickel-chromium alloy wire used for heating elements in small electric furnaces has a maximum rating of 10 watts/sq inch for a furnace temperature of 2000°F. Estimate the maximum power in watts which should be supplied to a wire of this alloy which is 30 ft long and 0.1 inch in diameter, if this wire were to be used to heat a furnace at 1800°F.

7–6 Laboratory furnace No. 1 consists of a rusty steel shell lined with a refractory material. This furnace is operated so that the inside surface temperature is 2000°F. Laboratory furnace No. 2 is identical with No. 1 except that the shell is made of polished aluminum sheet. Also it is operated with the same inside surface temperature as No. 1. Which furnace will have the hotter shell and why?

7–7 Figure 7–30 shows a cross section of a tube furnace normally operated with the furnace tube and contents maintained at 2540°F. Once the furnace is charged and brought to temperature, the heat content of the furnace tube and contents remains constant and the rate of heat flow into the furnace tube becomes negligible. In this steady state, the power input, which is 30 watts/sq inch of Globar surface, just balances the heat loss through the insulating wall. Emissivities are as follows: Globars, 0.65; furnace tube, 0.70; inside wall, 0.75. Ignore convection and conduction in the furnace gases.

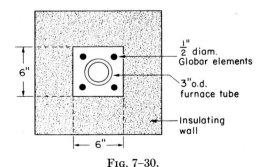

Fig. 7–30.

(a) By inspection of the sketch, estimate the fraction of the direct radiation from the Globars intercepted by the furnace tube and by the insulating wall, respectively. Ignore the small fraction intercepted by other Globars. (b) Calculate the steady surface temperatures of the Globars and of the inner surface of the insulating wall. [Hint: Make use of Eqs. (7–58) and (7–60).] (c) What would be the steady Globar temperature if the power input were kept the same and the furnace tube were removed from the chamber?

7–8 Two courses of firebrick ($k = 1.5$) in a furnace wall are poorly joined so that they are separated by a crack or air space averaging 2 mm in width. The brick temperature in the vicinity of the joint is estimated to be about 1000°F. Estimate the additional brick thickness, in inches, to which the thermal resistance of the joint is equivalent.

7–9 A 3-ft square duct carries 10,000 ft³/min (STP) hot air at 500°F, but the inside surface temperature of the duct is only 400°F. A measurement of the air temperature is attempted, using a thermocouple in a $\frac{1}{2}$-inch diameter, rusty, iron protection tube inserted to the center of the duct. Estimate the temperature reading of the thermocouple.

7–10 A steel pipe of 1-inch outside diameter and $\frac{1}{2}$-inch inside diameter carries 150 lb water/min through a furnace enclosure in which the gas and wall temperatures are 2100°F. (a) Calculate the length of pipe in which the water would be heated from an inlet temperature of 90°F to an outlet temperature of 110°F. (b) Calculate the inside and outside surface temperatures of the pipe at the section where the water temperature is 100°F.

7–11 A bare copper wire 0.46 inch in diameter carries 500 amp d.c. The wire crosses a room in which the temperature is 70°F. Calculate the operating temperature of the wire.

7–12 1000 lb/hr of gases from a smelting operation are to be cooled from 1000°F to 400°F in preparation for subsequent by-product recovery. The cooler will consist of an inverted U assembled from sheet-metal duct (circular cross section) 1 ft in diameter, and will be located outdoors where the ambient temperature is 70°F. Estimate the length of cooling duct required. The specific heat of the gases is 0.3 Btu/lb-°F, and other properties are substantially the same as for air. Neglect gas radiation inside the duct.

7–13 A vertical furnace wall is constructed of silica brick, with a total area of 400 sq ft and a thickness of 15 inches. The furnace interior is at 3000°F and the surroundings are at 80°F. Calculate (a) heat loss through this wall, Btu/hr, (b) fuel oil consumption equivalent to heat loss, barrels/day, for a fuel oil which furnishes 50,000 Btu/gal of available heat to the furnace chamber, and (c) the thickness of silica brick exposed to temperatures above 2800°F.

7–14 Addition of a 3-inch layer of insulating material ($k = 0.12$) to the furnace wall of problem 7–13 is being considered. Calculate (a) the percentage reduction in heat loss which should be attained by adding the insulation, (b) the daily fuel saving, if the fuel oil costs 3 cents/gal, and (c) the thickness of silica brick exposed to temperatures above 2800°F after adding insulation.

7–15 A chimney is 300 ft high and tapered from an inside diameter of 40 ft at the top to 50 ft at the bottom. The wall, constructed of brick ($k = 1.4$), varies gradually in thickness from top to bottom, as follows: 48 inches thick at the bottom, 30 inches at 100 ft, 24 inches at 200 ft, and 18 inches at the top. The chimney receives an estimated 1,000,000 ft³/min of gases at 500°F and 730 mm pressure (gas volume measured under actual conditions). It may be assumed that the gases have substantially the same properties as air. Estimate the temperature of the gases at the top of the stack, on a still day with outside air at 60°F.

7–16 A silica tube passes through a furnace kept at a uniform interior temperature of 2000°F (see Fig. 7–31). The tube is of $\frac{3}{8}$-inch inside diameter, $\frac{5}{8}$-inch outside diameter. Air at 70°F enters the tube at the rate of 10 lb/hr. At position P, indicated by the dotted line in the sketch, the

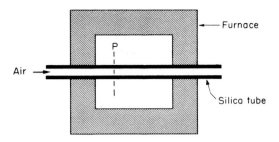

Fig. 7–31.

air temperature is 500°F. Estimate the temperature of the air stream 1 inch to the right of P. Neglect convection flow of heat to the outside surface of the silica tube.

7–17 A laboratory furnace is being designed for carrying out an exothermic gas-solid reaction on a continuous basis. The gaseous and solid reactants at room temperature (70°F) are to be fed continuously at a uniform total rate (for all reactants) of 10 lb/hr. The reaction products are to be discharged continuously at 1200°F. The heat of reaction is 1000 Btu/lb of total reactants. The average specific heat of the reaction products is 0.15 Btu/lb-°F.

The reaction chamber is to be a vertical Inconel cylinder, 4 inches in diameter by 3 ft high, with suitable auxiliary facilities for continuous feed and discharge. There will be good mixing in the process, so that the whole Inconel cylinder will be at substantially 1200°F, the temperature of the reaction products.

The Inconel cylinder will be insulated on the outside with a refractory material for which $k = 0.3$ and $\epsilon = 0.8$. Will a 2-inch layer of insulation (making total furnace diameter = 8 inches) cut the heat loss to the point where the furnace can be operated autogenously? Ignore end effects and make other reasonable assumptions if necessary.

7–18 A vertical laboratory tube furnace to operate at 2100°F is to be constructed as follows (see Fig. 7–32): An alundum tube 15 inches long by 2 inches outside diameter is wound with Kanthal-A resistance wire. The alundum tube is held centrally by two transite disks in a cylindrical

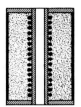

Fig. 7–32.

aluminum shell 15 inches long by 10 inches in diameter. The space between tube and shell is filled with crushed insulating brick.

For design purposes, the heat loss from this furnace may be regarded as radial heat flow out through concentric cylinders, and the end losses may be allowed for in the calculations by adding 40% to the length of the furnace.

If No. 16 B & S (0.05-inch diameter) wire is used for the heating element and the spacing between turns is made equal to the wire diameter, (a) what will be the voltage and current requirements of the furnace? (b) Will the resistance wire be overloaded? (c) Would you consider another wire size, either smaller or larger? Why or why not?

Data

Kanthal-A wire: Resistivity at working temperature, 147 microhms per cm cube (1 microhm = 10^{-6} ohm) or 885 ohms per circular mil ft. Maximum permissible surface load at 2100°F furnace temperature, 10 watts/inch².

Insulating material: Estimated thermal conductivity, 0.1 Btu-ft/hr-ft²-°F.

Aluminum: Thermal conductivity, 120 Btu-ft/hr-ft²-°F. Emissivity, 0.05.

7–19 A four-pass silicon-carbide tube recuperator similar to that shown in Fig. 5–11 receives 10,000 lb/hr of hot flue gases ($c_p = 0.25$) at 80°F. At each pass the air flows through 60 silicon carbide tubes ($k = 10$) in parallel. Each tube is 4 inches inside diameter, 6 inches outside

diameter, and has a working length of 3 ft. The heat transfer coefficient h from flue gases to outside surface of the tubes is estimated to be 3 Btu/hr-ft²-°F. For the estimates to be made below, the heat loss may be ignored and the recuperator may be regarded as a simple countercurrent heat exchanger.

Estimate the following, for air flow rates of 7000 lb/hr and 9000 lb/hr, respectively: (a) efficiency limits, %, (b) over-all coefficients (U), measuring air properties at 500°F, (c) over-all coefficients, measuring air properties at 1000°F, (d) q/(log mean Δt), (e) air preheat temperatures (use trial-and-error method, assuming various answers and making heat balance calculations of q/(log mean Δt) until the value indicated in part (d) is found), (f) q, Btu/hr, (g) over-all thermal efficiency of the heat exchanger, (h) relative efficiencies, and (i) pressure drops in the air stream per pass, assuming air flow at constant temperature equal to its top preheat temperature.

7–20 A regenerator receives hot flue gases at 2500°F and cold air at 60°F. The flue gases leave at 1300°F and the air is preheated to 2000°F. An estimated 15% of the heat given up by the flue gases is heat loss to the regenerator surroundings, and the rest (85%) is recovered in the preheated air. It may be assumed for estimating purposes that $c_p = 0.3$ for flue gases and $c_p = 0.25$ for air, independent of temperature. Estimate (a) over-all thermal efficiency, efficiency limit, and relative efficiency for this heat exchange operation.

Suppose now that the depth of the regenerator were doubled in such a way as to double the heat exchange area while keeping constant the over-all heat transfer coefficient U (Btu/hr-ft²-°F). The quantities and entering temperatures of the flue gases and air will be kept the same. Estimate for the enlarged regenerator (b) air preheat temperature, (c) over-all thermal efficiency and relative thermal efficiency.

CHAPTER 8

UNSTEADY HEAT FLOW

Heating and cooling are common steps in metallurgical processes, and often quantitative calculations of heating and cooling rates or of times required to accomplish certain temperature changes are useful. For example, when any cold material is placed in a furnace to be heated for melting, smelting, or other high-temperature processing, a definite period of time is required for the material to absorb heat and reach temperature. Heating up a cold furnace is another common process of unsteady heat transfer. When an ingot or other metal shape is hot rolled or hot worked, the time interval available for mechanical working obviously depends on the temperature range suitable for working and on the time consumed by the metal in cooling through this range. Checker brick in a regenerator and a bed of lump ore being heated or cooled by a stream of gas represent other types of unsteady heat flow systems.

Some of the most extensive and successful metallurgical applications of the principles of unsteady heat flow have been in the field of heat treatment of metals, particularly the heat treatment of steel.

Unsteady heat flow involves the same heat flow mechanisms as steady heat flow: conduction, convection, and radiation. The basic laws and empirical relations governing these mechanisms were presented in Chapter 7. However, rigorous applications of these principles to systems in which temperatures and heat contents along the heat flow path vary with time generally lead to more or less complicated differential equations which

only a specialist can pursue to a solution. Accordingly, this chapter will deal primarily with engineering methods of estimating unsteady heat flow. These methods require certain rather gross simplifying assumptions, but nevertheless have proven quite adequate for most practical purposes.

Simple heating and cooling of solid objects. The simplest kind of heating or cooling process to deal with quantitatively consists of taking a body at an initially uniform temperature t_1 throughout and placing it in surroundings at a uniform temperature t_2. Also, it will be assumed that the surroundings remain at a constant temperature t_2 throughout the heating or cooling process, so that t_2 is also the final uniform temperature of the object. This process is commonly realized by taking an object from a furnace and allowing it to cool in the room, in which case t_1 is the room temperature and t_2 is the furnace temperature. Most practical heating and cooling problems fall into this category, and the bulk of the theoretical relations to be developed in subsequent sections are limited to the type of process just described. When initial and final temperatures in the object are not uniform, or when the temperature of the surroundings varies with time, calculations are likely to be too complex and unwieldy for ordinary engineering application.

A further simplification which will be made in developing some of the principles of unsteady heat flow in the following pages will consist of limiting the theoretical discussion to homogeneous solid objects of simple and symmetrical geometric shapes. For example, if the object is a sphere, the temperature t at a given time in the heating or cooling process will be the same for all points at a given distance from the center of the sphere. Thus, the heating or cooling process for a given sphere can be specified completely by giving temperature t as a function of only two variables, time θ and distance from the center r.

For the simple unsteady heat-flow systems described above, the path of the heat flow can be divided into two major parts:

(1) *Conduction within the solid.*

(2) *Heat transfer between the surface of the solid and the surroundings by convection and radiation.*

One or the other of these two parts of the path may contribute most of the thermal resistance and thus may control the process, but under many conditions it is necessary to consider the joint effects of both parts of the heat flow path.

In view of the simplifications necessary to obtain a quantitative theoretical treatment, it will be apparent that rigorous and precise prediction of temperature behavior in heating and cooling will not be possible for many real systems. However, the relations found will serve well for estimating purposes and, perhaps of greater importance, will bring out the important variables in unsteady heat flow and their most significant interrelations.

Newton's law of heating or cooling; good conductors. The heating and cooling of bodies which are good conductors, such as metals, are often controlled primarily by the conditions affecting heat transfer from the surface to the surroundings. That is, the thermal resistance to heat flow within the body is small compared to the thermal resistance at the surface, so that during heating or cooling the temperature difference from

the inside to the outside of the body is small compared to the temperature difference between the surface and the surroundings. Accordingly, the behavior of many real bodies is correlated quite well by ignoring entirely the resistance to heat flow within the body. Assuming that the temperature t of the body at any time θ during the heating and cooling process is the same for all positions in the body, the rate of heat loss q to the surroundings at any given instant is given by the product of the mass (or volume $V \times$ density ρ), the specific heat c_p, and the rate of change of temperature with time, $dt/d\theta$:

$$q = -V\rho c_p \frac{dt}{d\theta}. \qquad (8\text{--}1)$$

The rate of heat flow from surface to surroundings can also be represented by the following relation from Chapter 7:

$$q = hA(t - t_2). \qquad (8\text{--}2)$$

In Eq. (8–2), h is the coefficient of heat transfer from surface to surroundings (convection and radiation combined), A is the surface area, t is the surface temperature of the object and is the same as the interior temperature for the case under consideration, and t_2 is the constant temperature of the surroundings. Combining Eqs. (8–1) and (8–2) and solving for $dt/d\theta$:

$$\frac{dt}{d\theta} = \frac{hA}{V\rho c_p}(t - t_2). \qquad (8\text{--}3)$$

Equation (8–3) states simply that the rate of heating or cooling of an object is proportional to the difference in temperature between the object and the surroundings. Assuming further that the heat transfer coefficient h and the specific heat c_p do not vary with temperature, Eq. (8–3) is readily integrated to obtain an equation giving the variation of temperature with time:

$$\log_{10}\frac{(t-t_2)}{(t_1-t_2)} = -\frac{hA}{2.3V\rho c_p}\theta, \quad (8\text{--}4)$$

or

$$\frac{t-t_2}{t_1-t_2} = 10^{-\frac{hA}{2.3V\rho c_p}\theta} = e^{-\frac{hA}{V\rho c_p}\theta}. \quad (8\text{--}5)$$

In these equations, t_1 is the initial temperature of the object when $\theta = 0$, t_2 is the surroundings temperature or final object temperature when $\theta = \infty$, and t is the temperature at any time θ. Equations (8–4) and (8–5) are statements of Newton's law of heating or cooling, which, as the derivation shows, is based on the assumption of a constant temperature throughout the object at any time and a constant heat transfer coefficient at the surface. For these conditions it is seen [Eq. (8–4)] that the time required to accomplish a given temperature change [that is, to obtain a given value of $(t-t_2)/(t_1-t_2)$] is inversely proportional to the heat transfer coefficient h and to the area-to-volume ratio A/V of the object, but directly proportional to the heat capacity per unit volume $c_p\rho$. Equations (8–4) and (8–5) are very useful for estimating purposes, but care must be taken to be sure that neglecting thermal resistance within the object does not introduce serious errors. The effects of thermal resistance within the object will be considered later.

EXAMPLE 1

A long piece of steel shafting 1 inch in diameter and initially at 70°F is inserted into a furnace at 2000°F. In 5 min the steel reaches a temperature of 1600°F. Using Newton's law, estimate the times required to reach (a) 1500°F, (b) 1950°F, (c) 1995°F.

SOLUTION

For the heating process described, $t_1 = 70°F$ and $t_2 = 2000°F$. Also when $t = 1600°F$, $\theta = 5$ min $= 0.083$ hr. Substituting these values in Eq. (8–4), we can solve for the expression $(hA)/(2.3V\rho c_p)$:

$$\frac{hA}{2.3V\rho c_p} = -\frac{1}{0.083}\log\frac{1600-2000}{70-2000}$$
$$= 8.2 \text{ hr}^{-1}.$$

Accordingly, for the heating process under consideration, Newton's law can be expressed:

$$\theta = -\frac{1}{8.2}\log\frac{t-t_2}{t_1-t_2}.$$

When $t = 1500°F$,

$$\theta = -\frac{1}{8.2}\log\frac{1500-2000}{70-2000} = \underline{0.0715 \text{ hr or } 4.3 \text{ min}}$$

Similarly, when $t = 1950°F$, we find $\theta = \underline{0.193 \text{ hr or } 11.6 \text{ min}}$; when $t = 1995°F$, we find $\theta = \underline{0.315 \text{ hr or } 18.9 \text{ min}}$. Thus we see that about 20 min should be taken to bring the piece substantially to furnace temperature.

It might be noted that the given data can be combined with other data to calculate h, the heat transfer coefficient from furnace to steel. Since we found

$$\frac{hA}{2.3V\rho c_p} = 8.2 \text{ hr}^{-1},$$

and since
$$A/V = \frac{\pi DL}{\frac{\pi D^2 L}{4}} = \frac{4}{D} = \frac{4}{\frac{1}{12}} = 48 \text{ ft}^{-1},$$

$$\rho \text{ for steel} = 500 \text{ lb/ft}^3,$$

$$c_p \text{ for steel} = 0.14 \text{ Btu/lb-°F}.$$

Therefore,
$$h = \frac{8.2 \times 2.3 \times 500 \times 0.14}{48} = 27.5 \text{ Btu/hr-ft}^2\text{-°F}.$$

EXAMPLE 2

Steel ingots $24 \times 24 \times 60$ inches are removed from a soaking pit at 2200°F and sent to a rolling mill. The minimum temperature at which rolling is satisfactory for this steel is 1900°F. Estimate the time available for rolling before the steel must be reheated, assuming $h = 15$ Btu/hr-ft²-°F for the cooling of the steel in the open.

SOLUTION

The time available for hot working is the time required to cool from 2200° to 1900°F in surroundings at 70°F, and is estimated by solving Eq. (8–4) for θ.

$$t = 1900°\text{F},$$

$$t_1 = 2200°\text{F},$$

$$t_2 = 70°\text{F},$$

$$h = 15 \text{ Btu/hr-ft}^2\text{-°F},$$

$$A = \frac{2 \times 24 \times 24 + 4 \times 24 \times 60}{144}$$

$$= 48 \text{ sq ft, assuming heat loss occurs from all ingot surfaces,}$$

$$V = \frac{24 \times 24 \times 60}{1728} = 20 \text{ cu ft},$$

$$\rho = 500 \text{ lb/ft}^3,$$

$$c_p = 0.14,$$

$$\theta = -\frac{2.3 V \rho c_p}{hA} \log \frac{t - t_2}{t_1 - t_2}$$

$$= -\frac{2.3 \times 20 \times 500 \times 0.14}{15 \times 48} \log \frac{1900-70}{2200-70}$$

$$= \underline{0.3 \text{ hr or 18 min.}}$$

For many purposes, it is helpful to measure temperature in terms of a dimensionless variable, known as *relative temperature*. Designating relative temperature as Y,

$$Y = \frac{t - t_2}{t_1 - t_2},\qquad (8\text{--}6)$$

in which t is the actual measured temperature at a given time, t_1 is the initial temperature of the object at $\theta = 0$, and t_2 is the surroundings temperature or final object temperature when $\theta = \infty$. In any given complete heating or cooling process, Y decreases from 1 at zero time to approach 0 as the object temperature approaches that of the surroundings. The relative temperature might be thought of also as the ratio of the temperature change yet to be accomplished for equilibrium to the total temperature change in the complete process from the start until equilibrium. It should be noted that Y is dimensionless, and therefore its value is independent of the choice of temperature units for t, t_1, and t_2, provided all three are in the same units.

In terms of relative temperature, Newton's law can be written:

$$\log_{10} Y = - \frac{hA}{2.3 V \rho c_p}\,\theta,\qquad (8\text{--}7)$$

or

$$\frac{d \log_{10} Y}{d\theta} = - \frac{hA}{2.3 V \rho c_p}.\qquad (8\text{--}8)$$

From these equations it is seen that for a process obeying Newton's law, a plot of the logarithm of relative temperature ($\log_{10} Y$) against time (θ) will be a straight line of slope $-(hA)/(2.3 V \rho c_p)$ passing through the origin ($\log Y = 0$ or $Y = 1$, $\theta = 0$). Thus, a semilogarithmic plot is especially convenient for heating and cooling data. This is illustrated in Fig. 8–1 (a) and (b) where two sets of cooling data and one set of heating data are shown both as semilogarithmic plots of $\log Y$ vs. θ and as conventional plots of t vs. θ. A useful feature of the semilog plot is that for processes obeying Newton's law the entire heating or cooling "curve" can be drawn as a straight line if one point is measured, that is, if Y is known at any one time after the start of the process.

Fourier's equation of unsteady heat conduction. Whereas in steady heat-flow systems temperature is a function of position only, in unsteady systems temperature is a function

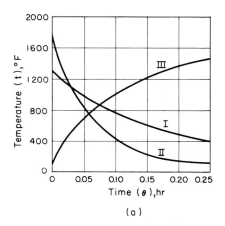

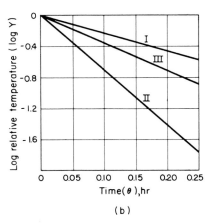

FIG. 8–1. Heating and cooling curves following Newton's law.

of two variables, position and time. If solid objects with some degree of symmetry are considered, position can be measured by a single variable s which represents, for example, distance from the center of a sphere or from the axis of a cylinder. For these cases, Fourier's law of heat conduction can be written as a partial differential equation:

$$\left(\frac{\partial t}{\partial \theta}\right)_s = \frac{k}{c_p \rho}\left[\left(\frac{\partial^2 t}{\partial s^2}\right)_\theta + \left(\frac{d \ln A}{ds}\right)\left(\frac{\partial t}{\partial s}\right)_\theta\right]. \quad (8\text{--}9)$$

A derivation of this relation was given in Chapter 7, Eqs. (7–2) to (7–6). Also it was pointed out that the physical properties of the solid are represented by the quantity $(k)/(c_p \rho)$ which is defined as the thermal diffusivity α.

Calculations of the thermal behavior of a heating or cooling object, when the process is dominated by resistance to heat flow within the object, require a solution of Eq. (8–9) for the boundary conditions (t_1, t_2) and geometry involved. In the more general case, where both surface and interior resistances to heat flow must be considered, Eq. (8–9) must be solved using Eq. (8–2) as one of the boundary conditions. The details of solving Eq. (8–9) for various object shapes and boundary conditions belong to the subject matter of advanced mathematics courses. However, numerical solutions have been published in the form of tables and graphs, which are readily used in working out engineering problems. These numerical solutions are most easily presented as relationships between dimensionless variables, so definitions of these variables will be given in the following section. Then in subsequent sections some of the numerical solutions and their applications to engineering systems will be presented.

Dimensionless variables of unsteady heat flow. The tabular and graphical presentation of relations between dimensionless variables of unsteady heat flow serves a similar purpose to uses of dimensionless variables in correlating empirical data on fluid friction (Chapter 6) and convection heat flow (Chapter 7). The principal physical variables of unsteady heat flow include: temperature t, time θ, position s, size, area A and volume of object V, density ρ, heat transfer coefficient h, and conductivity k. However, when these variables are arranged in dimensionless groups, it is found that only four dimensionless variables are needed ordinarily to describe unsteady heat flow. These are relative temperature Y, relative position n, relative time X, relative surface resistance m. *Relative temperature* has already been defined [Eq. (8–6)]. The definitions of the other three variables can now be given for certain specific geometric objects.

Figure 8–2 shows three common geometric shapes of solids of importance in heating and cooling problems. The behaviors of these

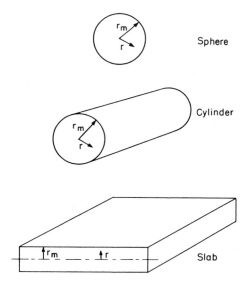

Fig. 8–2. Simple geometric shapes and position variables in unsteady heat flow.

simple shapes serve as good approximations of the behaviors of many other practical shapes. Thus the sphere represents the behavior of bodies approximately the same dimensions in all directions, the infinite cylinder represents bodies with one long dimension, and the infinite plate represents bodies with two long dimensions and one short dimension. In each case, the size is given by the distance r_m from the center to the surface. For a plate, r_m is the half-thickness; and for a cylinder or sphere, r_m is the radius. Similarly, owing to the symmetry of those shapes, position is given by the distance r from the center, since all points at a given distance from the center will have the same behavior during heating or cooling. *Relative position* (*n*) is given by r/r_m. Thus, relative position always varies from 0 at the center of the piece to 1 at the outer surface.

The dimensionless variable *relative time* (*X*) is defined in terms of the thermal diffusivity α, actual time θ, and radius or half-thickness r_m, and is equal to $\alpha\theta/r_m^2$. Thus, for a given value of relative time, the actual time is proportional to the square of the size of the object and inversely proportional to the thermal diffusivity. This definition of relative time thus implies that the actual time in hours to accomplish a given temperature change varies directly with the square of the size of the object and inversely as the thermal diffusivity if the shape and other factors are kept constant. These implied proportionalities are, however, strictly true only for heating and cooling objects with negligible surface resistance to heat flow, for which the thermal behavior is governed by resistance to heat flow within the object.

The fourth of the dimensionless criteria, the *relative surface resistance* (*m*), is given by the expression k/hr_m. This number affords a direct comparison of the resist-

ance to heat flow from the surface to the surroundings ($1/h$) with the resistance to heat flow within the object (r_m/k). For the conditions under which resistance to heat flow within the object is small compared to resistance to heat flow from the surface — that is, the conditions under which Newton's law is valid — *m* becomes very large, approaching ∞. At the other extreme, when the chief resistance to heat flow is within the solid object and surface resistance is negligible, *m* becomes very small, approaching 0. The ratio h/k (with dimensions of ft^{-1}) is sometimes referred to, especially in connection with heat treatment, as the "severity of quench."

Utilizing the above definitions and remembering also that $\alpha = k/c_p\rho$, Newton's law as given by Eqs. (8–7) and (8–8) can now be put into the following form:

$$\log Y = -\frac{1}{2.3}\frac{Ar_m}{V}\frac{1}{m}X. \qquad (8\text{–}10)$$

In Eq. (8–10), $Ar_m/V = 1$ for infinite plates, 2 for infinite cylinders, and 3 for spheres, as is easily shown by expressing A and V in terms of r_m. For simplicity in the subsequent discussion, Eq. (8–10) may be abbreviated to

$$\log Y = -N_eX; \qquad (8\text{–}11)$$

differentiating,

$$\frac{\partial \log Y}{\partial X} = -N_e. \qquad (8\text{–}12)$$

In these equations N_e is a dimensionless constant to be called the *Newton constant*. When Newton's law is applicable, comparison of Eqs. (8–10) and (8–11) shows:

$$N_e = \frac{1}{2.3m} \text{ for infinite plates,}$$

$$N_e = \frac{2}{2.3m} \text{ for infinite cylinders,} \qquad (8\text{–}13)$$

$$N_e = \frac{3}{2.3m} \text{ for spheres.}$$

Later on we will find Eqs. (8–11) and (8–12) useful for systems in which Newton's law is not valid. For these systems we will find N_e still a constant under certain conditions, but having values different from those given as Eq. (8–13). Regardless of whether or not Newton's law is valid, we will define $-N_e$ by Eq. (8–12) as the slope of a line or curve of log Y plotted against X.

Heating and cooling of plates, cylinders, and spheres. Numerical solutions of the Fourier equation (8–9) for plates, cylinders, spheres, and other geometries are given in a number of places in the literature and in a variety of graphical and tabular forms. Among the most convenient presentations are the Gurney-Lurie charts,* which take the form illustrated by Fig. 8–3. Log relative temperature ($\log_{10} Y$) is plotted against relative time (X), and each curve corresponds to a given pair of values of relative surface resistance (m) and relative position (n). Thus, when the value of m is available, the Gurney-Lurie chart gives directly the heating or cooling curves for various positions in the body. Actual temperature (t) and time (θ) are easily found by calculation from Y and X, respectively, knowing t_1, t_2, α, and r_m. Conversely, if a heating or cooling curve is known, the value of m can be determined by comparison of the data with the curves for various values of m on the Gurney-Lurie chart. These charts have a variety of other applications as well, and are widely used for engineering calculations.

Certain general characteristics of the Gurney-Lurie charts can be seen clearly in Fig. 8–3. In the first place, most of the data, in fact, all the data after a time interval at the start of heating or cooling

(roughly, after $X = 0.5$), are represented by straight lines. Since the chart is a semi-logarithmic plot, it follows, therefore, that these data fit Eq. (8–12), $-N_e$ representing the constant slopes of the straight lines. The values of N_e, however, are not given in general by Eq. (8–13), which was derived from Newton's law. On the other hand, all the lines for a given value of relative surface resistance (m) are parallel and have the same value of N_e in Eq. (8–12). In other words the slope N_e is a function of m only and does not depend on n.

Considering now any one group of parallel lines for a given value of m in Fig. 8–3, in general these lines do not extrapolate through the origin (log $Y = 0$ or $Y = 1$, $X = 0$) and therefore do not fit Eq. (8–11). Comparing the positions of the lines with those of lines through the origin fitting Eq. (8–11), the lines representing thermal behavior at positions toward the center of the object ($n \rightarrow 0$) show a time lag compared to Eq. (8–11), whereas positions toward the outside pass through a given temperature sooner than Eq. (8–11) would indicate. Accordingly, we may represent the straight lines on the Gurney-Lurie charts by the following equation:

$$\log_{10} Y = - N_e(X - \lambda), \qquad (8\text{–}14)$$

in which λ may be considered as a *lag coefficient*, with positive values for positions near the center of the object and negative values for positions near the surface.

The heavy dashed line in the group of lines for $m = 0.5$ represents the average temperature of the body at any time, which in this case is the temperature at a relative position (n) of 0.57. Thus, for this line, Y_{av} represents not only relative temperature but also relative heat content of the body as a whole.†

---

* Gurney, H. P., and Lurie, J., "Charts for Estimating Temperature Distribution in Heating and Cooling Solid Shapes." *Ind. Eng. Chem.*, **15**, 1170–1172 (1923).

---

† Assuming c_p constant.

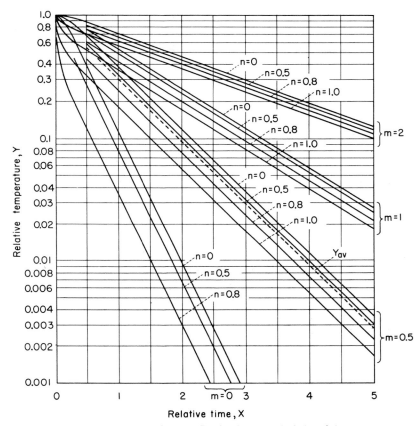

FIG. 8–3. Gurney-Lurie chart for infinite slabs.

That is, $Y_{av} = \dfrac{H - H_2}{H_1 - H_2}$, in which H is the heat content of the body at a given time, H_1 is the initial heat content when the body is at the initial uniform temperature t_1, and H_2 is the heat content at the final equilibrium or surroundings temperature t_2. One further observation can now be made in Fig. 8–3; that is, the dashed line showing Y_{av} as a function of X extrapolates substantially to the origin. Accordingly, as a convenient and very useful working basis, we may represent the *average* behavior of a heating or cooling object by a simple equation equivalent to Newton's law except for the value of N_e:

$$\log_{10} Y_{av} = - N_e X. \qquad (8\text{--}15)$$

This slightly arbitrary assumption of a zero lag coefficient for the average temperature behavior of various bodies greatly simplifies many engineering calculations without introducing errors of any engineering significance.

Much of the subsequent discussion and calculations will be based on Eqs. (8–14) and (8–15), which represent a compromise between the greater mathematical rigor of the Gurney-Lurie charts and the simplicity of form of Newton's law of heating or cooling. In view of the assumptions required even for the rigorous presentation, for example, the assumption that h is independent of temperature, it is unlikely that the simplification results in any significant loss of accuracy for most engineering calculations. However, it should be kept in mind that Eq. (8–14) and relations based on Eq. (8–14) cannot be used safely, even as approximations, to describe or predict heating or cooling behavior during the first instants of the process.

If problems are encountered requiring rigorous numerical solutions of Fourier's equation in terms of the dimensionless variables, the tables prepared by T. F. Russell* are more useful than the Gurney-Lurie charts for the initial period of heating or cooling; still other forms of solutions are available to meet specific needs.

Effect of surface resistance. Figure 8–4 shows the relationships between N_e and m for large plates, long cylinders, and spheres, obtained by integrations of the Fourier equation. The dashed curves for long bars of square cross section, cubes, and cylinders with length equal to diameter were located by estimation and interpolation. When the relative surface resistance m is known, this chart can be used to find N_e for use in Eqs. (8–11), (8–12), (8–14), and (8–15) to calculate heating and cooling curves for the conditions represented by straight lines on the Gurney-Lurie charts of log Y vs. X. Conversely, if experimental heating or cooling curves are plotted as log Y vs. X and give straight lines, the measured values of their slopes ($-N_e$) can be used with Fig. 8–4 to find the proper values of relative surface resistance m for the systems studied.

---

*See J. B. Austin, *Heat Flow in Metals.* Cleveland: American Society for Metals, 1942.

At high values of relative surface resistance, the curves in Fig. 8–4 are asymptotic to straight lines of slope -1, and the equations of these asymptotes were given as Eq. (8–13). That is, these asymptotes correspond to Newton's law derived on the assumption of uniform temperature and negligible resistance to heat flow within the object. It is evident that Newton's law [Eqs. (8–4), (8–7), (8–11), (8–13)] affords an accurate basis for calculation of heating and cooling curves if m is greater than 10, and affords a good basis of estimation for values of m as low as 1 or 2. Accepting an accuracy of $\pm 10\%$ in engineering estimates of heating or cooling times or rates, this means that Newton's law can be used when $k > 2hr_m$. Accordingly, if $k > 2hr_m$, the rates of heating or cooling (based on actual time units, not on X) are substantially independent of the numerical value of the conductivity k.

At low values of m, N_e reaches a maximum so that the curves of Fig. 8–4 are asymptotic to horizontal lines. The maximum values of N_e correspond to conditions of *negligible surface resistance* for which cooling or heating rates and temperature distribution are determined solely by heat conduction rates in the body. This condition is approached fairly closely for values of m of 0.1 and below. Thus, when $h > 10k/r_m$, heating and cooling rates are independent of the value of h. For example, if a layer of refractory material for which $k = 0.9$ Btu-ft/hr-ft^2-°F and $r_m = 0.5$ ft is to be heated, the rate of heating would be substantially the same whether h were 20 or 50, and attempts to increase the heating rate by increasing h above 20 or so would be rather futile.

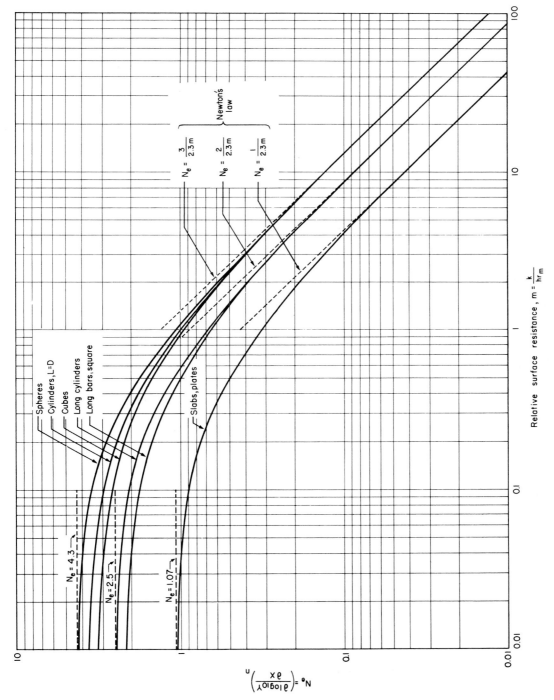

Fig. 8-4. N_e vs. relative surface resistance (m) for simple geometric shapes.

EXAMPLE 3

In example 2 on page 262, calculations of the cooling behavior of a steel ingot were made, assuming Newton's law. For this system, estimated values of h, k, and r_m are:

$$h = 15 \text{ Btu/hr-ft}^2\text{-°F},$$

$$k = 25 \text{ Btu-ft/hr-ft}^2\text{-°F},$$

$$r_m = 1 \text{ ft.}$$

Accordingly,
$$m = \frac{k}{hr_m} = \frac{25}{15 \times 1} = 1.67.$$

Reference to Fig. 8–4 indicates that the assumption of Newton's law did not introduce any significant error into the calculation of the cooling time.

EXAMPLE 4

A $3 \times 3 \times 12$-inch brick initially at 100°F is placed in a furnace at 1100°F. The thermal conductivity of the brick is 0.8 Btu-ft/hr-ft²-°F and h is estimated to be 10 Btu/hr-ft²-°F. Also, $\rho = 120$ lb/ft³ and $c_p = 0.22$ Btu/lb-°F. (a) Estimate the time required for the average temperature to reach 1000°F. (b) Compare this time with the time required by a piece of copper of the same dimensions to reach an average temperature of 1000°F under the same conditions.

SOLUTIONS

(a) First the relative surface resistance can be calculated for the brick:

$$m = \frac{k}{hr_m} = \frac{0.8}{10 \times 1.5/12} = 0.64.$$

From the curve in Fig. 8–4 for long bars we find when $m = 0.64$, $N_e = 0.9$. Accordingly, Eq. (8–15) for the average relative temperature becomes

$$\log Y_{\text{av}} = -N_e X = -0.9X.$$

The desired value of Y_{av} can be calculated:

$$Y_{\text{av}} = \frac{1000-1100}{100-1100} = 0.1.$$

Substituting this value in the above equation,

$$X = 1.1.$$

Since
$$X = \frac{\alpha\theta}{r_m^2} \quad \text{and} \quad \alpha = \frac{k}{c_p\rho},$$

$$\theta = X\frac{r_m^2 c_p\rho}{k} = 1.1\frac{\left(\frac{1.5}{12}\right)^2(0.22)(120)}{0.8}$$

$$= \underline{0.6 \text{ hr or about 35 min.}}$$

(b) Repeating the calculations for a piece of copper ($k =$ about 200),

$$m = \frac{k}{hr_m} = \frac{200}{10 \times \frac{1.5}{12}} = 160.$$

Reference to Fig. 8–4 shows that this value of m falls off the chart to the right indicating that Newton's law should be used. Accordingly, assuming a long bar and ignoring end effects,

$$N_e = \frac{2}{2.3m} = \frac{2}{2.3 \times 160} = 0.0055.$$

For $Y_{av} = 0.1$, Eq. (8–15) gives $X = 180$. Note that in *dimensionless time units*, the copper takes many times longer to heat than the refractory material of part (a). Assuming for the copper that $k = 200$ Btu-ft/hr-ft²-°F, $\rho = 500$ lb/ft³, and $c_p = 0.1$ Btu/lb-°F, the actual time θ can be calculated from X:

$$\theta = 180 \frac{\left(\frac{1.5}{12}\right)^2 (0.1)(500)}{200}$$

$$= \underline{0.7 \text{ hr or about 40 min.}}$$

The results for (a) and (b) show surprisingly little difference between the heating behaviors of the refractory material and of the copper under the conditions given. This is because in both cases the rate of heating is governed mainly by resistance to heat transfer at the surface. The gain in heating rate which might have been expected as a result of the much higher thermal conductivity of the copper is a little more than counterbalanced by the fact that the piece of copper requires more heat for the specified temperature rise than the refractory in the proportions of the volume specific heats $(c_p\rho)$, which are 50 Btu/ft³-°F and 26.4 Btu/ft³-°F, respectively.

Temperature distribution and time lag within heating and cooling objects. As shown by the Gurney-Lurie charts and as indicated by Eq. (8–14), interior temperatures will lag behind average temperatures given by Eq. (8–15), whereas surface-temperature change will be in advance of average temperature. These time differences remain constant throughout the later stages of heating or cooling (after $X =$ about 0.5, for straight-line portions of Gurney-Lurie charts) so that all points in the object have the same temperature-time behavior, except for the lag. It should be kept in mind, however, that these statements do not apply at the very start of heating and cooling. Thus, if all points in the body start at the same temperature, which is the case usually considered, a short time interval up to $X =$ about 0.5 or less is required to establish the kind of dynamic equilibrium represented by the straight lines on the Gurney-Lurie charts. After this state is reached, the relative temperature Y_1 at a given position remains in a constant ratio to the relative temperature Y_2 at any other given position. Thus, it is convenient to describe the temperature distribution in a heating or cooling object in terms of the variation of the ratio Y/Y_{av} with position ratio n, Y_{av} being the average body temperature. Figures 8–5, 8–6, and 8–7 are graphs of Y/Y_{av} against n

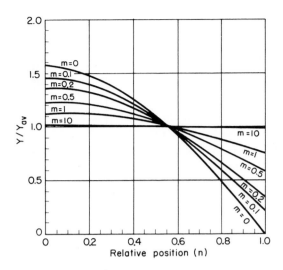

FIG. 8–5. Infinite slabs: variation of temperature with position for different values of relative surface resistance.

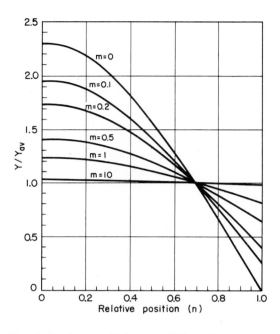

FIG. 8–6. Long cylinders: variation of temperature with position for different values of relative surface resistance.

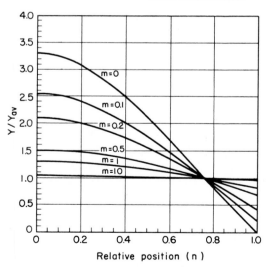

FIG. 8–7. Spheres: variation of temperature with position for different values of relative surface resistance.

for various values of relative surface resistance m.* These graphs show clearly that for high values of m, corresponding to behavior approaching Newton's law, Y/Y_{av} is not far from unity throughout the body, signifying substantially uniform temperature. On the other hand, for low values of m the temperature changes most rapidly with distance near the surface, and relatively large temperature gradients are obtained from the center to the outside in heating or cooling. Under conditions of negligible surface resistance ($m = 0$), the temperature at the surface of the object (at $n = 1$) is equal to that of the surroundings ($Y = 0$) at all times, and all the temperature drop is inside the object.

---

* The graphs represent the results of calculations based on Eq. (8–9).

The average relative temperature Y_{av} at any time X is found easily by Eq. (8–15), using Fig. 8–4 to find the appropriate value of the Newton constant N_e. From the value of Y_{av} calculated this way, the value of Y at any desired position can then be found by using the values of Y/Y_{av} interpolated from Figs. 8–5, 8–6, or 8–7.

Combining Eqs. (8–14) and (8–15) to eliminate X, we can express the time lag coefficient λ in terms of Y/Y_{av} as follows:

$$\lambda = \frac{1}{N_e} \log \frac{Y}{Y_{av}}. \qquad (8\text{–}16)$$

With this relationship, then, the information plotted in Figs. 8–5, 8–6, and 8–7 can also be used to evaluate the time lag. The numerical values of λ are converted to actual time lags in the same way that values of X are converted to actual time, that is, by multiplying by r_m^2/α.

Example 5

In example 4 on page 270 the times required to bring the average temperature of a brick and a piece of copper of the same shape to 1000°F in a furnace at 1100°F were estimated. Estimate for both objects (a) center and surface temperatures when average temperature reaches 1000°F, (b) times required for center and outside temperatures of brick to reach 1000°F, and (c) center temperature in brick when outside temperature is 600°F.

Solutions

For estimating purposes it can be assumed that the temperature distributions will be similar to those in a long cylinder, and Fig. 8–6 can be used. However, it should be noted that the square section, strictly speaking, does not have the symmetry necessary to enable position to be specified only by distance from the central axis.

(a) From the previous calculations, $m = 0.64$ for the brick and $m = 160$ for the copper. Interpolating in Fig. 8–6,

outside surface ($n = 1$): $Y/Y_{av} = 0.7$ for brick, 1.00 for copper,
center ($n = 0$): $Y/Y_{av} = 1.35$ for brick, 1.00 for copper.

Converting relative temperatures to actual °F, we find that when the average brick temperature is 1000°F, the center temperature will be about 965°F and the outside surface will be about 1030°F. For the piece of copper, it appears that the temperature will be within 1°F of 1000°F all the way from the center to the outside.

(b) For the brick, the lag coefficients are calculated by Eq. (8–16), with $N_e = 0.9$:

$$\text{outside surface:} \quad \lambda = \frac{1}{0.9} \log 0.7 = -0.17,$$

$$\text{center:} \quad \lambda = \frac{1}{0.9} \log 1.35 = +0.125.$$

Converting λ into actual time units, we find the time lag of the outside surface:

$$\text{lag} = \lambda \frac{r_m^2 c_p \rho}{k} = -0.17 \frac{\left(\frac{1.5}{12}\right)^2 (0.22)(120)}{0.8}$$

$$= -0.09 \text{ hr or about } -5 \text{ min.}$$

Combining this result with the previous calculation of the time required for the average temperature to reach 1000°F, we find the outside temperature will reach this temperature about 0.5 hr or 30 min after the piece is placed in the furnace. Similar calculations for the center indicate that the center will reach 1000°F about 39 min from the start of heating.

(c) When $t = 600$°F, $Y = \frac{100-600}{100-1100} = 0.5$, and $n = 1$. From Fig. 8–6, $Y_{n=1}/Y_{av} = 0.7$ and $Y_{n=0}/Y_{av} = 1.35$. Accordingly, when the outside temperature is 600°F,

$$Y_{n=0} = \frac{1.35}{0.7} \times 0.5 = 0.96.$$

The corresponding actual temperature is 140°F, just 40°F above the starting temperature of 100°F. However, this estimate must be considered very rough, because it involves the assumption that Eq. (8–14) is valid for an early stage in the heating process.

Heating and cooling rates. Continuing to deal with behavior after the body reaches the dynamically steady temperature distribution with constant time lags from one position to another, it follows directly from the definition of N_e [Eq. (8–12)] that the cooling rate or heating rate is proportional to the relative temperature, or

$$\left(\frac{\partial Y}{\partial X}\right)_{m,n} = Y \left(\frac{\partial \ln Y}{\partial X}\right)_{m,n} = -2.3 N_e Y. \quad (8\text{–}17)$$

This relation holds for all positions in the body. Thus, perhaps contrary to general expectations, the heating or cooling rate through a given temperature is no greater on the outside of a body than in the center. Moreover, at any given instant (after $X =$ about 0.5), the inside of a body is actually changing temperature more rapidly than the outside because the inside is lagging behind the outside and is further from the final temperature. Converting Eq. (8–17) from dimensionless units, the actual cooling rate is given by

$$\frac{\partial t}{\partial \theta} = -\frac{2.3 N_e \alpha}{r_m^2} (t - t_2). \quad (8\text{–}18)$$

As shown in Fig. 8–4, N_e approaches a maximum value dependent only on geometry as the surface resistance is decreased. Accordingly, applying Eq. (8–18), we may conclude that the maximum cooling or heating rate obtainable through a given temperature range under conditions of negligible surface resistance is proportional to the quotient α/r_m^2 for the body. When the process is controlled by surface resistance (Newton's-law conditions), this quotient is less significant and Eq. (8–3) shows more clearly the properties of the system which determine heating and cooling rates:

$$\frac{dt}{d\theta} = \frac{hA}{V\rho c_p}\,(t - t_2).\qquad(8\text{–}3)$$

As has been pointed out repeatedly, the quantitative relationships so far presented are not valid at the very start of a heating or cooling process. Fortunately, most heating and cooling problems do not require consideration of the initial stages of the heating or cooling. However, in studying the process of quenching metals for heat treating, especially hardening steel, the initial period is likely to be the most important. In fact, variations in hardness from inside to outside of quenched steel objects are common and result from large variations in cooling rates from center to outside surface during the initial period of cooling, although in the later stages of cooling, as pointed out above, the cooling rates through a given temperature should be the same in all parts of a symmetrical body.

Hardening of steel has been correlated with various criteria of the cooling process, including rate of cooling through a specified temperature, time of cooling through a specified temperature range, and others. Probably the most successful of these correlations, developed by Grossmann, Asimow, and Urban,* is based on the theory that the hardness of a given steel is fixed by the actual time required for cooling from the moment of quenching to a temperature halfway between the quenching temperature and the temperature of the cooling medium. In terms of the variables defined earlier, this "half-temperature time" is the time θ

---

* See *Hardenability of Alloy Steels.* Cleveland: American Society for Metals, 1939.

from the start of cooling when $Y = 1$ to $Y = 0.5$. This method of studying quenching involves the use of special charts representing solutions of the Fourier equation in terms of the dimensionless variables, and covering the initial period of the cooling during which cooling rates vary with position in the body.

Evaluation of relative surface resistance. Probably the most difficult variable to evaluate in heating and cooling problems is the relative surface resistance, m. More specifically, the difficulty lies in the evaluation of the heat transfer coefficient h governing the heat flow between the surface and the surroundings. Methods of estimating h for simple convection and radiation systems were discussed in some detail in Chapter 7 and those methods serve satisfactorily in many unsteady heat-flow calculations. Often, however, a satisfactory estimation of h by those methods is difficult, because of lack of data or because of the complexity of the heating or cooling conditions. Also, because in unsteady heat-flow calculations it is usually necessary to assume h constant and independent of temperature, which is not true in fact, the question naturally arises as to how to choose a single or average value of h for the entire temperature range of a heating or cooling process. Under these conditions it may be desirable to estimate h for the system under consideration from experimental heating or cooling data.

A convenient procedure for determining h in a given system consists of measuring the heating or cooling curve experimentally at a given position (usually at or near the surface) in a specimen more or less representative of the pieces whose heating or cooling is to be considered. From these data Y and X can be calculated and a graph made of log Y against X. If the usual assumptions of unsteady heat-flow calculations are reasonably well fulfilled, a substantial part of the data should determine a straight line on this graph. Measurement of the slope of this line gives the Newton constant N_e [Eq. (8–14)]. With this value of N_e available, Fig. 8–4 can be used to read off the proper value of m. With m, the thermal conductivity k, and the half-thickness or radius r_m all known, $h = k/mr_m$.

If the problems under consideration in a given furnace or cooling medium always involve heating or cooling the same material, it may be easier to utilize the experimental measurements to determine a surface coefficient H (also known as severity of quench), defined as follows:

$$H = \frac{h}{k} = \frac{1}{mr_m}. \qquad (8\text{–}19)$$

With a given material and given heating or cooling conditions, H once determined should be the same for all similar objects so that m in each case is found simply as $1/Hr_m$. Table 8–1 gives values of H in ft^{-1} representative of practices in heating and cooling steel.

The methods of Grossmann, Asimow, and Urban referred to previously for correlating steel hardening with quenching behavior involve determination of H for a given kind of quench from hardness vs. position measurements on a series of quenched specimens of different sizes.

TABLE 8-1

Typical Surface Coefficients (H) in Heating and Cooling Steel*

Heating:	
Radiant-tube furnace	0.6
Slab-heating furnace	1.2
Soaking pit	0.3
Resistance furnace	0.4
Cooling and quenching media, not agitated:	
Air	0.25
Oil	3.6
Water	12.0
Brine	25
Strong water spray:	75-150

*After J. B. Austin, Heat Flow in Metals. Cleveland: American Society for Metals, 1942.

In principle, m can be estimated from temperature vs. position measurements made at a given instant during the heating or cooling. If temperature-distribution data are plotted as log Y against n (relative position), m can be determined by finding the data in Figs. 8–5, 8–6, or 8–7 (depending on geometry of object) which best match the shape of the experimental curve of log Y vs. n. This procedure, however, does not appear to be of general utility, since it is usually easier to measure the temperature-time relation at a given position rather than the temperature-position relation at a given time.

For many estimating purposes it is unnecessary to obtain an accurate value of m; instead, it is necessary only to establish whether the body is being heated or cooled under conditions of *negligible surface resistance* on the one hand, or under conditions for which *Newton's law* is a good approximation on the other hand. Thus, if a rough estimate indicates that $m < 0.1$, heating and cooling calculations adequate for most needs can be based on the assumption of negligible surface resistance ($m = 0$, N_e values given by horizontal asymptotes in Fig. 8–4). On the contrary, if a rough estimate indicates $m > 1$, calculations can be based entirely on Newton's law, ignoring temperature differences within the body. For example, referring to the conditions described in Table 8–1, it can be seen that in heating steel or air-cooling steel, $m \ (= 1/Hr_m) > 1$ even for pieces a few feet in thickness, so that Newton's law is a useful approximation. In fact, $m > 1$ and Newton's law is useful generally for heating and air-cooling all metals. Owing to the larger values of H obtained in quenching into liquids, however, Newton's law is not generally applicable to quenching. In severe quenches of moderate-sized pieces, conditions of negligible surface resistance are approached. Thus, in a brine bath, $m < 0.1$ if the half-thickness or radius r_m is greater than 0.4 ft or 5 inches. Clearly, however, with as good a conductor as steel, conditions of negligible surface resistance are reached only by very severe quenches on relatively large sections. With much poorer conductors such as furnace refractories and other nonmetals, conditions of negligible surface resistance ($m < 0.1$) are more common.

Temperature changes in furnaces and other complex systems. The quantitative relations considered in previous sections have dealt primarily with the behavior of single bodies with simple, symmetrical shapes. For such bodies, estimations of unsteady heat flow were based on data obtained by integrations of the Fourier partial differential equation. However, when furnaces and other complex structures are considered, the mathematical difficulties become too great for the engineer to attempt solutions of the differential equations. Accordingly, for complex structures a rather more empirical approach must be adopted. The primary basis of this approach is the assumption that relative

temperature Y and time θ are related exponentially for complex systems in much the same way as has already been shown for simple bodies. That is, we will represent heating and cooling data generally for all kinds of systems by the empirical equation

$$Y = e^{-\frac{\theta - \theta_{\text{TR}}}{\theta_{\text{CAP}}}}, \qquad (8\text{--}20)$$

in which Y, the relative temperature, is defined as before in terms of the temperature t, the initial steady temperature t_1, and the final steady temperature t_2; θ is time; and θ_{TR} and θ_{CAP} are constants with dimensions of time. θ_{TR} is the *transfer lag* and θ_{CAP} is the *capacity lag*. Another useful form of this relationship is

$$\log_{10} Y = -\frac{1}{2.3} \frac{\theta - \theta_{\text{TR}}}{\theta_{\text{CAP}}}. \qquad (8\text{--}21)$$

Inspection of this equation shows that it corresponds to a straight line on a plot of log Y vs. θ, the slope of the line being $-1/(2.3\theta_{\text{CAP}})$ and the intercept on the θ axis being θ_{TR}. These characteristics are illustrated further in Fig. 8–8, where two sets of heating data obeying Eqs. (8–20) and (8–21) are each plotted by the conventional t vs. θ

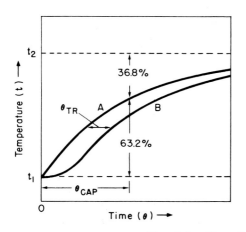
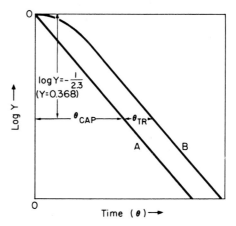

FIG. 8–8. Capacity lag and transfer lag.

method and by the log Y vs. θ method. Curve A represents a system for which the transfer lag θ_{TR} is zero, so that the straight line in the semilogarithmic plot passes through the origin (log $Y = 0$, $\theta = 0$). Instead of defining the capacity lag θ_{CAP} in terms of the slope of this straight line, a more meaningful but still equivalent definition is to consider the capacity lag θ_{CAP} as the time required for Y to change from 1 to e^{-1}, that is, the time required to accomplish 63.2% of the total temperature change. Defined in this way, θ_{CAP} is a convenient measure of the rate of response of a system to temperature changes. A sluggish, slowly responding system will have a large value of θ_{CAP} and a quickly responding system will have a low value of θ_{CAP}. Another characteristic of many complex systems is that different parts of the system start to respond to change at different times, just as the center of a body of simple geometric shape lags behind the outside surface in any heating or cooling process. This type of lag results from resistance to heat flow interposed between the point whose temperature change is being considered and the point or surface where the temperature change is initiated. In Fig. 8–8, curve B in each plot shows the heating behavior of a system with the same capacity lag (same slope) as curve A, but in addition having a transfer lag θ_{TR}. Thus, it can be seen that the time required to accomplish 63.2% of the total temperature change is the sum of the capacity lag θ_{CAP} and the transfer lag θ_{TR}. Under some conditions, as will be seen later, θ_{TR} can be negative.

Figure 8–9 gives t vs. θ and log Y vs. θ plots for data obtained in an experimental study of the temperature response of a laboratory electric furnace. The four experimental runs were conducted as follows:

(A) Starting with a cold furnace (26°C), the power was turned on at $\theta = 0$ and the furnace was allowed to heat with constant power input; $t_1 = 26$°C, $t_2 = 890$°C.

(B) Starting with the furnace operating at a steady temperature of 900°C, the power input was increased at $\theta = 0$ and the furnace was allowed to heat toward the new steady temperature; $t_1 = 900$°C, $t_2 = 1130$°C.

(C) Starting with the furnace operating at a steady temperature of 875°C, the power was shut off entirely at $\theta = 0$; $t_1 = 875$°C, $t_2 = 26$°C.

(D) Starting with the furnace at a steady temperature of 910°C, the power input was reduced at $\theta = 0$; $t_1 = 910$°C, $t_2 = 720$°C.

In the semilogarithmic graph of these data, both heating curves A and B coincide within experimental error, and the data after $\theta = 2$ hr are represented by a straight line. From the location and slope of this line, the capacity lag of the furnace, θ_{CAP}, is estimated to be 3.4 hr and the transfer lag θ_{TR} is -1.9 hr. The rather large negative transfer lag results from the fact that the thermocouple and heating elements were in direct sight of each other within the furnace enclosure so that the measured temperature response occurred in advance of the average temperature response of the furnace structure as a whole. The two cooling curves, C and D, both show a negative transfer lag of about 0.5 hr, but the capacity lag for cooling to room temperature (curve C) is about 3.8 hr as compared to about 2.5 hr for simply cooling to a lower furnace temperature (curve D).

Presumably different parts of the furnace structure and contents will show about the same capacity lags as the furnace interior, but will show quite different transfer lags depending on the thermal resistance between the part in question and the source of heat.

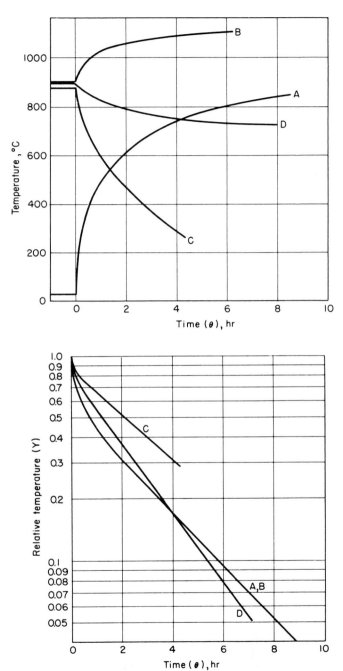

FIG. 8-9. Temperature response of laboratory electric furnace.

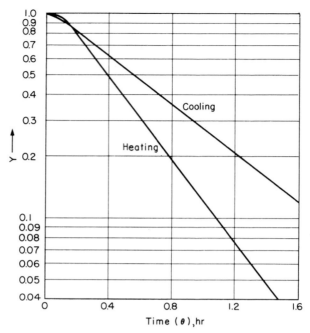

FIG. 8–10.   Temperature response within a furnace charge.

Thus it is to be expected that the outer surface of the furnace wall will have an appreciable positive transfer lag as compared to the negative transfer lag of the furnace interior.

Figure 8–10 gives the results of temperature measurements at the center of a crucible packed with fine ore. First the cold crucible was placed in a hot furnace and allowed to reach a steady temperature ($t_1 = 25°C$, $t_2 = 870°C$). Then the crucible was removed from the furnace and allowed to cool to room temperature ($t_1 = 870°C$, $t_2 = 25°C$). For the heating, $\theta_{TR} = 0.1$ hr or 6 min and $\theta_{CAP} = 0.43$ hr or 26 min. For the cooling, $\theta_{TR} = 0.05$ hr or 3 min and $\theta_{CAP} = 0.73$ hr or 44 min.

Factors affecting capacity lag and transfer lag. Comparisons of the empirical equations by which capacity and transfer lags are defined with the equations used earlier to represent Newton's law and to represent solutions of the Fourier equation throw some further light on the nature of capacity and transfer lags. For a simple body of symmetrical shape, comparison of the empirical Eq. (8–21) with Eq. (8–7) for Newton's law shows that the two relations are equivalent if

$$\theta_{CAP} = \frac{c_p \rho V}{hA} \quad \text{and} \quad \theta_{TR} = 0. \quad (8\text{–}22)$$

Since the resistance to heat flow within an object obeying Newton's law is negligible, there is no transfer lag from one part of the object to another. On the other hand, the capacity lag for a Newton's-law system is the total heat capacity of the system ($c_p \rho V$), or the total number of Btu required to raise the temperature 1°F, multiplied by the total surface resistance to heat flow [$1/(hA)$]. This close relation between the capacity lag and the total heat capacity accounts for the name capacity lag.

For systems whose behavior is not adequately described by Newton's law, we can compare Eq. (8–21) with Eq. (8–14), which is the general equation for heating and cooling behavior used previously to express the results of integration of the Fourier differential equation. Equations (8–14) and (8–21) become equivalent if

$$\theta_{\text{CAP}} = \frac{b}{2.3N_e}\left(\frac{r_m}{kA}\right)(c_p\rho V)*$$

and

$$\theta_{\text{TR}} = b\lambda\left(\frac{r_m}{kA}\right)(c_p\rho V)*, \qquad (8\text{–}23)$$

in which $b = Ar_m/V$ is a dimensionless shape factor with values of 1 for large

---

* The derivations are as follows: First, combining Eqs. (8–14) and (8–21),

$$-\log Y = N_e(X - \lambda) = \frac{\theta - \theta_{\text{TR}}}{2.3\theta_{\text{CAP}}}. \qquad (8\text{–}23a)$$

By the definitions of X and α,

$$X = \frac{\alpha\theta}{r_m^2} = \frac{k\theta}{c_p\rho r_m^2}. \qquad (8\text{–}23b)$$

Substituting Eq. (8–23b) in Eq. (8–23a),

$$\frac{N_e k}{c_p\rho r_m^2}(\theta) - N_e\lambda = \frac{1}{2.3\theta_{\text{CAP}}}(\theta) - \frac{\theta_{\text{TR}}}{2.3\theta_{\text{CAP}}}.$$
$$(8\text{–}23c)$$

If this equality is to be maintained for all values of the variable θ, the coefficients of θ on both sides of the equation must be equal and the constant terms must be equal:

$$\frac{N_e k}{c_p\rho V r_m^2} = \frac{1}{2.3\theta_{\text{CAP}}}, \text{ and } N_e\lambda = \frac{\theta_{\text{TR}}}{2.3\theta_{\text{CAP}}}. \quad (8\text{–}23d)$$

Solving for θ_{CAP}, and letting $r_m = bV/A$,

$$\theta_{\text{CAP}} = \frac{c_p\rho V r_m^2}{2.3N_e k} = \frac{b}{2.3N_e}\left(\frac{r_m}{kA}\right)(c_p\rho V). \qquad (8\text{–}23e)$$

With this expression for θ_{CAP}, θ_{TR} is found:

$$\theta_{\text{TR}} = 2.3\theta_{\text{CAP}}N_e\lambda = b\lambda\left(\frac{r_m}{kA}\right)(c_p\rho V). \qquad (8\text{–}23f)$$

---

plates, 2 for long cylinders, and 3 for spheres. In general, N_e depends on the geometry of the system and on the relative surface resistance, as was demonstrated previously for simple, symmetrical objects. Moreover, under conditions of negligible surface resistance, where heating behavior is governed by thermal resistance within the body, N_e becomes a constant determined only by the geometry of the system. Therefore, Eq. (8–23) shows that under conditions of *negligible surface resistance* the capacity lag θ_{CAP} is proportional to the product of the total heat capacity of the system $(c_p\rho V)$ and a term measuring the thermal resistance to heat flow within the system (r_m/kA). The transfer lag θ_{TR} is proportional to the same two general properties of the system, but λ, the proportionality constant, is a function of relative position as well as of system geometry and relative surface resistance, as was brought out earlier for the simple, symmetrical bodies.

One simple application of the above-discussed principles to furnace design lies in the use of lightweight insulating bricks in place of standard firebrick for furnace walls. Suppose, for example, a firebrick is compared with a porous brick made of the same material (same c_p), but the porous brick has $\frac{1}{4}$ the thermal conductivity and $\frac{1}{2}$ the density of the firebrick. For a wall section of given thermal resistance and given heat loss to the surroundings, the porous brick would have $\frac{1}{4}$ of the thickness and $\frac{1}{8}$ of the weight of the firebrick, accordingly $\frac{1}{8}$ of the heat capacity. Thus, the substitution of the insulating brick in this case should afford a very substantial decrease in the capacity lag of the system. On the other hand, if an equal volume of porous brick were substituted for the firebrick, the thermal resistance of the wall would be quadrupled while the heat capacity was halved and the net result would be some increase in the capacity lag of the system along with the major decrease in heat loss.

The addition of insulation to the outside of existing furnaces increases both thermal resistance and total heat capacity, and therefore increases the capacity lag of the furnace and the heating-up and cooling-down times.

Applications to temperature measurement. Strictly speaking, a temperature-measuring device such as a thermometer or a thermocouple measures only its own temperature, and thus measures the temperature of the surroundings only after it reaches equilibrium with the surroundings. When a

thermocouple is placed in a system, its temperature response is similar to that of any other body and involves both capacity and transfer lags. For example, a typical bare thermocouple in air will have a capacity lag θ_{CAP} of the order of a half a minute and no transfer lag. Placing the thermocouple in a protection tube will increase the capacity lag and also will introduce transfer lag. These lags become quite important when the attempt is being made to measure temperature in a system where the temperature of the system itself varies with time. Also, lags of this kind directly affect the response and general characteristics of devices used for automatic temperature control.

Supplementary References

Austin, J. B., *Heat Flow in Metals.* Cleveland: American Society for Metals, 1942.

Brown, George Granger, et al., *Unit Operations.* New York: John Wiley, 1950.

Kern, Donald Q., *Process Heat Transfer.* New York: McGraw-Hill, 1950.

McAdams, W. H., *Heat Transmission.* New York: McGraw-Hill, 1942.

Perry, John H., *Chemical Engineers' Handbook.* New York: McGraw-Hill, 1949.

Schack, A., *Industrial Heat Transfer* (translated from German by H. Goldschmidt and E. P. Partridge). New York: John Wiley, 1933.

Trinks, W., *Industrial Furnaces*, Vol. I. New York: John Wiley, 1950.

Problems

8–1 A metal object originally at 100°F is placed in a furnace at 1600°F and is observed to reach 1000°F after 15 min heating. Prepare graphs giving the estimated heating curves, as follows: (a) log Y vs. θ (hr) and (b) t (°F) vs. θ (hr). (c) and (d) Show on the same graphs the estimated heating curves for heating under the same conditions an object of the same shape and same material, but three times as large in linear measurements.

8–2 A cylinder of solid copper is taken from a furnace and allowed to cool in a room where the temperature is 80°F. The temperature of the copper is 930°F just 5 min after removal from the

furnace and is 800°F 10 min after removal from the furnace. (a) Estimate the temperature of the copper at the time of leaving the furnace. (b) What probable error would you ascribe to the estimate of part (a)? (c) Estimate the copper temperature 40 min after it is taken out of the furnace.

8–3 Two lengths of $1\frac{1}{2}$-inch steel shafting are removed from a furnace at 1600°F. One piece is cooled in the room where the temperature is 80°F and the other is quenched in a large tank of water at 60°F. Compare the times required for these pieces to cool to where they just feel warm to the touch.

8–4 A metal cylinder initially at 70°F is placed on end in a furnace chamber kept at 2000°F. Estimate the time required for the cylinder to reach its melting point, for each of the following cases: (a) lead cylinder, 4 inches in diameter by 3 ft long, emissivity (ϵ) = 0.6; (b) copper cylinder, 4 inches in diameter by 3 ft long, emissivity (ϵ) = 0.6; (c) copper cylinder, 8 inches in diameter by 4 ft long, ϵ = 0.6; (d) aluminum cylinder, 4 inches in diameter by 3 ft long, ϵ = 0.15.

8–5 Steel ingots $24 \times 24 \times 60$ inches in size are removed from a soaking pit at 2200°F and allowed to cool in the open. Estimate the temperature difference from center to outside surface when the outside surface is at 1900°F. See examples 2 and 3 for additional data on this cooling process.

8–6 A sphere 1 ft in diameter is cooled from 2100°F to 100°F under conditions of negligible surface resistance. Estimate (a) the temperature gradient (°F/inch) near the surface when the average temperature is 700°F, (b) the center temperature when the average temperature is 700°F, (c) the cooling rate (°F/min) at 700°F, if the thermal diffusivity is 0.05 ft²/hr, and (d) the cooling rate at 700°F if the thermal diffusivity is 0.5 ft²/hr.

8–7 A $9 \times 2\frac{1}{2} \times 2\frac{1}{4}$-inch insulating brick at a uniform temperature of 1800°F is removed from a furnace and allowed to cool in the room (70°F). For this material, $k = 0.12$ Btu-ft/hr-ft²-°F, $c_p = 0.21$ Btu/lb-°F, and $\rho = 25$ lb/ft³. Estimate (a) center temperature when outside surface reaches 100°F, (b) time required for outside surface to reach 100°F, and (c) time required for the center to reach 300°F. Also make the best estimate you can, with readily available data, of (d) the center temperature when the outside temperature is 400°F.

8–8 Standard $9 \times 4\frac{1}{2} \times 2\frac{1}{2}$-inch bricks are to be heated to 2500°F in a furnace, and it is estimated that the heat transfer coefficient from the furnace to the exposed brick surface is 20 Btu/hr-ft²-°F. For these bricks, $k = 0.8$ Btu-ft/hr-ft²-°F, $c_p = 0.21$ Btu/lb-°F, and $\rho = 125$ lb/ft³. Estimate the capacity lags and maximum transfer lags in hr if the bricks are carefully stacked in tall, separated columns (a) 9×9 inches at the base and (b) 18×18 inches at the base.

8–9 Estimate the capacity lags for bare-wire thermocouples in measuring air temperatures of 400–600°F, under the conditions outlined below. The metal wires used in making the couples have the following average properties: $c_p = 0.1$ Btu/lb-°F, $\rho = 525$ lb/ft³, $k = 30$ Btu-ft/hr-ft²-°F.

(a) Wire size, 0.125 inch; still air.
(b) Wire size, 0.125 inch; velocity, 20 ft/sec.
(c) Wire size, 0.125 inch; velocity, 100 ft/sec.
(d) Wire size, 0.025 inch; velocity, 20 ft/sec.

CHAPTER 9

PHASES IN PYROMETALLURGICAL SYSTEMS

Pyrometallurgical processes as a general rule are carried out in systems of several phases. Multiphase systems are the rule because each of the unit processes has the objective of making a separation of metals and other chemical elements from each other; these separations are accomplished by providing conditions which cause the elements to pass into two or more different phases removable as separate products. Each unit process of extracting metals involves a characteristic combination of kinds of phases; in fact, it is the kinds of phases dealt with that principally determine the apparatus and technique for a given unit process and thus distinguish that unit process from any other.

At the high temperatures characteristic of pyrometallurgical processes, certain of the phases often produced are very different in properties from the kinds of phases familiar to the chemist working with systems not far from room temperature. Thus, liquid metals, liquid slags, and liquid mattes might almost be considered as phases peculiar to the metallurgist. At the same time the extractive metallurgist handles these phases in such abundance and variety that he must be as familiar with them as the chemist is familiar with gases, solids, water, and water solutions at room temperature. Accordingly, the purpose of this chapter is to introduce some of the most important features and general properties of those phases peculiar to high-temperature pyrometallurgical processes.

Much of this introduction to the phases in pyrometallurgical systems will be based on information in constitution diagrams. For the most part, the principles on which these diagrams are based will be taken for granted, since they are treated quite fully in texts on physical chemistry and physical metallurgy.

Typical multiphase systems

Some of the most common combinations of phases characteristic of high-temperature unit processes are shown schematically in Fig. 9–1. Each of the systems shown is representative of processes carried out in extracting a number of metals, and the slag-metal and gas-metal systems in particular are important for nearly all the common metals. In refining liquid metals [Fig. 9–1(a)], the general procedure is to melt the metal, add reagents, and then control the temperature and chemical conditions so that the impurities are removed from the liquid metal and collected in a separate phase. As indicated in the figure, the phase in which the impurities are collected may be a gas, a liquid slag, or a solid precipitate. In smelting [Fig. 9–1(b)], a mixture of ores, fluxes, and other materials (for example, reducing agents, scrap, old slag, etc.) is heated to a temperature high enough so that everything becomes molten. Usually this yields two liquid phases, one a crude liquid metal or liquid matte which collects the bulk of the valuable metal and the other a liquid slag which collects in a single waste product the earthy gangue constituents of the ores and the other charge constituents not soluble in the metallic product. A few smelting processes yield more than two liquid phases.

In retorting [Fig. 9–1(c)], which is used only for relatively volatile metals such as zinc and magnesium, the valuable metal is formed by the reaction of solid metal oxides with solid reducing agents and leaves the retort as a gas. In distillation of volatile metals [Fig. 9–1(d)], of course, the characteristic phases are gaseous and liquid metal.

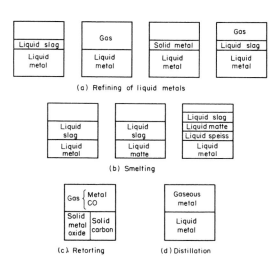

(a) Refining of liquid metals

(b) Smelting

(c) Retorting   (d) Distillation

FIG. 9–1. Typical multiphase metallurgical systems.

The broad objective in conducting a unit process is to accomplish certain separations, especially of valuable metals from waste, and this is achieved by controlling the distribution of the metals and other elements between the two or more phases formed as products of the unit process. In any one system these distributions are determined by a variety of chemical reactions, within phases (homogeneous reactions) and between phases (heterogeneous reactions), and the detailed studies of these reactions for individual unit processes make up the major part of Vol. II of this text. Thus, the succeeding portions of this chapter deal primarily with the constitution and general properties of the important kinds of phases as a background for the more specific and detailed physico-chemical treatment to follow in Vol. II.

Metals and metal solutions

Melting points and vapor pressures of pure metals. When the processes of extracting different metals are compared, it becomes evident immediately that the melting points and vapor pressures of the metals have a considerable effect on the choice of extraction processes. For example, metals of relatively low melting points and low vapor pressures, such as lead, tin, antimony, copper, gold, and silver, are commonly smelted and refined by processes which yield liquid metal as a principal phase. High-melting metals such as molybdenum and tungsten, on the other hand, are generally made by processes such as gaseous reduction of oxides which yield a solid metal powder. Metals with high vapor pressures (low boiling points) such as zinc, cadmium, arsenic, and mercury are most commonly extracted by procedures which yield gaseous metal at some stage. Besides these over-all effects on choice of process, it is further obvious that melting points and relationships of temperature to vapor pressure will affect many details of the procedures of extracting and refining.

Table 9–1 lists the common metals in four groups according to their melting points. Although the melting points of the metals represent only one of several factors determining the kinds of extraction processes used, examination of this classification shows that it places together several subclasses of metals which belong together on other grounds. Thus, the low-melting metals include the alkali metals, K, Na, and Li; the volatile metals Cd and Zn; and

TABLE 9-1

Classification of Common Metals According to Melting Points and Volatility

(Each group in approximate order of decreasing volatility)

	Melting point		Boiling point	
I. Low melting points (<600°C)	°C	°F	°C	°F
Hg	-39	-38	357	675
K	63	146	774	1425
Cd	321	610	765	1409
Na	98	208	892	1638
Zn	419	787	907	1663
Li	186	367	1372	2500
Bi	271	520	1420	2590
Pb	327	621	1744	3170
Sn	232	449	2270	4120
II. Medium melting points (600-1100°C)				
As	(814)*	(1497)*	(610)**	(1130)**
Mg	650	1202	1107	2025
Sb	630	1166	1440	2625
Ca	850	1560	1487	2710
Al	659	1217	2057	3740
Ag	961	1762	2210	4010
Cu	1084	1983	2600	4710
Au	1063	1945	2970	5380
III. High melting points (1100-1600°C)				
Mn	1244	2271	2150	3900
Si	1430	2605	2300	4170
Be	1280	2340	2400	4350
Cr	1550	2822	2500	4530
Ni	1452	2645	2730	4950
Fe	1535	2795	2740	4970
Co	1490	2714	2900	5250
IV. Very high melting points (>1600°C)				
Ti	1725	3140	3260	5900
V	1750	3180	3400	6150
Pt	1770	3220	4250	7680
Zr	2100	3810	----	----
Ta	2850	5160	----	----
Mo	2625	4760	4800	8670
W	3410	6170	5930	10700

*Melting point measured under pressure.
**Sublimation at 1 atm.

the group Bi, Pb, and Sn whose extraction processes have much in common. Similar sub-classes can be noted in the other three groups.

Within each of the four groups of Table 9–1, the metals are arranged in order of increasing boiling points, and thus in approximate order of decreasing volatilities. However, the boiling points are only roughly indicative of the tendencies to volatilize under actual process conditions, and a better comparison of metal volatilities can be made by reference to Fig. 9–2. This chart shows the vapor-pressure to temperature relation-

ships for most of the metals listed in Table 9–1, plotted according to the convenient coordinates log p vs. the reciprocal of absolute temperature $(1/T)$. This method of plotting is useful because the plot for each metal is substantially a straight line and because the complete ranges of temperature and pressure which have importance in extractive metallurgy can be shown on a single graph.

The temperature scale in Fig. 9–2 is divided arbitrarily into the same four temperature ranges which were the basis of the melting-point groupings in Table 9–1.

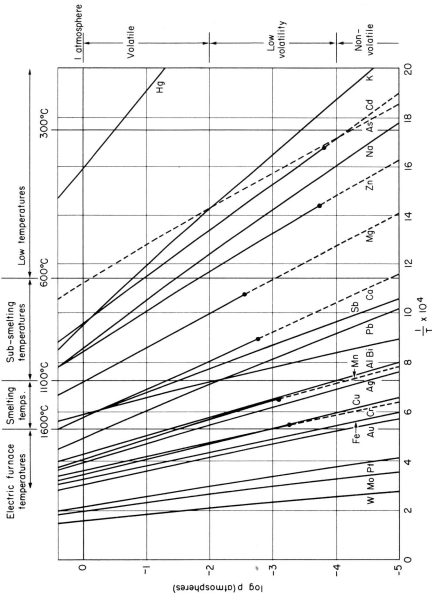

Fig. 9–2. Vapor pressures of common metals (solid lines represent data for liquid metals; dashed lines for solid metals).

Few nonaqueous metallurgical reactions are carried out below about 600°C (1100°F), so this temperature region is arbitrarily called *low temperatures*. The temperature region from 600 to 1100°C (1100–2000°F) covers a variety of processes, but is not high enough for most smelting and slag-making processes, so is designated *sub-smelting temperatures*. The temperature range from 1100°C to 1600°C (2000–2900°F) encompasses most of the common smelting steps in which slags are formed as major products, so is designated as *smelting temperatures*. Most of the processes in this temperature range are carried out in fuel-fired furnaces. Above 1600°C or so, ordinary fuel-firing techniques become inefficient (poor yields of available heat), so that electric furnaces have to be used. At *electric-furnace temperatures*, various difficulties are encountered, for example, in supplying heat and providing materials of furnace construction, which make processes in this range relatively expensive and consequently less common than processes at lower temperatures.

The vapor-pressure scale in Fig. 9–2 is also divided to indicate relative volatility of the metals. Since pressure equipment is rare in metallurgical processes, the metals are handled wholly as gases above the temperatures at which their vapor pressures become one atmosphere (boiling or sublimation points). In the pressure range labeled *volatile*, arbitrarily taken from 0.01 to 1 atm, a liquid or solid metal may evaporate in quite appreciable quantities during ordinary handling, especially, of course, at vapor pressures approaching 1 atm. The vapor-pressure range from 10^{-2} to 10^{-4} (roughly 10 mm to 0.1 mm Hg) is designated *low volatility*. In this range little evaporation occurs at ordinary pressures but the liquid and solid metals may be evaporated readily in commercial vacuum equipment.

Vapor pressures below 10^{-4} atm for most practical purposes represent conditions of *nonvolatility*.

Comparing now the positions of the lines for the various metals in Fig. 9–2, we can anticipate certain practical consequences of the vapor-pressure to temperature relationships. Magnesium and the metals to the right of magnesium on the diagram are readily vaporized at sub-smelting or lower temperatures and all these metals are commonly produced or refined by distillation or retorting processes which are based on their high volatilities. Also, it is evident that, except for As, all these metals can be melted and handled as liquids without too great a tendency to volatilize, provided the temperatures of the liquids are kept low. The next four metals toward the left, Ca, Sb, Pb, and Bi, have low volatilities at sub-smelting temperatures, but become quite volatile at higher smelting temperatures. The next group of seven, from Al to Au, and including Fe and Cu, the two metals which have been produced in largest tonnages of all the metals, do not enter the volatile range at ordinary smelting temperatures, although vaporization losses from liquid Al, Mn, and Ag are observed at relatively high smelting temperatures. Finally, the group Pt, Mo, and W represent the metals which can be considered as nonvolatile at all ordinary operating temperatures.

Gaseous metals. Gaseous metals are not essentially different from other gases in their behavior during metallurgical processes. In general, the metals volatilize as monatomic gases, and can be considered as obeying the ideal-gas relations well enough for all ordinary metallurgical calculations.

Solid metals. Physical metallurgy deals in detail with properties of solid metals, and of course the extractive metallurgist must become familiar with the properties of his product and with metal behavior in such stages of extraction as yield or treat metal in the solid state. In view of the extensive and intensive coverage of the solid state in physical metallurgy, there is no need in this chapter to go beyond pointing out that the extractive metallurgist often produces, handles, and processes solid metals and therefore must know a great deal about them.

Liquid metals. The majority of the common metals are smelted under conditions which yield liquid metal as the main valuable product. Also, pyrometallurgical refining processes generally are carried out on metals in the liquid state. We might almost regard the constant occupation and preoccupation with liquid metals as a hallmark of the extractive metallurgist, although probably it would be more accurate to include liquid slags with liquid metals in making such a distinction. In particular, liquid metals serve as reacting media and solvents for reactants in much the same way that water serves in many low-temperature chemical processes. Chemical reactions involving liquid metals as reactants and as solvents for other reactants will be considered in some detail in Vol. II in connection with each of the unit processes which have liquid metal as an important phase.

The viscosities of liquid metals in general are not much greater than the viscosity of water. For example, just above the melting point, iron has a viscosity of 0.04 poise compared to 0.01 poise for water at ordinary room temperatures. In fact, the kinematic viscosity (viscosity divided by density), which in some ways is a better comparative measure of fluid behavior, is less for liquid metals than it is for water. Thus, from a simple fluid-flow point of view, liquid metals behave very much like water, and can be

poured, agitated, flowed through orifices and ducts, and pumped in the same way except for the major difficulties with construction materials and moving parts at the high temperatures.

The surface tensions of liquid metals are much higher than for water. Thus, whereas water at room temperature has a surface tension of 72.8 dynes/cm, mercury at room temperature has a surface tension of 465 dynes/cm, lead 452 dynes/cm at its melting point, and metals of higher melting points such as copper, gold, iron, etc., have surface tensions above 1000 dynes/cm. With these high surface tensions should be coupled the fact that liquid metals do not wet the refractory surfaces with which they come into contact. Thus, their behavior with respect to refractory containers and furnace walls is well represented by the behavior of mercury in glass and is totally unlike water in glass. One important result of this behavior is that metals are readily held in containers with fairly porous walls, without much tendency for the metal to enter and penetrate capillary openings through the walls. Of course, if the metal is under some pressure the surface forces preventing penetration may be overcome and fairly rapid flow can then take place because of the low viscosity of the liquid metal.

Liquid metals as a group characteristically are good solvents for each other and for many nonmetallic elements as well. As a result, the crude metal produced as a liquid product of a smelting process often contains a variety of impurities in solution in amounts ranging from small fractions of a percent up to several percent. In refining such a crude metal, removal of each dissolved impurity can become a problem unto itself, requiring special steps and chemical treatments. We label many of these steps with words prefixed by de-. For example, dephosphorization, desulfurization, and de-oxidation of iron refer to removal of P, S and O; dezincing, desilverizing, and debismuthizing refer to processes of removing these elements from liquid lead.

The over-all solubility relationships and other important features of liquid-metal solutions are best seen by reference to the constitution diagrams for binary systems of specific metals and specific solutes. When solubilities of chemical compounds are to be considered rather than single chemical elements, it usually is necessary to consider the dissolution process as a more or less complex chemical reaction. For a given metal, the important binary solutes may be classified as follows:

(1) Other metals.
(2) Oxygen.
(3) Other gases:
    (a) Simple gases, such as hydrogen and nitrogen;
    (b) Compound gases, such as water and carbon monoxide.
(4) Other nonmetals, such as carbon, sulfur, and phosphorus.

These four classes will be considered briefly in the following sections.

Metal-metal solutions. In the present state of knowledge of liquid-metal solutions, it appears that no simple rules can be formulated as guides to the degree of solubility or insolubility of various metals in each other. The existence of "free electrons" and other special characteristics of metals seem to limit the applicability of solution concepts developed in the study of water solutions and nonmetallic solutions in general. However, for virtually all the pairs of metals whose solutions and mutual solubilities are of importance in metal extraction, constitution diagrams are available which give

quantitative solubility data and show the ranges of temperature and composition for which the various phases are stable. Excellent compilations of such data are given in *Der Aufbau der Zweistofflegierungen* by M. Hansen and in the *ASM Metals Handbook*. A complete and systematic survey of these data is beyond the scope of this text. However, in the following paragraphs a number of representative systems are described briefly and the relations of these data to the behavior of liquid metals in smelting and refining processes are indicated.

First, a few instances of *substantial insolubility* may be considered. A number of these systems are of considerable practical interest. When liquid metal A does not dissolve B in appreciable amounts, an extraction process yielding liquid A as a product should not present difficulties in keeping B out of the product, even if the raw materials contain substantial percentages of B. Also, when solid metal B does not dissolve appreciably in liquid metal A, the possibility exists that B may be used as a container for liquid A. Both of these conditions are well illustrated by the Pb-Fe system. Iron is substantially insoluble in liquid lead, so that iron-free liquid lead is readily produced by reduction of raw materials high in both iron and lead. Moreover, many of the steps of lead refining are conducted in iron or steel kettles. Other liquid metals dissolving negligible amounts of iron include bismuth, silver, mercury, alkali metals (Na, K, Li), and alkaline-earth metals such as barium and calcium. The solubility of iron in liquid magnesium is low enough so that liquid magnesium is handled commercially in iron pipes and containers, but the percent dissolved (0.02–0.04%) is not negligible as far as some special uses of magnesium are concerned. Other high melting-point metals which, like iron, are substantially insoluble in a number of metals of low melting points are tungsten, molybdenum, cobalt, and tantalum.

The second and fairly numerous group of binary metal systems which might be considered is the group showing *limited liquid solubilities*, and characterized by constitution diagrams with regions of two liquids, or *miscibility gaps*. A system typical of this group is the Zn-Pb system, represented in Fig. 9–3. The two metals are substantially insoluble in each other in the solid state and form two immiscible liquids at temperatures above the melting points of both metals. Just at its melting point (327°C), liquid lead dissolves 0.5% Zn but this solubility increases fairly rapidly with increasing temperature up to nearly 800°C, when the two metals become miscible in all proportions. This solubility-temperature relationship is utilized every day in one of the principal steps of lead refining. That is, zinc is added to lead as a precipitant for silver, and is first dissolved to the extent of 1 or 2 percent at temperatures well above the melting point of lead. Then the lead is slowly cooled almost to the melting point, so that a Zn-Ag alloy and the excess Zn precipitate out and can be removed as a crust. As the constitution diagram in Fig. 9–3 shows, this operation leaves 0.5% Zn in solution in the lead even if the greatest possible care is taken with the separation

of the solid zinc crust from the lead. Accordingly, further refining steps are required to recover the 0.5% Zn for re-use and to yield lead sufficiently low in zinc to meet market specifications. A further example of limited solubility in lead refining is exemplified by the Cu-Pb system shown in Fig. 9–4. Except for the large difference in melting points of the two pure metals and other differences of degree, this system shows the same over-all features as the Zn-Pb system: insolubility in the solid state and a two-liquid miscibility gap. Crude lead as smelted may carry in solution several percent copper derived from the primary raw materials. As Fig. 9–4 shows, most of this copper is precipitated by cooling the lead almost to its melting point. However, even at 327°C the lead still retains 0.06% Cu, a percentage too small to show in the diagram but too large to be tolerated in subsequent steps; hence, sulfur or other reagents may be used to reduce the copper percentage by another order of magnitude. The Fe-Sn system is quite similar to the Cu-Pb system insofar as the liquidus is concerned, but the solubility of iron in tin is only about 0.003% Fe just above the melting point of tin. Thus iron is satisfactorily removed from liquid tin by cooling to near the melting point and removing the precipitate. Other examples of binary systems showing two immiscible liquid phases are

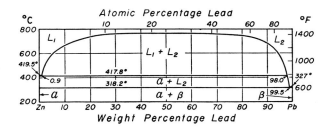

FIG. 9–3. Zn-Pb system (ASM, *Metals Handbook*).

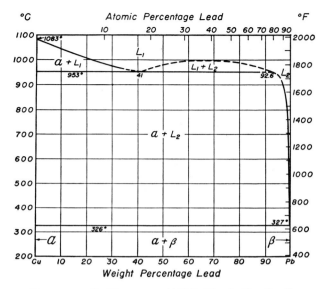

FIG. 9–4. Cu-Pb system (ASM, *Metals Handbook*).

the Ag-Ni, Al-Bi, Al-Cd, and Al-Pb systems.

As a third group of systems for which limited solubilities in the liquid state may influence metal-extraction processes, we may consider binary systems which, although not forming two immiscible liquids, do form *eutectics with small percentages of one of the components*. Generally systems of this kind are formed between metals having substantially different melting points, and the eutectic composition and temperature are close to the composition and melting point, respectively, of the lower melting-point component. Thus, in the Al-Fe system shown in Fig. 9–5, a eutectic is formed with only 1.7% Fe so that this percentage represents the solubility of iron in aluminum at temperatures not far from the melting point of aluminum. The diagram shows that the iron solubility increases fairly rapidly with temperature until at and above the melting point of iron the metals are miscible in all proportions. Also, it should be noted that

the solid phases which are in equilibrium with a liquid solution are not pure iron but instead are various solid solutions of Fe and Al. Systems of this kind, with low solubilities of the higher-melting metal in the lower-melting metal at temperatures near the lower melting point include the Ag-Pb, Ag-Sn, Al-Be, Al-Co, Al-Mn, Al-Sn, Ni-Sn, Ni-Zn, Bi-Cu, and Fe-Sb systems. Also, many binary metal systems with mercury are of this kind, mercury dissolving only small percentages of the metal at the freezing point of mercury ($-39°C$) but dissolving rapidly increasing percentages with increasing temperature. This behavior is illustrated in the constitution diagram of the Al-Hg system given in Fig. 9–6. Another kind of system showing limited solubility of the higher-melting in the lower-melting metal and therefore belonging in this group is characterized not by a eutectic, but by a *peritectic reaction occurring near the lower melting point*. The Cu-Fe system shown

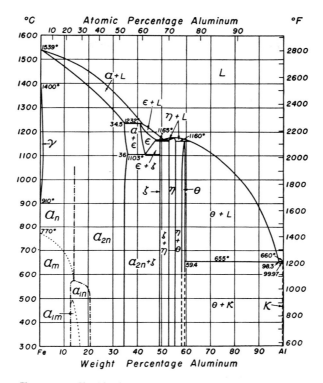

Fig. 9–5.   Fe-Al system (ASM, *Metals Handbook*).

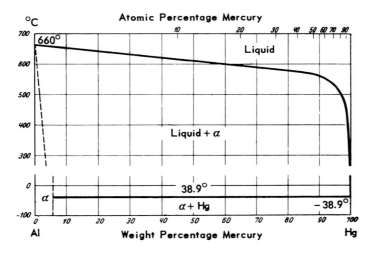

Fig. 9–6.   Al-Hg system (after Hansen, *Der Aufbau der Zweistoff-legierungen*).

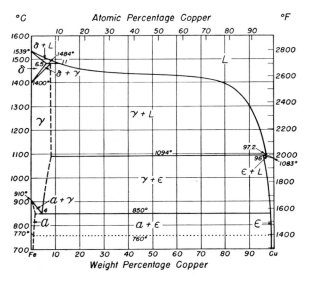

FIG. 9–7.   Fe-Cu system (ASM, *Metals Handbook*).

in Fig. 9–7 illustrates this behavior. This diagram brings out the fact that liquid copper formed in a smelting process at 1200 or 1300°C can carry in solution at the most a very few percent of iron, and if more metallic iron is formed in the process it will necessarily be as a separate solid phase.

Formation of one or more *stable solid intermetallic compounds* or *stable solid solutions* is a significant feature of some binary metal systems, because such solid phases often show strong tendencies to precipitate from various liquid-metal solutions. The Bi-Mg system (Fig. 9–8) shows such a behavior, the stable solid corresponding about to the composition Bi_2Mg_3 (85 weight % Bi). A practical application of this relationship is the use of Mg to precipitate Bi from liquid lead under conditions for which pure Pb and Bi are miscible in all proportions. The fact that Zn forms relatively stable solid solutions with Ag, Au, Cu, and other metallic elements similarly is to be correlated with the fact that zinc

is an efficient precipitant of these metals from solutions in liquid lead or liquid bismuth.

Previous paragraphs have dealt largely with various special instances of limited solubility in liquid metals. Such data show the maximum amounts of a given metal impurity which a liquid metal might carry in solution at a given temperature and show also the extent to which the impurity might be precipitated or separated by cooling the liquid to near the freezing point. However, when all the common metal-in-liquid-metal solutions handled in extractive metallurgy are considered, it is found that the great majority are produced and handled under conditions where the solvent metal and solute metal are *miscible in all proportions*. Thus in iron metallurgy, at temperatures above the melting point of iron, all the dissolved metals commonly present in iron and steel or added as alloying elements or reagents, such as Mn, Si, Ni, Co, Cr, V, W, Mo, Al, and Cu, are either miscible in all

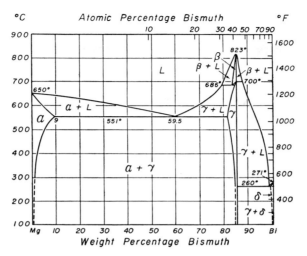

FIG. 9–8. Mg-Bi system (ASM, *Metals Handbook*).

proportions with iron or are so soluble that no solubility limitations need be considered in dealing with liquid iron and steel. Similarly, the liquid copper produced in a copper converter contains the metals As, Sb, Pb, Ni, Zn, Fe, Ag, Au, etc., in percentages determined largely by the amounts present in the raw materials and by the chemical conditions during smelting, and not limited in any way by solubility considerations at the smelting temperature. Thus, taking a broad over-all view, liquid metals are generally good solvents for other metals and, therefore, the liquid metals produced in various extractive processes tend to be complex solutions carrying major or minor amounts of many other metals derived from the ores and raw materials. Moreover, even at the lowest temperatures at which the solvent metal will remain liquid, the solubilities of the metallic impurities are usually so large that chemical reactions must be resorted to in order to purify the solvent metal.

Although emphasis in the above discussion has been placed on the liquid regions of the constitution diagrams, it should be stated that other regions of the constitution diagrams also become important in some extraction processes. Some of these will be referred to in later chapters in Vol. II, especially in connection with solidification of metals and with processes of refining liquid metals.

Oxygen-metal solutions. Oxygen is almost universally present in practical metallurgical systems, either free or combined, and is often a principal reagent or reactant in metallurgical reactions. The solubility of oxygen in most metals appears to be quite low, at any rate so low that few quantitative data are available. In some processes involving liquid metals, however, even very low concentrations of dissolved oxygen probably are significant in the functioning of the process. In lead refining, for example, a number of impurities are removed from solution in lead by oxidation with air, PbO, or other oxidizing agents. The solubility of oxygen in liquid lead has been measured to be of the order of $10^{-3}\%$ O at 500°C, but in spite of this low value it is possible that the reaction mechanism of oxidation with air involves dissolution of oxygen followed by reaction of dissolved oxygen with dissolved impurity.

The liquid metals which are known to dissolve relatively large quantities of oxygen are iron, nickel, cobalt, copper, and silver. Iron and copper are of particular interest because the dissolved oxygen has the leading role in many of the reactions occurring in the refining of these common metals. The subsequent discussion, therefore, refers primarily to these two metals.

The dissolution of oxygen in liquid iron or copper is fundamentally different from the dissolution of oxygen in water and in other common liquids at room temperature. In water solution, oxygen is dissolved molecularly from the gaseous state and thus is present as a molecular dispersion of O_2 molecules in the solvent. Also, the equilibrium quantity dissolved at a given temperature is proportional to the partial pressure of O_2 in the gas phase (Henry's law), and the solubility decreases with increasing temperature. On the other hand, when oxygen dissolves in liquid iron or liquid copper, a definite chemical reaction occurs so that the oxygen is no longer present as O_2 molecules. The solubility is not proportional to the partial pressure of O_2, and increases rather than decreases as the temperature is raised. Our knowledge of the structure of liquid metals is not sufficient to describe exactly how the oxygen is present, but the available quantitative information on the properties of the solution is well correlated by the assumption that the oxygen is dissolved as oxide, so that the dissolution of gaseous oxygen might be described:

O_2 (gas) + 4Cu (liq)

        $\longrightarrow$ 2Cu_2O (solution in Cu);

O_2 (gas) + 2Fe (liq)

        $\longrightarrow$ 2FeO (solution in Fe).

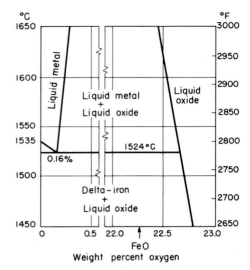

Fig. 9–9. Portion of Fe-O phase diagram showing solubility of oxygen in liquid iron.

Thus, there is no difference between a solution prepared by reacting O_2 with liquid metal and one prepared by dissolving the metal oxide in the metal, and we may refer to the solution interchangeably either as a solution of oxygen or as a solution of oxide.

A portion of the Fe-O constitution diagram showing the range of stability of oxygen solutions in liquid iron is given in Fig. 9–9. At any temperature above the melting point of iron, the maximum oxygen content of the liquid iron is reached when a separate phase of liquid iron oxide is formed. The saturation oxygen content increases with increasing temperature, from 0.16% O (0.72% FeO) at 1524°C to 0.23% O at 1600°C and 0.34% O at 1700°C. It should be noted that the liquid oxide which forms when the oxygen solubility in liquid iron is exceeded is not a stoichiometric oxide, but in fact contains a little more oxygen than would correspond to the formula FeO. Two other features of the diagram are of interest: first, the solubility of oxygen in solid iron is

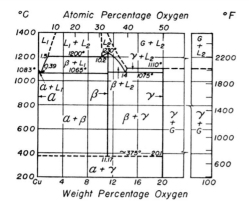

FIG. 9–10. Cu-O system (ASM, *Metals Handbook*).

too small to be shown and, second, the solubility of oxygen in liquid iron is great enough to lower the freezing point several degrees.

The diagram showing oxygen solutions in liquid copper (Fig. 9–10) is broadly similar to that for the oxygen-iron solutions. However, up to 1200°C the phase which precipitates on addition of an excess of oxygen is solid Cu_2O, and at 1065°C there is a Cu-Cu_2O eutectic. Above 1200°C saturation of the liquid copper with oxygen produces an immiscible liquid phase, an oxide phase with somewhat less oxygen than would correspond to the formula Cu_2O. Although the oxygen-solubility figures in % O do not appear large, they represent quite substantial solubilities when expressed in other units. Thus, the 1.5% O solubility at 1200°C corresponds to about 13.5% Cu_2O and also means that one volume of pure liquid copper at 1200°C will dissolve nearly 500 volumes of O_2 at 1200°C and 1 atm.

The oxygen content of liquid metals is readily increased (until saturation) by contact with air and oxidizing agents. Also, it is readily decreased by treating the liquid metal with reducing agents or by adding so-called deoxidizers which react with the dissolved oxygen to precipitate the oxygen

from solution as insoluble oxides or to eliminate the oxygen in the form of a gaseous oxide.

Solutions of gases in metals. The fact that gases dissolve in both liquid and solid metals and, moreover, dissolve in quantities large enough to have quite radical effects on metal properties and behavior is at first surprising to the nonmetallurgist, and in fact only in comparatively recent years have metallurgists themselves become fully aware of some of the effects of gases in metals. Some of the most significant recent advances in this field have focussed attention particularly on the evolution of dissolved gases during the solidification of metals, for in many cases such gas evolution has proved to be the most important single factor in determining the structure and properties of the cast metal.

As was pointed out for oxygen, the dissolution of gases in metals is not in general a simple molecular solution but partakes of the nature of a chemical reaction. Thus, it is not surprising to find that gases which are chemically inert, like the rare gases helium, argon, etc., do not dissolve in metals. Likewise, nitrogen is inert to some metals (e.g., copper) but dissolves readily in others (e.g., iron), with which it is also known to react chemically. Inert gases are useful in handling liquid metals when it is desired to displace gases in the atmosphere over the metal and thus to avoid gas dissolution entirely.

The solubilities of diatomic gases, especially H_2 and N_2, have been measured in a considerable number of different metals and alloys, mainly by the method of Sieverts. Sieverts' method involves bringing the liquid or solid metal into equilibrium with a measured quantity of the gas in a closed system, and then measuring the equilibrium gas pressure and the volume of the gas

298 METALLURGICAL ENGINEERING

dissolved. Since the quantity dissolved is measured as a volume of gas, the procedure is quite sensitive and is suitable for determining solubilities which represent small fractions of a weight percent hardly detectable by ordinary chemical analysis. The original work of Sieverts and subsequent work by Sieverts and others have established that the *solubility of a diatomic gas in a metal at a given temperature is proportional to the square root of the partial pressure of the gas.* This relation is known as Sieverts' law, and is to be contrasted with Henry's law for molecular gas solutions at room temperature according to which solubility is proportional to the first power of the pressure. Also, again in sharp contrast to the gas-in-water type of solutions, the solubilities of diatomic gases in metals generally increase with increasing temperature.

Figure 9–11 summarizes the experimental data on the solubility of H_2 in both solid and liquid copper at temperatures from 500 to 1500°C. At the left, the solubility is expressed in parts per million (ppm or % × 10^4) by weight and it is seen that in terms of

weight units the solubilities are very small. However, the scale at the right giving % by volume, based on gas volumes measured at 1 atm and at the melting point of copper, gives a much better measure of the effects of dissolved gas. The volumes of dissolved gas are generally commensurate with the volume of metal, and at higher temperatures and pressures become several times greater than the volume of metal. The data show clearly a common feature of gas-metal systems — at the melting point the solubility in the liquid metal is considerably greater than in the solid. Thus, liquid copper originally in equilibrium with H_2 at 1 atm and at 1083°C will on solidification evolve about 150% of its own volume of H_2. Even gas evolution of much lower magnitude during freezing leads to very porous castings and has radical effects on the structure and properties of the solid metal.

Figure 9–12 shows the solubilities of H_2 and N_2 in liquid iron and in the various allotropic forms of solid iron. It is interesting to note that gamma iron dissolves both these gases more readily than either alpha or delta iron. Also, the solubility of N_2 in gamma iron is unusual in that it decreases with increasing temperature.

Under many circumstances the gas in contact with the metal or the gas evolved from solution in a metal will contain compound gases such as CO, H_2O, SO_2, H_2S, etc. In some systems, in fact, the equilibrium solubilities of these gases at various partial pressures can be measured by the Sieverts technique already mentioned for diatomic gases. However, the term solubility may be somewhat misleading when applied to compound gases because the dissolution process is clearly a chemical reaction in which the gaseous compound loses its identity entirely. For example, H_2O dissolving in copper gives dissolved H and dissolved O

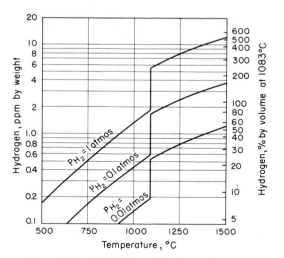

FIG. 9–11. Solubility of hydrogen in copper.

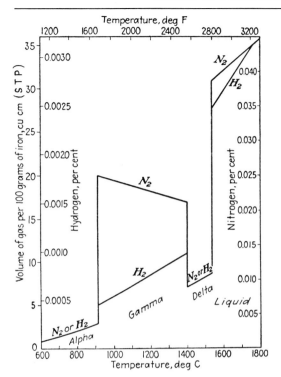

FIG. 9–12. Effect of temperature on the solubilities of hydrogen and nitrogen at 1 atm pressure in iron (AIME, *Basic Open-Hearth Steelmaking*).

just as if H_2 and Cu_2O were dissolved separately. The reaction can be written:

$$H_2O \text{ (gas)} \rightleftharpoons 2H \text{ (solution)} + O \text{ (solution)}.$$

In the reverse reaction, H_2O is evolved as a result of this chemical combination of dissolved H and dissolved O. Similarly, the following reaction is of great importance in steelmaking:

$$CO \text{ (gas)} \rightleftharpoons C \text{ (dissolved)} + O \text{ (dissolved)}.$$

Reactions of this kind, involving dissolution or evolution of compound gases, are considered in detail in Vol. II.

Solutions of other nonmetals in metals. Carbon, sulfur, and phosphorus are the principal nonmetallic elements, apart from the gaseous elements already considered, commonly dissolved in liquid metals. Crude liquid iron usually contains all three of these elements, and their removal and control represent a major part of the job of making steel from the iron. The liquid pig iron made in the blast furnace is saturated with carbon because it is made in contact with solid carbon (coke) and in addition carries in solution all the phosphorus and a large part of the sulfur which enter the blast furnace in the ore, coke, and other raw materials. Similarly, solutions of carbon in other metals have to be considered, because carbon enters most metallurgical processes as a fuel, reducing agent, or combined in the products of combustion. Sulfur is a common constituent of most base-metal ores and also may be dissolved in metal from the combustion products of many fuels.

As with other solutes considered previously, much of the useful information on solubility limits, ranges of stability of solution, and nature of precipitating phases for the elements in this group is available in the form of constitution diagrams.

The Fe-C diagram (Fig. 9–13) is familiar to all metallurgists, but attention is usually focussed on the portions of the diagram dealing with the solid phases. However, the diagram brings out several things of obvious bearing on the extractive metallurgy of iron. Liquid iron saturated with carbon, as it is in the blast furnace, will have on the order of 4 to 5% C in solution and will remain liquid at temperatures as low as 1135°C, about 400°C below the melting point of pure iron. As carbon is removed from solution during steel-making processes, the freezing point of the liquid is raised toward that of pure iron, and this must be taken into account by keeping the furnace temperature high enough to avoid freezing the metal.

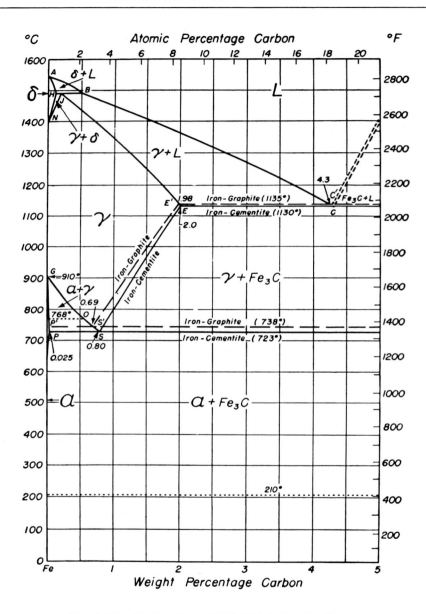

FIG. 9–13.   Fe-C system (ASM, *Metals Handbook*).

The Cu-C system shows a very low solubility of carbon in liquid copper, sharply contrasting with the behavior of carbon in iron. According to Bever and Floe, the carbon content of graphite-saturated liquid copper varies from about 0.0001% at 1100°C to 0.003% at 1700°C. The smallness of these figures, however, should not be taken as signifying that the dissolution of carbon in copper can be entirely ignored in the practical handling of liquid copper. Stoichiometric calculations show that 0.001% C, for example, is equivalent to a volume of CO which measured at 1 atm pressure and at the freezing point of copper is about 75% of the volume of the liquid copper.

The binary systems of carbon with many of the metals of relatively high melting points, such as Co, Ni, Mo, W, V, Cr, Zr, Ti, and Mn, show pronounced chemical affinities of these metals for carbon, and definite tendencies to form carbides and to form both liquid and solid solutions. These tendencies affect the processing of these metals very fundamentally in some cases, and may, for example, prevent the use of carbon as reducing agent or prohibit the use of graphite crucibles for melting. In metals of relatively low melting points, such as the common base metals Pb, Zn, Sn, Cd, Bi, and Hg, the solubility of carbon is probably a very small fraction of a percent, but good data are not available and there is little indication that carbon dissolution has any noteworthy effects on the behavior of these metals.

Sulfur is a common solute in liquid iron and liquid copper, from which it generally must be separated in the course of extracting and refining iron and copper. As with oxygen, the solutions are often conveniently regarded as solutions of metal sulfide in metal, rather than sulfur in metal, because at liquid-metal temperatures sulfur itself is a gas, and thus holding substantial percentages of sulfur in liquid-metal solution requires strong chemical bonding of some kind between the sulfur and the metal. Figure 9–14 is the constitution diagram for the Fe-S system, which shows that iron and sulfur form a complete range of liquid solutions from pure iron to solutions containing more sulfur than FeS. Also it is to be noted that Fe and FeS form a eutectic at 988°C and 31% S. The Cu-S system (Fig. 9–15) is characterized by a two-liquid miscibility gap, with liquid copper and liquid copper sulfide coexisting in equilibrium, but the

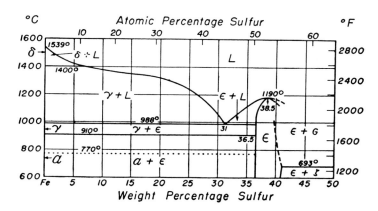

FIG. 9–14. Fe-S system (ASM, *Metals Handbook*).

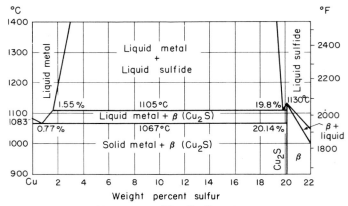

FIG. 9–15.   The Cu–S system.

sulfur solubility in liquid copper is quite large for all practical purposes, exceeding 1% at temperatures above the melting point of copper.

Sulfur, like oxygen, is relatively insoluble in many metals, especially in some common metals of low melting points.   Liquid lead and liquid bismuth, for example, dissolve very little sulfur just above their melting points, but several hundred degrees higher dissolve very substantial percentages.   Sulfur appears to be substantially insoluble in the liquid metals zinc, cadmium, aluminum, and gold.

Complex solutions; reactions in liquid metals.   So far we have considered in elementary fashion only a few representative binary solutions in which liquid metals serve as solvents.   In practice, liquid metals are likely to contain at the same time several dissolved elements, not infrequently over a dozen.   Under these conditions caution must be used in applying information about individual solutes obtained from binary constitution diagrams for the solute with the solvent metal.   Various kinds of interactions are encountered between different solutes, many of these corresponding roughly to interactions familiar in aqueous solutions.

Particularly important and useful are reactions leading to the formation of a new phase, such as precipitation reactions and gas-evolution reactions.   A few of these have been mentioned previously in passing, for example, precipitation of silver from lead with zinc, precipitation of copper from lead with sulfur, and formation of compound gases such as CO, H_2O, and SO_2 from the elements dissolved in liquid metals.   Another common type of interaction between elements dissolved in liquid metals is that they often affect each other's solubility even though no compounds may be formed.

The physical chemistry of liquid-metal solutions has in recent years attracted the attention of both chemists and metallurgists as a fruitful field of research.   As will be seen in Vol. II, quantitative physico-chemical studies of solutions of various elements and combinations of elements in liquid iron and liquid copper have contributed substantially to the development of steelmaking and copper-refining practices.   However, much remains to be done before our knowledge of metal solutions and reactions in metal solutions will approach that for aqueous solutions at room temperature.

Slags

What are slags? Although slag is a very common word in the metallurgist's vocabulary, it is applied to such a variety of things that an exact but brief definition is difficult. Perhaps the best definition can be obtained by summarizing the attributes and features which all slags have in common. *In the first place,* they are common liquid products of pyrometallurgical processes, especially the unit processes of simple smelting, blast furnace smelting, converting, and refining of liquid metals. In these processes, the complete fusion of all the starting materials gives two or more liquid products which stratify into separate layers on standing, with the slag, the liquid of lowest specific gravity, forming the layer which is skimmed off the top. *Second,* the slag from a given operation is generally the waste product or less valuable product and thus carries from the process a substantial portion of the substances not desired in the valuable product. This implies that the slag has a minimum content of the particular metal being extracted, so that most of the valuable

metal is recovered in a different product such as a liquid metal or matte. *Third,* the major components of a slag are generally compounds such as oxides, fluorides, silicates, phosphates, borates, and the like, although these compounds tend to lose their individual identities in the slag. This distinguishes slags from liquid metals which are primarily elemental metal, from mattes which are liquid sulfides, and from speisses which are mainly liquid arsenides. *Fourth,* at the temperatures at which they are produced in metallurgical processes, slags are complex single-phase solutions, though they may carry in suspension particles and globules of other phases. Solidified slags, on the other hand, tend to crystallize into a number of different solid phases.

The primary function of most slags is to collect and carry from the process in a convenient form various constituents and impurities present in the charge but not desired in the valuable-metal product. Tabulated below are a few examples of processes in which the slag serves as the principal outlet for the substances listed:

	Valuable product	*Major slag components*	*Charge components eliminated in slag*
Matte-smelting copper ores	Copper matte	Iron oxides, SiO_2	CaO, MgO, Al_2O_3, SiO_2, Fe
Converting copper matte	Blister copper	Iron oxides, SiO_2	Fe, SiO_2
Lead blast furnace	Crude lead	CaO, iron oxides, SiO_2	CaO, MgO, Al_2O_3, SiO_2, Fe, Zn
Lead softening	Soft lead	PbO, Sb_2O_3, As_2O_3	As, Sb, Zn
Iron blast furnace	Pig iron	CaO, MgO, Al_2O_3, SiO_2	CaO, MgO, Al_2O_3, SiO_2
Steelmaking, basic process	Liquid steel	CaO, SiO_2	Si, P, (S), (Mn)
Steelmaking, acid process	Liquid steel	Iron oxides, SiO_2	Si, (Mn)

In some cases, no difficulties are encountered in collecting all of a given charge component in the slag, provided only that a manageable liquid slag is formed and that the charge component has a reasonable physical opportunity to enter the slag. Thus, CaO, MgO, and Al_2O_3 are very common constituents of furnace charges, but always leave in the slag because they are insoluble in other phases and are not readily reduced to metal or changed to forms soluble in other phases. Similarly, in smelting nonferrous metals, conditions rarely are reducing enough to reduce SiO_2 to metal, so that substantially all the SiO_2 in the charge almost automatically goes into the slag. When other charge constituents are considered, however, collection in the slag is by no means as automatic, and special conditions of temperature, slag composition, and atmosphere may have to be provided to drive as much as possible into the slag and retain as little as possible in the liquid metal or other valuable product. This is true, for example, of Sb and As in lead softening, of P and S in steelmaking, and of most charge constituents in smelting and refining processes apart from CaO, MgO, Al_2O_3, and SiO_2.

Another function of slags, actually quite closely related to the function of collecting impurities, is that they often serve as the medium or solvent for important reagents and reactants used to treat the liquid metal. In fact, the slag as a whole might often be regarded as a refining reagent. Thus oxidation is often accomplished by making an oxidizing slag, in which the oxidizing agent is a dissolved oxide such as PbO in lead refining, or Fe_2O_3 in other cases. These agents oxidize impurities in the liquid metal and the resulting oxides then escape as gas or are dissolved by the slag.

Under some circumstances, a slag may be used primarily as a protective blanket to prevent or retard reaction of the underlying phase with the atmosphere.

Practical requirements for slags. Making a satisfactory slag which is readily formed and removed from the furnace and which fulfills the functions essential to the metallurgical process is often a difficult problem and often requires a compromise between a number of conflicting factors. The most important of these factors are indicated briefly below.

(1) *Metallurgical requirements.* Since the primary function of the slag is generally to serve as the outlet for charge components not wanted in the valuable product, the slag should collect and dissolve the maximum possible percentages of these components. At the same time, it should dissolve and carry mechanically the least possible amounts of the valuable metals in the charge. There is always some loss of metal in the slag and in fact some types of slags always require reprocessing to recover this metal.

(2) *Minimum cost.* In smelting ores, only rarely do the slag-forming constituents occur naturally in the right proportions to form a satisfactory slag. Thus, it is necessary to purchase reagents and fluxes which combine with the slag-forming components of the ore to give a slag composition of acceptable properties. Also, the slag is discharged molten with a high sensible-heat content, so that part of the fuel cost can logically be assigned to slag-making. The slag losses of valuable metal might also be expressed in dollars and cents as part of the slag cost. Other factors being equal, most slag costs increase directly with the quantity of slag made.

(3) *Formation and free-running temperature.* Smelting slags are formed by fusing together mixtures of solid substances which individually are often quite infusible at

ordinary smelting temperatures. Thus, CaO and SiO_2 separately have melting points far above usual smelting temperatures but are major constituents of many slags formed at temperatures as low as 1100 or 1200°C. The mechanism of forming a liquid slag is somewhat complex, but under given conditions it is clear that this process often places a lower limit on operating temperatures. Thus, the free-running temperature or minimum temperature at which the slag forms and flows freely from the furnace is often the critical temperature of the process in the sense that critical temperature was defined in the previous chapter on combustion and utilization of heat.

(4) *Viscosity and fluidity.* The normal method of separating slag consists in gravity settling to obtain a supernatant slag layer and then gravity flow of the slag out through a tap hole or other specially provided channel. Both the gravity stratification and the flow from the furnace require slag fluidity. Thick, viscous slags tend to carry particles of the metallic phase in suspension. Slags which do not flow freely are skimmed manually in some processes.

(5) *Specific gravity.* Since slag is separated in the furnace by gravity stratification, it must be enough lighter than the metallic phases so that slag particles will rise readily to the top of the metallic phase and particles of the other phases will drop readily out of the slag phase. This requirement is not often critical because liquid oxides and silicates as a rule have specific gravities considerably lower than metals, metal sulfides, and other metallic phases.

(6) *Corrosiveness.* As will be seen in more detail in the next chapter on refractory materials, the principal materials of furnace construction are composed of oxides like silica, alumina, magnesia, lime, etc., which are also principal slag-forming oxides. Thus,

care must be taken in the choice of refractory materials coming in contact with the slag; at the same time the limitations of the refractory materials can have considerable effect on the slag compositions which can be made without excessive corrosion of the furnace walls.

Typical slag compositions; fluxes. The principal components of metallurgical slags are the oxides which are most abundant as components of rocks in the earth's crust: SiO_2, CaO, MgO, FeO, Fe_2O_3, Al_2O_3, MnO, and P_2O_5. These oxides naturally enter metal extraction processes in large quantities as part of the gangue associated with the valuable ore minerals. At the same time, when fluxing materials must be added to the charge for the primary purpose of obtaining a slag of satisfactory composition, raw materials made up of these same oxides are likely to be used as fluxes because of their wide availability and cheapness. Thus, high-SiO_2 rock, limestone, and iron ore are three of the most common fluxing materials.

Generally speaking, a given unit process is likely to be characterized by a fairly narrow range of slag compositions which meet the chemical requirements of the process and at the same time satisfactorily fulfill the various practical requirements discussed in the preceding section. Table 9–2 gives some typical slag analyses from a variety of processes. Although the slag compositions for a given process usually fall into a recognizable pattern, the data show very large variations from one process to another. In terms of weight percent, a major constituent of one slag might be a minor constituent of another. However, it should be noted that silica is present in substantial percentages in virtually all the slags listed, so that it is not unreasonable to conclude that most metallurgical slags are silicate slags. On the other hand, no other single

TABLE 9-2

Typical Slag Compositions

Process	Weight percentages										
	SiO_2	CaO	FeO*	MnO	MgO	Al_2O_3	S	Cu	Pb	ZnO	P_2O_5
I. Nonferrous smelting processes											
Matte smelting copper ores (a)	37.3	4.7	46.0			6.9	1.1	0.44			
Matte smelting copper ores (b)	38.3	20.3	22.3		2.1	10.9	0.14	1.1			
Converting copper matte	26.2	0.7	58.5			4.9	1.5	2.93			
Lead blast furnace (a)	26.8	8.1	36.8		4.9	5.7	3.0	0.3	1.6	12.5	
Lead blast furnace (b)	35.0	20.5	28.7		1.7	3.4	1.1		0.9	5.6	
II. Ferrous metallurgy											
Iron blast furnace	34.7	44.8	0.5	1.5	3.2	14.6	1.4				
Basic open-hearth process	18.0	44.0	14.1	11.4	7.0	1.6	0.3				2.8
Basic Bessemer process	16.6	44.1	9.8	6.4	4.5	2.1					16.5
Acid open-hearth process	55.6	0.7	28.9	10.2	0.1	4.2					0.02
Acid Bessemer process	63.2	0.7	18.0	14.7	0.3	2.8	0.01				0.01

*Total iron content expressed as FeO.

oxide constituent appears in any way essential to slags — some slags have virtually no lime or magnesia, others have virtually no iron oxides, etc. In special instances, the slag may carry large percentages of unusual constituents. Thus, lead blast furnace slags can carry large amounts of ZnO and basic steelmaking slags may be high in P_2O_5. The smelting slags (e.g., matte smelting, iron blast furnace) have generally low contents of the valuable metal being smelted so that they can be discarded without serious loss. However, other slags (e.g., copper converter) usually contain too much metal to be discarded and hence are re-treated.

Chemical properties of oxide slags. Only a beginning has been made in developing the science of slag chemistry; the chemical properties of slags are not nearly as well understood as are the corresponding properties of aqueous solutions. One major factor in the slow progress of slag chemistry is the difficulty of making quantitative measurements at the high temperatures required for liquid slags. However, a great deal of empirical information is available from accumulated practical data on slags, so that

certain rather general features of slag chemistry can be distinguished. Thus, the concepts of basicity and acidity clearly refer to important chemical properties of the slag although we are unable to define and measure them in a scientifically satisfactory fashion. Likewise, oxidizing power and reducing power are general slag properties which play important roles in most slag-producing processes. Apart from these very general properties, present knowledge of slag chemistry consists primarily of data on specific reactions of individual slag constituents, such as the slag-metal reactions in steelmaking. Moreover, the metallurgist operating a given process usually knows empirically at least the over-all effects of changes in slag composition, and thus in a given situation may be able to predict the chemical effects of, for example, increasing SiO_2 or increasing CaO in the slag.

The concepts of *slag basicity* and *slag acidity* are only tenuously analogous to the ordinary ideas of basicity and acidity of aqueous solutions. Qualitatively, for both kinds of solutions, certain oxides are considered to be basic or base-forming, others are considered acidic, and still others are considered amphoteric. Also it is recognized in both cases that there are all degrees of basicity and acidity. Bases and acids tend to react with each other to form salts and analogous compounds. However, in slags we do not have hydrogen ions, hydroxyl ions, and water, and the reactions involving these, as a simple basis of correlating acid-base behavior.

Silica is the principal acid oxide in metallurgical slags, which is of course why they are often thought of as silicates (that is, as salts of silicic acid). In fact, silica is such a common denominator that silica content alone in weight percent serves as a rough criterion of acidity. Slags high in SiO_2 are acid; slags low in SiO_2 are basic (see Table 9-2). It might be noted that geologists adopt a similar point of view in classifying rocks. Silica is the common acidic constituent of rocks, and is combined with many different basic oxides to form silicate minerals.

Lime, magnesia, and ferrous oxide are the common basic oxides in slags, but other metal oxides occasionally are present in sufficient quantities to be considered major basic constituents. For example, MnO is an important base in some steelmaking slags. Lime is usually considered as the strongest base among these common oxides, and this characteristic together with the natural abundance of limestone and dolomite make these materials especially prominent as basic fluxes. Accordingly, strongly basic slags, such as those in basic steelmaking processes (Table 9-2), are often high-lime slags.

Alumina is a common slag constituent which cannot be classified unequivocally either as an acid or as a base. Towards silica and other acid oxides, alumina acts as a base, forming aluminum silicates, for example. Towards lime, on the other hand, alumina acts as an acid, forming calcium

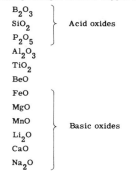

TABLE 9-3

Relative Basic Strengths of Oxides

(Oxides arranged in order of increasing basicity and decreasing strength of metal-oxygen bonding)

aluminates. Other oxides behave in similar fashion, and thus are often called amphoteric. A more constructive point of view, however, consists of regarding basicity and acidity as relative and not absolute properties. Then the common oxides can be arranged in the form of a basicity series, in order of increasing basicity, as illustrated in Table 9-3. Within such a series, any one oxide acts as a base to oxides above and as an acid to oxides below. Some kind of a quantitative scale of basicity might be imagined to go with this listing, just as the electromotive series of the metals follows the standard electrode potentials. Thus, oxides far apart in the series, such as SiO_2 and CaO, show a clear-cut acid-base differentiation. On the other hand, oxides close together in the series show little difference in basicity and in fact might well shift places somewhat depending on what is selected as a quantitative criterion of basicity. The order in Table 9-3 is based on theoretical calculations of the relative strengths of the metal-oxygen bonds in the oxides, the most acid oxides having the strongest bonds.

The broad significance of slag basicity to the metallurgist is that this variable seems to affect markedly most of the chemical reactions of slags. Some reactions require basic slags to proceed at all, some require acid slags, and in others control of the extent of reaction is exercised by control of slag basicity. The best-known correlations of slag basicity with metallurgical behavior are in steelmaking, where the primary classification of processes is according to basicity and acidity. Thus, we have the basic open-hearth process and the acid open-hearth process, the basic Bessemer process and the acid Bessemer process, each serving quite different purposes. Perhaps the most striking difference between acid and basic steelmaking is that phosphorus can be removed from liquid steel into basic slags but cannot be removed into acid slags. The classification of slags and processes is also extended to the refractory materials of which furnaces are constructed. Basic slags and basic processes are carried out in basic furnaces or furnaces lined with basic refractories, because acid refractories such as silica brick will not withstand the corrosive action of basic slags. Similarly, of course, acid slags are best contained by acid refractories.

Many different empirical measures of slag basicity or acidity have been proposed and used as guides in adjusting slag compositions to process requirements. These figures are calculated in various ways from slag analyses, and hence do not afford a real measure of a chemical property. In spite of this failing, though, when the behavior of a group of slags all of the same type is considered, very often these empirical figures make possible simple and useful correlations. The figure of this kind most widely used in the past has been the *silicate degree*, defined as follows:

$$\frac{\text{silicate}}{\text{degree}} = \frac{\text{mols oxygen in } SiO_2}{\text{mols oxygen in basic oxides}}.$$

In calculating this ratio, CaO, MgO, FeO, MnO, and, in fact, most of the simple metal oxides are considered as bases, and no distinction is made as to basic strength among these oxides. Al_2O_3 is often ignored in the calculation, that is, considered neutral, but may under some circumstances be counted as a base.

Example

Calculate the silicate degree of a slag having the following analysis:

32.9% SiO_2, 34.2% FeO, 16.9% CaO, 2.0% MgO, 3.0% Al_2O_3, 8.5% ZnO.

Solution

In calculating the silicate degree of this slag, we will arbitrarily consider as bases all the oxides except SiO_2. However, it should be recognized that for some purposes, the Al_2O_3 and ZnO might be considered neutral and, therefore, might be omitted from the bases in the calculation. Also, it is assumed implicitly that the iron is all present as FeO, whereas in actuality both ferrous and ferric iron are usually present.

Taking 100 lb of slag as a basis,

$$\text{lb mols O in } SiO_2 = \frac{2 \times 32.9}{60} = 1.10;$$

$$\text{lb mols O in bases} = \frac{34.2}{72} + \frac{16.9}{56} + \frac{2.0}{40} + \frac{3 \times 3.0}{102} + \frac{8.5}{81}$$
$$= 1.02.$$

Therefore, silicate degree $= 1.10/1.02 = \underline{1.08}$.

For crystalline metal silicates and silicate minerals, the ratios of oxygen in silica to oxygen in bases can usually be represented as ratios of small whole numbers, readily seen by inspection of the chemical formulas. Thus, the mineral fayalite ($2FeO \cdot SiO_2$) has a silicate degree of 1; wollastonite ($CaO \cdot SiO_2$) has a silicate degree of 2. A nomenclature sometimes used both on minerals and on slags is essentially the same as classification according to silicate degree, but involves assigning names to various silicate degrees, as follows:

Silicate degree	Name	Typical mineral formulas
$\frac{1}{2}$, $\frac{2}{3}$, etc.	Subsilicates	$4RO \cdot SiO_2$
		$3RO \cdot SiO_2$
		$R_2O_3 \cdot SiO_2$
1	Monosilicate	$2RO \cdot SiO_2$
		$2R_2O_3 \cdot 3SiO_2$
1.5	Sesquisilicate	$4RO \cdot 3SiO_2$
2	Bisilicate	$RO \cdot SiO_2$
		$R_2O_3 \cdot 3SiO_2$
3	Trisilicate	$2RO \cdot 3SiO_2$

These terms are used to describe slags as well as crystalline silicates, even though stoichiometric chemical compounds of the types indicated may not exist in liquid slags. That is, the expression "monosilicate slag" means simply a slag with a silicate degree of substantially unity. Still other expressions used to indicate the silica-to-base relationship include "orthosilicate" for a silicate degree of 1, implying that the substance is a salt of orthosilicic acid (H_4SiO_4); and "metasilicate" for a silicate degree of 2, corresponding to metasilicic acid (H_2SiO_3).

Since it has been found that slag basicity is a primary factor in most of the refining reactions of steelmaking, metallurgists in this field have been particularly active in devising various measures of basicity and in attempting to correlate them with the extent of desulfurization, extent of dephosphorization, and other criteria of metallurgical performance. Probably over a dozen basicity ratios have been used by various workers, all based on chemical analyses of the slags. Some of the simplest and most used of these are called V-ratios. One common V-ratio is simply: (weight % CaO)/(weight % SiO_2). In another V-ratio, the weight % P_2O_5 is added to the weight % SiO_2 in the denominator. Simple ratios of this kind are indeed helpful guides in process control, but as more and more chemical data are becoming available, the scientific shortcomings of these ratios as basicity measures are becoming more and more evident.

Another general property of slags often ranking in importance with basicity and acidity is the behavior of the slag in oxidation-reduction reactions. Thus the metallurgist may say, for such and such a purpose "the slag must be a strongly oxidizing slag," or for another purpose "the slag must be strongly reducing." Oxidation-reduction reactions form the backbone of many metallurgical processes in which slags are major phases. In such reactions, the slag may carry in solution one or more of the reactants and reaction products.

One type of oxidation-reduction reaction frequently encountered in both ferrous and nonferrous extraction and refining processes is based on the fact that iron exhibits two valences in the slag, ferric and ferrous. Examples of reactions of this type are given below:

$$Fe_2O_3 \text{ (slag)} + Fe \text{ (liquid metal}$$
$$\text{or matte)} \longrightarrow 3FeO \text{ (slag)}; \quad (9\text{-}1)$$

$$2Fe_2O_3 \text{ (slag)} + Si \text{ (liquid iron)} \longrightarrow$$
$$4FeO \text{ (slag)} + SiO_2 \text{ (slag)}; \quad (9\text{-}2)$$

$$Fe_2O_3 \text{ (slag)} + C \text{ (liquid iron)} \longrightarrow$$
$$2FeO \text{ (slag)} + CO \text{ (gas)}; \quad (9\text{-}3)$$

$$CO_2 \text{ (gas)} + 2FeO \text{ (slag)} \longrightarrow$$
$$Fe_2O_3 \text{ (slag)} + CO \text{ (gas)}; \quad (9\text{-}4)$$

$$O_2 \text{ (air)} + 4FeO \text{ (slag)} \longrightarrow$$
$$2Fe_2O_3 \text{ (slag)}. \quad (9\text{-}5)$$

For iron-bearing slags, the ferric-to-ferrous relationship found by chemical analysis is a rough indication of oxidation-reduction behavior. Except for a few relatively pure laboratory slags, available data are insufficient to set up a quantitative scale of oxidizing power. Considering reactions (9-1) to (9-3) on the one hand and reactions (9-4) and (9-5) on the other, it is readily seen that slags containing iron oxides may act as oxygen carriers. Thus, the Fe_2O_3 consumed by slag-metal reactions may be constantly replenished by slag-gas reactions with the net result that constituents of the metal are oxidized by oxygen carried by Fe_2O_3 across the slag from the gas phase.

In a second group of oxidation-reduction reactions, the slag carries in solution an oxidizing agent, particularly an oxide of a metal which enters the metal phase when reaction occurs. Examples are:

$$3PbO \text{ (slag)} + 2Sb \text{ (liquid lead)} \longrightarrow$$
$$3Pb \text{ (liquid lead)} + Sb_2O_3 \text{ (slag)}; \quad (9\text{-}6)$$

$$Cu_2O \text{ (slag)} + Fe \text{ (liquid copper)} \longrightarrow$$
$$2Cu \text{ (liquid copper)} + FeO \text{ (slag)}; \quad (9\text{-}7)$$

$$2FeO \text{ (slag)} + Si \text{ (liquid iron)} \longrightarrow$$
$$2Fe \text{ (liquid iron)} + SiO_2 \text{ (slag)}. \quad (9\text{-}8)$$

Here again, the chemical analysis giving the percentage of oxidizing agent in the slag serves as a rough guide to the oxidizing power of the slag. With this percentage must be taken into account the strength of the oxidizing agent, or its ability to furnish oxygen. Reactions of this type involved in specific unit processes will be considered in more detail in Vol. II.

Some slags contain no substances which participate in oxidation-reduction reactions under the conditions in which the slags are produced. Thus, oxides like CaO, MgO, and Al_2O_3 are very stable and remain inert to oxidation-reduction reactions. Under some conditions, especially in ferrous metallurgy, SiO_2 may be reduced [for example, by reverse of reaction (9-8)], but SiO_2 is inert to the common oxidation-reduction reactions in nonferrous metallurgy. Accordingly, slags composed mainly of CaO, MgO, Al_2O_3, SiO_2, and other relatively inert compounds may be neither oxidizing nor reducing but entirely neutral as far as the process reactions are concerned.

Structure of liquid slags. Physical chemists and physicists correlate and explain many of the properties of gases, crystalline solids, electrolytic solutions, and organic liquids by referring to the structure of these phases and the forces and energies binding the structure together. At the present time, however, we do not yet have an understanding of the structure of molten slags sufficient to account quantitatively in any detail for slag properties. Various concepts of slag structure have been used in metallurgical literature and these are discussed briefly in the following paragraphs.

One simple approach is to consider slags as molecular solutions of the various component oxides, in other words, as intimate mixtures of SiO_2, CaO, FeO, Fe_2O_3 molecules, and the like. Chemical analyses of slags are conventionally reported as percentages of oxides, and this practice naturally lends some illusion of reality to the view that slags contain simple oxide molecules as fundamental structural units. In stoichiometric calculations, also, it is usually convenient to regard the oxides as the slag components and to write chemical equations for reactions between the oxides. The fact that these customs give correct stoichiometric results is in no way evidence for the existence of oxide molecules; no substantial evidence exists that slags are molecular solutions of the simple oxide molecules.

Until recently the most prevalent views of slag structure have been based on the idea that slags are molecular solutions, but that the constituent molecules are more or less complex compounds formed between the oxides. Most of these compounds represent combinations of basic oxides with acid oxides or, even more specifically, are thought of as silicates, aluminates, ferrites, and the like. The formulas for these molecules and the names of the corresponding compounds generally correspond to specific minerals or crystals whose existence is well established in the solid state. For example, an iron silicate slag of silicate degree close to unity is thought of as consisting primarily of the compound $2FeO \cdot SiO_2$ and might be called a fayalite slag because this formula represents the composition of the mineral fayalite. Postulated molecular structures of this kind have been and still are useful as a basis of correlating data, and afford a fairly flexible approach because of the wide range of choice of compound formulas usually possible in a given situation. Moreover, the assumption

that molecules of metal silicates exist in slags, the molecules having compositions of known minerals, can well lead to useful conclusions, because the stability of the solid minerals certainly suggests that a corresponding association or structure in the liquid state will have some tendency to exist, even though not as complex molecules. In this connection it should be pointed out that molecules are not ultimate structural units in the solid minerals either, although the minerals correspond in composition to the molecular formulas.

As evidence of various kinds accumulates, it appears that we must regard common slags as ionic liquids, that is, as liquids in which ions are important structural units. Perhaps the most direct indication of the existence of ions in oxide slags is that their electrical conductivities in the systems so far measured are of the correct orders of magnitude for ionic liquids but not for molecular or electronic liquids. In fact, slag conductivities expressed on an equivalent basis are about the same as conductivities of dilute aqueous salt solutions, indicating for slags a relatively high degree of dissociation. However, we cannot apply our knowledge of the structure of aqueous solutions to slags to any great extent because in slags there is no ever-present solvent like water, whose special behavior is so predominant in determining the behavior of electrolytic solutions. Ionic theories of the structure of liquid slags are in some instances more consistent with physico-chemical equilibrium data than molecular theories. In other instances, both theories fit equally well or equally poorly, depending on the assumptions made in applying them. The over-all status of the ionic theory at the present time seems to be that a strong case can be made for the position that oxide slags are in general highly dissociated ionic liquids, but the details of the structure and of the interactions between ions in the structure are not understood well enough to permit quantitative predictions and correlations of slag properties.

Accepting the theory that ions are important structural units in liquid slags, some further qualitative ideas and confirmatory information can be obtained by analogy from the structural characteristics of solid oxides, silicate minerals, glasses, and fused salts. The simple basic oxides like CaO, MgO, FeO, and MnO in the crystalline state form lattices similar to that of NaCl, in which the units are ions, Ca^{++}, Mg^{++}, Fe^{++}, Mn^{++}, and $O^=$. Each metal ion is shared by six oxygen ions and also each oxygen ion is surrounded by six metal ions in a continuously repeated geometric structure. When the oxide is melted to form a slag, the principal structural change probably is the loss of long-range order. That is, the liquid oxide still consists of the same ions and each ion probably tends to surround itself with six ions of the opposite sign, thus retaining the short-range order of the solid. On the other hand, when the ions beyond the immediate neighbors of a given ion in the liquid are considered, the ordered geometric arrangement characteristic of the crystal is not retained in the liquid. This conception of liquid-oxide structure implies a high degree of ionic dissociation in the sense that the fundamental structural units are ions, but at the same time the individual ions are not freely-moving independent bodies because they are chemically bonded to near neighbors in much the same way as they are in crystals.

As already mentioned, acidic oxides are characterized by greater bond strengths than basic oxides (see Table 9–3), and hence do not ionize simply into metal ions and oxygen ions but rather tend to form complex oxy-ions such as SiO_4^{4+}, PO_4^{3-}, BO_3^{3-}, etc.

Studies of the structures of solid silica and silicate minerals indicate broadly how acidic oxides might behave in liquid slags. The common structural unit in these solids is the SiO_4 group, which may be visualized in space as a tetrahedron with a silicon in the center and oxygens at the four corners. In solid silica (e.g., quartz) these tetrahedra are all linked together so that each oxygen is shared by two tetrahedra, to obtain the over-all composition SiO_2. In the silicate minerals these tetrahedra are present in several forms: (a) discrete orthosilicate ions (SiO_4^{4-}), each a single tetrahedron, (b) $Si_2O_7^{6-}$ ions in which one oxygen atom is shared by two SiO_4 tetrahedra, (c) two oxygen atoms of each SiO_4 tetrahedron are shared with other tetrahedra to obtain complex rings or long chains such as

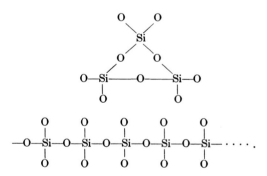

(d) three oxygens of each tetrahedron may be shared to obtain sheets, and (e) sharing of all four oxygens leads to the structures of solid silica already mentioned. From these features of crystalline silicates, it is reasonable to expect that silica in liquid slags will be present in the form of silicate ions of perhaps more than one kind and that there will be a tendency to retain the tetrahedral SiO_4 grouping characteristic of all the mineral silicates. Thus, the following simple ions are probably constituents of silicate slags:

$$\left[\begin{array}{c} \text{O} \\ | \\ \text{O}-\text{Si}-\text{O} \\ | \\ \text{O} \end{array}\right]^{4-} , \left[\begin{array}{cc} \text{O} & \text{O} \\ | & | \\ \text{O}-\text{Si}-\text{O}-\text{Si}-\text{O} \\ | & | \\ \text{O} & \text{O} \end{array}\right]^{6-}$$

Though metasilicate ions ($SiO_3^=$) are not found in minerals, the possibility of their existence in liquid slags should not be ruled out.

Probably the ionic form of silica will vary in liquid slags, depending as in solid minerals on the availability of oxygen or $O^=$ furnished by the basic oxides. In slags high in SiO_2 and low in $O^=$, relatively large ions in the form of chains and rings seem likely because of the fact that the silicate tetrahedra have to share their oxygens with each other. This tendency to form chains, rings, sheets and the like accounts for the high viscosity of strongly silicious slags. With increasing $O^=$ derived from increasing contents of basic oxides, these large ions may be broken up by reactions such as the following:

$$2\left[\begin{array}{ccc} \text{O} & \text{O} & \text{O} \\ | & | & | \\ \text{O}-\text{Si}-\text{O}-\text{Si}-\text{O}-\text{Si}-\text{O} \\ | & | & | \\ \text{O} & \text{O} & \text{O} \end{array}\right]^{8-} + \text{O}^=$$

$$\rightleftharpoons 3\left[\begin{array}{cc} \text{O} & \text{O} \\ | & | \\ \text{O}-\text{Si}-\text{O}-\text{Si}-\text{O} \\ | & | \\ \text{O} & \text{O} \end{array}\right]^{6-}$$

and

$$\left[\begin{array}{cc} \text{O} & \text{O} \\ | & | \\ \text{O}-\text{Si}-\text{O}-\text{Si}-\text{O} \\ | & | \\ \text{O} & \text{O} \end{array}\right]^{6-} + \text{O}^=$$

$$\rightleftharpoons 2\left[\begin{array}{c} \text{O} \\ | \\ \text{O}-\text{Si}-\text{O} \\ | \\ \text{O} \end{array}\right]^{4-}$$

As is the case in dealing with aqueous solutions, the chemical equations for reactions involving slag constituents can be written either in ionic form or in molecular form. For example, the oxidation of ferrous iron with CO_2 can be represented:

$$2FeO \text{ (slag)} + CO_2 \text{ (gas)} \longrightarrow$$
$$Fe_2O_3 \text{ (slag)} + CO \text{ (gas)},$$

$$2Fe^{++} \text{ (slag)} + 4O^= \text{ (slag)} + CO_2 \text{ (gas)} \longrightarrow$$
$$Fe_2O_5^{4-} \text{ (slag)} + CO \text{ (gas)},$$
or

$$2Fe^{++} \text{ (slag)} + CO_2 \text{ (gas)} \longrightarrow$$
$$2Fe^{+++} \text{ (slag)} + O^= \text{ (slag)} + CO \text{ (gas)}.$$

In our present state of knowledge, no sound basis exists for considering one way of writing the equation better than another. All are equivalent as far as stoichiometric, thermodynamic, and other calculations independent of reaction mechanism are concerned.

Ionic theories of slag structure show particular promise as a foundation for understanding and eventually perhaps measuring quantitatively the property of slag basicity. This property appears to be closely related to the concentration of oxygen ions ($O^=$) in the liquid slag. Thus, the relative basic or acidic strength of oxides appears to be directly related to their relative abilities to furnish $O^=$ when dissolved in slags. A strong base such as CaO may dissociate completely into metal cations and oxygen ions. On the other hand, a strong acid such as SiO_2 has virtually no tendency to dissociate into silicon and oxygen ions, but on the contrary actually consumes oxygen ions when dissolved in the melt (e.g., $SiO_2 + 2O^= \rightarrow SiO_4^{4-}$). Further development of these concepts of basicity and acidity, founded on ionic theories, is likely in the not too distant future.

TABLE 9-4

Melting Points of Oxide Components of Slags

Compound	Mineral or crystal name	Formula	Melting point	
			°C	°F
Alumina	corundum	Al_2O_3	2050	3720
Antimony trioxide		Sb_4O_6	655	1210
Beryllia	bromellite	BeO	2530	4580
Boric oxide		B_2O_3	450	842
Calcium oxide (lime)		CaO	2570	4660
Cuprous oxide		Cu_2O	1230	2245
Iron oxides	wüstite	FeO_x	1371	2500
	magnetite	Fe_3O_4	1597	2907
	hematite	Fe_2O_3	1455*	2651*
Lead oxide (litharge)		PbO	890	1634
Magnesia	periclase	MgO	2800	5070
Nickel oxide		NiO	1990	3610
Phosphorus pentoxide		P_2O_5	358**	676**
Silica	cristobalite	SiO_2	1728	3142
Titania	rutile	TiO_2	1830	3330
Zinc oxide		ZnO	1975	3590

*Decomposes into magnetite and oxygen at 1 atm.
**Sublimes without melting at 1 atm.

Slag constitution diagrams. Most of the common slag-forming oxides have melting points well above the operating temperatures of the processes in which slags are made. Melting points of the oxides most frequently found in slags are given in Table 9–4. The five oxides which constitute the bulk of smelting slags are underlined. These data indicate that fusible oxide slags can be made only by combining two or more oxides to obtain mixtures which are liquid at temperatures far below the melting points of the pure oxides. For the simpler slags, especially those composed primarily of only two or three oxides, an understanding of the melting and freezing behavior and data on fusion temperatures can be obtained from published constitution diagrams. Even for complex slags, data of considerable practical value usually can be obtained by study of the binary and ternary diagrams for various simple combinations of the oxide components of the complex slags.

The CaO-SiO_2 diagram, which is particularly important to a number of slags in ferrous metallurgy, is shown in Fig. 9–16. In this binary system the composition range in which liquid slags are possible at any reasonable furnace-smelting temperature is relatively narrow and falls in the vicinity of the composition of wollastonite, which is a bisilicate (or metasilicate) composition, 48.3 weight % CaO, 51.7 weight % SiO_2. A monosilicate mixture forms a very stable solid with melting point above 2130°C, and CaO itself is substantially infusible. Mixtures of CaO and SiO_2 high in SiO_2 do not form homogeneous slags, but at temperatures above the melting point of SiO_2 form two immiscible liquids. The constitution diagram also shows the solid phases or mineral silicates which are to be found in solidified calcium silicate slags.

The MgO-SiO_2 system is represented by a constitution diagram quite similar to that for the CaO-SiO_2 system. Accordingly, for

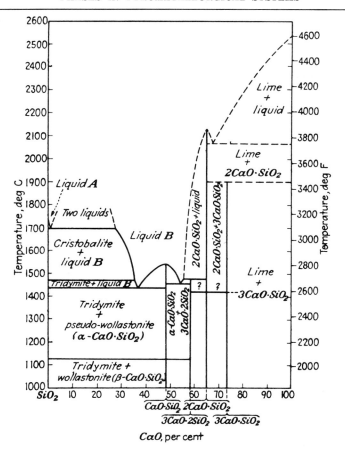

FIG. 9–16. CaO-SiO₂ system (AIME, *Basic Open-Hearth Steelmaking*).

many purposes it is satisfactory to assume that the behavior of MgO in slags is substantially the same as that of CaO, on a mol for mol basis. Thus dolomitic limestone (MgCO₃ + CaCO₃) when more available may be substituted for ordinary limestone (CaCO₃) as a flux without greatly affecting the slag chemistry. The assumption of equivalence of MgO and CaO is often used because in most metallurgical slags the MgO content is relatively small. However, if large percentages of CaO are replaced by MgO, differences in basicity and other slag properties must be considered.

Iron oxides and silica are the major constituents in copper-smelting slags and occur together in substantial percentages in many other slags as well. Strictly speaking, all iron silicate slags are ternary slags since even the simplest contain both ferrous and ferric iron. Under reducing conditions, however, the content of ferric iron is small enough so that the slags may be regarded as ferrous silicates as a good working approximation. This approximation is common practice, not always justified, and is furthered in some cases by the fact that the analytical chemist may report total iron

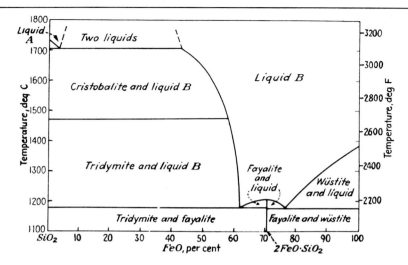

FIG. 9–17.   FeO-SiO₂ system (AIME, *Basic Open-Hearth Steelmaking*).

content as equivalent percent FeO.   Figure 9–17 represents the constitution of iron silicate slags in equilibrium with solid iron crucibles.   This diagram is not a true binary constitution diagram, but because the Fe_2O_3 percentages in the slags are small it can be regarded as a binary FeO-SiO_2 diagram for many purposes.   Actually, for the slags in iron crucibles, which represent the closest possible approach to liquid ferrous silicates, the proportion of ferric iron ranges from about one-tenth of the total iron in high-iron slags down to less than one-fiftieth in melts with 30% or more silica.   In comparison with the simple lime-silica slags, the iron silicates have freezing points several hundred degrees lower.   Also, since the iron oxide itself melts at only 1371°C, the range of possible compositions of liquid slags at ordinary smelting temperatures extends all the way to 0% SiO_2.   As in the CaO-SiO_2 system, two immiscible liquids are formed in systems high in SiO_2.   One stoichiometric compound is formed on freezing, fayalite; this is a monosilicate or orthosilicate.   Accordingly, iron silicate slags

with silicate degrees near unity are often called fayalite slags.

Although it is recognized that changing the ratio of ferric to ferrous iron in iron silicate slags has very large effects on slag chemistry, especially on the oxidizing power, we do not yet have complete data on the constitution of ternary FeO-Fe_2O_3-SiO_2 slags.   Figure 9–18 shows the composition range of the ternary melts at 1300°C.   The melt field is bounded by four curves representing, respectively, (a) melts saturated with solid iron, (b) melts saturated with solid silica, (c) melts saturated with magnetite, and (d) melts saturated with solid wüstite.   Comparison of Fig. 9–18 with similar data for other temperatures shows that while the SiO_2 solubility is relatively insensitive to temperature changes, the solubility of Fe_2O_3 (or magnetite) in the slag increases very markedly with increasing temperature.

Just as MgO behaves much like CaO, so MnO behaves in a general way like FeO and can be considered as replacing an equivalent amount of FeO, as long as the MnO

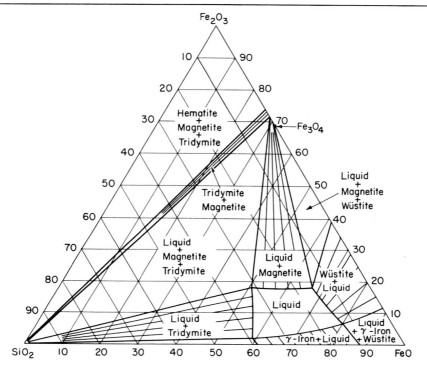

FIG. 9–18. Isothermal section of FeO-Fe₂O₃-SiO₂ system at 1300°C.

content does not become large in proportion to the FeO.

Several processes yield silicate slags with substantial percentages of both calcium and iron oxides. Figure 9–19 is an over-all view of the liquidus surface of the FeO-CaO-SiO₂ ternary system which indicates the composition ranges of interest in slagmaking as a rough valley extending diagonally across the diagram from the middle of the CaO-SiO₂ edge to the FeO-SiO₂ edge near the FeO corner. The slags of lowest freezing points fall in the vicinity of 45% FeO, 20% CaO, and 35% SiO₂, which represent about the relative proportions of those components in some lead blast furnace slags.

The Al₂O₃-SiO₂ binary system is noted more as a basis for refractory materials than as a slag-forming system, and, therefore, is

discussed in Chapter 10. Alumina, however, is present in metal ores as part of the gangue, and, therefore, is very often a secondary constituent of smelting slags. Only rarely is high-alumina material used as a flux, and then generally only for the lack of lime or iron fluxes. However, CaO and Al₂O₃ form a narrow range of slags liquid below 1500°C in which the Al₂O₃ may be thought of as the acid oxide. The CaO-Al₂O₃-SiO₂ system is of particular importance for iron blast furnace slags, having a ternary eutectic at 1170°C, 23.25% CaO, 14.75% Al₂O₃, and 62.0% SiO₂. In some nonferrous smelting slags, too high a content of Al₂O₃ is said to cause difficulties because of the tendency of aluminate spinels (e.g., MgO·Al₂O₃, ZnO·Al₂O₃) to precipitate or form infusible solid phases.

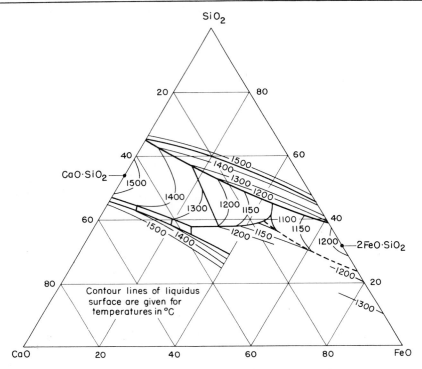

FIG. 9–19. A portion of the liquidus surface in the FeO-CaO-SiO₂ system (after Bowen, Schairer, and Posnjak).

Many other constitution diagrams are available in the literature for both silicious and nonsilicious slag systems. A great deal of careful work has been done on silicate systems particularly, because these systems are vital to geochemists and geologists in accounting for the silicate minerals which make up a large part of the earth's crust.

Slag viscosity. The previous discussion of practical requirements for slags brought out the fact that the gravity separation from other phases and subsequent slag handling place some definite requirements and limitations on the physical properties of slags, especially on the viscosity. The experimental measurement of slag viscosity at high temperatures is difficult, so that few quantitative data are available. However,

these data and ordinary visual observations show clearly that silicate slags characteristically are relatively viscous liquids. Thus, whereas liquid metals are similar to water and have viscosities of a few hundredths of a poise, an average smelting slag is more like castor oil with a viscosity of a few poises, roughly one hundred times greater than the viscosity of liquid metals. Slags considered to be very viscous might have viscosities of a few hundred poises.

Some notion of the effect of viscosity on gravity separation in a furnace can be had by estimating the settling velocities of metallic particles. To take a numerical example, we can estimate the velocity of fall of a 1-mm sphere of liquid metal of specific gravity 8 through a slag of viscosity

1 poise and specific gravity 3. Stokes' law for the terminal velocity of a sphere under viscous flow conditions is given by:

$$u = \frac{g(\rho - \rho_F)D^2}{18\mu}$$

in which u is terminal velocity in cm/sec, g is acceleration of gravity in cm/sec^2, ρ is particle density and ρ_F is fluid density in gm/cm^3, D is particle diameter in cm, and μ is viscosity in poises. Applying this relation to the data given, we find a settling velocity of about 3 cm/sec, which suggests no difficulty in making a slag-metal separation. However, if the slag were as viscous as 100 poises, the velocity of the 1-mm particle would be on the order of 0.03 cm/sec, roughly 1 ft in 15 min, and some difficulties might be expected in making a clean separation. The effects of specific gravity and particle size can also be estimated by the Stokes' law equation.

Viscosity depends on both temperature and slag composition, and under some conditions may be very sensitive to small changes in these variables. Also, the apparent slag viscosity may be increased by the presence of suspended or precipitated solid particles. Sometimes the effects of these variables can be correlated with the help of the slag-constitution diagram. Viscosity increases with decreasing temperature in general, but as the freezing point or liquidus temperature is approached in a given slag, the viscosity may change more rapidly. This change will continue as the temperature drops below the freezing point, either because of precipitation of the solid phase or simply as an accompaniment of supercooling. Because of this effect, it may be helpful to regard the viscosity as a function of the degree of superheat of the liquid, that is, of how far the slag is heated above its freezing point or liquidus temperature.

Enough measurements have been made to show that changes in slag composition have large effects on viscosity and that increases in specific components may have rather specific effects, but the data are as yet insufficient to establish any correlations of over-all applicability. The only relationship which appears general is that increasing SiO_2 content causes increasing viscosity. This relationship is readily understood, in a qualitative way at least, in terms of the behavior of silicate groups and ions in the slag structure discussed in a previous section.

Although viscosity has been so far considered principally as a physical property of a liquid affecting gravity separation, particle settling, and flow, it is a property which also has considerable significance in other ways for metallurgical processes involving slags. In a broad way, the magnitude of the viscosity is an indication of other magnitudes such as diffusion constants and reaction-rate constants. Thus, a viscous or sluggishly-moving slag is also likely to be more sluggish in chemical reactions, in mixing, and in transferring heat.

Slag formation. The constitution diagrams considered previously show the phases present when various oxide mixtures are brought to *equilibrium* under given conditions. In determining these diagrams great care was taken to reach equilibrium, and the experimental mixtures in some cases were held for many hours at constant temperature to insure completion of slow changes. Such precautions are especially essential in silicate systems where reactions and formation of new phases may proceed very slowly. For example, it is very easy to cool a high-silica liquid slag so that no crystallization occurs and the resulting solid is a glass, which should be considered as a super-cooled liquid. The change from liquid slag to glass as temperature decreases represents a gradual and continuous change in properties over a wide temperature range and has little relation to the equilibrium freezing behavior shown by the constitution diagram. The same thing is true of the process of slag formation. A mixture of solid oxides held just above the liquidus temperature shown by the constitution diagram might take many hours to form the homogeneous liquid indicated by the constitution diagram. Accordingly, the conditions required in practice for forming slags may differ quite materially from the conditions indicated by equilibrium diagram. Usually the process temperature necessary to form a reasonably homogeneous and satisfactory slag is higher than the liquidus temperature by 100°C or more.

The minimum temperature required to form a slag which discharges freely from the furnace is sometimes called the "free-running temperature" of the slag. This temperature is a somewhat arbitrary practical criterion for the process as a whole rather than a reproducible temperature property of the slag. Moreover, it may depend on the temperature-viscosity relation for the slag as well as on the conditions of formation. To the furnace operator, however, the free-running temperature may be very significant as a minimum satisfactory operating temperature or as the critical furnace temperature. In the blast furnace, the slag-forming oxides move downward from the smelting zone as soon as they are fused into a free-running slag, so that the free-running temperature of the slag more or less automatically is a major factor in determining the temperature of the smelting zone of the furnace. This leads to control of furnace temperature by control of slag composition.

One method of measuring the formation temperature of a slag consists of preparing a mixture of the solid oxides with a little binder (e.g., dextrine) and pressing it in the form of a narrow cone. This cone is then heated slowly under visual observation, and the furnace temperature at which the cone melts down into a pool of slag is determined. For a given mixture composition, the formation temperature depends on particle size of the solid oxides, on intimacy of mixing, heating rate, and other experimental variables, just as it does in a full-scale process. Accordingly, for previously untried slag compositions, experiments of this kind may be quite useful in anticipation of full-scale processing. Formation of slags from mechanical mixtures of relatively infusible solid oxides may require temperatures several hundred degrees above the freezing point of the final homogeneous slag indicated by the constitution diagram. On the other hand, if the solid mixture contains some solid slag previously fused, this material will serve to start the process of slag formation at a much lower temperature. For a continuous process in which a slag pool is maintained at all times, slag formation involves mainly dissolution of new slag constituents in the pool.

Mattes and other products

Liquid metals and slags are the most common liquid products of smelting and refining processes, but in a few processes other liquid phases such as mattes and speisses are produced, either in place of liquid metal as a primary valuable product or in addition to a primary liquid-metal product. It is interesting to note that slags, metals, mattes, and speisses are mutually insoluble so that various combinations of two- and three-phase systems are possible. Under certain conditions even four liquid phases may be present, in four separate liquid layers.

Mattes are liquid solutions of metal sulfides. The most common major constituents are copper, iron, and nickel sulfides, although other base metals and precious metals may be present in substantial amounts. Mattes are the principal metallic products in smelting copper and copper-nickel sulfide ores, and represent important by-products and intermediate products in other smelting and refining processes of nonferrous metallurgy, especially lead metallurgy. Table 9–5 gives chemical analyses of a number of commercial mattes and shows the type of process in which each matte is made.

The simplest mattes which can be prepared in the laboratory are binary liquid solutions of one metal and sulfur, usually approximately the composition of a simple sulfide, such as Cu_2S, FeS, PbS, ZnS, etc. When a metal forms more than one sulfide (e.g., Cu_2S, CuS; FeS, FeS_2), the higher sulfide tends to decompose, before melting, into the lower sulfide and sulfur gas so that the simple mattes prepared by ordinary techniques will be close to the lower sulfides in composition. In fact, at the temperatures of molten mattes, the sulfur/metal ratio may be below the stoichiometric ratio for even the lower sulfide because of the

tendency of the matte to volatilize sulfur. When the sulfur/metal ratio is below stoichiometric, the matte may be regarded if desired as a solution of (metal) + (metal sulfide). The composition ranges in which binary mattes can exist are shown in general by the metal-sulfur or metal-metal sulfide constitution diagrams.

Copper and iron are the metals most commonly present in mattes, and the constitution diagrams for the Cu-S and the Fe-S systems have already been presented in connection with the discussion of solutions of sulfur in liquid metals. Figure 9–15 shows that copper-sulfur mattes exist at a given temperature over a range of S/Cu ratios from a little below the stoichiometric ratio for Cu_2S (Cu-saturated mattes) to well above Cu_2S. However, if the liquid has only slightly more sulfur than corresponds to the composition Cu_2S, the vapor pressure of sulfur gas over the liquid becomes so large that the excess sulfur is lost by volatilization. Accordingly, the Cu-S matte produced in copper smelting, known as "white metal," has substantially the composition of Cu_2S and is often considered as Cu_2S. As shown in Fig. 9–14, iron and sulfur form mattes with a relatively wide range of possible sulfur/iron ratios. At

TABLE 9-5

Chemical Analyses of Mattes

Matte	Weight percent								Ounces/ton	
	Cu	Fe	S	Pb	Zn	Ni + Co	As	Sb	Ag	Au
Copper mattes: (a)	22.0	41.4	24.5	1.6	4.2					
(b)	35.7	33.9	26.3						1.4	0.02
(c)	45.9	26.0	25.2							
(d)	68.0	8.0	21.4							
(e)	78.7	0.7	19.5							
Copper-nickel matte	24.8	0.5	22.0			52.8				
By-product mattes, lead blast furnace:(a)	5.9	48.8	20.0	9.5	5.4		1.5			
(b)	56.4	1.2	14.4	23.2			2.9			
(c)	61.3	1.0	15.8	15.8			2.9	0.4	37.2	0.16
Soda matte (contains sodium sulfide), dross smelting	11.1	18.6	19.8	15.1	8.5	2.6				

temperatures below the melting point of iron, the lower limit of the S/Fe ratio is reached with mattes in equilibrium with and saturated with solid iron, but above the melting point of iron a continuous solution field extends all the way to pure iron. As with Cu-S melts, the sulfur vapor pressure increases rapidly with increasing sulfur so that melts appreciably above FeS in sulfur content are not stable unless held in pressure apparatus to avoid volatilization of sulfur. Figure 9–14 also shows some interesting features of the behavior of the solid iron sulfides, pyrrhotite and pyrite. Pyrrhotite is represented by a solid-solution field, with minimum sulfur at lower temperatures corresponding to the composition of FeS (36.5% S). The iron sulfide which melts congruently at 1190°C has more sulfur than the composition FeS. Pyrite (FeS$_2$) decomposes at 693°C and 1 atm to give pyrrhotite and sulfur gas.

Reference to other base metal-sulfur binary constitution diagrams shows that the formation of binary liquid mattes over a considerable range of sulfur/metal ratios is common. In particular, as with both copper and iron, the sulfur/metal ratio can be lower than that for the lowest stoichiometric sulfide, which means that mattes in general readily dissolve excess metal. Several systems, including Ag-S, Sb-S, and Sn-S, show two-liquid miscibility gaps similar to the one in the Cu-S system. Another characteristic which can be seen in most of the base metal-sulfur binary systems is that the freezing points of the metal sulfides are not especially high. Typical values for various sulfides are given in Table 9–6. As these data indicate, matte-forming temperatures are generally below slag-forming temperatures, so that no difficulty is experienced in making liquid mattes in ordinary smelting and refining processes. The chief exception

is that materials high in Zn and S under some conditions form relatively infusible and troublesome accretions.

A matte containing two metals and sulfur should strictly speaking be considered as a ternary solution. Thus, a complete representation of the constitution of copper-iron mattes would require a ternary diagram for the system Cu-Fe-S, and, as a matter of fact, some of the major features of this ternary diagram have been determined and are available in the literature on the subject. However, most mattes can be represented approximately in composition as mixtures of simple stoichiometric sulfides, and a simpler approach can be made. Following this procedure, the constitution of copper-iron mattes may be considered in terms of the diagram for the Cu$_2$S-FeS system shown in Fig. 9–20. This diagram does not represent a true binary system, but actually is a section through the ternary system. Figure 9–20 shows that liquid cuprous and ferrous sulfides are miscible in all proportions at smelting temperatures, and the liquid of eutectic composition has a freezing point just below 1000°C. The constitutional relationships for several other sulfide mixtures are similar, but without the relatively wide ranges of solid solubility. For example, Cu$_2$S and PbS, both with melting points just over 1100°C, form a eutectic at about 50% Cu$_2$S with a eutectic temperature of 540°C. Similarly, FeS and PbS show a eutectic composition of about 26% FeS with a eutectic freezing point of 782°C.

Since at smelting temperatures most of the base-metal sulfides are soluble in each other, they tend to collect in a single matte phase of complex composition if the proportions of metal to sulfur are in the right range. However, when a matte phase is produced in the presence of other phases, such as liquid metals, slags, or speisses, different

TABLE 9-6

Melting Points of Metal Sulfides

Compound	Mineral	Formula	Melting point	
			°C	°F
Antimony sulfide	stibnite	Sb_2S_3	546	1015
Arsenic sulfide	orpiment	As_2S_3	300	572
Bismuth sulfide	bismuthinite	Bi_2S_3	747	1377
Cadmium sulfide	greenockite	CdS	(1750)*	(3182)*
Cuprous sulfide	chalcocite	Cu_2S	1131	2068
Ferrous sulfide	pyrrhotite	FeS	1190	2175
Lead sulfide	galena	PbS	1114	2037
Manganese sulfide	alabandite	MnS	1530	2786
Nickel sulfide	millerite	NiS	797	1467
Potassium sulfide		K_2S	912	1674
Silver sulfide	argentite	Ag_2S	842	1548
Sodium sulfide		Na_2S	978	1792
Zinc sulfide	sphalerite	ZnS	(1645)*	(2993)*

*Melting points at about 100 atm. CdS and ZnS volatilize without melting when heated at 1 atm.

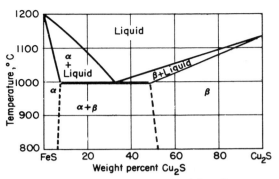

FIG. 9-20. The Cu_2S-FeS system (after Carpenter and Hayward).

metals are distributed in different ways between the various liquid phases, and the distribution of a given metal will depend on chemical conditions in the system. Thus, whenever a matte is produced it tends to collect the bulk of the copper if enough sulfur is present to form Cu_2S. On the other hand, in lead smelting the lead largely enters the liquid metal and enters the matte only to a minor extent, although it is enough to justify subsequent recovery of lead from the matte. Similarly, in copper smelting the bulk of the copper is collected in the matte but the iron enters only to the extent that sulfur is available after satisfying the copper, the balance of the iron going to the slag. Of great practical importance is the fact that the precious metals gold and silver dissolve readily in mattes, so that these valuable metals are efficiently collected and saved in smelting processes yielding mattes as the primary metallic products. When both liquid matte and liquid metal are made together, however, the major portions of the precious metals tend to enter the metal phase in preference to the matte.

Commercial mattes also dissolve elements other than metals and sulfur. For example, copper mattes contain oxygen in solution, sometimes described as dissolved magnetite because of indications that magnetite crystallizes out from the mattes on cooling. Selenium and tellurium also tend to dissolve in copper mattes, and though present only in small percentages must eventually be separated from the copper as the matte undergoes further processing to make commercial copper.

The solidification of mattes containing several metals, sulfur, and possibly dissolved oxides as well is a complex crystallization process usually yielding a mixture of several solid phases. Microscopic examinations of polished sections usually show solid metal (e.g., copper), solid sulfides identical with known sulfide minerals, oxides (e.g., magnetite crystals), and other phases not readily identified. In commercial mattes, these phases crystallized from liquid mattes are to be distinguished from solids carried mechanically in suspension in the liquid matte as it is discharged from the furnace. Generally speaking, however, the components of mattes are characterized by limited solubilities in the solid state. This behavior can be made the basis of separation processes, in which the solidified matte is crushed and then treated by such mineral-dressing processes as flotation and magnetic separation. In order to accomplish the most effective separations by this procedure, the matte composition and cooling conditions are regulated so that the metals to be separated have the opportunity to segregate completely into crystals of a reasonable size. This process has recently assumed great importance in the metallurgy of copper-nickel ores.

Little is known quantitatively about the physical properties of liquid mattes. Liquid mattes are quite fluid, probably having viscosities of the same order as liquid metals. Also, some mattes are known to be good electrical conductors, even approaching the conductivities of liquid metals and suggesting the presence of free electrons in the melt structure.

Speisses are liquid solutions of arsenides, or arsenides and antimonides, and thus are produced only when the materials being treated contain large quantities of arsenic, or arsenic and antimony, more than can be carried readily in solution in the metal or matte. Accordingly, speisses are not made and handled in the large quantities and wide variety of processes involved in liquid metals, slags and mattes.

Drosses are heterogeneous products skimmed or drossed from the surface of liquid metals during refining or simply after melting. Unlike slags, mattes, and speisses, they are not formed as single liquid phases, but instead are mixtures of precipitated solid and liquid compounds with substantial proportions of mechanically-trapped mother metal from the underlying metal bath.

Supplementary References

AIME, Committee on Physical Chemistry of Steelmaking, *Basic Open-Hearth Steelmaking*, revised ed. New York: AIME, 1951.

ASM, *Metals Handbook*. Cleveland: American Society for Metals, 1948.

Bever, M. B., "Gases in Cast Metals." *Iron Age*, **161**, 90–94, April 22, 1948.

Bragg, W. L., *Atomic Structure of Minerals*. Ithaca, N. Y.: Cornell University Press, 1937.

Chipman, J., and Chang, L., "The Ionic Nature of Metallurgical Slags. Simple Oxide Systems." *AIME J. of Metals*, **1**, 191–197, 1949.

Data on Chemicals for Ceramic Use. National Research Council, Bull. 107, 1943.

Grant, N. J., and Chipman, J., "The Chemistry of Metallurgical Slags." *AISI Yearbook*, 469–486, 1949.

Hansen, M., *Der Aufbau der Zweistofflegierungen*. Berlin: J. Springer, 1936.

Kelley, K. K., *Contributions to the Data on Theoretical Metallurgy. III. The Free Energies of Vaporization and the Vapor Pressures of Inorganic Substances*. U. S. Bureau of Mines, Bull. 383, 1935.

Marsh, J. S., *Principles of Phase Diagrams*. New York: McGraw-Hill, 1935.

Preston, E., "The Structure and Constitution of Glass." *J. Soc. Glass Tech.*, **26**, 82–107, 1942.

Smithells, C. J., *Gases and Metals*. London: Chapman & Hall, 1937.

Smithells, C. J., *Metals Reference Book*. New York: Interscience, 1949.

Stull, D. R., "Vapor Pressure of Pure Substances. Inorganic Compounds." *Ind. Eng. Chem.*, **39**, 540–550, 1947.

Wells, A. F., *Structural Inorganic Chemistry*. Oxford: Clarendon, 1950.

Zachariasen, W. H., "The Atomic Arrangement in Glass." *J. Am. Chem. Soc.*, **54**, 3841–3851, 1932.

CHAPTER 10

REFRACTORY MATERIALS

Construction of furnaces for pyrometallurgical processes requires refractory materials which will withstand the high temperatures, severe chemical actions, and other conditions characteristic of the processes. A wide variety of service conditions is encountered, not only in going from one process to another but also in different parts of the furnace used for any one process. In some applications, the chief requirement may be infusibility at extreme temperatures. In others, resistance to chemical attack by a certain kind of slag may be the most important factor. Mechanical properties, such as compressive strength at high temperatures or resistance to abrasion and dust erosion, play major roles in some applications. Thermal conductivity should be low for some uses, high for others. For most individual applications, a combination of these and other requirements has to be considered. Accordingly, a wide variety of refractory materials has been developed to meet the metallurgist's needs, and the refractory manufacturers continually work with the metallurgist in an effort to develop better materials for specific purposes.

To the metallurgical engineer, proper choice and use of refractory materials is often very vital to the success or failure of a process. In the past, some otherwise attractive processes have failed or been replaced primarily because materials available for apparatus construction would not stand up under the process conditions. Many of the virtually standard smelting and refining processes for common metals, such as the iron blast furnace, steelmaking, copper smelting, zinc retorting, and others, have refractory requirements which account for substantial proportions of total metal-producing costs. Accordingly, proper choice of refractories, design of furnaces, and maintenance and repair of refractories during furnace operation represent a constant challenge to the engineer attempting to achieve minimum operating costs.

The properties of the refractory materials affect various aspects of the way the process itself is conducted. Chemically the oxides of which the commonest refractories are composed are the same as the oxides which are the most important constituents of metallurgical slags. Thus, the refractory walls of the furnace are in many cases to be numbered among the reacting phases in the system and are not at all inert in the way a glass beaker is inert to reactions in aqueous solutions. It may be necessary, for example, to control temperature, gas composition, or slag composition for the process in order to avoid or to minimize reactions with the refractories. Flame temperatures in fuel-fired reverberatory furnaces are limited by what the roof arch will stand without failing. In most metallurgical processes, operation at atmospheric pressure is taken for granted because gas tightness cannot be achieved with refractories at high temperatures. Refractory-material properties directly limit furnace shapes and dimensions and affect all the mechanical manipulations of metallurgical processes, such as charging, tapping, skimming, blowing, etc.

In this chapter the important properties of refractory materials for metallurgical applications are defined and the common refractory materials are classified and discussed in terms of these properties. Finally, methods of furnace construction are taken up briefly and a few metallurgical furnaces are described to illustrate how knowledge of refractory properties is utilized.

Refractory properties

The service behavior of refractories is determined largely by the skill and judgment exercised in matching refractory properties to specific service conditions. In particular it is essential to consider the various ways in which the process conditions are likely to cause failure. Accordingly the list of properties considered in selecting and using refractories includes not only straightforward physical and chemical properties, such as density, specific heat, thermal conductivity, and chemical analyses, but also other complex properties which relate primarily to the various mechanisms of failure. Following is a list of the most important refractory properties and other attributes which concern the user of refractory materials.

I. Composition and structure
  (a) Chemical analysis
  (b) Structure, texture, and phase constitution
II. Service properties
  (a) Fusion and softening temperatures
  (b) Resistance to chemical attack
  (c) Expansion and shrinkage
  (d) Spalling resistance
  (e) Load-bearing capacity
  (f) Abrasion resistance
III. Other physical properties
  (a) Thermal conductivity
  (b) Cold crushing strength and modulus of rupture
  (c) Specific gravity, porosity, permeability
  (d) Specific heat
  (e) Electrical conductivity
IV. Costs
  (a) Materials
  (b) Fabrication
  (c) Maintenance
  (d) Indirect and other costs

Composition and structure. In the final analysis, the important service properties and physical properties are determined by the chemical composition and by the structure and phase constitution of the refractory body. Composition is usually expressed by giving the chemical analysis in percentages of the component oxides as % SiO_2, % MgO, % Al_2O_3, % FeO, etc. In many instances, of course, the individual oxides are not present as such but are chemically combined with other oxides. Composition affords the primary basis of classifying refractory materials as acidic, basic, and neutral, and the common names for most materials refer to the major oxide constituents, as for example, silica brick, high-alumina brick, chrome refractories, etc. Also, it is often found that constituents present in minor percentages have serious effects on service properties, especially under certain extreme furnace conditions. Accordingly, a complete chemical analysis is one of the first essentials if the behavior of a given material as a refractory is to be understood.

The fabrication of a refractory body starts generally with granular or particulate materials, and by various steps and procedures these materials are finally incorporated in a single coherent body, such as a brick or a furnace bottom, with some mechanical strength. Starting with given raw materials, the procedure of forming the refractory body can be varied in many ways to obtain a variety of final structures with widely different properties. Further changes in structure occur during service, accompanied by further changes in properties. In the study of refractories the macrostructure and microstructure, and the correlation of structure with properties, play much the same roles as the corresponding studies of structure and properties in the field of physical metallurgy. Thus, when current developments in common refractory bodies are examined, in many cases it is found that the important improvements in properties are primarily the result of improvements in structure rather than in composition.

The structure of a typical refractory body involves solid grains and crystals in a matrix or groundmass. The matrix may be a glass, a finely crystalline mass, or a mixture, and acts as a cement or bond for the primary grains of refractory material. In general, the matrix does not entirely fill the space between refractory grains, so that the structure as a whole is porous. This basic structure is subject to wide variations in relative amounts of primary refractory grains and matrix, in sizes and size distribution of the primary grains, in porosity, etc. To obtain this type of structure, the refractory body is formed from a natural or blended mixture of a granular and relatively pure primary refractory constituent with other constituents which combine to form the matrix when the body is heated. The formation of the bond generally is accomplished by firing or heating to a temperature high enough to form a liquid of the bonding constituents. Depending on the temperature and time of heating, more or less of the primary refractory constituent may dissolve in this liquid. When the body is cooled, this liquid may simply become a glassy bonding phase or it may crystallize. Thus, the groundmass or bond of a refractory body is the less refractory portion of the structure and is the portion where liquid formation occurs first when the body is in service at high temperatures. Actually, a glass is an undercooled liquid whose viscosity decreases continuously with increase in temperature; hence, the high-temperature mechanical behavior of many refractory bodies is best understood by thinking of the bond as a highly viscous liquid. In this way a gradual deterioration of properties with increasing service temperature is accounted for by increases in the proportion of liquid and decreases in its viscosity as the temperature is raised, in contrast to the abrupt and complete failure of a pure solid when its melting point is reached.

In view of the fact that the bond is the place where an ordinary refractory body starts to fail at high temperatures, it is found that the bodies which remain most rigid at extreme temperatures are those composed of relatively pure crystalline compounds with a minimum of bonding constituents. By controlling the size distribution of the pure refractory material so that the grains pack closely and by using special forming methods, the amount of bonding material can be minimized. As the amount of bond is decreased, firing temperatures must be increased and it is likely that the rigidity of the body depends more and more on growth of interlocking crystals and on solid-solid sintering phenomena. In this connection, special refractory bodies of outstanding properties, so far limited mainly to laboratory crucibles and the like, have been fabricated of high-purity refractory oxides without the addition of any bonding material at all other than a chemical binder to hold the body together until the strength is developed in firing.

Equilibrium diagrams are essential tools in interpreting and predicting structures of refractory bodies and structural changes during firing and service. They show the important relationships between composition and melting point. However, it must be kept in mind that refractory bodies are often not in a state of chemical equilibrium, so that the phase constitutions indicated by equilibrium diagrams are not attained. One common example, of course, is the existence of a glass as a metastable phase under conditions for which the equilibrium diagram shows only crystalline phases.

Because of the many important relationships between structure and properties, the direct microscopic study of refractory structures often affords valuable information and facilitates the diagnosis of refractory failures. Both the petrographic examination of thin sections by transmitted light and the examination of polished sections by reflected light are useful.

Service properties and their measurement. Refractory bodies fail in service in a variety of ways and from a variety of causes, and not infrequently the mechanism of failure is complex and poorly understood. However, certain prime causes of failure are readily recognized, such as softening and fusion, chemical attack by slags, thermal spalling, mechanical failure under load, and abrasion. To facilitate the choice of materials for different combinations of service conditions, various laboratory tests have been devised which measure more or less quantitatively the service conditions under which a given body can be used without failure. Some of these tests have been quite satisfactory and have been standardized in the industry, so that results obtained by different laboratories, by manufacturers, and by consumers are comparable and reproducible and can be made the basis of specifications. The principal standard tests are described in *Manual of ASTM Standards on Refractory Materials* published by the American Society for Testing Materials, and in the subsequent discus-

sion these standard tests will be referred to by the ASTM code number. However, satisfactory laboratory tests have not been found as yet to measure some of the important service properties, for example, resistance to slag attack, and reliance must be placed on a theoretical understanding of the mode of failure, on previous operating experience, or on actual trials in the full-scale furnace operation.

Fusion and softening temperatures logically come first in the list of service properties, since the first requirement for a refractory material is that it will withstand the high temperature at which it is to be used. The word refractoriness is often, though perhaps loosely, used in referring to the ability of a material to withstand high temperatures. Pure crystalline solids with congruent melting points, such as most of the simple oxides, fail abruptly by completely melting to a liquid at a fixed temperature, the melting point. Most refractory bodies, however, do not behave so simply. As the temperature is raised for a body consisting of a crystalline phase and a glassy phase, the glass, which is itself essentially a very viscous supercooled liquid, gradually becomes more fluid and may also increase in quantity by dissolving some of the crystalline material. Changes in the mechanical properties of the material, softening, and eventually fusion may occur gradually over a considerable temperature range, so that specification of any one temperature as the softening or fusion temperature is quite arbitrary and very dependent on the testing procedure and on the kind of observations made to detect softening and fusion.

The most widely used method of measuring softening behavior at high temperatures has been the determination of the Pyrometric Cone Equivalent (PCE), by the procedure standardized as ASTM Test C–24.

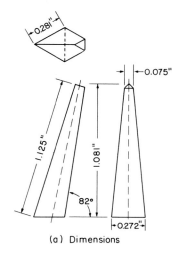

(a) Dimensions   (b) Method of comparison

FIG. 10–1.   Pyrometric cones (ASTM).

In determining the PCE, the sample is pulverized to minus 65 mesh and then formed into a test cone having the shape and dimensions shown in Fig. 10–1. Water and a small amount of dextrine or glue act as binders in making the cones. These cones are mounted on an inert plaque along with a series of Standard Pyrometric Cones with which the behavior of the sample is to be compared. The plaque is then heated at a specified rate in a furnace with uniform temperature distribution and preferably a neutral or oxidizing atmosphere. Softening of a cone is indicated when the cone bends over until its tip touches the plaque. The PCE of the test sample is the number of the Standard Pyrometric Cone which most nearly corresponds in time of softening to the test cone. Although this procedure really measures not softening temperatures, but softening behavior in comparison with standard materials during a heating cycle of controlled time and temperature, the PCE's can be converted to approximate softening temperatures, since the behavior of the standard cones has been carefully measured.

Table 10–1 gives the temperatures corresponding to the end points of the standard cones measured under known heating rates.

Depending on the material being tested and on the application to be made of the material, various other ways of measuring softening behavior may be useful. Instead of depending on Standard Pyrometric Cones, the behavior of the test cone may be observed while temperature is measured with an optical pyrometer. In other cases, the most significant test of softening temperature may be the temperature at which deformation starts under a fixed load. For a pure oxide or other crystalline compound, of course, melting occurs sharply at a single characteristic temperature.

Resistance to chemical attack, especially by slags, is all-important in the choice of materials for many metallurgical uses, since even with the most suitable materials corrosion is one of the prime causes of refractory failure. The common refractory oxides are the same oxides as those forming slags, so that only very rarely is it possible to find a refractory which is insoluble in the slag.

TABLE 10-1

Temperature End Points of American Pyrometric Cones*

Heating rates: 150°C per hour for cones 022 to 20
100°C per hour for cones 23 to 38

| No. of | End point | | No. of | End point | |
cone	°F	°C	cone	°F	°C
022	1121	605	10	2381	1305
021	1139	615	11	2417	1325
020	1202	650	12	2435	1335
019	1220	660	13	2462	1350
018	1328	720	14	2552	1400
017	1418	770	15	2615	1435
016	1463	795	16	2669	1465
015	1481	805	17	2687	1475
014	1526	830	18	2714	1490
013	1580	860	19	2768	1520
012	1607	875	20	2786	1530
011	1661	905	23	2876	1580
010	1643	895	26	2903	1595
09	1706	930	27	2921	1605
08	1742	950	28	2939	1615
07	1814	990	29	2984	1640
06	1859	1015	30	3002	1650
05	1904	1040	31	3056	1680
04	1940	1060	32	3092	1700
03	2039	1115	**32-1/2	3137	1725
02	2057	1125	33	3173	1745
01	2093	1145	34	3200	1760
1	2120	1160	35	3245	1785
2	2129	1165	36	3290	1810
3	2138	1170	37	3308	1820
4	2174	1190	38	3335	1835
5	2201	1205	**39	3389	1865
6	2246	1230	**40	3425	1885
7	2282	1250	**41	3578	1970
8	2300	1260	**42	3659	2015
9	2345	1285			

*Fairchild, C. O., and Peters, M. F., "Characteristics of Pyrometric Cones." J. Am. Ceram. Soc. 9, 701-743, 1926.

**Not included in the tests of Fairchild and Peters. The temperatures given are approximate.

Rather the problem is to minimize the rate of slag attack. Many factors affect rate of slag attack, so that in any given case a careful study of the mechanism of the attack may be required to find the best means of overcoming slag action or of bringing it within tolerable limits. No simple laboratory tests have been developed which afford a reliable prediction of the rate of slag corrosion in full-scale furnaces, so that in the end operating experience affords the only safe guide.

When a slag remains in contact with a refractory wall for any period of time, the action between the two is generally not confined to a geometric interface separating them but instead usually results in the development of several transitional zones from the interior of the homogeneous slag to the unaffected interior of the wall. These zones vary greatly in nature and thickness from one case to another, but it is what happens in these zones that determines the rate of corrosion of the wall. In a typical case, a liquid zone is formed on the slag side by the slow dissolution of refractory in the slag immediately adjacent to the wall. Preferably the liquid formed by dissolution of the refractory in this zone should have a higher melting point than the slag and a relatively high viscosity so that it will be retained at the surface as a deterrent to rapid reaction. The effectiveness of such a barrier is also related to the fact that low diffusion constants and low reaction-rate constants in general accompany high viscosity. With such factors as melting points, viscosities, diffusion constants, and reaction-rate constants playing major roles in the action of this liquid zone, it is further apparent that slag temperatures and temperature gradients at and through the wall will have large effects on the rate of corrosion. Within the wall other zones may develop from such processes as slag penetration of pores, selective dissolution of certain refractory constituents, growth of crystals, and the like. These phenomena also may have some part in the ultimate failure of the refractory by slag action.

The point of view that slag action is minimized by choosing a refractory whose dissolution in the slag increases the slag melting point and viscosity and produces corresponding changes in such related properties as diffusion constants is consistent with general results of metallurgical experience. Typical instances are the use of acid refractories to hold acid slags and the use of basic refractories to hold basic slags. When, for example, a high-SiO_2, acid slag is placed in contact with a basic refractory, excessive slag action and rapid corrosion are likely to occur because the liquids formed by solution of basic refractories in acid slags generally have lower melting points and lower viscosities than the original slags.

Chemical attack is involved in a variety of refractory failures other than direct slag action. Erosion by dust-laden gases is accelerated by reactions, often slagging reactions, between the dust and the refractory. In steelmaking processes, iron oxides in significant quantities may condense from the gas phase onto furnace wall and roof surfaces. Reactions may occur at joints between dissimilar refractories, or between bricks and cement. Under special circumstances failure may result from the action of carbon monoxide decomposing to deposit carbon in the interior of the brick.

Expansions and shrinkages of several kinds must be considered in accounting for service behavior of refractories. In the first place, there is the ordinary expansion with increasing temperature and contraction with decreasing temperature characteristic of all solids and measured by the coefficient of thermal expansion. Also, there are the dimensional changes occurring under loads, both elastic and plastic deformations. Simple thermal expansion and elastic deformation are *reversible*, in that the solid returns to original dimensions when the original temperature or loading is restored. Plastic deformation or flow under load, on the other hand, is *irreversible* and causes permanent dimensional changes. In dealing with refractory bricks and other bodies, however, still other types of expansion and shrinkage often play the chief roles. Particularly important are the volume changes accompanying allotropic changes in solid phases and the volume changes associated with changes in the struc-

ture and phase constitution of the body. Some of the allotropic changes are reversible, occurring at a definite temperature just like melting and boiling; others are so sluggish that even at favorable temperatures they do not go to completion during the life of the refractory. Structural changes, such as crystallization from the glass phase or dissolution of crystals in the glass phase, are generally slow and irreversible, starting during fabrication and continuing often in some degree through the entire service life of the brick.

The coefficient of expansion of refractory bodies varies as a rule with temperature, so that it is most convenient to show expansion behavior by means of expansion curves like those in Fig. 10–2. These curves show total reversible linear expansion in percent from room temperature to the temperatures indicated as abscissas. The actual coefficient of thermal expansion at any temperature is proportional to the slope of the curve at that temperature.

One of the functions of the process of firing a brick, besides the function of forming a ceramic bond, is to complete insofar as is practicable the irreversible and permanent volume changes. When most refractory raw materials are formed and then fired, the body shrinks. The greater the firing temperature, the greater is the irreversible shrinkage. This change is generally associated with the formation of glass and liquid which pulls the structure together, but in some materials crystal growth and sintering phenomena may be important in causing shrinkage. Since excessive shrinkage leads to difficulties of various kinds, raw materials characterized by high firing shrinkage are often prefired before incorporation in the mix for a refractory body or are mixed with other materials having low firing shrinkages. The prefired material added for this

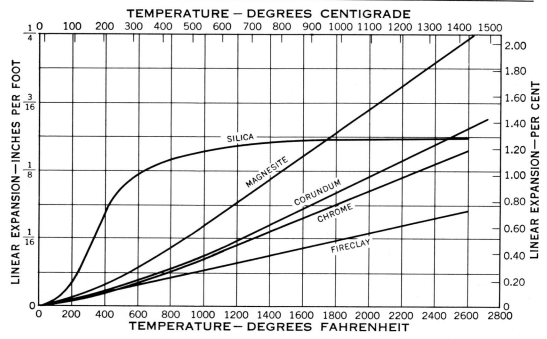

FIG. 10–2.   Reversible-expansion curves for common refractory bricks.

purpose is called "grog," and in many cases consists simply of crushed brick. A few materials expand instead of shrinking when first fired. Expansion may result from allotropic changes in crystal structure (as with SiO_2) or from other causes such as gas evolution.

ASTM Test C–113, "Reheat Change of Refractory Brick," affords a standard procedure for measuring irreversible changes which occur as the result of heating alone. Specific heating schedules are used for different types of refractories, and the change is reported as a percentage change in length.

Spalling of a refractory brick or other body involves fragmentation and fracture occurring in the vicinity of the surface in such a way that the fragments or spalls are lost, progressively exposing new surface and eventually causing failure. Though spalling is a common cause of failure, its mechanisms

are not well understood. It is clear, however, that spalling resistance not only is a function of refractory composition but also depends very much on structure and constitution and, therefore, for a given class of materials is subject to some control in the process of brick fabrication. Also, the causes of spalling are recognized so that a great deal can be done in furnace design and operation to minimize conditions leading to spalling failure.

The prime cause of spalling is that stresses are set up in the brick by temperature gradients. The temperature gradients cause unequal expansion in different parts of the brick, and if the temperature gradient and coefficient of thermal expansion are both large enough this differential expansion will generate stresses large enough to cause failure by shear or tension. Any ordinary furnace wall in steady operation, of course, has a temperature gradient from inside to outside and this gradient undoubtedly generates thermal stresses in the wall, but not great enough to cause fracture. Much larger temperature gradients and thus much larger stresses are generated during heating and cooling, as can be seen by applying the principles of unsteady heat flow discussed in Chapter 8. In particular, heating and cooling produce high temperature gradients near the surface. High stresses may result also when the heating or cooling passes through a temperature at which an allotropic change occurs, with a sudden volume change in a small range of temperatures. Spalling resulting from those factors is known as thermal spalling and the term *spalling resistance* usually refers primarily to resistance to thermal spalling. The standard test for spalling resistance consists of heating and rapidly cooling test panels through a fixed number of cycles, then determining the weight loss by spalling as a measure of spalling resistance. Details of this test are given as ASTM Test C–38.

Mechanical spalling is caused by external mechanical forces acting on a brick. If a wall is built without adequate provision for thermal expansion, the bricks in some parts of the wall may be overloaded mechanically and may thus tend to fracture by spalling.

Another cause of spalling under some service conditions is changes in brick structure in zones extending only part way into the brick. Slag penetration and prolonged exposure to high temperatures and chemical action may alter the coefficient of thermal expansion and other physical properties in the affected zone enough to cause spalling away from the unaffected zones or to produce a greater sensitivity to thermal spalling.

Load-bearing capacity at high temperatures is a critical property in the choice of a refractory material for a roof arch which is supported by a constant compressive force applied at the sides. The bricks in furnace walls must, of course, support the weight of bricks and other parts above, but high-temperature load-bearing capacity is not often critical here because in ordinary furnace design the bulk of the weight is carried by cooler parts of the wall. The nature of failure under load is closely related to several of the service properties already discussed: softening temperature, expansion and shrinkage, and spalling resistance. In particular, the softening temperature represents a condition at which the load-bearing capacity has become so small that the material in the form of a small test cone deforms under its own weight. Under an applied load the same kind of deformation and flow occurs at a lower temperature, corresponding to a higher viscosity in the glassy phase than that at the softening temperature. Moreover, time is likely to be an important variable in this kind of failure under load. The rate of deformation or creep over weeks or months under service conditions may in the end determine the suitability of the material for a given load-bearing application.

Although load-bearing capacity and maximum operating temperatures under load are usually limited by softening and plastic-flow behavior, other types of failure under load are encountered occasionally. When the material does not readily deform plastically, for example, because of too small a proportion of glassy phase in the structure or because the temperature is too low for the glass to flow, failure may occur by shear right across the brick. Uneven compressive loading and pinching may cause failure through spalling, if the structure is unable to accommodate itself by plastic flow.

The standard test for behavior of refractories under load at high temperature, ASTM Test C–16, consists of heating test specimens according to a fixed time-temperature schedule under a compressive load of 25 psi. The lengths of the specimens are measured at room temperature before and after the heating, and the deformation is expressed as percent of the original length. If the brick fails by shear during the heating schedule, the temperature of failure is reported. Although this test does not measure in any way the magnitude of the load which can be applied to a refractory under working conditions, the test results nevertheless have been found to serve well as comparative measures of load-bearing capacity and as the basis for specifications and quality control.

Abrasion resistance and mechanical toughness must be provided in furnace parts which are continually or frequently in contact with moving solid materials or objects. In various kinds of vertical shaft furnaces, which are characterized by handling of lump feed, abrasion is especially severe. Rotary kilns are another common apparatus characterized by severe abrasion. Gases carrying suspended solids may afford an abrasion problem, especially in places where the gas velocity is high or directed toward the surface. Though refractories have been developed and continue to be developed for good abrasion resistance, there is no generally accepted, quantitative measure of the property which might be determined in a laboratory test. Here again previous service performance is likely to be the metallurgists' guide in the choice of materials.

Other physical properties. In addition to the service properties just discussed, which relate primarily to causes of refractory failure, the user of refractory materials finds need for data on a number of the familiar general physical properties of solid materials.

The property of *thermal conductivity* was discussed in detail in Chapter 7 and should require no further explanation at this point. Likewise, its importance as a property of refractories should be obvious; in some applications low conductivity is advantageous while in others high conductivity is desirable. ASTM Test C–201 can be used to measure thermal conductivity of refractory materials. Typical values of this property for common materials are given in the Appendix.

Cold *crushing strength* and *modulus of rupture*, as determined by ASTM Test C–133, give some indication of how brick will withstand handling and shipping, and mechanical loading at low temperatures. Also, these mechanical properties serve as some measure of the extent to which the ceramic bond has been developed in firing and sometimes may be correlated with certain structure-sensitive service characteristics of the brick.

The *bulk density* of a refractory body is its weight divided by total volume including pores, and is the figure used in various engineering calculations of weight and volume of refractory structures. *True specific gravity* or *true density* refers to the solid refractory, exclusive of pores. *Apparent specific gravity* is measured by immersion in a fluid which penetrates the open pores of the body, and thus refers to the solid refractory including closed pores. *Porosity* and *apparent porosity* (open pores) are readily calculated from these specific gravities. Standard procedures of measurement and calculation are given in ASTM Tests C–134 and C–135. These properties tell much about the structure of a refractory body, and are usefully correlated with other properties such as thermal conductivity, resistance to slag penetration and attack, mechanical strength, phase constitution, etc.

Heat capacity and *specific heat* data for refractory materials and for many of their pure oxide components are given in the Appendix.

Electrical conductivity is primarily of interest in connection with materials used in electric furnaces. It should be noted that two fairly common refractory materials, graphite and silicon carbide, are relatively good conductors and are very useful as resistance heating elements.

Costs. In the final analysis, costs should be the determining factors in the choice of refractories for metallurgical processes. However, this criterion is more easily stated than practiced because of the difficulties in establishing and predicting some of the important costs. Low first cost for refractory materials is necessarily one of the prime considerations in building large furnaces, and this accounts for the fact that clay refractories and high-silica refractories represent well over 90% of the total tonnage of refractories used by the metallurgical industry. On the other hand, in metallurgical furnaces there are many places where clay or silica refractories are entirely unsatisfactory and many other places where specific properties of more costly materials justify their use. Thus, in any one furnace a wide variety of refractories may be used, each being chosen to meet the requirements of a particular zone or section of the furnace structure; the bulk of the furnace structure may still be made up of cheaper standard materials such as firebrick or silica brick.

A detailed analysis of refractory costs is outside the scope of this text, but to illustrate the complexity of the problem in some cases, one specific service might be considered briefly; namely, the choice of materials for constructing the roofs or arches of large, high-temperature, fuel-fired, reverberatory furnaces, such as basic open-hearth and copper-smelting furnaces. Silica brick was universally used for roof construction in these furnaces at one time and still is predominant in this service because its relatively high load-bearing capacity at high temperatures makes possible a wide span of sprung arch, supported only at the sides. In recent years, however, increasing numbers of magnesite roofs have been installed. Magnesite bricks are more expensive to start with. Also, they have poor load-bearing capacity at high temperatures, insufficient for the construction of wide arch spans under compression, so that an elaborate steel structure must be provided to suspend the roof, brick by brick. Still another less desirable property is their high thermal conductivity, resulting in higher heat losses. In spite of the added first cost and other disadvantages, substitution of suspended magnesite roofs for sprung silica roofs has proved to be an over-all economy in many instances. The items of cost savings, which overbalance the items of increased cost, result from the greatly increased roof life and increased furnace campaign without shutdown for repairs, decreased repair and maintenance, and increased furnace efficiency and capacity possible with the higher flame temperatures and harder furnace driving tolerated by the magnesite roof. The establishment of this kind of a cost comparison for a given operation obviously requires a careful full-scale trial over a long period of time.

Metallurgical refractories

Classification. One basis long in use of classifying refractory materials is closely related to the classification of slags in terms of basicity and acidity, which was discussed in Chapter 9. Following this usage, materials high in SiO_2 are considered acid, those high in CaO or MgO are basic, and such oxides as Al_2O_3, Cr_2O_3 and non-oxides like C and SiC are considered to form neutral refractories. As a rough working generalization, acid refractories are used in contact with acid slags and basic refractories in contact with basic slags, while neutral refractories are thought of as resistant to both acid and basic slags. This point of view extends further to the classification of processes, such as the acid Bessemer and the basic Bessemer processes, the adjectives acid and basic referring both to the slags and to the refractory materials used in the respective processes. However, as better knowledge of the chemistry of slags and refractories has accumulated, it has shown that these generalizations, though retaining much qualitative validity, do represent a considerable oversimplification.

A convenient classification of refractory materials, based largely on chemical composition, is as follows:

    I. Silica-alumina refractories
        (a) High silica
        (b) Fireclay
        (c) High alumina
   II. Basic refractories
        (a) Magnesite and dolomite
        (b) Chrome, chrome-magnesite, and magnesite-chrome
        (c) Forsterite
  III. Insulating refractories
   IV. Special refractories
        (a) Carbon and graphite
        (b) Silicon carbide
        (c) Zircon and zirconia
        (d) Laboratory ware
        (e) Miscellaneous

All the types listed have important metallurgical applications, but, very roughly, the above tabulation is in the order of increasing specialization of application, decreasing tonnage of materials used, and increasing unit cost.

Behavior of silica. Silica has three crystalline forms, *quartz, tridymite,* and *cristobalite,* and also is readily made into a non-crystalline glass, *vitreous silica* or *silica glass.* Each of the three crystalline varieties in turn exists in high- and low-temperature modifications. In all these forms, as in all the silicate minerals, the structural unit is the SiO_4 tetrahedron with a Si atom in the center and four O atoms at the corners. Also, the composition SiO_2 requires that each O atom be shared by two SiO_4 tetrahedra, so that each tetrahedron is linked to four others around it. Thus the differences between the various structures lie in the arrangements according to which the tetrahedra are linked together and in their relative geometrical locations in space in the solid. The three primary crystalline forms and the glass differ from each other in the scheme of linkage, so that transformation from one to the other requires the breaking of some linkages and the formation of new linkages. On the other hand, the changes from high- to low-temperature modifications, called *inversions,* represent a less radical structural change, a geometrical rearrangement of the tetrahedra without breaking any linkages. This is illustrated in Fig. 10–3, which shows schematically the difference between low quartz (α) and high quartz (β).

A further complication in the behavior of silica, besides the multiplicity of crystalline modifications, is the fact that many of the changes from one variety to another tend to be slow and sluggish. Accordingly, at a

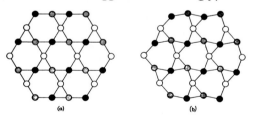

FIG. 10–3. The relation between (a) β-quartz and (b) α-quartz. Silicon atoms only are shown (after Bragg, *Atomic Structure of Minerals,* Cornell University Press, 1937).

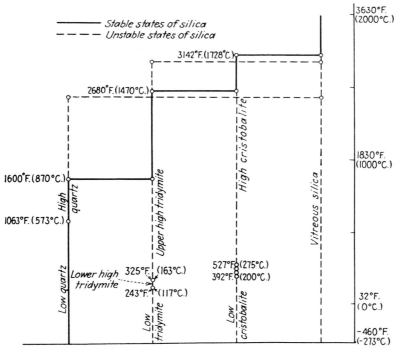

FIG. 10–4. Relations between the modifications of pure silica (after Sosman).

given temperature, though only one form is stable and at equilibrium thermodynamically, several of the other forms may exist metastably without, for practical purposes, any appreciable change to the more stable form. Thus quartz, tridymite, cristobalite, and silica glass can all exist indefinitely at room temperature even though quartz is the only truly stable structure at room temperature. Figure 10–4 summarizes the stability relationships between the various forms of silica. The solid line indicates the stable form at each temperature, and shows that starting with α-quartz at room temperature or below and increasing the temperature, the successive stable phases are α-quartz, β-quartz, tridymite, cristobalite, and finally liquid silica. In ordinary heating and cooling of silica, however, the silica behavior does not follow this relationship, except that the α-β inversion at 575°C generally occurs quickly and reversibly as long as either form of quartz is present. When quartz is heated, then, the inversion occurs at 575°C but the silica tends to remain as β-quartz above 870°C. If the temperature is raised to 1250°C or above, the quartz tends to transform directly into cristobalite, although under certain favorable conditions it may change directly into tridymite. At higher temperatures the change from quartz to cristobalite becomes fairly rapid. Also, at temperatures up to 1470°C, the cristobalite tends to change slowly to tridymite. In the range from 1470°C up to the melting point 1728°C any prolonged heating yields cristobalite, regardless of the starting phase. Once liquid silica, cristobalite, or tridymite is formed, each structure tends to be retained on cooling, so that when a furnace structure containing an SiO_2 phase is cooled from operating temperature, the structure at operating temperature is largely retained. However, each of the crystalline varieties undergoes the inversions at the temperatures indicated in Fig. 10–4 on heating and cooling, these inversions occurring rapidly enough to be observed during all ordinary heating and cooling cycles. Vitreous silica is a colorless, transparent glass with many desirable properties for laboratory and research uses, and is reasonably stable at temperatures up to about 1200°C. At higher temperatures, however, it devitrifies and loses its mechanical strength and other desirable properties too rapidly to be useful for more than a short period such as a single laboratory test or a single experimental run.

For refractories made up largely of solid SiO_2, like silica brick, and for other refractories in the SiO_2-Al_2O_3 series containing substantial proportions of solid SiO_2, these structural transformations are significant because they are accompanied by changes in properties. Of particular importance are the changes in density, as can be seen by referring to the data in Table 10–2. Large expansions occur when quartz transforms to either tridymite (17.2% increase in volume) or cristobalite (14.3% increase in volume). In high-silica refractories this large expansion must be provided for in manufacture, because the silica in the raw materials is in the

form of quartz. Obviously it is necessary that this expansion be completed insofar as possible under carefully controlled conditions during the manufacture of the brick. If the brick installed in a furnace were to contain large proportions of quartz, the dimensional changes on heating would be too large to allow for by the usual methods, and most seriously, the temperature gradients in the furnace structure and across single bricks would lead to unequal rates of transformation and unequal expansions, resulting in large stresses, spalling, and failure.

The density changes accompanying the inversions, though not as large as those when quartz transforms to tridymite or cristobalite, are of serious concern because they occur rapidly as the silica passes through the inversion temperature. Thus refractory bodies high in silica are particularly prone to spalling if heated and cooled repeatedly through 527°F (275°C), which is the inversion temperature of cristobalite.

Silica brick and other high-silica refractories. The high-silica refractories are made primarily from quartzite of relatively high purity, usually about 98% SiO_2. In some applications, natural sandstone or ganister may be cut and the blocks used in furnace construction, or silica sand may be used directly in making furnace bottoms. For most purposes, however, the manufactured silica brick is the standard form for furnace construction. Silica brick is essentially an aggregate of crystalline silica, with just enough other oxides present to form a ceramic bond. Starting with a relatively pure quartzite, about 2% of lime will be added to form the silicate bond. At high temperatures the bonding phase will become liquid and may increase in volume through dissolution of silica. Accordingly, to secure best mechanical properties of the brick at high temperatures and to obtain a brick us-

TABLE 10-2

Densities of Crystalline Silica and Density Changes Accompanying Inversions

	Room-temperature density, g/cc	Density change on inversion*
Quartz	2.65	0.02
Tridymite	2.26	0.01
Cristobalite	2.32	0.04

*Density decreases in passing from low- to high-temperature modification.

able to near the melting point of silica, care must be taken to control or minimize the percentages of the oxide impurities in the raw material which tend to flux the silica and form excessive liquid under working conditions. Thus, the difference between poor and good quality silica brick is often due to a fraction of a percent difference in impurity content.

In manufacturing silica brick, the quartzite is crushed, and may then be washed or cleaned to eliminate impurities. The crushed rock is ground with addition of water and lime to make a mud which is then pressed into bricks. The bricks are dried, fired, and cooled, according to a carefully regulated time-temperature schedule. The firing, which takes many days and reaches temperatures over 2600°F, transforms the quartz into tridymite and cristobalite and develops the bond. The brick expands about 10% by volume during firing, and is fired to the point where little or no further permanent volume change will occur during use. The finished brick will be roughly 45 to 60% tridymite, 30 to 40% cristobalite and up to 5% residual quartz. In service these proportions will change further depending on operating temperature, but the further transformations are accomplished without appreciable volume changes.

The properties of silica brick which are most important in its applications will be largely apparent from the preceding discussions of general refractory properties and of the behavior of pure SiO_2. In the first place, silica is the principal acid refractory for containing acid slags, and conversely it does not resist satisfactorily the corrosive action of basic slags. Owing to its cheapness, good mechanical strength, and ability to withstand temperatures up to 3000°F or so under load, silica brick is a common material of general furnace construction, and may be so used even in basic processes for furnace parts not coming in contact with basic slags. The good load-bearing capacity to near the melting point of silica is a unique characteristic among the common refractories, which in general start to soften and flow under load at temperatures far below those at which they actually melt to form a liquid. Silica brick is unique also in that its volume is nearly constant as the temperature varies above about 1000°F (see Fig. 10–2). The combination of light weight, high-temperature load-bearing capacity, and constancy of volume make silica brick the outstanding material for constructing furnace roofs and sprung arches of wide spans. In all the uses of silica brick, special care must be taken in heating and cooling in the temperature range below 1200°F (650°C), because the brick are very sensitive to spalling in this range. This spalling results from the large volume change accompanying the inversion of cristobalite. Thus, this material is not well suited for constructing intermittent furnaces and other applications involving frequent and rapid heating and cooling through a low temperature range. On the other hand, at temperatures above a red heat (above 1200°F), silica brick are virtually immune from spalling because of the very low coefficient of thermal expansion at high temperatures. This high-temperature spalling resistance is valuable in furnaces such as steelmaking furnaces, which undergo rapid temperature changes during the operating cycle. Back from the hot face, of course, the brick temperatures fall into the range of the silica inversions, but the temperature changes occur more slowly without large temperature gradients.

Table 10–3 gives typical compositions and summarizes some of the common properties of silica brick. The bulk density is the lowest of the common refractories (excluding insulating materials). Comparison with

TABLE 10-3

Typical Properties and Composition of Silica Brick

Bulk density:	100-115 lb/ft^3	
True specific gravity:	2.30-2.38	
Apparent porosity:	20-30%	
Cold crushing strength:	1000-3000 psi	
Thermal conductivity at 2000°F:	0.9-1.2 Btu-ft/ft^2-hr-°F	
	Conventional	Super duty
Temperature of failure in load test, 25 psi:	3000°F	3070°F
Chemical analyses (weight %):		
SiO_2	95.63	96.33
Al_2O_3	0.75	0.28
CaO	2.60	2.74
Fe_2O_3	0.75	0.56
Na_2O	0.04	0.04
K_2O	0.15	0.04
TiO_2	0.08	0.03

Table 10–2 shows that the true specific gravity is close to that of cristobalite. The principal and important difference between the chemical analyses of conventional and superduty brick is the percentage of Al_2O_3. The resulting improvement of about 70°F in refractoriness may appear small, but is very significant in certain applications (for example, roofs of open-hearth furnaces) where the refractory determines the top working temperature of the furnace.

The SiO_2-Al_2O_3 system. The binary SiO_2-Al_2O_3 system is the basis of a whole series of refractories, ranging in composition from the high-silica materials already discussed to high-alumina and pure-alumina materials, and including the clay refractories which are the most important group of all in quantity and variety of use. The constitution diagram for this system in Fig. 10–5 shows the phases present in equilibrium at various temperatures. Although in many cases the refractory body does not attain equilibrium and also contains impurities other than SiO_2 and Al_2O_3, this diagram nevertheless represents a good starting point for consideration of the SiO_2-Al_2O_3 refractories.

At the SiO_2 end of the diagram it can be seen that small additions of Al_2O_3 markedly lower the temperature at which the refractory becomes entirely liquid, only 5.5% Al_2O_3 being required to reach a eutectic temperature of 1595°C (2903°F) as compared to the melting point of pure SiO_2, 1724°C (3135°F). Thus, at the high-SiO_2 end of the system, compositions in the vicinity of the eutectic composition are not particularly desirable for refractories. In silica brick, the effect of small percentages of Al_2O_3 in increasing the proportion of liquid at operating temperatures is even more serious than the binary diagram indicates, because with the CaO added to form the bond a ternary CaO-Al_2O_3-SiO_2 eutectic can be formed which is liquid at only 1165°C (2129°F). Accordingly, the quartzites used in making silica brick generally contain under 1% Al_2O_3. For highest quality silica brick, a limit of 0.5% or so may be placed on the total Al_2O_3, TiO_2, and alkali oxides which contribute to the formation of a liquid phase at relatively low temperatures (see Table 10–3).

Below the eutectic temperature (1595°C), Al_2O_3-SiO_2 mixtures up to 71.8% Al_2O_3 form at equilibrium two solid phases, silica (cristobalite or tridymite, depending on temperature) and the crystalline compound mullite ($3Al_2O_3 \cdot 2SiO_2$). A pure clay with Al_2O_3 and SiO_2 present in the proportions 40:60, for example, would give about 55% mullite and 45% SiO_2. Such a mixture on heating above 1595°C becomes mullite plus a liquid phase, the proportions of liquid increasing with increasing temperature until the mixture becomes completely liquid a little above 1800°C. Other compositions in the range 5.5 to 71.8% Al_2O_3 behave similarly, but at any given temperature above 1595°C the ratio of liquid to mullite decreases with increasing percentage Al_2O_3. Also, the temperature of complete melting increases with increasing Al_2O_3. The liquid formed at or not far above the

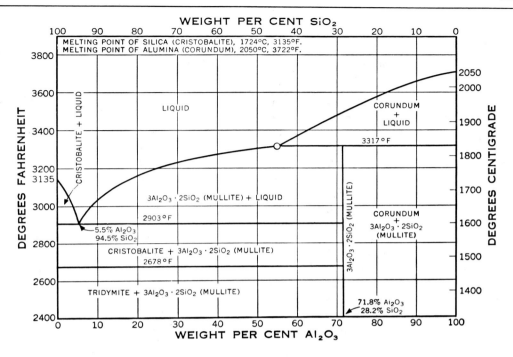

FIG. 10–5. SiO$_2$-Al$_2$O$_3$ system (Harbison-Walker, *Modern Refractory Practice*).

eutectic temperature is very viscous and may partake of the nature of a glass more than of a common liquid. On cooling, this liquid tends to remain as a glass. In fact, all the crystallization and phase changes indicated by the diagram are sluggish under most circumstances.

Mullite itself is quite refractory, being stable up to 1825°C (3317°F), at which temperature it decomposes to form corundum (Al$_2$O$_3$) and liquid. For compositions with 71.8% Al$_2$O$_3$ or more this represents the lowest temperature at which any liquid is formed, under equilibrium conditions. Corundum, or pure alumina, is the most refractory solid in the SiO$_2$-Al$_2$O$_3$ system, melting at 2050°C (3722°F). Thus, in a broad way, the refractoriness of SiO$_2$-Al$_2$O$_3$ refractories increases with alumina content over the entire composition range from the eutectic with only 5.5% Al$_2$O$_3$ up to pure Al$_2$O$_3$.

Fireclay and firebrick. Fireclay refractories are the principal materials of construction of metallurgical furnaces. Because they are the cheapest and most widely available of all refractory materials, they are universally used in all types of furnaces except for places where very high temperatures, severe chemical attack by slags and other agents, or other service conditions make necessary the use of other materials in spite of their higher costs. Many types of metallurgical furnaces are constructed wholly of fireclay refractories; even for a furnace such as a basic open hearth, where the conditions within the furnace chamber proper are too severe for fireclay materials, the tonnage used in constructing the furnace and its auxiliary parts exceeds the total tonnage of all the other refractories used.

Clay is not a single mineral or specific mixture of minerals, but is a material formed in the earth's crust by prolonged weathering and other geochemical processes acting on rocks high in feldspar. These processes break up the original rock structure and yield a very fine grained mixture or clay composed primarily of hydrous aluminum silicate minerals, of which kaolinite ($Al_2O_3 \cdot 2SiO_3 \cdot 2H_2O$) is typical. The clay mixture may include also hydrated aluminum minerals such as gibbsite [$Al(OH)_3$] and diaspore ($HAlO_2$), quartz (SiO_2), and various other minerals derived from the original rock and containing additional elements like the alkalis, sodium and potassium, magnesium, iron, titanium, etc. In terms of the chemical analysis, silica, alumina, and combined water are generally the major constituents. The fine-grained structure is an important characteristic, most clays being made up of grains in the range 10^{-4} to 10^{-1} mm in size and containing large proportions of particles below 10^{-2} mm in size. It is the combination of fine particle size and specific properties of the plate-like clay minerals which accounts for the workability and plastic properties common to clays and so vital in their various uses. Clays differ very considerably among themselves in structure, workability, plasticity, particle-size distribution, and mineralogical composition, and these differences lead to classification of clays into flint clays, plastic clays, kaolins, ball clays, and the like.

Fireclays, as might be expected, are simply clays that have some degree of refractoriness when fired. Using the standard pyrometric cone test of softening temperature mentioned earlier as a criterion, fireclays generally have PCE's of 19 (1515°C) or above. It has been proposed that clays with PCE's from 19 to 26 (1515–1595°C) be called "low heat duty" fireclays and those with PCE's of 27 (1605°C) or above be called "refractory" fireclays.

The refractoriness depends very much on the content of impurities, the purest clays approaching kaolinite ($Al_2O_3 \cdot 2SiO_2 \cdot 2H_2O$) in composition having PCE's up to 34 and 35 (1755–1775°C). Kaolinite probably is the principal mineral in fireclays and analyzes 39.5% Al_2O_3, 46.5% SiO_2, and 14.0% H_2O. Elimination of all the water by firing gives a composition of 45.9% Al_2O_3 and 54.1% SiO_2. Free silica, present in many clays in substantial amounts, changes these proportions materially, and clays analyzing more than 70% SiO_2 (calcined basis) are called silicious fireclays. On the other hand, clays from some sources contain substantial proportions of diaspore or gibbsite and are classed as aluminous fireclays. In terms of composition of the finished refractory brick, the dividing line between fireclay refractories and high-alumina refractories is generally placed at about 50% Al_2O_3. Table 10–4 gives chemical analyses of a representative group of fireclays. Another basis of classifying fireclays, apart from the chemical composition, is illustrated in the tabulation and involves the terms "flint clay" and "plastic clay." A flint clay as mined is a hard and rocklike material, quite resistant to weathering, and requires crushing and grinding to develop plasticity. A plastic clay, on the other hand, is soft and easily worked into a plastic mass.

The manufacture of bricks and other shapes from fireclays involves a number of steps, principally (1) blending and mixing, (2) tempering with water, (3) forming or molding, (4) drying, and (5) firing. Although superficially this may appear to be a simple procedure, actually it represents a complex art with many factors and variables at each stage which influence the nature and properties of the final product. Thus firebrick is not a single product of standardized properties but rather is made with many different combinations of properties to meet the

TABLE 10-4

Chemical Analyses of Typical Fireclays*

	New Jersey siliceous	Penna. flint	Penna. plastic	Penna. semi-plastic	Ohio flint	Ohio plastic	Missouri plastic	Missouri flint	Diaspore	Bauxite	Georgia kaolin	Maryland flint
SiO_2	77.7%	43.1%	44.6%	46.2%	45.1%	54.3%	56.8%	42.7%	20.5%	18.9%	45.9%	56.1%
Al_2O_3	15.7	39.4	37.3	36.3	34.1	27.0	26.6	39.7	60.9	52.5	37.0	33.3
Fe_2O_3	0.5	1.3	1.2	1.6	2.3	2.4	1.6	1.1	0.6	4.4	1.1	0.6
TiO_2	----	2.0	2.2	2.2	2.1	2.0	1.7	2.2	3.1	2.7	1.3	----
CaO	trace	0.1	0.3	0.2	0.25	0.2	0.4	0.1	0.4	3.2	0.3	0.2
MgO	0.8	0.1	0.4	0.3	0.5	0.1	0.2	0.3	0.3	0.4	0.2	0.1
Alk.	trace	0.2	1.6	1.7	1.4	1.6	0.1	0.2	1.3	0.4	0.8	----
Ign. loss	5.6	13.6	12.7	11.5	12.8	9.3	12.1	73.8	13.1	17.9	12.7	9.7
PCE**	27-28	33-34	32-33	32	32-33	29	30	34-35	40	34	33-34	33

*After J. R. Coxey, Refractories. State College, Penna.: Pennsylvania State College, 1950.
**Pyrometric Cone Equivalent.

specific requirements of different services. Obviously, the compositions and properties of the fireclays used as raw materials will be major factors in determining the properties of the product, but in addition much can be done in the manufacturing to develop and augment specific properties. For example, several different forming methods are used, including hand molding, extrusion of a thick mud, power pressing, and slip casting. For some purposes, especially to obtain a denser and harder brick, a vacuum may be applied during forming to remove air. Another variable is the quantity of grog, or prefired fireclay, used in the mix. Still another factor is the control of the particle-size distribution of the raw materials going into the mix. In particular, it may be important to blend various sizes of grog and other nonplastic raw materials in optimum proportions to give high packing density.

In the firing process, the kaolinite and other minerals in the mix are decomposed chemically, and the crystalline phases indicated in the equilibrium diagram (Fig. 10-5) tend to develop, particularly mullite and silica (cristobalite or tridymite). Also, the firing develops a glassy phase which bonds the mass together and accounts in large measure for the hardness and mechanical strength of the fired brick. The proportions of glass present in the final brick vary a great deal, depending on the nature of the mix and on the firing conditions (temperature and time). In general the brick are not held at the firing temperature long enough to attain the equilibrium constitution shown by the SiO_2-Al_2O_3 equilibrium diagram. Instead, firing temperatures and times are chosen to give the brick a desired set of properties such as density, cold mechanical strength, load-bearing capacity at high temperatures, dimensional changes on reheating, spalling resistance, etc. The relationships between these properties and the structure produced by firing are rather complex and are best left to the ceramic specialist.

Fireclay brick are classified primarily according to refractoriness, as *super duty, high heat duty, intermediate heat duty,* and *low heat duty.* The ASTM gives standard specifications for these four classes, as follows (ASTM, C-27).

Super duty:

PCE not lower than No. 33 on fired product. Less than 1% shrinkage in reheat test to 1600°C (2910°F). Less than 4% weight loss in spalling test, brick preheated to 1650°C (3000°F)

High heat duty:

PCE not lower than Nos. 31–32 *or* less than 1.5% deformation in load test at 1350°C (2460°F)

Intermediate heat duty:

PCE not lower than No. 29 *or* less than 3% deformation in load test at 1350°C (2460°F)

Low heat duty:

PCE not lower than No. 19

Within each of these four classes, the properties of the brick still vary over a wide range, so that for many services additional specification is necessary to insure obtaining the desired combination of properties. A single refractory manufacturer may supply several brands of brick in a given heat duty classification, each designed for certain kinds of service. Special brands of firebrick are made for the various parts of the iron blast furnace, where several kinds of severe service conditions have to be met. Also, other brands are made for service in rotary kilns and shaft furnaces where abrasive resistance is particularly important. Table 10–5 gives typical compositions and properties for the various classes of firebrick along with similar data for the high-alumina materials discussed in the following section.

High-alumina refractories. As indicated by the SiO_2-Al_2O_3 equilibrium diagram, the

TABLE 10-5

Typical Properties and Compositions of Fireclay and High-Alumina Brick

A. Physical Properties		
	Fireclay	High-alumina
Bulk density, lb/ft^3	120-150	130-175
Apparent porosity, %	10-30	20-30
Cold crushing strength, psi	1000-6000	2000-10,000
Thermal conductivity at 2000°F Btu-ft/ft^2-hr-°F	0.8	0.8-1.4

B. Composition and True Specific Gravity						
	Chemical analyses, percent				True specific gravity	Pyrometric Cone Equivalent
	SiO_2	Al_2O_3	TiO_2	Other oxides		
Fireclay						
Super duty	49-53	40-44	2.0-2.5	3-4	2.65-2.75	33-34
High duty, aluminous	51-60	35-40	1.7-3.3	3-6	2.60-2.70	31½-33
High duty, silicious	65-80	18-28	1.0-2.0	2-6	2.40-2.45	28-31
Intermediate duty	57-70	25-36	1.3-2.1	4-7	2.55-2.65	29-31
Low duty	60-70	22-33	1.0-2.0	5-8	2.55-2.65	19-26
High Alumina						
50% class	41-47	47.5-52.5	2.0-2.8	3-4	2.75-2.85	34-35
60% class	31-37	57.5-62.5	2.0-3.3	3-4	2.90-3.05	35-37
70% class	20-26	67.5-72.5	3.0-4.0	3-4	3.15-3.25	36-38
80% class	11-15	77.5-82.5	3.0-4.0	3-4	3.35-3.45	38-39
90% class	8-9	89-91	0.4-0.8	1-2	3.55-3.65	39-40
99% class	0.5-1.0	98-99	trace	0.6	3.70-3.90	41½

refractoriness of silica-alumina materials increases as the alumina contents are raised above those of the standard fireclays. Mullite, containing 71.8% Al_2O_3 and 28.2% SiO_2 and with an incongruent melting point of 1825°C (3317°F), is a principal crystalline constituent of most of the high-alumina refractories. Corundum, or pure Al_2O_3, with a melting point of 2050°C (3720°F), is the other important crystalline constituent and, of course, predominates in the materials with the highest alumina contents. However, as with fireclay refractories, equilibrium is not attained in firing, so that the phases actually present in high-alumina refractories depend on the raw materials used and on the manufacturing process and thus may include variable amounts of silica (cristobalite or tridymite), glass, and other phases similar to those in fired fireclays. The high-alumina refractories are generally classed according to content of Al_2O_3, and are available as 50%, 60%, 70%, 80%, 90%, and 99% Al_2O_3. Table 10–5 shows how the refractoriness of silica-alumina materials increases with alumina content for the various brands of high-alumina brick furnished by the Harbison-Walker Refractories Company. With increasing alumina, other properties are also improved so that high-alumina materials have replaced fireclay and other materials at many points of severe service in metallurgical plants, in spite of their greater cost. One of the properties of importance to metallurgists is the good resistance to slag attack, especially by basic slags and by slags high in lead oxide, alkali-metal oxides, or other compounds very destructive to fireclay refractories.

Another property of high-alumina brick which leads to its choice over fireclay brick for some applications is good load-bearing capacity at high temperatures.

High-alumina refractories of similar overall compositions may differ in structure and properties because they are made from different raw materials. Diaspore clays are blended with good quality fireclays to make some of the high-alumina refractories. In others, especially of 60% Al_2O_3 and higher, fused alumina (corundum) of high purity is used in substantial proportions. The 99%-Al_2O_3 bricks consist almost entirely of fused alumina, and, owing to the low content of fluxing oxides, approach pure crystalline corundum in melting point and some other properties. Still another type of raw material is accounted for by the group of minerals, sillimanite, kyanite, and andalusite. These three minerals all correspond to the formula $Al_2O_3 \cdot SiO_2$, which is 62.9% Al_2O_3 and 37.1% SiO_2. When properly heat treated these minerals are converted largely into mullite. Theoretically, complete conversion leads to a mixture of 87.7% mullite and 12.3% silica.

Basic refractories. The chief needs for basic refractories are in processes where resistance to basic slags is essential, such as the basic open-hearth process. In addition, however, some major applications are accounted for by the ability of common basic refractories to withstand higher temperatures than other common materials such as fireclay brick and silica brick.

Magnesia (MgO) is a principal constituent of most basic refractory materials, but in many of these it is combined or mixed with substantial proportions of other oxides, especially chromic oxide, iron oxide, lime, alumina, or silica. The high content of MgO accounts to a considerable extent for the high refractoriness of basic refractories. Pure crystalline MgO, known as periclase, has one of the highest melting points of all the refractory oxides, about 2800°C (5070°F). However, owing to the fluxing action of impurities and of substances added to form a bond, the commercial refractories high in MgO fail at much lower temperatures, especially under load.

Since high-magnesia refractories have in the past been manufactured starting with magnesite (magnesium carbonate) as the chief raw material, the custom has been established of referring to them as magnesite refractories, for example, "magnesite brick," even though no magnesite can be present after the first firing or heating. The mineral magnesite occurs in relatively pure deposits, and the first step in treatment, generally carried out at the mine, is the burning of the magnesite to eliminate CO_2 and give MgO. For refractory purposes, the magnesite is dead-burned at relatively high temperatures (above 1500°C), so that the product becomes sintered and hard with partial fusion of impurities and development of periclase crystals. Iron oxide often present as an impurity aids in the dead-burning process, forming a bond between periclase crystals, and if not present in sufficient amount for this purpose may be added in the form of iron ore or mill scale before burning. For chemical and other nonrefractory purposes, magnesite is "caustic-calcined" at a lower temperature which gives a powdery product of much greater chemical reactivity. In recent years, increasing quantities of magnesia have been produced from seawater and brines, and though magnesium carbonate is not involved, the hard-burned magnesium oxide and the products made from it are still called magnesite refractories. Chemical analyses of several dead-burned magnesites are given in Table 10–6. It can be seen that silica is a common impurity, along with iron, lime, and small amounts of alumina.

One of the useful characteristics of magnesium oxide is that it combines with iron oxides to form relatively refractory solid phases. In the crystal structures of many common minerals, magnesium ions and ferrous ions are often interchangeable, both having the same charge and about the same size. Accordingly, magnesia and ferrous oxide (wüstite) are thought to form a complete series of solid solutions ranging in melting point from that of magnesia (2800°C) down to that of wüstite (1371°C). Likewise, MgO can replace FeO in magnetite to form the compound magnesioferrite ($MgO \cdot Fe_2O_3$), which has a melting point of about 1750°C compared to 1597°C for magnetite (Fe_3O_4). At high temperatures magnesioferrite and magnetite form solid solutions with intermediate melting points. These phenomena explain the good resistance of high magnesite refractories to attack by iron oxides, and show how a magnesite

TABLE 10-6

Chemical Analyses of Dead-Burned Magnesite*

Source	MgO	SiO_2	Al_2O_3	Fe_2O_3	CaO	Ignition loss
Chewelah, Washington						
regular	81.6	6.9	1.5	4.6	3.3	0.1
flotation	88.3	4.9	1.5	2.2	3.0	0.1
Sea water (California)	84.5	5.1	0.3	7.3	2.3	0.1
	90.3	5.1	0.7	1.3	1.8	0.7
Austria	83.0	5.8	1.7	4.0	5.0	0.2
Czechoslovakia	88.9	0.6	0.5	6.5	2.4	0.9
Russia	92.5	1.5	0.5	1.6	2.2	0.6
Manchukuo	92.0	3.7	1.0	1.5	1.6	0.1

*After D. M. Liddell, Handbook of Nonferrous Metallurgy. New York: McGraw-Hill, 1945.

refractory can absorb more than its own weight of iron oxides without fluxing and failure at smelting temperatures. Also, the formation of iron oxide, magnesia solid solutions of good refractoriness is utilized in the sintering and bonding of magnesite refractories, iron ore or mill scale often being added for this purpose if the magnesite does not already contain enough iron oxide as an impurity.

Magnesite refractories are attacked by high-silica or acid slags, and also should not be used in direct contact with acid refractories such as silica brick and fireclay. The phase diagram for the MgO-SiO_2 system in Fig. 10–6 shows that magnesia and silica, even in the absence of other oxides, form liquid melts at temperatures as low as 1543°C. However, a substantial percentage of silica may be present in magnesite refractories without seriously impairing their refractoriness because of the formation of the compound forsterite, $2MgO \cdot SiO_2$, which itself has a melting point of 1890°C. Through the reaction of silica to form this compound, silica present as an impurity in magnesite raw materials helps to form a ceramic bond in much the same way as the iron oxide forms a magnesioferrite bond. As a matter of fact, basic refractories are now successfully used in which forsterite itself is the primary refractory phase.

In addition to its unique and useful behavior toward iron oxides and its ability to form the highly refractory forsterite with limited quantities of silica, magnesia combines with still other common oxides to form refractory phases of relatively high melting points. As shown in Fig. 10–7, magnesia and alumina form the stable compound $MgO \cdot Al_2O_3$, which is the mineral spinel, and the lowest temperature at which a liquid is formed is 1925°C, corresponding to the spinel-alumina eutectic. Bricks, crucibles, and other prod-

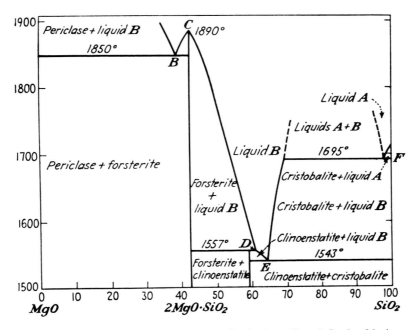

FIG. 10–6. MgO-SiO_2 system (AIME, *Basic Open-Hearth Steelmaking*).

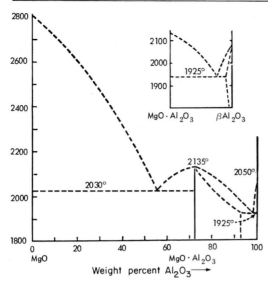

FIG. 10–7. MgO-Al₂O₃ system (after Hall and Insley, *J. Am. Ceram. Soc.*).

ucts have been made with the composition of spinel and are said to possess valuable properties as refractories. Magnesia also forms a refractory, spinel-type compound with chromium oxide, $MgO \cdot Cr_2O_3$. The binary system of lime-magnesia is thought to be a simple eutectic-type system, with a eutectic temperature of about 2200°C (4200°F) and no compounds or solid solutions. However, lime present as an impurity in magnesite bodies under some conditions forms silicates and ferrites which affect the properties of the body.

Dead-burned magnesite is the major starting material for several refractory products with important metallurgical applications. In the first place, large quantities of the dead-burned grain are used directly in furnaces, without fabrication into bricks. A common method of making a monolithic furnace bottom consists of spreading dead-burned magnesite over the hearth in a thin layer and then sintering or burning-in this layer to a dense monolithic mass by heating to a high temperature (for example, 2900°F). Usually some slag, iron ore, or other material is mixed with the magnesite to aid the sintering process and to form a bond. A working bottom of the desired thickness is built up by repeatedly sintering in one thin layer at a time.

Several kinds of magnesite bricks are used in furnace construction. Some are bonded by firing, as are silica brick and fireclay brick, starting with mixtures of dead-burned magnesite with sufficient fluxing oxides present in the mixture or deliberately added to form a ceramic bond of the desired properties. Burned magnesite brick are dense (about 170 lb/ft³) and very resistant to basic slags and high temperatures, but have the disadvantage of poor spalling resistance, a high coefficient of thermal expansion (see Fig. 10–2), poor load-carrying ability at high temperatures, and a tendency to shrink irreversibly in use. Some of this shrinkage may be due to continued formation of crystalline periclase from the "amorphous" magnesium oxide first formed by decomposition of carbonate (or hydroxide), and some of the shrinkage is associated with the continued fusion and development of the liquid or bonding phase. Accordingly, the shrinkage of magnesite brick in service is minimized by increasing the percentage of magnesia and firing at higher temperatures. On the other hand, some of the disadvantages of fired magnesite brick appear to have been overcome by the use of chemically bonded, unburned brick and by the development of combinations of magnesite with chrome refractories. In forming the unburned brick, the chemical bonding agent is incorporated in the mix and develops a bond during the drying. Further changes occur in the structure after the brick is placed in service. A further modification with definite advantages in some applications is to encase the unburned brick in soft sheet steel on three or more sides. In service, the iron exposed to the interior of the furnace oxidizes and forms a strong magnesioferrite bond between adjacent bricks. Thus, a wall constructed of metal-encased brick becomes substantially monolithic and shows better spalling resistance than a wall of ordinary magnesite brick. The unoxidized steel in the cooler parts of the wall adds to the mechanical strength of the wall.

Lime (CaO), like magnesia, has a very high melting point (2570°C, 4660°F) and is similar to magnesia in general chemical behavior. However, there are no satisfactory furnace refractories in which lime is the principal constituent because lime slakes so readily with water and even slakes rapidly in the air, forming the hydroxide and the carbonate. Calcined dolomite, containing CaO and MgO roughly in molecularly equivalent amounts, also slakes readily in the air and thus tends to disintegrate on exposure at ordinary temperatures. In spite of this shortcoming, dolomite is used in large quantities as a basic refractory, especially in maintaining the bottoms and banks of basic open-hearth furnaces, where in normal operation it is not exposed to the slaking action of the atmosphere. For this application, the dolomite may be simply crushed and used raw without calcining, or it may be burned to eliminate CO_2 and to initiate the sintering and bonding reactions. The granular dolomite, raw or calcined, is simply thrown into place in the furnace without cooling, and at the furnace temperature quickly becomes sintered and bonded to the old furnace refractories.

Chrome ores have in recent years assumed considerable importance as raw materials for basic refractories, particularly in various combinations with magnesite. At one time, however, chrome refractories were used in furnace construction primarily because they were considered as "neutral." Actually, chrome brick resists both basic and acid slags quite well and is the only common refractory with this property. For this reason, it is used in furnace construction to separate basic and acid refractories, for example, magnesite and fireclay, which tend to react and flux each other if laid in direct contact. Except for the fact that chrome refractories do not withstand the destructive action of iron oxides, they are used in much the same way as magnesite, that is, as general-purpose basic refractories. In such use they have had the advantages, compared with magnesite, of slightly lower cost, better resistance to fluxing with silica, and lower thermal conductivities.

The major mineralogical constituent of chrome ores is chromite, a double oxide of the spinel group fitting the general formula $RO \cdot R'_2O_3$ (like spinel, $MgO \cdot Al_2O_3$; hercynite, $FeO \cdot Al_2O_3$; magnetite, $FeO \cdot Fe_2O_3$; and magnesioferrite, $MgO \cdot Fe_2O_3$). The basal formula for chromite is $FeO \cdot Cr_2O_3$, but in the chromite of commercial chrome ores a substantial part of the FeO is replaced by MgO and part of the Cr_2O_3 is replaced by Al_2O_3. Thus, the principal mineral species in chrome ores might be described as a solid solution of the spinel type containing the oxides FeO and MgO in varying proportions and the oxides Cr_2O_3 and Al_2O_3 in varying proportions, but with total FeO + MgO molecularly equivalent to total Al_2O_3 + Cr_2O_3.* Typical analyses of chrome ores used for refractories are given in Table 10–7. Ores with higher percentages of Cr_2O_3 (above 48%) and higher ratios of chromium to iron

---

* A portion of the iron (probably a small portion) may be present as Fe_2O_3, and therefore should be included with the Al_2O_3 and Cr_2O_3 in the spinel stoichiometry.

TABLE 10-7

Chemical Analyses of Refractory Chrome Ores*

Source	Cr_2O_3	SiO_2	Al_2O_3	FeO	MgO	CaO	Ignition loss
Cuba	31.8	6.3	27.2	13.0	18.5	0.6	2.2
Cuba	36.9	2.5	28.2	13.5	17.1	0.3	0.8
Greece	40.0	4.6	21.9	15.2	16.7	0.7	1.4
India	46.7	3.8	17.3	18.8	13.1	---	2.0
Philippines	33.9	4.4	29.5	12.8	17.9	0.2	1.3
Rhodesia	45.4	8.2	13.4	15.8	15.1	---	3.1
Russia	41.4	7.5	18.2	14.4	16.5	---	3.8

*After D. M. Liddell, Handbook of Nonferrous Metallurgy. New York: McGraw-Hill, 1945.

are used to make ferrochromium. For refractory purposes, the Al_2O_3 and Cr_2O_3 combined should be at least 60%, but the Cr_2O_3 content can be as low as 30 to 35%. Silicate minerals, especially magnesium silicates, are the main impurities in chrome ores, but if not present in too large percentages help to form a silicate bond on firing. Higher silica percentages are not desirable, however, because the magnesium silicates occurring as impurities tend to have magnesia:silica ratios corresponding approximately to the formula $MgO \cdot SiO_2$ and therefore are not highly refractory (see Fig. 10–6). That is, the spinel crystals in chrome refractories have very high melting points, so that the temperatures of failure of chrome refractories are determined mainly by the character and amounts of the silicate impurities which enter the ground mass or bonding portion of the fired chrome ore or chrome brick.

In the last fifteen years or so, constantly increasing use has been made of various combinations of chrome ores and deadburned magnesite, so that this group of refractories is now produced in larger tonnages than the straight chrome materials. In some ways, the mixing of chrome and magnesite combines the respective advantages of the two separate materials. Mixtures in which the chrome ore predominates are called "chrome-magnesite"; those in which

the magnesite predominates are called "magnesite-chrome." Some of the benefits of adding magnesite to chrome ore can be readily understood in the light of the preceding discussion. Thus, the added magnesia combines with the relatively nonrefractory magnesium silicates already present as impurities in chrome ores to form forsterite in the bonding mass surrounding the grains of chrome spinel, and this gives a more heat-resistant product with better load-bearing capacity at high temperatures than that made from the chrome ore alone. Magnesia added to chrome ores also tends to replace FeO in the spinel structure, probably with some increase in the stability of the spinel. Also, the added magnesia combines with Fe_2O_3 in a stable solid solution, as discussed previously, and this action, too, may be beneficial if the material is exposed to iron oxides or if the brick is exposed to oxidation-reduction processes which oxidize the FeO in the spinel.

Chrome ores and chrome-magnesite mixtures are utilized in a variety of ways in furnace construction. Large quantities are used in the raw or unfired state, in ramming and patching mixtures, which are bonded into monolithic structures in place in the furnace. Bricks may be fired to develop a desired structure during manufacture or they may be of unfired, chemically-bonded manufacture. The unfired brick are often steel-encased. These variations in methods of using basic refractories, and other variations in the individual steps of manufacture, such as in the method of pressing and forming brick, the sizing of the raw materials, firing temperatures, etc., account for large differences in properties among materials of substantially the same chemical composition.

Insulating refractories. Low thermal conductivity and good insulating properties are achieved primarily in materials of high, well-distributed porosity and low bulk density, without much regard to chemical composition. For insulating metallurgical furnaces, low thermal conductivity must be combined with the ability to withstand high temperatures without decomposition or structural changes. The fact that low bulk density and light weight go hand-in-hand with low thermal conductivity and high porosity is, of course, very advantageous in certain types of furnace construction. To the extent that these light-weight materials can be used in place of standard refractories, the furnace foundations, steel structural parts, door mechanisms, and other supporting parts can be of lighter construction. Another advantage of using light refractories, especially in intermittently operating furnaces, is that they have lower heat capacities per unit volume. The heat-storage capacity for a wall of given dimensions is directly proportional to the bulk density, so that less heat is required to bring a furnace constructed of insulating refractories to operating temperature. These advantages, however, cannot be realized to any great extent in the construction of furnaces for most large-scale processes of extractive metallurgy because the high porosity and low bulk density of insulating refractories are also associated with poor mechanical properties such as strength, load-bearing ability, and abrasion resistance, poor resistance to slag action and

other chemical attack, and poor spalling resistance. Accordingly, where these requirements must be met within the furnace, insulating materials are added as additional layers on the outside of the furnace for the main purpose of saving heat.

High-temperature insulation is available in several different forms, each having its own particular fields of use. These forms include loose granular or earthy material, insulating bricks, and castable materials mixed with water and poured, tamped, or troweled into place like concrete. Diatomaceous earth is a naturally occurring high-silica material which with suitable admixtures and processing is made into all these forms. Another useful raw material is vermiculite, a micaceous material which exfoliates and expands manyfold on heating. Insulating firebrick is widely used, both as backing for regular dense brick and, where furnace and process conditions permit, as the material for constructing entire furnaces. These bricks are made from fireclays much like ordinary firebrick, but with procedures modified to give a light, porous product. A common procedure involves adding to the mix an organic material like sawdust which burns out during firing to leave the desired degree of porosity. The ASTM has set up a classification of insulating firebrick (ASTM C–155) as Groups 16, 20, 23, 26, and 28, these numbers representing the maximum temperatures to which the bricks should be exposed in 100's of degrees Fahrenheit. Table 10–8 gives the properties of insulating bricks furnished by one manufacturer. Examination of these data shows that low bulk density and low thermal conductivity are properties somewhat contradictory to high refractoriness and good mechanical properties.

Adding insulation to the exteriors of existing furnaces has several consequences, in addition to the reduction of heat loss, and

TABLE 10-8

Properties of Insulating Firebrick*

Properties	Number of brick					
	K-16	K-20	K-23	K-26	K-28	K-30
Bulk density, lb/ft^3	18.8	25.3	25.9	41.6	42.5	50.5
Thermal conductivity, Btu-ft/hr-ft^2-$^\circ$F						
Mean temp.: 500°F	0.058	0.075	0.070	0.122	0.119	0.163
1000°F	0.079	0.098	0.091	0.153	0.147	0.208
1500°F	0.103	0.122	0.112	0.193	0.177	0.262
2000°F	---	---	0.138	0.258	0.212	0.338
Cold crushing strength, psi	48	95	117	154	148	308
Reheat shrinkage, ASTM C-210 $\{$ $^\circ$F	1550	1950	2250	2550	2750	2800
%	< 0.1	< 0.1	0.6	0.4	0.6	0.5

*Courtesy of the Babcock and Wilcox Co.

these should be considered with care before the insulation is added. Heat-flow calculations (Chapter 7) show that an insulating layer which accomplishes a substantial reduction in heat loss automatically will account for a substantial portion of the total temperature drop from the furnace interior to the surroundings. This means that the original refractory wall will be subjected to a much higher average temperature after it is insulated, and this may seriously affect its service. Under severe conditions, in other words, adequate refractory life may require that the outer portion of the refractory be kept cool, so that good air cooling is needed instead of insulation. Another consequence of getting most of the temperature drop into the insulating layer, of course, is that the inner surface of the insulation is itself subjected to a temperature near that of the furnace interior, and the insulating material must be chosen to meet this condition.

Special refractories. A number of refractory materials have rather limited applications in full-scale metallurgical operations but have outstanding properties for certain specialized kinds of service. These properties are so vital in some places that the use of special refractories is justified even though their first cost may be several times that of

any of the standard materials already discussed. Also, in small-scale work with valuable metals and metal products and in most laboratory work at high temperatures, especially analytical work and research, best practice is generally to use the materials with the best physical and chemical properties for the job, with only secondary regard to cost.

Silicon carbide (SiC), well known by the trade name carborundum, has several valuable properties to make it probably the most widely used of the special refractories. Silicon carbide is made from silica and carbon in an electric furnace. The crystalline furnace product is crushed and sized, and the relatively pure granular material serves as the starting point for fabricating bricks, muffles, crucibles, and many special shapes. Fireclay is usually added as a binder and the mixture is shaped and fired much like other refractories. Silicon carbide does not fail by fusion or softening at any temperatures attained in metallurgical furnaces, but is little used above about 1650°C (3000°F) because of chemical decomposition. In air, and in the presence of other oxidizing agents, carbon is lost as carbon monoxide, and silicon is oxidized or lost by volatilization. This decomposition occurs slowly even at temperatures as low as 1300°C (2370°F). Resistance to attack by both acid and basic slags is good, so that silicon carbide is considered a neutral refractory, but decomposition occurs through oxidation of the refractory and may be quite rapid in contact with some metal oxides and oxidizing slags.

The thermal conductivity of silicon carbide is of the order of ten times that of fireclay brick. This is, of course, a pronounced advantage in construction of apparatus requiring heat flow across the refractory. Accordingly, retorts with charge inside and heat source outside, muffles, and recuperator tubes are often fabricated of carborundum. The relatively high electrical conductivity, combined with refractoriness and good mechanical properties, is utilized in the well-known globar-rod resistance-heating elements which have been successful in a variety of furnace sizes and designs. Another field in which silicon carbide refractories are used is in furnace parts for which a premium is placed on exceptionally good mechanical properties, particularly abrasion resistance, hardness, strength, low thermal expansion, and spalling resistance.

Graphite stands well above other common refractory substances in sheer refractoriness and infusibility. Pure graphite does not melt, but vaporizes at about 3600°C or above. The chief limitation, unfortunately a controlling factor in many applications, is that even at low furnace temperatures (for example, 700°C or above), the carbon reacts with oxygen, many metal oxides, and oxidizing slags, to be consumed as CO or CO_2. Under reducing conditions, graphite is quite inert to most liquid metals, and to liquid mattes and slags which do not contain readily reduced oxides. Accordingly, graphite crucibles are much used for melting metals. However, it should be kept in mind that carbon is quite soluble in liquid iron and that

it also dissolves or reacts with a few other metals in percentages sufficient to have major effects on metal properties. Another outstanding property of graphite, which accounts for its extensive use as heating resistors and electrodes in extreme high-temperature electric furnaces, is its good electrical conductivity. The thermal conductivity also is high. By suitable manufacturing methods, carbon and graphite can be made into refractories possessing excellent mechanical properties, including good resistance to thermal shock and freedom from spalling.

Several forms of carbon are used in making refractories, including coke, natural graphite, and artificial graphite made in the electric furnace. Carbon blocks and carbon brick, fabricated from crushed coke, using tar or pitch as a binder and firing under reducing conditions, are used sometimes in constructing hearths for iron blast furnaces and for other furnaces operated at high temperatures under reducing conditions. Crucibles made of natural graphite with plastic fireclay as binder, roughly half graphite and half clay by weight, are commonly used for melting metals. For the many specialized uses of carbon where the various desirable properties of pure graphite, its high refractoriness, electrical conductivity, dimensional stability, and good mechanical properties under severe service, are needed to the fullest extent, artificial graphite is the primary material. To make this product, amorphous carbon in various forms (for example, lumps of anthracite coal or electrodes molded of crushed coke and tar) is treated in special electric furnaces at high temperatures (3000°C) such that the carbon crystallizes to graphite or "graphitizes" and the impurities are lost by volatilization. One of the valuable features of massive graphite made in this way is that it is very easily machined to obtain accurately-dimensioned objects and shapes which would be very difficult if not impossible to fabricate from any other refractory material.

Much of the small-scale high-temperature work of the metallurgical laboratory is conducted in furnaces and apparatus made of various refractory materials already discussed, particularly fireclay, silica, alumina, magnesia, silicon carbide, and graphite. In addition, quite a number of materials are utilized effectively in the laboratory which are not used on a large scale, because of high cost or other limitations. These include such materials as high-temperature porcelains, fused mullite, zircon, beryllia, thoria, and even certain sulfides and carbides. In fabricating combustion tubes, boats, crucibles, pyrometer protection tubes, and other small parts for the laboratory, desirable properties can be developed to a high degree by using high-purity raw materials and fabrication procedures which are too expensive to be considered on a large scale or simply are not readily adapted to making large shapes. Thus silica glass, or pure fused quartz, which is available as a clear transparent glass and also in other translucent and less expensive forms, is very useful in laboratory research at high temperatures, but is not adapted to any large-scale apparatus. Similarly, crucibles and tubes made of pure alumina, using special forming methods and very high firing temperatures, are substantially impermeable and gastight.

Metallurgical furnace construction

Bricks, blocks, and other shapes. The standard brick used in the largest quantities for furnace construction is the "9-inch straight," which is a rectangular brick measuring $9 \times 4\frac{1}{2} \times 2\frac{1}{2}$ inches. In addition there are many other sizes and shapes which are used in sufficient quantities to justify mass production and standardization. Actually the 9-inch straight is one of a number of standard $9 \times 4\frac{1}{2} \times 2\frac{1}{2}$-inch sizes, as shown in Fig. 10–8. All the refractory materials discussed previously which are used in large quantities are available in these standard shapes. Another similar group of standard shapes is the $9 \times 4\frac{1}{2} \times 3$-inch group, the group designation representing the dimensions of the 9-inch straight. Many other sizes and shapes are standardized, especially in the fireclay refractories, and some sizes and shapes are made primarily for a single kind of furnace, such as cupola blocks, blast furnace bottom blocks, rotary-kiln blocks, etc.

Mortars and cements. Mortars and cements are used to form the joints in laying brickwork. Most of these materials are purchased dry and then mixed with water to the proper consistency before use. The resulting batter is applied by troweling, dipping, or pouring. For refractory brickwork, it is usually desirable to make the joints as thin as possible, with substantially brick-to-brick contact. In some common types of construction, the bricks are laid without mortar and the joints are sealed by grouting or by inter-brick bonding reactions which occur when the furnace is placed in operation at high temperature. The proper choice of mortar and the technique of making the joints are very important factors in the service behavior of a refractory structure, calling for as much care as is exercised in the choice of the refractory brick itself.

One of the important properties of the mortar is, of course, the Pyrometric Cone Equivalent (PCE) or softening temperature, which is measured by ASTM Test C–24. Shrinkage during drying and firing should be minimized. The mechanical strength of the bond formed by the mortar is another signif-

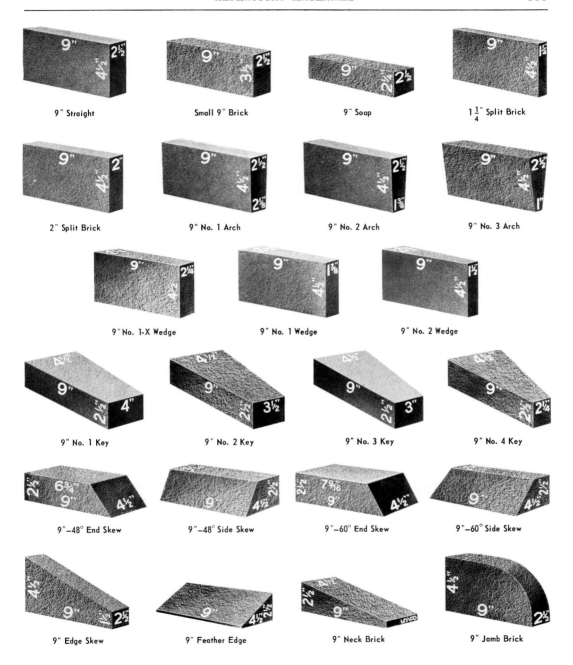

FIG. 10–8. Standard $9 \times 4\frac{1}{2} \times 2\frac{1}{2}$-inch sizes of refractory bricks (courtesy of Harbison-Walker Refractories Company).

icant property (ASTM Test C–198). For obvious reasons, mortar materials are selected to have physical and chemical characteristics compatible with the bricks they are used with. Accordingly, they are usually made of the same class of refractory raw materials as the bricks. Fireclay mortars, often plastic bonding clays with prefired or calcined clays added in large proportions to reduce firing shrinkage, are used with firebrick. Similarly, silica, alumina, deadburned magnesite, and chrome ores form the bases for mortars to be used with bricks of these respective classes. However, some of the mortars and high-temperature cements, for example, those with a chrome-ore base, serve satisfactorily to bond several different classes of brick and also can be used to make neutral joints between bricks of different classes which tend to react with each other at high temperatures.

Mortars and cements, irrespective of raw-material base, are of two broad types: *hot-setting* and *air-setting*. Hot-setting mortars form a strong bond only after firing, and make a ceramic bond essentially the same as that developed in firing a brick or other refractory body. Such mortars in general form a strong joint only near the hot surface of a wall and act mainly as filler in the outer parts of the wall which never reach bonding temperature. Since the more refractory mortars will require higher temperatures to develop strong bonds, a balance between refractoriness and bond-forming capacity is desirable to match the conditions of service of the mortar. Air-setting mortars and cements contain additions such as sodium silicate which develop a strong bond and impart a set merely on drying at room temperature. When the structure is heated, a ceramic bond develops in the hotter portions of the wall just as with hot-setting mortars, while the chemical bond continues to hold the brick together in the colder parts of the structure.

Plastics, ramming mixtures, and castables. Furnace linings and furnace hearths can often to advantage be monolithic and formed in place instead of being laid with bricks. The refractory plastics and ramming mixtures are used for this purpose. These materials are mixed with water and then rammed into place and into the desired shape in the cold furnace. An air-setting bond is used with many of these plastics to afford strength at low temperatures. After careful drying and firing, the monolithic structure is ready for service. Entire hearths of large furnaces can be built up, but care must be taken in all steps to insure that the final structure does not form deep cracks and shrink away from the sides of the furnace. Virtually all the common refractory materials are available or can be prepared in mixtures suitable for use in forming monolithic hearths and linings and for patching holes in brickwork. The fireclay-base materials known as plastic firebricks are especially convenient in repairing firebrick structures and are available in mixtures which after firing in the furnace have properties approaching those of good-quality firebrick. High-silica furnace bottoms are prepared using silica sand or ground ganister with enough fireclay added to obtain a workable mixture and to form a bond on heating. Plastics and ramming mixtures of the basic refractory materials magnesite and chrome ore are also widely used, and the chrome-base plastics in particular have a variety of uses because of their resistance to slag erosion and their chemical neutrality to other types of refractories, both acid and basic.

The castables are mixed and poured like concrete and, in fact, are also called refractory concretes. This is, of course, a convenient technique of forming special shapes, but is used mainly for low and moderately low temperature applications because the materials added to obtain the set, such as Portland cements, act as fluxes and reduce the refractoriness of the mixture.

Elements of furnace structures. Each of the metallurgical unit processes has its own special furnace requirements based on such aspects of the process as operating temperatures, chemical conditions, heat flow, materials handling, and operating manipulations. The details of furnace design, important though they are to successful and profitable operation of a given process, represent an art beyond the scope of this book and a field in which the metallurgist must lean heavily on the experience and knowledge of ceramists and other specialists. However, to complete this introduction to refractory materials and their metallurgical uses, brief consideration will be given to certain essential features of furnace construction which are common to most metallurgical furnaces.

Broadly speaking, there are two types of furnace construction, one involving building the refractories as a thick lining into a steel shell and the other involving masonry or brick construction usually with exterior structural-steel binding to hold the furnace together. Some furnaces combine these two types of construction. The shell-type construction is almost essential in furnaces which are rotated, tipped, or otherwise moved about and also is a common feature of most vertical shaft furnaces. On the other hand, a supporting shell is not necessary for stationary furnaces of masonry construction, for example, reverberatory furnaces, and such furnaces are built in the open with an exterior structure of I-beams, tie rods, and other structural-steel members sufficient to bind the furnace together, and in many cases also to support the roof and keep the roof arch under compression.

The copper converter shown in Fig. 10–9 is a simple example of the shell-and-lining type of construction. A common size is a cylinder 13 feet in diameter by 30 feet long, with a capacity of about 75 tons of liquid matte. In this apparatus large volumes of air are blown through liquid copper matte (copper and iron sulfides) at 1100–1300°C to produce iron oxides and sulfur dioxide. Silica is added as flux to form a slag with the iron oxide. The converter is in the form of a horizontal cylinder and is mounted on rollers so it can be rotated into different positions for blowing, charging, skimming slag, and pouring. Magnesite is the common refractory material for lining copper converters primarily because the slags are basic and high in iron oxides. Historically, the process was originally attempted in converters with acid linings and was unsatisfactory largely because of the short life of the refractory. The converter shown in the figure is lined with magnesite brick, using special wedge shapes to form a stable cylindrical lining. The lining is 18 inches thick (when new) over most of the surface, but can be made somewhat thinner in the upper portions where erosion is less severe. A 1- or 2-inch layer of crushed dead-burned magnesite is used between the brick and the shell to afford some flexibility and cushioning to take care of differential expansion and contraction. Common practice in some smelters is to maintain a protective coating of magnetite over the brick, this coating being formed by a special blowing of low-grade matte with insufficient flux to slag the iron oxides. Under favorable conditions a magnesite-lining will last many months, with occasional repairs in the zones of severe erosion, particularly around the tuyères.

The reverberatory matte-smelting furnace shown in Fig. 10–10 illustrates many of the common features of large masonry furnaces with exterior structural-steel binding. The matte-smelting process carried out in this

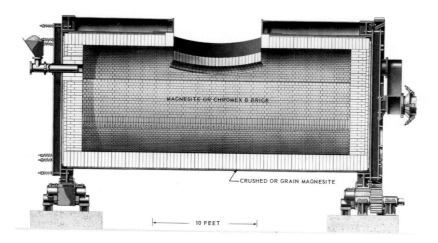

CRUSHED OR GRAIN MAGNESITE

MAGNESITE OR CHROMEX B BRICK

10 FEET

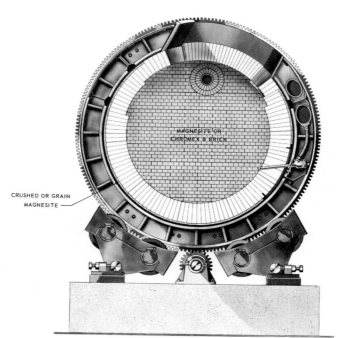

MAGNESITE OR CHROMEX B BRICK

CRUSHED OR GRAIN MAGNESITE

FIG. 10–9.   Copper converter (courtesy of Harbison-Walker Refractories Company).

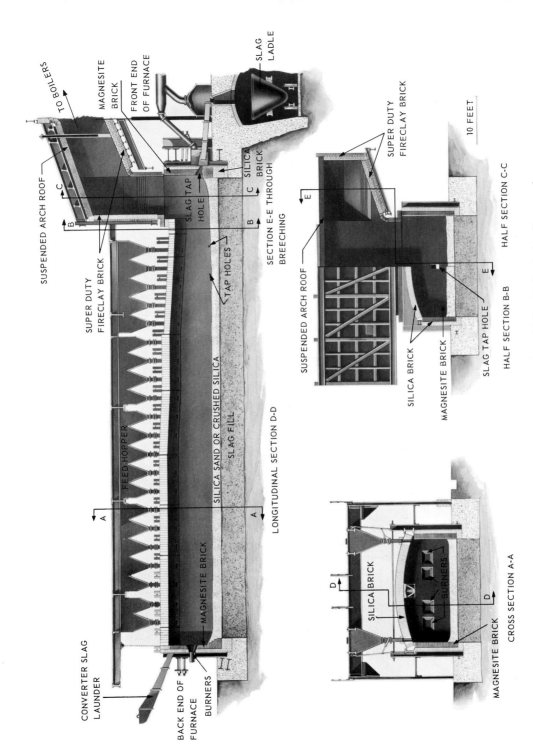

CONVERTER SLAG
LAUNDER

SUSPENDED ARCH ROOF

TO BOILERS

MAGNESITE
BRICK

FRONT END
OF FURNACE

SUPER DUTY
FIRECLAY BRICK

SILICA
BRICK

SLAG TAP
HOLE

SECTION E-E THROUGH
BREECHING

SLAG
LADLE

SUPER DUTY
FIRECLAY BRICK

10 FEET

HALF SECTION C-C

SUSPENDED ARCH ROOF

SILICA BRICK

MAGNESITE BRICK

SLAG TAP HOLE

HALF SECTION B-B

FEED HOPPER

TAP HOLES

SILICA SAND OR CRUSHED SILICA
SLAG FILL

MAGNESITE BRICK

LONGITUDINAL SECTION D-D

BACK END OF
FURNACE

BURNERS

SILICA BRICK

BURNERS

MAGNESITE BRICK

CROSS SECTION A-A

FIG. 10-10. Reverberatory smelting furnace (courtesy of Harbison-Walker Refractories Company).

359

furnace was described briefly in Chapter 1, pages 6-9, and this description should be referred to for an understanding of the service conditions to which the refractories are subjected.

The furnace construction, of course, starts with the *foundation*, which must be well designed to carry the large load of furnace and contents, because foundation instability can have very grave and costly consequences. When the hearth rests directly on the foundation, concrete is not a satisfactory material of foundation construction because it loses strength at temperatures as low as 500°C. Accordingly a common procedure is to pour a slag foundation of considerable thickness in a location picked with due regard for the load-carrying power of the ground below the slag fill. The vertical steel buckstays, usually I-beams or channels spaced 2 to 4 ft apart all around the furnace, must be solidly based to take the thrust of the furnace and sometimes to carry the weight of the roof or other parts of the furnace itself, so they may be set in concrete before the slag is poured, as shown in Fig. 10-10.

Another common type of construction, especially for smaller furnaces, involves a conventional foundation of reinforced concrete or steel, with a ventilated furnace bottom. That is, the furnace bottom is laid on steel plate which is supported and separated from the main foundation by a series of parallel I-beams. This arrangement was shown in Fig. 5-1 (page 110). The space between the beams is left unobstructed so that the bottom is kept reasonably cool by natural ventilation and the foundation is not overheated.

The *hearth* of the matte-smelting furnace shown in Fig. 10-10 represents one of a considerable variety of hearth constructions, and consists of silica sand rammed in with a small amount of binder and then sintered into place. Such a bottom is built up after the walls and roof are complete. Another type of bottom is built up of successive layers of bricks, as shown in Fig. 5-1. When the furnace is to contain liquid metal or matte, the bottom must be a tight monolith as in Fig. 10-10, or an inverted arch as in Fig. 5-1 to prevent its floating up in the heavy liquid. Another type of bottom, especially common in steelmaking furnaces, involves several layers of brick followed by a monolithic working hearth made of a plastic refractory rammed into shape and fired, or built up by sintering on successive thin layers of a mixture such as burned grain magnesite and slag. In many ways, the design and building of hearths and bottoms for smelting and melting furnaces is a difficult art. Not only must the bottom be tight against penetration of liquid metal and against floating up, but also it must withstand severe corrosive and erosive actions. In batch processes the bottom may be subjected to severe thermal shocks and to mechanical shocks and abrasion during charging, and at the same time a smooth bottom contour may be necessary to obtain clean drainage of the liquid metal at the end of treatment. Once made, a good bottom is often relatively permanent and with proper maintenance and care may last through several rebuildings of other parts of the furnace.

Furnace *walls* are constructed of several courses of brick to a thickness depending on the severity of the service conditions. In some furnaces, the entire wall is of one kind of brick. In other furnaces, it has been found most economical to use composite wall constructions, with more refractory and more expensive materials on the inside, backed up by cheaper firebrick or silica brick, sometimes with insulating brick as the very outside layer. Also, the type of refractory used may be varied from top to bottom

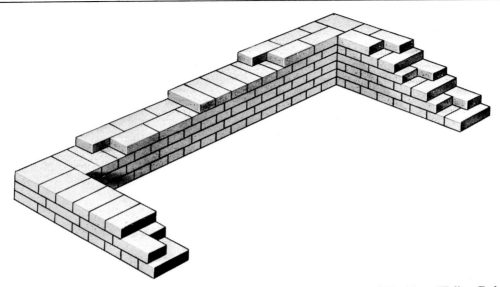

FIG. 10–11.   Alternate-header-and-stretcher wall construction (courtesy of Harbison-Walker Refractories Company).

of the wall and along the length of the wall to meet different conditions. The walls shown in Fig. 10–10 are of magnesite backed up with silica brick. In other matte-smelting furnaces, the magnesite is carried up to just above the slag line, where chemical attack is most severe, and silica brick is used for the remainder. There are several different ways of laying the brick so that the wall is bonded or tied together in a stable structure. Figure 10–11 shows how 9-inch walls are built up of alternate "header" and "stretcher" courses of standard 9-inch bricks. For the "stretcher" courses, the long axes of the bricks are parallel to the wall surface and for the "header" courses the long axes are perpendicular to the wall surface. Not shown in the figure are vertical expansion joints which must be left in the walls at intervals ranging from 1 or 2 feet up to 20 or 30 feet, depending on the expansion characteristics of the brick and on the service conditions.

The most prevalent type of *roof* construction is the *sprung arch* (Figs. 10–10, 5–1). The essential features of one common type of sprung arch are shown in Fig. 10–12. Standard tapered arch or wedge bricks are used, often combined with straights to attain the desired curvature (see Fig. 10–8). Both the horizontal thrust and the weight of the arch are supported by the specially shaped skewback bricks, which are supported in turn by steel channels or buttresses running the entire length of each side of the furnace. The buttresses are supported by the vertical buckstays, only one pair of which is shown, and the horizontal thrust of the arch is taken up by the tie rods above the furnace between opposite buckstays.

Figure 10–13 (a) shows the geometry of a sprung arch and the chief dimensions. The conventional sprung arch is circular with one center of curvature common to top and bottom surfaces, as shown. The rise is commonly expressed in inches per foot of span

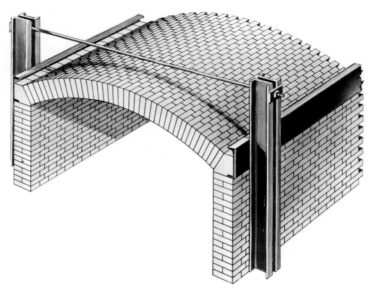

FIG. 10–12. Simple sprung arch (courtesy of Harbison-Walker Refractories Company).

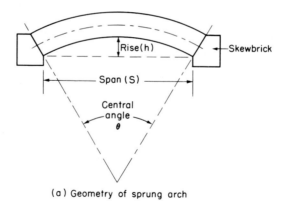

(a) Geometry of sprung arch

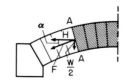

(b) Forces in a sprung arch

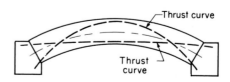

(c) Extreme positions of thrust curve

FIG. 10–13. Sprung arches.

(h/S), and ranges from 1 to 3 inches/ft. Specification of this value is equivalent to specifying the central angle (θ). For example, if h/S is 1.608 inches/ft, the central angle θ is 60° regardless of the span, and the arch is built up starting from 60° skew bricks, the inclined surfaces of which form an angle of 30° with the vertical. The standard 48° skews have inclined surfaces 42° from the vertical and thus correspond to a central angle of 84° and a rise of 2.3 inches/ft (see Fig. 10–8 for dimensions of 9-inch standard skews). These relationships follow from simple plane-geometric considerations, and it is readily shown that

$$\frac{h}{S} = 6\left[\frac{1 - \cos (\theta/2)}{\sin (\theta/2)}\right], \qquad (10\text{–}1)$$

where h/S is the rise in inches/ft and θ is the central angle.

Referring now to Fig. 10–13(b), the total force or thrust F acting across the surface AA between any two bricks in the arch is the resultant of two forces: (1) H, the horizontal thrust component applied by the skewbacks, which is the same in all parts of the arch, and (2) $W/2$, the weight of the shaded portion of the arch from the surface AA up to the center of the span. Expressing F, H, and W as lb/ft of running length of the arch,

$$F = \sqrt{H^2 + \frac{W^2}{4}}. \qquad (10\text{–}2)$$

At the center of the arch, $W/2 = 0$, and the thrust F is horizontal and equal to H. The maximum thrust occurs where the arch bears on the skews, and is given by Eq. 10–2 with W equal to the total weight of the arch in lb/running ft. The position and direction of the resultant force F in any part of the arch is indicated by the curve labelled "line of thrust," which for an arch of constant density and thickness is substantially a cate-

nary curve.* The angle α which the line of thrust makes with the horizontal at any position is determined by the force triangle at that position:

$$\tan \alpha = \frac{W}{2H}, \qquad (10\text{–}3)$$

in which $W/2$ is the weight of the arch between this position and the center, and H is the horizontal thrust, both in lb/running ft. Thus for a given arch, the location and shape of the line of thrust vary with the horizontal thrust H applied to the arch through the skewbacks. As H is increased, Eq. (10–3) shows that α at any position decreases, which means that the thrust curve becomes flatter. If it becomes so flat as to drop below the arch at the center, the arch will buckle. Conversely, if H is decreased, the thrust curve increases in curvature, and if the thrust curve then bends to move outside the arch, the arch falls in. Extreme positions of the line of thrust are shown in Fig. 10–13(c). Actually, to obtain a stable arch without a tendency for joints to open at the top or bottom, the line of thrust should pass through the middle one-third of arch thickness.

To estimate the magnitudes of the forces in an arch and the horizontal thrust which must be taken by the buttresses, buckstays, and tie rods, we can assume as an approximation that the thrust curve coincides with the curve representing the center line of the arch. On this assumption, the thrust angle α at the skewback will be half the central angle θ of the arch. Applying Eq. (10–3),

$$H = \frac{W_t}{2} \cot \frac{\theta}{2} \qquad (10\text{–}4)$$

and

$$F_t = \frac{W_t}{2} \csc \frac{\theta}{2}, \qquad (10\text{–}5)$$

---

* The student should demonstrate this to his own satisfaction.

in which W_t is the total weight of the arch from skewback to skewback. If F_t, as given by Eq. (10–5), represents the thrust on the skewbacks in lb/ft of arch length, the average pressure or stress P in psi is given by

$$P = \frac{F_t}{12t} = \frac{W_t}{24t} \csc\left(\frac{\theta}{2}\right), \quad (10\text{–}6)$$

in which t is the arch thickness in inches. The estimates made by Eqs. (10–4), (10–5), and (10–6) on the assumption that the thrust curve follows the center line serve only to

indicate orders of magnitude, because these forces vary over a considerable range as the line of thrust is allowed to move within the limits already mentioned. One approach to design of the arch and its auxiliary steelwork is to assume that the actual forces and stresses are 2 or $2\frac{1}{2}$ times those estimated by the above equations. Rigorous calculation of arch stresses is a very complex problem in mechanics, and the indeterminate nature of some of the conditions imposed on actual arch structures makes the value of more elaborate calculations doubtful.

EXAMPLE

The arch of a copper-smelting furnace is constructed of silica brick, and has a span (S) of 25 ft, a rise (h) of $27\frac{1}{2}$ inches, and a thickness (t) of 20 inches. Estimate (a) the horizontal thrust H in lb/ft of arch length, (b) the compressive stress at the center of the arch, and (c) the compressive stress at the skews.

The rise is given by

$$\frac{h}{S} = \frac{27.5}{25} = 1.10 \text{ inches/ft of span.}$$

Applying Eq. (10–1),

$$\frac{h}{S} = 1.10 = 6\left[\frac{1 - \cos(\theta/2)}{\sin(\theta/2)}\right].$$

Solution of this equation gives the central angle, $\theta = 41.5°$. Referring to Fig. 10–13, the radius of the lower arch surface is $S/(2 \sin \theta/2) = 25/(2 \sin 20.75°) = 35.3$ ft. The roof is $20/12$ or 1.66 ft thick, so the outer radius is 36.96 ft. Using these dimensions, the volume of silica brick in the arch per foot of length is found to be 42.5 cu ft. Taking the weight of the silica brick as 110 lb/ft³, we find the total weight W_t of the arch:

$$W_t = 42.5 \times 110 = 4675 \text{ lb/running ft.}$$

Assuming that the line of thrust corresponds closely to the center of the arch, the horizontal thrust is given by Eq. (10–4):

$$H = \frac{W_t}{2} \cot \frac{\theta}{2} = \frac{4675}{2} \cot 20.75°$$

(a)
$$H = \underline{6160 \text{ lb/running ft.}}$$

At the center of the arch, the total thrust equals the horizontal thrust, H, so that the average compressive stress is estimated to be

(b)
$$P = \frac{H}{12t} = \frac{6160}{240} = \underline{25.7 \text{ psi.}}$$

At the skews, by Eq. (10–5),

$$F_t = \frac{W_t}{2} \csc \frac{\theta}{2} = \underline{6600 \text{ lb/running ft,}}$$

which corresponds to a stress of

(c) $$P = \frac{F_t}{12t} = \frac{6600}{24} = \underline{27.5 \text{ psi.}}$$

Sprung arches of matte-smelting furnaces, open hearths, and other large high-temperature furnaces are universally constructed of silica brick, since no other common refractory material possesses the necessary high-temperature load-bearing capacity. Also of importance are the low bulk density of silica brick and its constancy of volume above about 650°C. For furnaces with smaller roof spans and lower operating temperatures, high-duty and super-duty fireclay bricks are widely used, and are preferable to silica brick in many batch and intermittent operations where the rapid temperature changes would cause serious spalling of silica brick. Under special conditions, high-alumina, forsterite, magnesite, or other basic refractories are used. Although their high refractoriness and resistance to slag and dust attack make magnesite refractories desirable for roof construction, several factors combine to render them unsuitable for sprung arches. These factors are: poor load-bearing ability at high temperatures, high density which makes the weight and thrust forces greater in direct proportion to the density, high coefficient of thermal expansion, and, in some of the common high-magnesite materials, poor spalling resistance. However, composite sprung arches of silica brick with basic brick sections at the shoulders where service conditions are most severe have been used. Also, complete arches of newer basic refractories of improved properties, such as high-fired chrome-mag-

nesite bricks, have proved successful in smaller furnaces.

In recent years, there has been a great deal of experimentation and development work with *suspended arches*, which do not require the light weight and high-temperature mechanical properties that so severely limit the choice of refractories for sprung arches. For copper reverberatory furnaces and open-hearth steelmaking furnaces especially, the operating conditions are so severe with respect to temperature and chemical attack of various kinds that silica roofs require much care and even under favorable conditions require rebuilding more often than the rest of the furnace. Suspended roofs of magnesite and various magnesite-chrome and chrome-magnesite refractories appear to have been quite successful in some installations at copper smelters and steel plants, where the relatively high initial costs for the refractory and its installation with suspending steelwork have been offset by the improved service life and the increased furnace production obtained by driving the furnace harder. Several forms of suspended-roof construction have been used, some curved like a sprung arch and others flat, but most of the methods of suspension involve special brick shapes, each brick being hung on a steel hangar of some kind suspended from a relatively elaborate framework over the top of the furnace. One method is illustrated in Fig. 10–14.

FIG. 10–14. View of one type of suspended arch (courtesy of General Refractories Company).

Supplementary References

AIME, Committee on Physical Chemistry of Steelmaking, *Basic Open-Hearth Steelmaking*, revised ed. New York: AIME, 1951.

ASTM, Committee C–8 on Refractories, *Manual of ASTM Standards on Refractory Materials*. Philadelphia: ASTM, 1948.

Coxey, James R., *Refractories*. State College, Penna.: Pennsylvania State College, 1950.

Demmerle, R. L., Birch, R. E., and Hendryx, D. B., "Manufacture of Basic Refractory Brick." *Ind. Eng. Chem.*, **40**, 1762–1772, 1948.

Etherington, H., *Modern Furnace Technology*. London: Charles Griffin, 1944.

General Refractories Co., *Refractories*. Philadelphia: General Refractories Co., 1949.

Harbison-Walker Refractories Co., *Modern Refractory Practice*. Pittsburgh: Harbison-Walker Refractories Co., 1950.

Liddell, Donald M., *Handbook of Nonferrous Metallurgy*. Vol. 1, *Principles and Processes*. New York: McGraw-Hill, 1945.

Norton, F. H., *Refractories*. New York: McGraw-Hill, 1949.

Norton, F. H., *Elements of Ceramics*. Cambridge: Addison-Wesley, 1952.

Trinks, W., *Industrial Furnaces*, Vol. I. New York: John Wiley, 1950.

APPENDIX

TABLE A-1

Atomic Weights and Atomic Numbers

Element	Symbol	Atomic No.	Atomic Mass	Element	Symbol	Atomic No.	Atomic Mass
Actinium	Ac	89	(227)	Molybdenum	Mo	42	95.95
Ag (See Silver)	Ag	47	107.880	Na (See Sodium)	Na	11	22.997
Aluminum	Al	13	26.97	Nb (See Columbium)	Cb	41	92.91
Americium	Am	95	(241)	Neodymium	Nd	60	144.27
Antimony, stibium	Sb	51	121.76	Neon	Ne	10	20.183
Argon	A	18	39.944	Neptunium	Np	93	(237)
Arsenic	As	33	74.91	Nickel	Ni	28	58.69
Astatine	At	85	(211)	Nitrogen	N	7	14.008
Au (See Gold)	Au	79	197.2	Osmium	Os	76	190.2
Barium	Ba	56	137.36	Oxygen	O	8	16.0000
Berkelium	Bk	97	—	Palladium	Pd	46	106.7
Beryllium, glucinum	Be	4	9.013	Pb (See Lead)	Pb	82	207.21
Bismuth	Bi	83	209.00	Phosphorus	P	15	30.98
Boron	B	5	10.82	Platinum	Pt	78	195.23
Bromine	Br	35	79.916	Plutonium	Pu	94	(239)
Cadmium	Cd	48	112.41	Polonium	Po	84	210
Calcium	Ca	20	40.08	Potassium, kalium	K	19	39.096
Californium	Cf	98	—	Promethium	Pm	61	(147)
Carbon	C	6	12.010	Praseodymium	Pr	59	140.92
Cerium	Ce	58	140.13	Protactinium	Pa	91	231
Cesium	Cs	55	132.91	Radium	Ra	88	226.05
Chlorine	Cl	17	35.457	Radon, niton	Rn	86	222
Chromium	Cr	24	52.01	Rhenium	Re	75	186.31
Cobalt	Co	27	58.94	Rhodium	Rh	45	102.91
Columbium, niobium	Cb	41	92.91	Rubidium	Rb	37	85.48
Copper, cuprum	Cu	29	63.54	Ruthenium	Ru	44	101.7
Curium	Cm	96	(242)	Samarium	Sm	62	150.43
Dysprosium	Dy	66	162.46	Sb (See Antimony)	Sb	51	121.76
Erbium	Er	68	167.2	Sn (See Tin)	Sn	50	118.70
Europium	Eu	63	152.0	Scandium	Sc	21	45.10
Fe (See Iron)	Fe	26	55.85	Selenium	Se	34	78.96
Fluorine	F	9	19.000	Silicon	Si	14	28.06
Francium	Fr	87	(223)	Silver, argentum	Ag	47	107.880
Gadolinium	Gd	64	156.9	Sodium, natrium	Na	11	22.997
Gallium	Ga	31	69.72	Strontium	Sr	38	87.63
Germanium	Ge	32	72.60	Sulfur	S	16	32.066
Gold, aurum	Au	79	197.2	Tantulum	Ta	73	180.88
Hafnium	Hf	72	178.6	Technetium	Tc	43	(99)
Helium	He	2	4.003	Tellurium	Te	52	127.61
Holmium	Ho	67	164.94	Terbium	Tb	65	159.2
Hg (See Mercury)	Hg	80	200.61	Thallium	Tl	81	204.39
Hydrogen	H	1	1.0080	Thorium	Th	90	232.12
Indium	In	49	114.76	Thulium	Tm	69	169.4
Iodine	I	53	126.92	Tin, stannum	Sn	50	118.70
Iridium	Ir	77	193.1	Titanium	Ti	22	47.90
Iron, ferrum	Fe	26	55.85	Tungsten, wolfram	W	74	183.92
K (See Potassium)	K	19	39.096	Uranium	U	92	238.07
Krypton	Kr	36	83.7	Vanadium	V	23	50.95
Lanthanum	La	57	138.92	W (See Tungsten)	W	74	183.92
Lead, plumbum	Pb	82	207.21	Xenon	Xe	54	131.3
Lithium	Li	3	6.940	Ytterbium	Yb	70	173.04
Lutecium	Lu	71	174.99	Yttrium	Y	39	88.92
Magnesium	Mg	12	24.32	Zinc	Zn	30	65.38
Manganese	Mn	25	54.93	Zirconium	Zr	40	91.22
Mercury, hydrargyrum	Hg	80	200.61				

368

TABLE A-2

Vapor Pressure of Water, 0° to 100°C

	Millimeters Mercury									
$^\circ$C	0	1	2	3	4	5	6	7	8	9
0	4.58	4.93	5.29	5.69	6.10	6.54	7.01	7.51	8.05	8.61
10	9.21	9.84	10.52	11.23	11.99	12.79	13.63	14.53	15.48	16.48
20	17.54	18.65	19.83	21.07	22.38	23.76	25.21	26.74	28.35	30.04
30	31.82	33.70	35.66	37.73	39.90	42.18	44.56	47.07	49.69	52.44
40	55.32	58.34	61.50	64.80	68.26	71.88	75.65	79.60	83.71	88.02
50	92.5	97.2	102.1	107.2	112.5	118.0	123.8	129.8	136.1	142.6
60	149.4	156.4	163.8	171.4	179.3	187.5	196.1	205.0	214.2	223.7
70	233.7	243.9	254.6	265.7	277.2	289.1	301.4	314.1	327.3	341.0
80	355.1	369.7	384.9	400.6	416.8	433.6	450.9	468.7	487.1	506.1
90	525.8	546.0	567.0	588.6	610.9	633.9	657.6	682.1	707.3	733.2
100	760.0									

TABLE A-3

High-Temperature Heat Contents and Heat Capacities

All data are taken from K. K. Kelley, Bulletin 476, U.S. Bureau of Mines (1949). This bulletin should be consulted for data on substances not listed below. Values of a, b, c, and d are given for the empirical equations:

$$H_T - H_{298} = aT + bT^2 + cT^{-1} + d, \qquad C_p = a + 2bT - cT^{-2}.$$

In these equations, T is in °K; H_T and H_{298} are in cal/mol (g cal/g mol, kg cal/kg mol, lb cal/lb mol); and C_p is in cal/mol-°C or Btu/lb mol-°F. The tabulated values of $H_t - H_{77}$ are in Btu/lb mol.

Metals — Empirical coefficients

	Al cryst	Al liq	Cu cryst	Cu liq	Fe α	Fe β	Fe γ	Fe δ	Fe liq	Hg liq	Hg gas	Mg cryst	Mg liq	Pb cryst	Pb liq	Zn cryst	Zn liq
a	4.94	7.00	5.41	7.50	3.37	10.40	4.85	10.30	10.00	6.61	4.969	6.14	7.40	5.82	6.80	5.35	7.50
b×10³	1.48	---	0.75	---	3.55	---	1.50	---	---	---	---	0.75	---	0.95	---	1.20	---
c×10⁻⁵	---	---	---	---	-0.43	---	---	---	---	---	---	0.78	---	---	---	---	---
d	-1605	330	-1680	-20	-1176	-4280	390	-4420	-180	-1971	13,055	-2159	-440	-1820	-838	-1702	-850
Fusion or transition pt.	931.7°K		1357°K		1033°K	1179°K	1674°K	1803°K		634°K		923°K		600.5°K		692.7°K	

Al (Fusion 931.7°K)

t,°F	H_t−H_77
200	720
400	1,960
600	3,270
800	4,640
1000	6,110
1200	7,580
1217(c)	7,700
1217(ℓ)	12,330
1400	13,600
1600	15,000
1800	16,400

Cu (Fusion 1357°K)

t,°F	H_t−H_77
200	720
400	1,940
600	3,190
800	4,460
1000	5,770
1200	7,110
1400	8,500
1600	9,900
1800	11,300
1983(c)	12,670
1983(ℓ)	18,290
2000	18,400
2200	19,900
2400	21,400

Fe (transition pts. 1033°K, 1179°K, 1674°K, 1803°K)

t,°F	H_t−H_77	t,°F	H_t−H_77
200	770	2200	19,500
400	2,090	2400	21,400
600	3,540	2553(γ)	22,880
800	5,150	2553(δ)	23,080
1000	6,880	2600	23,560
1200	8,880	2786(δ)	25,470
1400(α)	11,630	2786(ℓ)	32,130
1400(β)	11,630	2800	32,300
1600	13,720	2900	33,300
1662(β)	14,360		
1662(γ)	14,740		
1800	15,900		
2000	17,700		

Hg (Fusion 634°K)

t,°F	H_t−H_77
200	810
400	2,140
600	3,460
681(ℓ)	4,000
681(g)	29,170
800	29,800
1000	30,800
1200	31,700
1400	32,700
1700	34,200
2000	35,700
2500	38,200
3000	40,700

Mg (Fusion 923°K)

t,°F	H_t−H_77
200	740
400	2,000
600	3,320
800	4,710
1000	6,140
1200	7,600
1202(c)	7,610
1202(ℓ)	11,500
1400	13,000
1500	13,700

Pb (Fusion 600.5°K)

t,°F	H_t−H_77
200	790
400	2,130
600	3,490
621(c)	3,640
621(ℓ)	5,840
800	7,060
1000	8,410
1200	9,770
1400	11,100
1600	12,500
1800	13,900
1900	14,500

Zn (Fusion 692.7°K)

t,°F	H_t−H_77
200	750
400	2,020
600	3,360
787(c)	4,640
787(ℓ)	7,820
800	7,920
1000	9,410
1200	10,900
1400	12,400
1600	13,900
1700	14,700

TABLE A-3 (cont'd.)

Gases

t,°F	Rare gases*	CO	CO_2	CH_4	H_2	H_2O	H_2S	N_2	O_2	SO_2	SO_3
a	4.969	6.79	10.55	5.65	6.52	7.17	7.02	6.66	7.16	10.38	13.70
$b \times 10^3$	---	0.49	1.08	5.72	0.39	1.28	1.84	0.51	0.50	1.27	3.21
$c \times 10^{-5}$	---	0.11	2.04	0.46	-0.12	-0.08	---	---	0.40	1.42	3.12
d	-1481.6	-2105	-3926	-2347	-1939	-2225	-2257	-2031	-2313	-3683	-5417

$H_t - H_{77}$ Btu/lb mol

t,°F	Rare gases*	CO	CO_2	CH_4	H_2	H_2O	H_2S	N_2	O_2	SO_2	SO_3
200	610	860	1,160	1,120	850	1,000	1,020	860	870	1,220	1,600
400	1,610	2,270	3,160	3,120	2,250	2,640	2,710	2,260	2,320	3,320	4,490
600	2,600	3,700	5,340	5,440	3,650	4,340	4,520	3,680	3,820	5,570	7,690
800	3,590	5,170	7,650	8,050	5,050	6,100	6,390	5,140	5,380	7,940	11,100
1000	4,590	6,680	10,100	11,000	6,470	7,930	8,380	6,620	6,970	10,400	14,800
1200	5,580	8,230	12,600	14,100	7,890	9,820	10,400	8,150	8,610	12,900	18,600
1400	6,570	9,810	15,200	17,500	9,330	11,800	12,600	9,720	10,300	15,500	22,500
1700	8,070	12,200	19,200	23,000	11,500	14,800	15,900	12,200	12,800	19,400	28,600
2000	9,560	14,700	23,300	28,800	13,800	18,100	19,500	14,600	15,400	23,400	
2500	12,000	18,900	30,300		17,600	23,700	25,500	18,700	19,800	30,100	
3000	14,500	23,200	37,400		21,600	29,600		23,000	24,200	36,900	
3500	17,000	27,600	44,700		25,800	35,800		27,300	28,700	43,700	
4000	19,500	31,900	52,100		30,000	42,100		31,600	33,400	50,500	

*Ideal monatomic gases: A, He, Kr, Ne, Xe; also monatomic gaseous metals, Ag, Cd, Hg, Zn, and others, referred to standard state of ideal gas at 298°K (77°F).

TABLE A-3 (cont'd.)

Solid Oxides

	Al_2O_3	Cu_2O	CaO	FeO	Fe_2O_3	Fe_3O_4	MgO	PbO	ZnO
a	27.43	14.90	11.67	12.38	23.36	39.92	10.18	9.05	11.71
$b \times 10^3$	1.53	2.85	0.54	0.81	8.62	9.43	0.87	3.20	0.61
$c \times 10^{-5}$	8.47	---	1.56	0.38	3.08	10.01	1.48	---	2.18
d	-11,155	-4,696	-4,051	-3,891	-8,764	-16,098	-3,609	-2,983	-4,277
Transition temp.									

$H_t - H_{77}$, Btu/lb mol

$t, °F$	Al_2O_3	Cu_2O	CaO	FeO	Fe_2O_3	Fe_3O_4	MgO	PbO	ZnO
200	2,650	2,080	1,330	1,560	3,210	4,650	1,160	1,390	1,290
400	7,310	5,540	3,550	4,120	8,960	12,900	3,150	3,730	3,530
600	12,500	9,130	5,890	6,740	15,300	22,000	5,230	6,230	5,800
800	18,000	12,900	8,280	9,410	22,300	32,200	7,380	8,860	8,150
1000	23,700	16,700	10,700	12,100	29,600	42,900	9,640	11,600	10,600
1200	29,500	20,600	13,200	14,900	37,100	53,900	12,000	14,600	13,000
1400	35,500	24,700	15,600	17,600	44,600	65,000	14,400		15,600
1600	41,500	29,000	18,200	20,500			16,800		18,200
1800	47,600		20,800				19,300		20,800
2000	53,700		23,400				21,900		23,400
2200	59,800		26,000				24,400		26,100
2400	66,000		28,700				26,900		28,800
2600	72,200		31,400				29,400		
2800	78,400		34,100				31,900		
3000							34,500		

SiO_2

	quartz α	quartz β	cristobalite α	cristobalite β	tridymite α	tridymite β	glass
a	11.22	14.41	4.28	14.40	3.27	13.64	13.38
$b \times 10^3$	4.10	0.97	10.53	1.02	12.40	1.32	1.84
$c \times 10^{-5}$	2.70	---	---	---	---	---	3.45
d	-4,615	-4,455	-2,212	-4,696	-2,077	-4,395	-5,310
Transition temp.	848°K		523°K		390°K		

$H_t - H_{77}$

quartz

$t, °F$	$H_t - H_{77}$
200	1,450
400	4,060
600	6,970
800	10,100
1000	13,500
1067(α)	14,710
1067(β)	15,230
1200	17,400
1400	20,600
1600	24,000
1800	27,300
2000	30,700
2200	34,200
2400	37,600
2600	41,100
2800	44,700
3000	48,300
3100	50,100

cristobalite

$t, °F$	$H_t - H_{77}$
200	1,460
400	4,060
482(α)	5,240
482(β)	5,600
600	7,440
800	10,500
1000	13,800
1200	17,000
1400	20,300
1600	23,700
1800	27,000
2000	30,500
2200	33,900
2400	37,400
2600	40,900
2800	44,500
3000	48,100
3100	49,900

tridymite

$t, °F$	$H_t - H_{77}$
200	1,420
242(α)	1,950
242(β)	2,020
400	4,290
600	7,210
800	10,300
1000	13,500
1200	16,800
1400	20,100
1600	23,400
1800	26,800
2000	30,200
2200	33,700
2400	37,200
2600	40,700
2800	44,300
3000	47,800
3100	49,600

glass

$t, °F$	$H_t - H_{77}$
200	1,480
400	4,060
600	6,830
800	9,770
1000	12,900
1200	16,200
1400	19,500
1600	22,900
1800	26,400
2000	30,000
2200	33,600
2400	37,400
2600	41,300
2800	45,300
3000	49,300
3100	51,400

TABLE A-3 (cont'd.)

Other Substances

	C, graphite	CaCO₃, calcite	Cu₂S α	Cu₂S β	Cu₂S γ	FeS α	FeS β	FeS liq	Fe₃C α	Fe₃C β	Fe₃C liq	SiC	ZnS
a	4.10	24.98	19.50	23.25	20.32	15.20	10.95	16.00	19.64	25.62	30.60	8.93	12.16
$b \times 10^3$	0.51	2.62	---	---	---	---	1.90	---	10.00	1.50	---	1.50	0.62
$c \times 10^{-5}$	2.10	6.20	---	---	---	---	---	---	---	---	---	3.07	1.36
d	−1,972	−9,760	−5,665	−6,300	−4,275	−4,532	−1,857	−508	−6,745	−7,515	+740	−3,825	−4,137

Transition temp.: Cu₂S — 376°K, 623°K; FeS — 412°K, 1468°K; Fe₃C — 463°K, 1500°K.

C, graphite

t, °F	$H_t - H_{77}$
200	300
400	900
600	1,630
800	2,470
1000	3,390
1200	4,360
1400	5,360
1600	6,390
1800	7,470
2000	8,550
2200	9,640
2400	10,800
2600	11,900
2800	13,100
3000	14,200
3200	15,400
3400	16,700
3600	17,900

CaCO₃, calcite

t, °F	$H_t - H_{77}$
200	2,680
400	7,330
600	12,400
800	17,800
1000	23,300
1200	29,100
1400	35,000
1600	41,100
1700	44,200

Cu₂S

t, °F	$H_t - H_{77}$
200	2,400
217(α)	2,740
217(β)	4,390
400	8,640
600	13,300
662(β)	14,720
662(α)	15,080
800	17,900
1000	22,000
1200	26,000
1400	30,100
1600	34,200
1800	38,200
2000	42,300

FeS

t, °F	$H_t - H_{77}$
200	1,870
282(α)	3,110
282(β)	5,360
400	6,850
600	9,460
800	12,200
1000	14,900
1200	17,700
1400	20,700
1600	23,700
1800	26,800
2000	30,000
2183(β)	32,960
2183(ℓ)	41,360
2200	41,600

Fe₃C

t, °F	$H_t - H_{77}$
200	3,240
300	5,980
374(α)	8,080
374(β)	8,410
400	9,120
600	14,600
800	20,100
1000	25,700
1200	31,300
1400	37,000
1600	42,800
1800	48,700
2000	54,600
2200	60,600
2240(β)	61,760
2240(ℓ)	83,950
2400	88,800
2600	95,000
2800	101,100
2900	104,100

SiC

t, °F	$H_t - H_{77}$
200	920
400	2,560
600	4,430
800	6,480
1000	8,600
1200	10,900
1400	13,100
1600	15,500
1800	17,900
2000	20,500
2200	23,200
2400	25,900
2600	28,700

ZnS

t, °F	$H_t - H_{77}$
200	1,400
400	3,780
600	6,330
800	8,910
1000	11,500
1200	14,100
1400	16,600
1600	19,200
1700	20,500

TABLE A-4

Mean Specific Heats of Common Materials

These data are for engineering estimates only.

Units: $\dfrac{g\ cal}{g\text{-}{}^{\circ}C} = \dfrac{kg\ cal}{kg\text{-}{}^{\circ}C} = \dfrac{lb\ cal}{lb\text{-}{}^{\circ}C} = \dfrac{Btu}{lb\text{-}{}^{\circ}F}$

Materials	Temperature range	c_m	Materials	Temperature range	c_m
Air*	77-400°F	0.24	Metals, liquid:*	Above melting point	
	77-1000°F	0.25	Aluminum		0.26
	77-2000°F	0.27	Copper		0.12
	77-3000°F	0.28	Iron		0.18
Construction materials, miscellaneous:			Lead		0.03
brickwork, concrete, glass, granite,			Mercury		0.03
limestone, sand, stone		c. 0.2	Zinc		0.12
Fuels:			Refractory bricks:		
Coal		c. 0.3 (0.26-0.37)	Chrome	77-400°F	0.18
Coal tar		c. 0.4		77-1000°F	0.20
Coke (low ash)	77-500°F	0.24		77-1800°F	0.22
	77-1000°F	0.30		77-2600°F	0.23
	77-1500°F	0.34	Fireclay	77-400°F	0.20
	77-2000°F	0.36		77-1000°F	0.23
Fuel oils		c. 0.5 (0.4-0.5)		77-1800°F	0.26
Wood		c. 0.5 (0.45-0.65)		77-2600°F	0.29
			Magnesite	77-400°F	0.23
Matte, liquid:		c. 0.15		77-1000°F	0.25
Metals, solid:*	Room temperature			77-1800°F	0.28
				77-2600°F	0.29
Aluminum		0.22	Silica	77-400°F	0.21
Brass		0.09		77-1000°F	0.23
Copper		0.09		77-1800°F	0.26
Iron and steel		0.12		77-2600°F	0.29
Lead		0.03	Slags, liquid		c. 0.3
Magnesium		0.25	Water:		
Nickel		0.12	Ice	32°F	0.49
Platinum		0.03	Liquid		1.00
Silver		0.06	Steam (1 atmosphere)*	77-200°F	0.45
Zinc		0.09		200-400°F	0.45
				400-800°F	0.48

*Use Table A-3 for more precise calculations.

TABLE A-5

Standard Heats of Formation from the Elements at 298°K, 1 Atmosphere

(From Selected Values of Chemical Thermodynamic Properties, U. S. Bureau of Standards, 1949.)

$-\Delta H_f^o$ is given in kg cal/g mol. To obtain cal/mol (g cal/g mol, kg cal/kg mol, or lb cal/lb mol), multiply by 1000. To obtain Btu/lb mol, multiply by 1800.

	Substance	$-\Delta H_f^o$	Substance	$-\Delta H_f^o$	Substance	$-\Delta H_f^o$
OXIDES	Ag_2O	7.31	CuO	37.1	PbO (yellow)	52.07
	Al_2O_3 (corundum)	399.09	Cu_2O	39.84	Sb_2O_4 (cryst)	214
	As_4O_6 (monoclinic)	312.8	$Fe_{0.95}O$ (wüstite)	63.7	Sb_2O_5 (cryst)	234.4
	As_2O_5	218.6	Fe_2O_3	196.5	SiO (gas)	26.72
	BaO	133.4	Fe_3O_4	267.0	SiO_2 (quartz)	205.4
	BeO	146.0	HgO (red)	21.68	SiO_2 (cristobalite)	205.0
	Bi_2O_3	137.9	H_2O (liq)	68.32	SiO_2 (tridymite)	204.8
	CaO	151.9	H_2O (gas)	57.80	SiO_2 (glass)	202.5
	CdO	60.86	MgO	143.84	SnO_2	138.8
	CO (gas)	26.42 (29.6)*	MnO	92.0	SO_2 (gas)	70.96
			MnO_2	124.5	SO_3 (gas)	94.45
	CO_2 (gas)	94.05 (97.2)*	MoO_2	130.0	WO_3	200.84
	CoO	57.2	Na_2O	99.4	ZnO	83.17
	CrO_3	138.4	NiO	58.4		
	Cr_2O_3	269.7	P_2O_5	360.0		
SULFIDES	Ag_2S (α)	7.60	FeS (β)	21.35	NiS	17.5
	As_2S_3	35	FeS_2 (pyrite)	42.52	Ni_3S_2	43.4
	CaS	115.3	FeS_2 (marcasite)	36.88	PbS	22.54
	CdS	34.5	H_2S (gas)	4.82	Sb_2S_3 (black)	43.5
	CS_2 (gas)	-27.55	HgS (cinnabar)	13.90	Sb_2S_3 (orange)	36.0
	CuS	11.6	MnS (green)	48.8	ZnS (wurtzite)	45.3
	Cu_2S	19.0	MoS_2	55.5	ZnS (sphalerite)	48.5
	FeS (α)	22.72	Na_2S	89.2		
SULFATES	$BaSO_4$	350.2	$MgSO_4$	305.5	$PbSO_4 \cdot PbO$	282.5
	$CaSO_4$	342.42	Na_2SO_4	330.90	$PbSO_4 \cdot 2PbO$	344
	$CuSO_4$	184.00	$NiSO_4$	213.0	$PbSO_4 \cdot 3PbO$	403
	$FeSO_4$	220.5	$PbSO_4$	219.50	$ZnSO_4$	233.88
	H_2SO_4 (liq)	193.91				
CARBONATES	$BaCO_3$ (witherite)	291.3	$MgCO_3$	266	$NaHCO_3$	226.5
	$CaCO_3$ (calcite)	288.45	$MgCO_3 . CaCO_3$	556	$PbCO_3$	167.3
	$FeCO_3$ (siderite)	178.7	Na_2CO_3	270.3	$ZnCO_3$	194.2
HYDROXIDES	$Al(OH)_3$ (amorph)	304.2	$Ca(OH)_2$	235.8	$Mg(OH)_2$	221.00
	$Al_2O_3 \cdot H_2O$	471	$Fe(OH)_3$	197	NaOH	101.99
	$Al_2O_3 \cdot 3H_2O$	613.7				
HYDROCARBONS	CH_4	17.89	C_2H_4	-12.50	C_3H_8	24.82
	C_2H_2	-54.19	C_2H_6	20.24	$n-C_4H_{10}$ (gas)	29.81
MISCELLANEOUS	AgCl (cryst)	30.36	Fe_2SiO_4	343.7	NaF	136.0
	CaC_2	15.0	HCl (gas)	22.06	NH_3 (gas)	11.04
	Fe_3C	-5.0	NaCl	98.23	SiC	26.7
	FeSi	19.2				

*Heats of formation from "amorphous" C in fuels.

375

TABLE A-6

Thermal Conductivities k of Common Materials

$$\frac{\text{Btu} - \text{ft}}{\text{ft}^2 - \text{hr} - {}^{\circ}\text{F}}$$

Materials	Temperature, °F						
	70	200	400	800	1200	1600	2000
Air	0.015	0.018	0.023				
Construction materials:							
Brick, red	c. 0.4						
Concrete, cinder	0.2						
1:4 dry	0.44						
stone	0.54						
Pyrex glass	c. 0.6						
Rubber	c. 0.1						
Sand, dry	c. 0.2						
Sandstone	c. 1.0						
Wood	c. 0.1						
Graphite, longitudinal	95						
-100 mesh powder	0.1						
Insulating materials:							
Asbestos fiber, 36 lb/ft^3	0.09	0.11	0.12	0.13			
Diatomaceous earth powder, 18 lb/ft^3	0.039	0.042	0.048	0.061			
Insulating brick (see Table 10-8)							
Lampblack	0.04						
85% magnesia, 13 lb/ft^3	0.034	0.036	0.040				
Mineral wool, fibrous	0.023						
Metals, solid:							
Aluminum	120						
Brass	60						
Cast iron	30						
Copper	220						
Gold	170						
Lead	20						
Magnesium	90						
Mercury (liquid)	7						
Nickel	35						
Platinum	40						
Silver	240						
Steel, low carbon	25						
Steel, stainless	9						
Zinc	65						
Oils, miscellaneous hydrocarbons	c. 0.08						
Refractory materials:*							
90% alumina brick		1.70	1.56	1.34	1.19	1.13	1.10
Chrome brick			0.84	0.90	0.94	0.97	0.99
Fireclay brick		0.57	0.60	0.67	0.74	0.80	0.87
Graphite			61	42	33	26	18
Magnesite brick			3.0	2.8	2.4	2.0	1.7
Silica brick		0.57	0.63	0.75	0.88	1.00	1.13
Silicon carbide brick			17.5	14.6	12.5	9.6	7.9
Water	0.35	0.39					

*Thermal conductivities of refractories may differ very considerably from the representative values given here, because of differences in structure and porosity and because of changes occurring during service.

TABLE A-7

Total Emissivities (ϵ) of Various Surfaces

Abstracted largely from the compilation by H. C. Hottel, in McAdams, W. H., Heat Transmission, New York: McGraw - Hill, 1942.

Surface	Temperature °F	Total emissivity (ϵ)	Surface	Temperature °F	Total emissivity (ϵ)
Asbestos board, asbestos paper	100-700	0.95	rolled sheet steel, oxidized	70	0.66
Brick:			rough ingot iron	1700-2040	0.87-0.95
Red, building	70	0.93	rough steel plate	100-700	0.94-0.97
Refractory	1100-1800	0.7-0.9	liquid iron or steel		0.28
Silica	1800-2000	0.8-0.85	Lead, pure unoxidized	260-440	0.057-0.075
Metals:			gray oxidized	70	0.28
Aluminum, highly polished	400-1100	0.039-0.057	Platinum, pure polished plate	440-1200	0.054-0.104
rough plate	70	0.055	wire	440-2500	0.073-0.182
oxidized at 1100°F	400-1100	0.11-0.19	Painted and lacquered surfaces:		
Brass, highly polished	500	0.03	Black shiny lacquer, sprayed on iron	70	0.875
rolled plate	70	0.06	Black lacquer	100-200	0.80-0.95
oxidized at 1100°F	400-1100	0.60	Flat black lacquer	100-200	0.96-0.98
Copper, polished	200	0.02	White lacquer	100-200	0.80-0.95
heavily oxidized by heating	70	0.78	Oil paints, all colors	212	0.92-0.96
liquid	2000-2300	0.16-0.13	Aluminum paints and lacquers	212	0.27-0.67
Iron and steel:			Water	32-212	0.95-0.96
electrolytic iron, highly polished	350-400	0.052-0.064			
freshly machined	70	0.44			

USEFUL CONVERSION FACTORS

Extensive and unneccessary use of conversion factors should be avoided by care in planning problem solutions.
At the start of each problem solution, choose the system of units which for the data available will require the fewest conversions.

Length	1 ft = 30.48 cm 1 inch = 2.54 cm = 25.4 mm	Standard conditions (STP)	Temperature: $0^\circ C = 32^\circ F = 273^\circ K = 492^\circ R$ Pressure: 1 atm=760 mm Hg=29.92 inches Hg=14.7 psi Ideal gas volumes: 1 g mol = 22.4 liters 1 kg mol = 22.4 cu m 1 lb mol = 359 cu ft
Volume	1 cu m = 35.31 cu ft 1 cu ft = 7.48 gal (U.S.) 1 bbl = 42 gal (U.S.)		
Mass	1 kg = 2.205 lb 1 lb = 453.6 g 1 lb = 16 oz (avoir) = 14.583 oz (troy) = 7000 grains. 1 short ton (or net ton) = 2000 lb 1 long ton = 2240 lb 1 metric ton = 1000 kg = 2205 lb	Energy	1 g cal=4.183 joules (international) = 4.183 watt-sec 1 kg cal = 1000 g cal = 2.205 lb cal = 3.97 Btu 1 kw hr = 3415 Btu 1 Btu = 778 ft-lb
Density	$1 \text{ g/cm}^3 = 62.43 \text{ lb/ft}^3 = 8.345 \text{ lb/gal}$	Energy/Time (Power)	1 hp = 550 ft-lb/sec = 33,000 ft-lb/min 1 hp = 745.5 watts = 2546 Btu/hr 1 kw = 1.341 hp = 3415 Btu/hr
Pressure	1 atm = 760 mm mercury = 29.92 inches mercury 1 atm = 14.7 psi 1 inch water = 5.2 lb/ft^2	Viscosity	1 poise = 100 centipoises = 1 g mass/cm-sec 1 poise = 0.0672 lb mass/ft-sec = 242 lb mass/ft-hr
Temperature	$^\circ C = \dfrac{^\circ F - 32}{1.8}$ $^\circ F = 1.8(^\circ C) + 32$ $^\circ K = ^\circ C + 273$ $^\circ R = ^\circ F + 460 = 1.8 (^\circ K)$	Logarithms	e (base of natural logarithms) = 2.718 $\ln_e x = 2.303 \log_{10} x$ $e^n = 10^{n/2.303}$

INDEX

Abrasion resistance, 325–326, 334, 344, 351, 353
Absorptivities:
  surfaces, 220–239
  gas bodies, 221, 235–239
Acceleration of gravity (g), 144
Accumulation, 17, 20, 28, 30, 31
Acid Bessemer process, 308, 336
Acid open-hearth process, 308
Acid oxides, 307–309, 311–312
Acid refractories, 308, 326, 330–331, 335–336, 339, 347, 349
Acid slags, 306–309, 313, 330–331, 335–336, 339, 347, 349, 352
Adiabatic combustion temperature, 114–118, 119–120, 123, 140
Adiabatic compression, 181–183, 190
Adiabatic discharge coefficient, 185–187
Adiabatic expansion, 184
Adiabatic flame temperature, 114–118, 119–120, 123, 140
Adiabatic flow, 184–185
Adiabatic heat exchanger, 247–250
Adiabatic nozzles, 189–190
Adiabatic path, 45
Agitator, 18–20
Air adjustment, 113, 118, 122–123, 137- 142, 177
Air composition, 27
Air compression and compressors, 147–148, 179–183, 190
Air consumption in combustion, 111–115, 118, 121–123, 126, 137–142
Air leakage, 111–112, 121–123, 137–140, 173, 189
Aluminum, extraction from bauxite, 2–4
Amphoteric oxides, 307
Analyses, *see* type of analysis and kind of material analyzed
Arches, 109–110, 325, 333, 335, 339, 340, 361–366
Ash, 72–73, *see also* Proximate analyses
ASTM classifications and test procedures:
  fuels, 72–74, 76–77, 79, 89, 103–105, 107
  refractories, 328, 329, 332–334, 343–344, 351–352, 354
Atomic numbers, 12–13, Table A–1
Atomic weights and masses, 13, Table A–1

Atomization, 105–106
Autogenous processes, 54–55, 59–60, 67–69, 257
Automatic control, 9, 169, 173–174, 188, 282
Available heat, 119–129, 137–142, 197–198, 244
Average reciprocal density, 149, 183

Balances, *see* type of balance, such as Element balance, Heat balance, Energy balances, etc.
Base temperature, *see* Reference temperature
Basic Bessemer process, 308, 336
Basicity series, 307
Basic open-hearth process, 4, 308, *see also* Open-hearth steelmaking
Basic oxides, 307–309, 311–312
Basic refractories, 308, 326, 330–331, 335–336, 345–351
Basic slags, 330–331, 335–336, 339, 345, 349, 352
Batch operation, 15, 16, 30
Beds of solid particles:
  fluid flow, 160–162
  forced convection, 217–218
Beehive coke ovens, 85–86
Bends in pipes and ducts, 158–159, 166, 189
Bernoulli equations, 143, *see* Energy balances
Bessemer converter air supply, 178–179, 189–190
Bessemer processes, 303, 306, 308, 336
Blackbodies and black surfaces, 219–229
Blast furnace:
  air humidity, 126
  air supply, 178–179, 185–186
  charge calculation, 36–40, 41–42
  coke, 69, 70, 78, 80, 88, 89, 91
  critical temperature, 126
  gas, 37, 39–40, 61–65, 87, 90, 91, 118, 123, 136–137, 140
  heat balance, 55–56, 58, 60–66
  heat transfer, 213, 217, 254
  metallurgical balances, 28, 36–40, 61–63
  process, 4
  reactions, 61–62, 94
  refractories, 325, 344
  slags, 303, 306, 317, 320
  stoves, 133, 135
  tuyères, 185–186

Blowers, 178–183, 188–190
Boiling points of metals, 285–288
Bond and bonding phase, see Ceramic bond
Boyle's law, 14
Bricks, see Refractory bricks, Firebrick, Silica brick, Magnesite brick, etc.
Buckstays, 109–110, 360–361
Bulk densities:
    refractory materials, 326, 334, 339, 351, 364–365
    various materials, 21
Burners, 83–84, 98–101, 105–106
By-product coke ovens, 85–87, 107

Calcining:
    aluminum trihydrate, 3, 33–34, 67
    heat balances, 67, 68
    limestone, 68, 191, 254
Calorific intensity, 117
Calorific power:
    availability, 119–120
    coal, 73–78, 106–108, 118
    coke, 107
    definitions, 73–75
    Dulong's formula, 74, 75
    fuel oils, 105, 118
    gaseous fuels, 90–91, 107–108, 118
    recovery in gasification, 91–92, 97–98, 107
Calorimeters, 74, 105
Capacities of equipment and furnaces, 17–20, 121, 253–254
Capacity lag, 277–283
Capillary tubes, 157, 167, 188
Carbon balance, 112–114
Carbon deposition, 40–41, 331
Carbon in liquid metals, 299–301
Carbonization, see Coking
Carbon refractories, 353–354
Carborundum, see Silicon carbide
Castable refractory materials, 351, 356
Casting; 3, 4, 191, 254
Cements, 354, 356–357
Centrifugal fans, see Fans
Ceramic bond, 327, 331, 334, 338–340, 343, 346–350, 354, 356
Changes in state, thermodynamic, 44–53
Changes in state of aggregation, 44–50
Charcoal, 69, 70, 85, 94

Charge calculations, 31, 36–40, 41–42
Checkers, 133–134, 135–140
Chemical analyses, see type of analysis and material analyzed
Chemical bonding of refractories, 348, 350, 356
Chemical metallurgy, 1–11
Chimneys, see Stacks and chimneys
Chrome-magnesite refractories, 350–351, 365
Chrome ores, 349–351, 356
Chrome refractories, 336, 348–351
City gas, 91
Clays and clay refractories, 335, 340–345
Coal, 69–81, see also Pulverized coal
    agglomeration, 76–77
    anthracite, 69–71, 75–80
    bituminous, 69–71, 75–80, 88, 107
    calorific powers, 73–79, 106–108, 118
    classification, 72, 75–79, 106–107
    coking, 77–78, 80, 83, 85–88
    combustion and heat utilization, 140–141
    constitution, 71–72
    deposits and producing areas, 79–81
    destructive distillation, 83–88, 91–92, 107
    flame temperatures, 118
    gas, 90, 91, 107
    gasification, 91–98, 107–108
    grades, 78–79
    grate firing, 70, 78, 81, 94, 174
    preparation, 79–81
    proximate analyses, 72–73, 75–78, 106–107
    quality, 78–80
    rank and origin, 70–71, 75–80, 106–107
    rational analysis, 71–72
    storage, 81–83
    tar, 103, 106
    ultimate analyses, 73, 75, 106–107, 118
    weathering, 76–78
Coalification, 70–71, 75
Coefficients of discharge, 163–165, 171–172, 175, 184–190
Coefficients of expansion, 331–332, 339, 348, 365
Coke, 83–89
    blast furnace, 61–64, 69–70, 78, 80, 88, 89, 91
    breeze, 89, 97
    gasification, 68, 91–98
    ovens, 85–87
    properties, 70, 88–89
Coke-oven gas, 87, 90, 117–118, 136–137, 142

Coking:
  beehive ovens, 85–86
  by-product ovens, 85–87, 91, 107
  by-products, 85–87
  coal, 77–78, 80, 85–88
  heat balance, 107
  mechanism, 85
Combining volumes, 15
Combining weights, 7, 12–14
Combustion, 7–8, 109–142
  air supply, 111–115, 118, 121–123, 126, 137–142, 177, 188
  heat utilization, 114–142
  principal variables, 120–121
  products, see Flue gases
  stoichiometry, 33–36, 111–114, 140
Complex processes, thermal calculations, 51–52
Compressed air and blast, 146–147, 178–187, 189–190
Compression of gases, 178–187, 189–190
Compressive strength:
  coke, 89
  refractory materials, 325–326, 334
Compressors, 178–183
Concentric cylinders, radial heat flow, 198–202, 255–258
Concentric spheres, radial heat flow, 202
Conduction of electricity, 204–206, 240, 256
Conduction of heat, 191–206, see also Heat conduction systems, Thermal conductivities
Conductivity, see Thermal conductivity, Electrical conductivity
Conservation of elements, 12–14, 30
Conservation of energy, 7, 43–52, 109, 143, 145
Constant-pressure processes, 45–55, 74, 97
Constant-volume processes, 45, 67, 74
Constitution diagrams:
  mattes and metal-sulfur systems
    Cu-S, 302
    Cu_2S-FeS, 323
    Fe-S, 301
  metal-metal systems
    Al-Hg, 293
    Cu-Pb, 292
    Fe-Al, 293
    Fe-Cu, 294
    Mg-Bi, 295
    Zn-Pb, 291

  metal-nonmetal systems
    Cu-S, 302
    Fe-C, 300
    Fe-S, 301
  metal-oxygen systems
    Cu-O, 297
    Fe-O, 296
  oxide and silicate systems
    CaO-SiO_2, 315
    FeO-CaO-SiO_2, 318
    FeO-Fe_2O_3-SiO_2, 317
    FeO-SiO_2, 316
    MgO-Al_2O_3, 348
    MgO-SiO_2, 347
    SiO_2-Al_2O_3, 341
Contact resistances, 197, 206, 242, 255–256
Contraction of flow, 171–173
Convection heat flow, 191, 194, 207–218
  beds of solids, 217–218
  dimensionless numbers, 211–218
  forced convection, 207, 213–218, 256–258
  furnace exteriors, 230–231, 256–258
  heat exchangers, 239
  heat-transfer coefficients, 208, 211–218, 236, 240
  horizontal cylinders, 211–214, 217, 255–258
  natural convection, 207–213, 255–256
  pipes and ducts, 215–217, 235–238, 255–258
  plane surfaces, 212–214
  spheres, 217
  unsteady systems, 259–260
Convergent and convergent-divergent nozzles, 185–187
Converters:
  air requirement, 178–179, 189–190
  construction, 357–358
  tuyères, 185–186, 189–190
Converting, 3, 4, 51, 59–60, 303, 306, 357–358
Cooling, see Heating and cooling
Cooling rates, see Heating and cooling rates
Copper smelting:
  furnaces, 325, 335, 357–360, 364–365
  heat balances, 51, 54–56, 59–60
  mattes, 3, 6, 8–9, 321–323
  processes, 2–9, 294
  slags, 3, 6, 8–9, 303, 306, 315
  waste-heat recovery, 124
Corundum, 341, 345
Cosine law, 223–224

Critical pressure ratio, 186–187
Critical Reynolds numbers, 157–158
Critical temperatures, 118–129, 135, 137–142
Cristobalite, 336–340, 343, 345
Crushing strength:
    coke, 89
    refractory bodies, 325–326, 334
Cyclic operation, 16, 17
Cylinders, convection heat transfer, 211–214, 217, 255–258, *see also* Concentric cylinders

Dampers, 174–177
Dead-burned magnesite, 346–348, 350, 356, 357
Dead load, 20
Deformation under load, 329, 333–334, 344
Deoxidation, 290, 297
Dephosphorization, 289–290, 309
Desulfurization, 290, 309
Dew point, 27, 41, 140
Diaspore, 342, 345
Diatomaceous earth, 351
Diffusion, 205
Dilution methods of flow measurement, 168
Dimensional analysis:
    fluid flow, 152–154, 160
    forced convection, 214–215, 218
    natural convection, 211–213
Dimensional homogeneity, 152–154
Dimensionless numbers:
    convection heat flow, 211–218
    fluid flow, 152–155, 160–161, 171–172
    unsteady heat flow, 264–266
Dimensions, *see* Units and dimensions
Discharge coefficients, *see* Coefficients of discharge
Displacement, 17–20
Dissociation, effect on flame temperature, 116–117
Distillation, 284–285
Dolomite, 336, 349
Draft and draft control, 121–122, 168–178, 189
Drosses, 324
Ducts and flues:
    gas flow, 154–160, 189
    heat flow to walls, 235–239, 256–258
    pressure drop and size, 159–160, 189
Dulong's formula, 74, 75

Efficiencies of fans, blowers, and compressors,

178–179, 183, 190, *see also* Thermal efficiencies
Electrical conductivities:
    mattes, 324
    metals, 194
    refractory materials, 326, 335, 353
    slags, 311
Electrical conductors, temperature rise, 253, 256
Electrical work, 43, 45, 52, 54–57, 66–67
Electric furnaces, 129, 255–258, 287–288, 353
Elemental analyses, *see* Ultimate analyses
Element balances, 12–17, 20, 27–42, 112–114
Emissive power, 219–223
Emissivities (total):
    gases, 221, 235–239
    surfaces, 220–239, Table A–7
Endothermic reactions, 52, 54, 56, 97
Energy balances, 43–45, 53–68, 143–150, *see also* type of energy balance and kind of system
Energy content, 43–45, 66, 67, 144–146
Energy, kinds of, 43–44, 143–145, *see also* specific kinds of energy
Enlargement of flow, 171–173
Enthalpy, *see* Heat content
Entrance losses, 171–172
Equilibria in metallurgical reactions, 8
Equilibrium diagrams, *see* Constitution diagrams
Erosion, 325, 331
Errors in weighing, sampling, and analysis, 22–24, 30, 42
Eutectics, 292–293, 317, 322, 340–341
Excess air, 33–36, 41, 112, 113, 121–123, 126, 137–142
Exothermic reactions, 52, 54, 97–98
Expansion factors, 163, 185
Expansion and shrinkage of refractory bodies, 326, 331–333, 338–340, 343, 348, 353–354
Expansion work, 45–46, 49, 52, 66
Extraction of metals, 1–4

Fanning equations, 155, 158, 160–162
Fans, 177–180, 184, 189
Final state, 44, 45, 49–53
Firebrick, 335, 341–345, 354–356, 360, 365
Fireclays and fireclay refractories, 341–345, 347, 349, 351, 356, 365
Fire refining, 3, 4, *see also* Refining of liquid metals

Firing refractory bodies, 327, 331-332, 334, 339, 342–343, 345

First law of thermodynamics, 7, 43–46, 49, 52–55, 74, 97, 109, 114, 145, 146, 181

Fixed carbon, 72–73, *see also* Proximate analyses

Flame radiation, 239

Flames:
  pulverized coal, 82
  gaseous fuels, 98–101

Flame temperatures, 114–118, 126–127, 140, 239, 325

Flame velocities, 99–101

Flow of fluids, 8, 143–190, *see also* specific apparatus or flow system
  energy balances, 143–149
  kinds of energy, 143–145

Flow of heat, *see* Heat flow

Flow of materials through apparatus, 17–20, 42

Flow measurement and flow meters, 161–168, 186, 188, *see also* specific apparatus, such as Capillary tubes, Nozzles, Orifices, Pitot tubes, etc.

Flowsheets, 2–4

Flues, 110–111, *see also* Ducts and flues

Flue gases:
  analyses, 112–114, 140–141
  heat recovery from, 123–124, 129–142
  heat transfer to walls, 235–239
  measuring and sampling, 167
  temperatures, 119–130, 141

Fluid flow, 8, 143–190, *see also* Flow of fluids, name of specific apparatus or flow system

Fluidized beds, 161

Fluxes, 285, 304–306

Forced convection, 207, 213–218, 256–258

Forced draft, 177–179, 189

Formation temperatures of slags, 304–305, 320

Forsterite, 336, 347, 350, 365

Fourier's equations, 192–193, 263–264, 266–268

Fractional kinetic energy loss, 154, 158–159, 171–172, 174–177

Free convection, *see* Natural convection

Free electrons, 290, 324

Free-running temperatures of slags, 119, 304–305, 319–320

Freezing, heat flow in, 244–246

Freezing points, metal sulfides, 322–323, *see also* Melting points and Constitution diagrams

Friction, 144–145, 149–167, 170–176, 179, 186, 188–189, *see also* specific apparatus or flow system

Friction factors, 154–162, 171–176
  relations to Reynolds number, 155, 157–162, 171

Frictionless flow, 162–163, 181–185

Fuel consumption, 7–8, 111–114, 120–129, 135, 137–142, 173, 189, 197–198, 244, 257

Fuel oils, 69–70, 101–106
  burners, 105–106
  calorific powers, 105, 118
  combustion and heat utilization, 137–141
  compositions, 103, 118
  flame temperatures, 118
  grades, 103–105
  properties, 103–105

Fuels, 7–8, 69–108, *see also* specific fuels and fuel properties
  classification, 69–70
  combustion and heat utilization, 109–142

Furnace, *see also* type of furnace or process
  arches, 109–110, 325, 333, 335, 339, 361–366
  construction, 325, 351–366
  draft, 121–122, 168–177, 189
  foundations, 359–360
  gas displacement, 210–211
  hearths, 109–110, 204–205, 348, 356, 359–360
  heating and cooling, 259, 277–283
  heat losses, 197–198, 203–204, 206, 225–227, 230–231, 240–244, 255–258
  heat transfer, 225–229, 239, 255–258
  insulation, 281–283, 351–352
  openings, 170–173, 189
  pressure, *see* draft
  roofs, *see* arches
  smelting power, 120–121, 124, 141–142
  walls, 333, 360–361

Ganister, 338, 356

Gas analyses, 15, 26–27, 41

Gas cooler, 256–257

Gas displacement in furnaces, 210–211

Gas evolution from metals, 297–298

Gaseous fuels, 69–70, 89–101
  analyses, 90–91, 107–108, 118
  burners, 98–101
  calorific powers, 90–91, 107–108, 118

Gaseous fuels (*continued*)
  combustion and heat utilization, 140–142
  combustion mechanism and characteristics, 98–
    101
  constituents, 89–90
  flame temperatures, 118
  from coal and coke, 70, 91–98, 107–108
  preheating, 123–124, 129, 136–137
  types, 90–92
Gaseous metals, 285–288
Gases, flow of, 143–190, *see* specific apparatus or
    flow system
Gases in metals, 296–299
Gases, temperature measurement, 252–253, 256
Gasification of coal and coke, 67, 68, 70, 92–98,
    107–108
Gas laws, 7, 12, 14–15
Gas-metal reactions, systems, 284, 296–299, 310
Gas producers, 94–98, 107–108
Gas radiation, 234–239
Gas-slag reactions, 310, 313
Gas-solid processes, 10, 254, 257
Gas solubilities in liquid metals, 297–299
Gas volumes, calculation, 14–15
Gay-Lussac's law, 14
g_c, conversion factor, 144
g, acceleration of gravity, 144
Geometric mean area, 202
Gibbsite, 342
Glass, glassy state, 319, 327–328, 331, 333, 343, 345
Globar furnaces, 231–232, 256, 353
Graphite, 70, 71, 85, 336, 353–354
Grashof number, 211–212
Gravity separation, slags, 303, 305, 318–319
Gray bodies and surfaces, 219–222, 229–235
Grog, 332, 343
Gurney-Lurie charts, 266–268, 271

Head, 178–179
Heat, 43–68
Heat absorbed, 44–46, 49, 51–52, 66–68, 145
Heat balance, 7, 53–68
  calculation procedure, 57–66
  choice of reactions, 57–58
  difficulties, 60
  items, 53–54
  reference temperatures, 53, 59–60, 68
  relation to First Law, 53, 55

Heat balances:
  adiabatic combustion, 114–117
  blast furnace, 55–56, 58, 60–66
  calcining, 67, 68
  coking, 107
  converting, 51, 59–60
  gasification, 97–98, 107–108
  gas producer, 107–108
  heat exchangers, 130
  matte smelting, 54–56
  open-hearth system, 137–140, 142
  recuperators, 247–249, 258
  regenerators, 137–140, 142
  roasting zinc ore, 67–68
Heat capacities, 45–48, 60, 181–182, Table A–3
  ratio C_p/C_v, 181–182
Heat conduction systems:
  constant area: flat wall, insulated rod, 196–198,
    255–257
  furnace hearths, 204–205
  furnace walls, 197–198, 203–204, 206, 255–258
  radial flow, concentric cylinders, 198–202, 255–
    258
  radial flow, concentric spheres, 202
  series paths, 204–206
  unsteady, 259–260
Heat conductivity, *see* Thermal conductivity
Heat content, 45–50, 144–148, Table A–3
Heat duty classification (ASTM), 343–344
Heat evolved, compression of gases, 182–183
Heat exchangers, 129–135, 191, 239, 247–252, 258,
    see also Recuperators and Regenerators
  design calculations, 251–252, 258
  over-all coefficients, 249–252, 258
  thermal efficiencies, 129–130, 137–140, 251, 258
Heat flow, *see also* specific heat-flow mechanisms
    and systems
  as a rate-determining factor, 253–254
  complex paths, 230, 239–258
  steady, 191–258
  unsteady, 259–283
Heating and cooling, 259–283
  bricks, 270–271, 273–274, 283
  conditions of negligible surface resistance, 268–
    269, 273–274, 276–277, 281
  curves, 263, 266–268, 275–280, 282
  effects on refractories, 333, 337–339
  furnaces and other complex systems, 277–282

Heating and cooling (*continued*)
    good conductors, *see* Newton's law of heating
        and cooling
    plates, cylinders, spheres, 264–269, 271–273,
        281–283
    quenching metals, 275–277
    rates, 259–260, 268, 274–275, 283
    simplified process for solid objects, 259–260
    steel ingots, 262, 270, 283
    steel shafting, 261–262, 282
    temperature distribution, 260, 266, 271–274,
        276, 278, 283
    temperature gradients, 333, 338–339
    time lags, 266, 271–274, 277–283
Heating value, *see* Calorific power
Heat input, 43, 53–59, 64
Heat losses, 54–60
    furnaces, furnace walls, 197–198, 203–204, 206,
        230, 231, 240–244, 255–258
    gases in pipes and ducts, 215–216, 235–239,
        256–258
    insulated pipes, cylinders, 199–202, 255–258
    laboratory furnaces, 202–204, 255–258
    liquid metal in open cylinder, 232–234
    nonventilated hearths, 204–205
    radiation through furnace openings, 243–244
Heat of fusion, 48, 49, 52
Heat of transformation, 48–49, 52
Heat of vaporization, 48–49, 52, 54
    of water, 48–49, 65, 68, 73–75
Heat output, 43, 53–59, 64–66
Heat recovery from flue gases, 123–124, 129–142
Heats of combustion, *see* Calorific power
Heats of formation, 46, 49–52, 58, Table A–5
Heats of reaction, 46, 49, 50–68
    at high temperatures, 52, 59, 67, 68
Heats of solution, 46
Heat storage in furnace walls, 351
Heat transfer coefficients:
    convection, 208, 211–218, 236, 240
    heating and cooling bodies, 260–261, 265, 268,
        275–276
    over-all, local over-all, 249–252
    radiation, 229–231, 240
Heat treatment, 259, 275–277
Heat utilization in furnaces, 114–142
    open-hearth steelmaking, 135–140
    principal variables, 120–121

Henry's law, 296, 298
Heterogeneous reactions, 285
High-alumina refractories, 336, 340–345, 365
High-silica refractories, 335, 336, 338–340
Homogeneous reactions, 285
Humidity, 27, 41, 61–63
Hydraulic radius, 154–156
Hydrocarbons, 71, 89–91, 102–103
Hydrogen balance, 112–114
Hydrogen in metals, 297–299

Ideal gases, 14, 15, 147, 148, 181, 183
Ignition temperatures of fuels, 89, 98
Impact pressure, 165–166
Incomplete combustion, 116–117, 121–122, 126
Incompressible fluids, 149, 183
Inert gases, 297
Inflammability limits, 98–99
Initial state, 44, 45, 49–53
Input, 17, 20, 28
Insulating bricks, materials, 194, 206, 255, 281–
    283, 336, 351–352, 360
Insulation:
    aluminum foil, 246
    cylinders and pipes, 198–202, 255–258
    furnaces, 198–199, 206, 244, 255–258, 360
    spheres, equidimensional bodies, 202
Integrated processes, 2
Intercooler, 183
Intermetallic compounds, 294–295
Internal energy, *see* Energy content
Internal friction, *see* Viscosity
Inverse-square law, 223
Ionic theory of slags, 311–313
Iron and steel, flowsheet, 2, 4
Irreversible expansion, 331–332, 338–339
Irreversible heat exchanger, 247–250
Isothermal compression, 181–183, 190

Kaolinite, 342–343
Kinematic viscosity, 151, 289
Kinetic energy, 44, 144–145, 147, 154
Kinetic theory, 152, 195

Laboratory furnaces, 198–199, 202–204, 211, 231–
    232, 244, 255–258, 278–280
Laboratory refractories, 336, 354

Lags, lag coefficients, 19–20, 266–267, 271–274, 277–283
Laminar flow, 151, 157
Lead refining, 41, 290, 291, 295
Lead smelting, 321, 323
Lignite, 69–71, 75–80, 118
Liquid fuels, *see* Fuel oils, Tar
Liquid metals, 284–302
  as solvents, 289–302
  classes of dissolved impurities, 290
  surface tensions and viscosities, 289
  wetting refractories, 289
Liquids, flow of, 143–190
Lime as a refractory, 349
Load-bearing capacity, 326, 333–335, 339, 343–345, 348, 351, 365
Logarithmic mean area, 199–200
Louvres, 174

Magnesioferrite, 346
Magnesite brick, 335, 346–349, 357–361, 365
Magnesite-chrome refractories, 350–351, 365
Magnesite refractories, 336, 345–351
Manometers, 168–169, 188–189
Mass velocity, 155
Materials balances, 12–17, 20, 27–42
  blast furnace, 61–63
  combustion, 111–114
Matrix, 327
Mattes, 3, 6, 8–9, 40, 42, 320–324
Matte smelting, 3–9, 40, 42, 54–56, 303, 306, 320–324
Maximum flame temperatures, *see* Adiabatic combustion temperature
Maximum velocity in pipes, 166
Mean areas:
  arithmetic, 196, 199, 203
  geometric, 202–203
  logarithmic, 199–200, 203
Mean heat capacity, mean specific heat, 47–48, 67, Table A–4
Mechanical energy, 43, 52, 66, 67, 144–150
  compressing gases, 178–183, 189–190
Mechanical energy balance, 149–150, *see also* specific fluid flow systems and kinds of flow energy
Melting, 10, 118–120, 191, 244–246, 254

Melting furnace, 34–36
Melting points, *see also* Constitution diagrams
  metals, 285–286, Table A–3
  refractory materials, 328
  slag-forming oxides, 314
  sulfides, 322–323
Metallurgical balances, *see* Materials balances
Metallurgy, branches of, 1–4
Metals, thermal conductivities, 192
Metal sulfides, metal-sulfur systems, 301–302, 321–324
Metasilicates, 309
Mineral matter in coal, 72, 76–77, 79
Mineralogical analyses, 26, 41
Miscibility gaps, 291, 301, 322
Mixing, 17–20, 41
Models, 152, 215
Modulus of rupture, 326, 334
Moisture, moisture content:
  combustion air, 113, 121, 124–126
  effect on gas radiation, 235–238
  flue gases, 113
  fuels, 72–73, 113, 121, 124–126, *see also* Proximate analyses
  gases, 27, 33–35, 41
  solid materials, 21, 24
Mol, 13–15
Monochromatic absorptivities, emissivities, emissive powers, 219–223
Monolithic furnace structures, 348–350, 356, 360
Morgan ejector, 177–178
Mortars, 354, 356
Mullite, 340–341, 343, 345
Multiphase systems, 284–285, 320

Natural convection, 207–213, 255–256
Natural draft, 169–170, 189
Natural gas, 70, 90, 91, 117–118, 123, 140–141
Neutral refractories, 326, 335–336, 349, 353, 356
Newton constant, heating and cooling, 265–271, 273–274, 276, 281
Newton's law, acceleration, 144
Newton's law of heating and cooling, 260–263, 265–271, 273–274, 276, 280–283
Nitrogen balance, 112–114
Nozzles, 163–165, 184–187, 189–190
Nusselt number, 211–212, 215–217

Oil gas, 90, 91
Open-hearth steelmaking:
  fuels and heat supply, 135–142
  furnaces, 135–136, 335, 340, 341, 349, 365
  slags, 303, 306, 308, 309
Optical pyrometer, 222–223
Orifices, orifice meters, 163–165, 170–172, 184–185, 188–189
Orsat analyses, 26–27, 34, 41, 112–114, 140–141
Orthosilicates, 309
Output, 17, 20, 28
Over-all coefficients, 249–252
Oxidation of liquid metals, 296–297
Oxidation-reduction reactions, 309–310, 313
Oxide slags, *see* Slags
Oxides, melting points, 314
Oxygen balance, 112–114
Oxygen enrichment of combustion air, 118, 121, 126–128, 137–141
Oxygen-metal solutions, 290, 295–297

Parr's formulas, 76–77
Partial pressure, 15, 27
Paths of thermodynamic changes in state, 44–45, 180–182
Peat, 69–71, 75
Pebble heater, 131–133, 217–218
Periclase, 345–346, 348
Peritectic reactions, 292, 294
Permeability, 326
Petroleum, petroleum refining, 67, 70, 91, 102–103
Phase diagrams, *see* Constitution diagrams
Phases in metallurgical systems, 8–10, 284–324
Pig iron, 37–40, 61–64, 299–300
Pipes and ducts:
  bends and fittings, 158–159, 166, 189
  forced convection, 214–217, 235–240, 256–258
  friction, 154–160, 189
  pressure drops and sizes, 159–160, 189
  surface roughness, 155–157
  velocity distribution, 166–167, 188
Pitot tubes, 165–167, 188
Planck's law, 219, 222
Plastics, 356, 360
Poiseuille's equation, 157
Porosity:
  coke, 88–89

  effect on thermal conductivity, 193–194
  refractory materials, 326, 327, 334, 351
Ports, 170–172
Potential energy, 44, 144–145
Powdered coal, *see* Pulverized coal
Power consumption:
  air compression, 179–183, 190
  electric furnaces, 203–204, 231–232, 244, 255–258
  flow systems, 188–190
Prandtl number, 195, 211–212, 215–217
Preheating combustion air, gaseous fuels, 115–116, 118, 121, 123–124, 127, 129–142, 258
Pressure drops, 155–157, 183–190
Pressure energy, 144–146, 149, 169, 185–187
Pressure taps, 164–166
Pressure variation with height, 169
Primary air, 83–84, 100, 121
Probable error, 22–24, 42
Producer gas, 87–99, 107–108
  adiabatic combustion temperatures, 115–118
  available heat, 122–128
  combustion, 140
  open-hearth fuel, 136–137
Proximate analyses, 26, 72–73, 75–78, 88, 106–107
Pulverized coal, 7–8, 34–36, 70, 81–84, 125, 140
Pumps, 189
Pyrometric cone equivalents, pyrometric cones, 329–330, 342–344, 354
Pyrometry, *see* Temperature measurement

Quartz, 336–339
Quenching, 275–277

Radiant energy, 218
Radiation heat flow, 191, 194, 218–239
  any two surfaces, 223–224, 229–230
  coefficients, 229–231
  direct, between black surfaces, 223–225
  effect of refractory surfaces, 224–229
  emissivities, 220–239, Table A-7
  exterior furnace surfaces, 230–231
  flames and gases, 234–239
  flue gases to flue, 235–238
  furnace openings, 243–244
  gas bodies, 234–239
  geometric factors, 223–229
  Globars, 231–232, 256

Radiation heat flow (*continued*)
  gray surfaces, 229
  heat exchangers, 239
  interior furnace surfaces, 224–233, 239
  isothermal enclosures, 220–221
  liquid metal in cylindrical vessel, 232–234
  object in furnace, 225
  parallel plane surfaces, 224–225
  unsteady systems, 259–260
Radiation shields, 246–247, 253
Ramming mixtures, 350, 356
Ranks of coal and other solid fuels, 70–71, 75–80, 106–107
Rate of cooling, liquid iron in open cylinder, 232–234
Rates of flame propagation, 99–101
Rates of reaction, 6, 9, 82, 98, 101
Rates of treatment, 17–20, 239–240, 253–254
Rational analyses, 24, 26, 71–72
Reactions in liquid metals, 302
Reciprocal density, 149, 183
Recovery of calorific power, 91–92, 97–98, 107
Recuperators, 123, 129–132, 247–252, 258, 353
Reduction of oxides, 10, 40, 50–51, 57, 67, 68
Reference states, 47, 50–54, 57, 58, 64, 67, 68, 73–74, 97
Reference temperatures, 47, 50–53, 57–59, 67, 68
Refining of liquid metals, 3, 4, 10, 119, 284–285, 289, 291, 295–296, 302–304
Refining of metals, 1–4
Reflection of radiant energy, 218, 222, 225–228
Refractory brick, *see also* specific varieties of bricks, such as Firebrick, Silica brick, etc.
  fabrication, 326–327, 338–339, 342–343
  joints, 354–356
  sizes and shapes, 354–355
Refractory materials, 9, 325–367, *see also* specific materials and properties
  classification, 335–336
  composition and structure, 326–328
  costs, 326, 335
  properties, 325–335
  typical analyses, 340, 343–344, 346, 350
Refractory surfaces in radiation heat flow, 224–229
Regenerators, 16, 86–87, 123, 129–135, 137–142, 250–252
  gas flow, 210

heat balances, 137–142
heat flow, 249–252, 258
thermal efficiencies, 129–130, 137–142, 250–252, 258
Relative heat content, 266–267
Relative position, 264–267, 271–274, 276
Relative surface resistance, 264–277, 281
Relative temperature, 204, 263–283
Relative time, 264–271, 273–275, 281
Resistance furnaces, 203–204, 231–232, 244, 255–258, 353
Resistance to slag attack, 325–326, 328–331, 334, 339, 345–347, 349, 351
Retention time, 17–20, 42, 254
Retorting, 68, 284–285, 325, 353
Reverberatory furnaces, 6–9, 109–111, 141, 325, 335, 340, 357–366
Reversible compression, 181–183
Reversible expansion, 331–332
Reversible heat exchanger, 247–250
Reversible processes, 45
Reynolds numbers, 154–162, 165–166, 214–217
Roasting, 41
  heat balances, 54–56, 67–68
Rotameter, 167–168
Rotary kilns, 33–34, 334, 344
Russell tables, 268

Sampling, 20–25, 42, 72
Sandstone, 338
Secondary air, 83–84, 100, 121
Segregation process, 324
Self-expansion work, 145, 149, 179–183
Sensible heats, 47–48, 53–60, 64–68, Table A–3
Series paths of heat flow, 204–206, 230, 240, 246, 249, 255–258
Service properties of refractories, 326–334
Severity of quench, 265, 276–277
Shaft furnaces, 42, 334, 344
Short-circuiting, 17, 20
Shrinkage, 326, 331–333, 343, 348, 354
Sieverts' law, 297–298
Signs of ΔH and Q_p, 52–54
Silica-alumina refractories, 336, 338, 340–345
Silica brick, 335, 338–340, 360–361, 364–365
Silica glass, 336–338, 354
Silica, inversions and transformations, 336–339
Silica sand, 338, 356

Silicate degree, 308–309
Silicate minerals, 309, 312
Silicate slags, 305–309, 312–320
Silicon carbide, 131, 336, 352–354
Simple smelting, 6–10, 254, 284–285, 303
Size of sample, 23–24, 42
Sizes of pipes and ducts, 156–157, 159–160
Slags, 284–285, 289, 303–320
  acidity and basicity, 306–309, 313, 330–331, 335–336
  action on refractories, 305, 329–331, 335–336, 339, 345, 349, 352
  blast furnace, 37–38, 317–318
  chemical properties, 306–318
  component oxides, 303–307, 314
  constitution, 314–320, see also Constitution diagrams
  copper smelting, 3, 6, 8–9, 303, 306, 315
  formation and free-running temperatures, 119, 304–305, 319–320
  functions and practical requirements, 303–305
  gravity separation, 303, 305, 318–319
  ions, 311–313
  oxidizing and reducing power, 306, 309–310
  silicate degree, 308–309
  steelmaking, 303, 306, 308–309
  structures, 310–313
  typical analyses, 305–306
  viscosity and fluidity, 143, 305, 313, 318–320, 330–331
  V-ratio, 309
Slag-metal:
  reactions, 306, 310
  separation, 303, 305, 318–319
  systems, 284–285
Smelting, 2–9, 191, 284–285, 288–289, 304, 325
Smelting power, 120–121, 124
Softening temperatures of refractories, 326, 328–330, 333, 340, 342–344, 352, 354
Solid fuels, see name of specific fuel
Solidification as a heat flow process, 244–246, 254
Solidification of metals, 297
Solid metals, 288–289
Solid solutions, 294–295
Solubilities in liquid metals, 290–302
Spalling and spalling resistance, 326, 332–333, 338–339, 343–344, 348–349, 351, 353, 365
Special refractories, 336, 352–354

Specific gravities:
  coke, 88–89
  fuel oils, 103
  refractory materials, 326, 334, 338–340, 348
Specific heats, 47–48, 67, 326, 334, Table A–4
Specific surface, 161
Spectral emissivity, 222
Speisses, 320, 322, 324
Spheres, convection heat transfer, 213, 217
Spinels, 317, 347–350
Sprung arches, 361–365
Stacks and chimneys, 168–170, 175, 177, 189, 257
Stagnant layer, 197, 207–208, 213–215
Standard conditions, 14–15
Standard tests:
  fuels, 72–74, 76–77, 79, 89, 103–105, 107
  refractory materials, 328–329, 332–334, 343–344, 351–352, 354
Stanton number, 215
Static draft, 169–170, 189
Static pressure, 165–166
Steady heat flow, 191–258
Steady state, 15–17, 22, 30, 53, 92
Steam engine, 146, 148–149
Steam tables, 148
Steel-encased brick, 348–351
Steelmaking, 4, 299, 302–309, 325, 331, 339, 360, see also specific steelmaking processes
Stefan-Boltzmann law, 219–220
Stoichiometry, stoichiometric calculations, 7, 12–42, 109, 111–114, 140–142
Stokes' law, 318–319
STP, 14
Streamline flow, 155, 157–158, 161, 214
Structures:
  liquid slags, 310–313
  refractories, 326–328
  silicate minerals, 312
Sublimation points, 286–288
Sulfides, 301–302, 321–324
Sulfur in liquid metals, 299, 301–302, 321–324
Superficial velocity, 161
Surface resistance, see Relative surface resistance
Surface temperatures, 196–197, 240–243, 255–256, 271–275
Surface tensions of liquid metals, 289
Suspended arches, 365–366

System, 10, 11, 44, 53

Tar, 69–70, 87, 103, 105–108, 136
Temperature distribution, heating and cooling objects, 260, 266, 271–274, 276, 278, 283
Temperature measurement, 8, 60, 197, 222–223, 239, 252–253, 256, 282–283, 329–330
Temperatures of metallurgical processes, 287–288
Tests, *see* Standard tests and specific materials tested
Theoretical air, 33–36, 41, 112, 115, 118, 121–123, 126, 137–142
Thermal conductance, 240
Thermal conductivities, 192–196, Table A–6
   refractory materials, 325–326, 334–335, 349, 351–353
Thermal diffusivities, 193, 264–265, 283
Thermal efficiencies:
   heat exchangers, 129–130, 137–140, 142, 251, 258
   processes, 57, 58
Thermal resistivity, thermal resistance, 204–206, 229–230, 240, 249, 256
Thermochemical data, 45–50, 60
Thermochemistry and thermophysics, 7, 43–68
Thermocouples, thermometers, 252–253, 256, 282–283
Thermodynamic properties, 44, 45, 49, 54
Thermodynamic states, 44, 50
Time lags, *see* Lags, Capacity lags, Transfer lags
Time of residence, 17–20, 42, 254
Time of response, 19–20, 41, *see also* Lags, Capacity lags, Transfer lags
Time of treatment, 17–20, 42, 254
Total energy balances, 145–149, 182–183
Transfer lag, 277–283
Transition region, fluid flow, 155, 157–158
Traverse, 167
Tridymite, 336–340, 343, 345
Turbocompressors, 179
Turbulent flow, 155, 157–158, 161, 214–217
Tuyères, 185–187, 189–190

Ultimate analyses, 24, 26, 73–75, 88, 103, 107
Unit operations, 1–4
Unit processes, 1–11
   classification, 2, 8–10

engineering fundamentals, 6–8
   modes of operation, 15–20
Units and dimensions, 378
   energy balances, 144–145
   heat flow, 193
Unsteady heat flow, 192–193, 250–251, 259–283, *see also* Heating and cooling
Unsteady state, 15–17, 31, 53, 60

Vapor pressures:
   metals, 285–288
   water, Table A–2
Velocity distribution in pipes, 166–167, 188
Vena contracta, 164
Venturis, Venturi meters, 162–164, 178, 184–188
Vermiculite, 351
Viscosities:
   definitions, 151–152
   gases, 152
   glass, 327, 333
   liquid metals, 289
   mattes, 324
   relations to fluid friction, 151–154
   relations to other fluid properties, 195
   slags, 305, 313, 318–320, 330–331
Vitreous silica, 336–338, 354
Volatile matter, 72–73, *see also* Proximate analyses
Volatilities of metals, 285–288
V-ratio, 309

Waste-heat boilers, 67, 129, 141
Waste-heat recovery, 123–124, 129–142
Water, *see also* Moisture, Moisture content
   heat of vaporization, 48–49, 65, 68, 73–75
   reference state, 50, 53, 54, 57, 64, 73–74, 97
   vapor pressure, Table A–2
Water gas, 90, 91, 94–95, 97, 117–118, 123, 141
Weighing, 20–21, 24
White metal, 321
Wiedemann-Franz ratio, 194
Wood, 69–71, 75
Work, 43–44, 52, 66, *see also* Mechanical energy, Self-expansion work

Zircon and zirconia, 336